ITALY

Fossano

N85

Luc de
Serre-Ponçon

Barcelonnette

Cuneo

Limone
Piemonte

Digne

Mercantour National
Park

Baiardo

N85

PROVENCE-
ALPES-CÔTE D'AZUR

N202

Castellane

Breil-Sur-Roya

D6202

Ventimiglia

E80

N85

Des Préalpes d'Azur
Regional Nature Park

Contes

Menton

Verdon Gorge

Lac de
Sainte-Croix

Escragnolles

Monaco

Menton

Grasse

Nice Ville

Nice

MONACO

NICE-CÔTE
D'AZUR AIRPORT

N85

A8

Baie des Anges

EASTERN CÔTE D'AZUR

Draguignan

CANNES-
MANDELIEU
AIRPORT

Cannes

Antibes

Cannes

Puget-Sur-
Argens

Vidauban

A8

Saint-
Raphael

Fréjus

Golfe
de Fréjus

Plaine
des Maures

A57

Ste-Maxime

D559

Port-Grimaud

Golfe de
Saint-Tropez

Cogolin

Saint-Tropez

WESTERN CÔTE D'AZUR

D98

Mediterranean Sea

Hyères

TOULON-HYÈRES
AIRPORT

Îles d'Or

Île du
Levant

Îles
d'Hereys

Port-Cros

© MOON.COM

Contents

Gordes

Provence & the French Riviera

The southeast corner of France has been one of Europe's most popular travel destinations for over 150 years. It is a magnet for artists, writers, and holidaymakers who continue to be drawn by the remarkable beauty of its landscapes, the slow-paced life of its medieval villages, and the captivating glamour of its Mediterranean resorts.

In Provence, discover the major cities of Marseille, Aix-en-Provence, Arles, and Avignon, as well as the regional nature parks of the Luberon, Les Alpilles, and the Gorges du Verdon. Immerse your senses in the sweet-smelling lavender fields of Haute Provence, the pungent, earthy truffle markets of the northern Vaucluse, and the salty, sun-kissed beaches and calanques of the southern coast.

The French Riviera, essentially the coastal strip of Provence from Hyères to Menton on the Italian border, begins in colorful, artistic Nice and moves along the seafront to Antibes, Cannes, Saint-Raphaël, and the tourist hot spot of Saint-Tropez. Heading east, we travel through the villages and resorts of the Corniches toward the principality of Monaco.

The French art de vivre, a way of enjoying life's pleasures, culture, and food, is a big part of life in the south. The slow-paced experience is amplified by the poetic surroundings of Provence, the resplendent café terraces of Aix-en-Provence (known as the Paris of the South), the glorious arched Roman arena in Arles, and the vaulted splendor of the Palais des Papes in Avignon. On the Riviera are some spectacular examples of art deco and belle epoque, as well as stuccoed white palace hotels and state-of-the-art shopping malls and sports arenas.

Tourists come for the art, museums, food markets, and festivals, but outside the high season, life in Provence and on the Riviera is unexpectedly simple. Locals play card games on café terraces, chat about the Tour de France, and enjoy a game of boules in the village square—and they are always happy to play a visitor.

Grande Plage in Juan-les-Pins

10 TOP
EXPERIENCES

1 **Dining alfresco by the sea** and enjoying the best food Southern France has to offer (page 31).

2 Experiencing the wonder and reverence induced by the **Palais des Papes.** New technology has revived the splendor of touring this vast Gothic palace (page 43).

3 Wine-tasting in the northern Vaucluse, hopping between **Châteauneuf-du-Pape** (page 53) and **Vacqueyras, Gigondas,** and **Rasteau** (page 56) to discover the region's stellar reds.

4 Walking through the waving seas of the **lavender fields of Haute Provence,** with picturesque villages rising like islands and swarms of bees darting this way and that (page 171).

5 Exploring the **Gorges du Verdon,** especially via kayak. At 700 m (2,296 ft) deep and 25 km (16 mi) long, the gorge is an awe-inspiring demonstration of the power of nature and time (page 175).

6 Visiting the daring, futuristic **MuCEM,** a glass box containing a museum of Mediterranean culture, one of the many architectural projects transforming the reputation of Marseille (page 200).

7 Getting familiar with the Riviera's sublime **architecture,** from belle epoque villas to modernist beachside cabins (page 248).

8 Mountain biking and hiking the incredible geological giants of Southern France, from the striking **Mont Ventoux** (page 67) to the rust-red **Massif de l'Estérel** (page 364).

9 Lying on the sands of France's most famous beaches, such as **Plage de Pampelonne** near Saint-Tropez (page 383) or **La Grande Plage** in Juan-les-Pins (page 306).

10 Rolling the dice at the **Casino de Monte-Carlo** (or maybe just stepping inside) and imagining yourself as James Bond for an evening (page 403).

Planning Your Trip

WHERE TO GO

Avignon and the Vaucluse

The busy metropolitan outskirts of Avignon give way to a walled medieval interior. The main draw is the world-famous **Palais des Papes,** closely followed by the **Pont Saint-Bénézet,** whose arches famously stretch only halfway across the Rhône. Day trips to the **Pont du Gard** aqueduct and **Châteauneuf-du-Pape** vineyards are a must. Nearby **L'Isle-sur-la-Sorgue** is an antiques lover's mecca and a great base to explore north toward **Mont Ventoux.**

Arles and Les Alpilles

For the locals here, life is all about bull running and bullfighting. There are plenty of other reasons to recommend the region, including major Roman sites and the van Gogh art trail.

Saint-Rémy-de-Provence and **Arles** are filled with restaurants, art galleries, and boutique shops. Don't miss the surrounding countryside, which is dominated by rocky peaks and a pungent combination of pine forests, olive groves, and vineyards.

Aix-en-Provence, the Luberon, and the Gorges du Verdon

Time in Aix passes almost unnoticed. Most visits begin underneath the shady canopy of the plane trees on the wide, café-lined **Cours Mirabeau** and end in the old town amid a maze of cobbled streets and squares adorned with fountains. Don't miss a visit to **Cézanne's workshop** or the chance to see his work at the **Musée Granet.**

the Cours Mirabeau, Aix-en-Provence

The **Luberon Regional Nature Park** is a rural idyll, dotted with renowned villages such as **Gordes, Bonnieux,** and **Lourmarin.** Depending on the time of year, the fields will be full of melons, pumpkins, drifting blossoms, or vines sagging with fruit. The hills are criss-crossed with cycle and hiking routes and dotted with vineyards to visit. From late June until the end of July, the lure of lavender in full bloom calls visitors to the high plateaus of **Haute Provence.**

Farther east, the limestone **Verdon Gorge** takes your breath away. Cooler summer temperatures, great water sports, and traditional Provençal villages such as **Moustiers-Sainte-Marie** make the Verdon Regional Nature Park a popular destination in high summer.

Marseille and the Var Coast

Marseille is a loud, brash, in-your-face melting pot of cultures, founded by the Greeks, and, thanks to centuries of immigration, France's second-largest city. Sparkling new cultural centers, hidden neighborhoods, fishers offering the catch of the day, and restaurants offering vrai (true) bouillabaisse all compete for attention. To the east is the **Calanques National Park,** perhaps the most scenic, unspoiled stretch of Mediterranean coast remaining in France. Accessible largely on foot or by boat, the Calanques are a series of rocky inlets cut into the coastline. The **Var resorts** between **Bandol** and **Sanary-sur-Mer** are just as scenic as their more celebrated Riviera counterparts and a lot less busy.

Nice and Antibes

Nice, capital of the Côte d'Azur, is a fun, modern, sculpture-filled, eco-friendly city with a sleek tramway that allows direct, cheap travel to the city center or port from the airport. The promenade des Anglais, Nice's celebrated seafront walkway, is a constant flow of cyclists, Rollerbladers, honeymooners, hikers, and holidaymakers. Strolling along the promenade at dusk above the

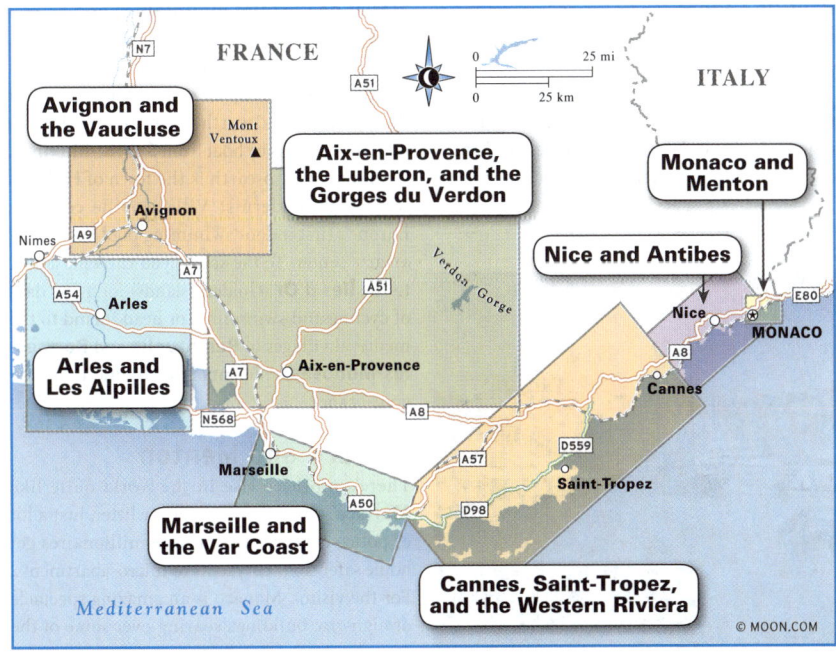

© MOON.COM

promenade des Anglais, Nice

pebbly beach is a magical experience. You could easily spend a month in Nice and still not see half of what's on offer.

West of Nice, **Cagnes-sur-Mer** was home to the artist Pierre-Auguste Renoir, and a whole village of writers and painters settled in neighboring **Haut-de-Cagnes.** Their legacy lives on in art studios and design workshops all over the medieval village.

Antibes has the largest pleasure-boat marina in Europe and hosts the Voiles d'Antibes regatta in June. It's a great place to spend a few days, with a Provençal market, interesting

Saint-Tropez

shops, lively nightlife, and the Musée Picasso occupying a fortified seafront château. Waterskiing was invented in **Juan-les-Pins,** Antibes's close neighbor, which also boasts some of the best beach bars on the Riviera and a famous annual jazz festival that has been running since 1960.

Cannes, Saint-Tropez, and the Western Riviera

Famed for its annual film festival, superyachts, nightlife, and sandy beaches, **Cannes** feels like a dreamed-up fantasy of a Riviera resort. Its seafront promenade, **La Croisette,** is lined with belle epoque villas and designer boutiques. At the port nearby, passenger ferries leave for the **Îles de Lérins,** perfect for a day among the pines and eucalyptus or tasting wines from the vineyard of a 5th-century monastery.

Farther west, **Fréjus** has some of France's best **Roman ruins** and has an unpretentious atmosphere—meaning cheaper hotels and restaurants. It's a great base for exploring the rust-colored **Massif de l'Estérel mountain range** on foot or by mountain bike.

Undeniably glamorous and often unbearably crowded, **Saint-Tropez** is the poster resort for the whole Riviera, worth visiting even if just for a couple of hours by boat from **Sainte-Maxime.**

More down-to-earth is the town of **Hyères,** where the modernist Villa Noaille contains Europe's first indoor swimming pool and is now an arts center. To the south, you can take a boat to the **Îles d'Or** (Golden Islands) for a few days of cycling and swimming, or head inland to the medieval villages of **Ramatuelle** and **Bormes-les-Mimosas,** redolent of the traditions of a bygone France.

Monaco and Menton

There's nowhere else in the world quite like **Monaco.** There's no graffiti or litter, just a lot of police officers making sure millionaires get home safely to their yachts or micro-apartments. For the visitor, Monaco is an amazing spectacle of high-rise buildings soaring over some of the world's most prestigious art galleries, museums, jewelers, and real estate. Everyone should pay a visit to the **Monte-Carlo Casino,** even if just on a morning tour.

Heading east toward Italy brings you to

Menton, the last town in France before Italy, boasting the country's sunniest climate, a breathtakingly beautiful seafront, a famous annual Lemon Festival, and a wealth of ornamental parks and gardens.

WHEN TO GO

Spring

As long as you don't want to spend all your time on the beach, spring is the best time to visit the South of France. The sun is usually shining, the temperature can reach the mid-20s Celsius (high 70s Fahrenheit), and the Provençal countryside looks its most beautiful. In early spring the fields fill with the wild blossoms of almond and cherry trees. Hotels, restaurants, and beach bars that closed for winter begin to open in **April** with a fresh coat of paint and cheerful staff. In **May,** there's the **Cannes Film Festival, Saint-Tropez's Les Bravades,** and the **Monaco Grand Prix,** some of the highlights of the Riviera year, but as it's not yet high season, hotel prices will generally be cheaper. By mid-June the lavender fields are coming into bloom.

Summer

The period between **mid-July** and **mid-August** is by far the busiest and hottest in the South of France. Sightseeing can be oppressive under the beating heat of the summer sun. Most tourists in these months get out and about early, doing as much as they can before lunch, and then spend the afternoon by the pool. Restaurants are busier, the weekend traffic on the autoroutes is terrible, and at the seaside there's hardly a spare inch of beach. The first two weeks in July and the last two in August are slightly quieter, and a good option for determined sunseekers who want to avoid the worst of the crowds. The first week in July is often the best time to see the **lavender fields** in full bloom.

Fall

September and **October** are some of the best

months to visit. Temperatures are still in the low 20s Celsius (low 70s Fahrenheit), so it's the perfect time of year for activities like cycling, hiking, and kayaking. The vines turn a magnificent range of russet and golden colors, and the roads are quiet, so it's easy to get from sight to sight.

Winter

If you love hiking and cycling, then winter is still a good time to visit. In **December,** the weather can still be mild, and Christmas lights and decorations go up at the beginning of the month. Special **Christmas markets** take place on an almost daily basis.

In **January** and **February,** it's best to base yourself in a city. By the seaside, it's not unusual for people to sunbathe in sheltered spots, and the brave even swim. Away from the sea, the towns and villages of Provence can be exceedingly quiet, and it gets quite cold, with nighttime temperatures often falling below 0°C (32°F).

lavender fields of Haute Provence

By **March** the countryside is slowly stirring into life, vines sprout with an urgent vigor, and the fruit trees begin to blossom. The markets fill up with traders, and village shops and restaurants reopen. Everyone seems happy as the tourist season begins.

BEFORE YOU GO

Passports and Visas

The latest visa requirements can be checked at the France Diplomatic website (www.france-visas.gouv.fr/en/web/france-visas/visa-wizard). The site offers a "visa wizard" that will quickly tell you your requirements. Here is a summary of the situation at time of writing:

Nationals from the **United States, Canada, the United Kingdom, Australia,** and **New Zealand** can enter France and stay for up to 90 days without a visa. Stays of more than 90 days require a visa and proof of income and medical insurance. Citizens of **EU member countries** who have a valid passport and national identity card can travel freely to France. **South African** nationals require a short-stay visa for visits up to 90 days, and a long-stay visa if the traveler plans to stay more than 90 days.

What to Pack

What should go into your suitcase very much depends on when you visit. In addition to weather-appropriate clothing and other necessities, don't forget a plug adapter for your electronic devices. European outlets take round two-prong plugs.

If You Have . . .

ONE WEEK

Pick a central base for three or four nights such as **L'Isle-sur-la-Sorgue** in the Vaucluse. From here you can explore the cities of **Arles** and **Avignon** and the **Luberon Nature Park.** For the remaining nights, use **Nice** as a base to explore the Riviera, catching the train that runs along the coast to visit iconic seaside resorts such as **Villefranche-sur-Mer.**

TWO WEEKS

Divide your time between the coast and inland. Once again, pick a central base. **Saint-Rémy-de-Provence** is a good alternative. Then stay in **Cassis** for a few nights to visit the **Calanques** and Marseille. Head east to **Saint-Tropez** and dress as glitzy as you can to strut your stuff portside, before rounding out the vacation with a stay in **Antibes** to explore the Riviera.

THREE WEEKS

Plan to stay a couple of nights in each of the **big cities:** Aix, Arles, Avignon, and Marseille. Explore the cities and the neighboring sights, such as Mont Sainte-Victoire near Aix, the Camargue near Arles, Châteauneuf-du-Pape and the Pont du Gard from Avignon, and the Calanques near Marseille. Spend the second week in the heart of the **Provençal countryside.** Choose among the villages of the Luberon, Les Alpilles Nature Park, northern Vaucluse, or inland Var, which all share a similar relaxed Provençal atmosphere. For the third week, head to the Riviera and meander eastward from Hyères to Saint-Tropez to Cannes and then Nice, finishing with a roll of the dice in Monaco's casino. If you are lucky you may win enough to pay for the whole trip.

TRANSPORTATION

Air

Most travelers will arrive at either **Marseille** or **Nice airport.** Both have low-cost terminals as well as regular scheduled flights to all major destinations in Europe.

For transatlantic passengers, the Marseille and Nice airports have direct flights with Air Canada and Air Transat to Montreal. Nice Airport also has direct flights with Delta/Virgin Atlantic from New York JFK airport, Philadelphia, and Atlanta. For other North American destinations, travelers must fly to common hubs such as Paris, London, or Amsterdam and then take a connecting flight to Marseille or Nice.

For Australians the quickest way to get to the South of France is a flight to Dubai. From Dubai, Emirates flies direct to Nice and Transavia offers limited service to Marseille. Alternatively, there are direct flights to Nice from Hong Kong.

There are no direct flights to the South of France from South Africa. South Africans should fly to London, Amsterdam, or Paris and pick up a connecting flight.

Nîmes Airport, a 45-minute drive from Avignon, is a low-cost hub with flights to and from Stansted near London. Farther into the Languedoc, **Montpellier Airport,** which is a 60-minute drive from Avignon, also offers flights to London, as well as other major European hubs.

Train

Provence is directly connected by a **TGV line** to Paris. Hourly TGV trains stop at the major cities of Avignon (under 3 hours from Paris), Aix-en-Provence, Marseille, and Nice. From Marseille, there's a slower coastal track that stops at all the major cities along the Riviera except Saint-Tropez.

Interrail (www.interrail.eu) offers 5-day, 7-day, and monthly **rail passes** for travel throughout Europe, as well as a pass for France alone, that may make sense if you plan to travel a lot by train, depending on the length of your trip. Non-UK and European residents should apply for a Eurail pass (www.eurail.com).

Bus

BlaBlaCar (www.blablacar.fr) and **Flixbus** (www.flixbus.fr) operate long-distance bus routes between all major towns in France. They are usually cheaper than traveling by train, but can be very slow, with lots of stops and waiting time. Buses have Wi-Fi, toilets, and reclining seats.

The Riviera is very well serviced by **local bus networks.** Each town has its own network (single journeys cost €1-3), with connections to the countryside beyond. Bus service on Sundays is usually poor or even nonexistent.

Car

The main route to Provence from the north is the **A7 autoroute,** which runs between Lyon, Avignon, and Aix-en-Provence. It is nicknamed the Autoroute du Soleil. During weekends in the summer months of July and August, it is advisable to find alternative routes because the stretch between Avignon and Lyon becomes one long traffic jam.

The A7 intersects with the **A8 autoroute** just west of Aix-en-Provence, from where it takes just under 2 hours to reach Nice. The A8 serves all the main towns and cities along the Riviera, passing alongside the Massif de l'Estérel between Saint-Raphaël and Cannes and the Massif des Maures west of Fréjus. It has regular **aires de répos** (rest stops with toilets and places to sit) and **aires de services** (petrol stations).

The **A9 autoroute** runs from just outside Avignon to the Spanish border near Barcelona.

Provence & the French Riviera

The itinerary recommended below is relatively ambitious. You will need a car, and it's intended as a happy combination of history and culture, blending inland Provence with a trip to the coast and mixing in plenty of stops that will satisfy gourmands. For some there may be too much travel from place to place, so adapt as necessary and, above all, don't forget to leave time to sit in a café, preferably with a pastis, and watch the world go by.

Day 1

Start in **Avignon.** An early TGV (bullet train) from Paris will get you there in time for lunch at **Le Carré de Palais** restaurant. In the afternoon, tour the **Palais des Papes** and cross the river to **Villeneuve-lès-Avignon** for a view back over the city and of the famous **Pont d'Avignon** (Pont Saint-Bénézet). Stay overnight in the **Clos Saluces** bed-and-breakfast and make it your base for the first three nights.

Day 2

Rise bright and early for breakfast in the verdant garden of Clos Saluces and then head out into the countryside around Avignon. History buffs will love a morning trip to the **Pont du Gard Roman aqueduct.** On the way back to Avignon in the afternoon, stop off at **Châteauneuf-du-Pape** to sample the luxurious, velvety red wines.

Day 3

Tour the hill villages of the **Luberon Nature Park.** Be sure to include Gordes, Goult, and Roussillon as stopping-off points. If you have time, go and see the **Abbaye Notre-Dame de Sénanque,** just outside Gordes. It is particularly worth the detour in early summer when the lavender is in bloom. Eat lunch on the terrace of **L'Orangerie** restaurant looking out from Gordes over the Luberon. In the afternoon head back to Avignon via **L'Isle-sur-la-Sorgue.** Walk along the banks of the **River Sorgue,** enjoying the town's waterwheels and browsing the renowned antiques shops.

Day 4

Head to **Arles** and prepare for some serious culture. Visit the Roman theater and Roman arena as well as the Fondation Vincent van Gogh. Then leap forward in time with a tour of the futuristic Frank Gehry skyscraper at **Luma Arles.** Drive to Aix-en-Provence and stay overnight in the **Hotel des Augustins.**

Day 5

In the morning explore **Aix-en-Provence** by following the **Cézanne walking trail.** Have lunch at the newly refurbished **Les Deux Garcons** on the Cours Mirabeau before enjoying a cultural afternoon visiting the **Musée Granet** and the **Hôtel Caumont.**

Day 6

Head north to the **Gorges du Verdon,** stopping at the **Plateau de Valensole** en route to see the lavender (late June-July). Wander the picturesque streets of **Moustiers-Sainte-Marie** before driving the head-spinning **Corniche Sublime** in the afternoon. If there is time, rent a paddleboat from one of the beaches on **Lac de Sainte-Croix.** Return to Aix.

Day 7

Head to Cassis, and either hike out into **Les Calanques** or catch a boat from the port. Enjoy a bouillabaisse portside either for lunch or as the sun dips below the horizon. If you have time, visit **Domaine Paternel** to pick up some of the finest white wine and rosé in Provence. Treat yourself and stay overnight in **Hotel Mahogany** on Plage du Bestouan.

Day 8

Head to **Sainte-Maxime,** a 1-hour, 40-minute journey along the coastal D98 road. There you can leave the car and board the 20-minute ferry

1: Pont du Gard Roman aqueduct **2:** the port of Cassis
3: Plage de Pampelonne, Saint-Tropez

across the water to Saint-Tropez. Have a drink in one of the bars overlooking the superyachts before a visit to **L'Annonciade** art museum on the harbor. Enjoy a seafood meal at **Le Petit Pointu** and a floodlit game of boules on the gravel courts of the **place des Lices,** and then end the evening in the 1st-floor bar of **Hotel Sube.**

Day 9

On Tuesdays and Saturdays there's a big market on Saint-Tropez's **place des Lices;** otherwise, order a slice of **Tarte Tropézienne** at the café of the same name. Take Zou bus 875 toward the medieval village of **Ramatuelle** (for 15 minutes) and follow the signs for **Plage de Pampelonne,** which runs parallel with the road for about 5 km (3 mi). Have lunch at a beach club and enjoy the stretch of soft white sand and the chance to swim.

Day 10

After a second night at the Hotel Sube, take the ferry back to Sainte-Maxime. From there it's only half an hour to **Cannes,** via a route that passes beneath the red rocks of the **Massif de l'Estérel.** Wander through the narrow streets of **Le Suquet** old town and walk up to the summit for great views of the **Baie de Cannes.** Return for a bouillabaisse at the **Bistrot Gourmand** and do some shopping along **rue d'Antibes** and **La Croisette.** Stay for a couple of nights at the stylish **Hotel Le Cavendish** near the railway station.

Day 11

Antibes is a perfect day trip from Cannes. Begin the day at the covered **Marché Provençal** in the old town before enjoying a drink while watching the locals (and tourists) at **Le Zinc.** Visit the **Musée Picasso** in the old town and the art and glassblowing studios on **boulevard d'Aguillon.** Relax on **Plage de la Salis** or take a boat tour from one of the operators at **Port Vauban,** the largest pleasure-boat marina in Europe. Head back to Le Cavendish for drinks and a meal at **Aux Bons Enfants.**

Day 12

Arrive in **Nice** early (it's only 30 minutes from Cannes by car or train) for the food and flower market along the **cours Saleya,** one block from the sea. Stroll through **Vieux Nice** (the old town) to the **MAMAC** contemporary art museum and its roof garden. Have a salade Niçoise at one of the many restaurants on the nearby **place Garibaldi.** In the afternoon, take the free elevator up to the castle ruins on the **Colline du Château** before a seafront walk along the **promenade des Anglais.** Stay at **Le Negresco** (expensive) or **Villa Victoria** a few blocks away.

Day 13

Leave the car and board bus number 5 behind the **Galeries Lafayette** department store bound for **Cimiez,** a historic neighborhood uphill from the old town. Get off at the first stop on the steep hill (a 10-minute ride) for the museum dedicated to Russian-born artist **Marc Chagall.** Continue on the same bus to the top of the hill, where there's a **Roman amphitheater** and, for even more culture, the **Musée Matisse,** former house of the French painter. Return to the center for supper along the lively **rue Dalpozzo.**

Day 14

From Nice, it's a 30-minute drive to **Monaco** on the A8 autoroute or a 21-minute train journey. Monaco's **Musée Océanographique** is one of the highlights of the Riviera. Visit the **Casino de Monte-Carlo** and play if you want to, or just sit outside at the **Café de Paris** for some people- and car-watching. Spend some quiet time among the plants and waterfalls in the **Jardin Japonais** and continue along avenue Princesse Grace for a Chinese dinner at **Song Qi** or an Italian meal at **Avenue 31.** Walk back up to the place du Casino and finish the tour with a late-night drink in the **Buddha Bar** … or even head back into the casino if you're feeling lucky.

Best of the Outdoors and Nature

Southern France is perfect for both serious outdoor enthusiasts and people who prefer a leisurely stroll. Those who love nothing better than filling their holiday with activities will be thrilled at the breadth of hiking, kayaking, biking, and bird-watching available here.

Hiking

- Hike to the top of **Mont Sainte-Victoire.** There are four exhilarating routes to choose from at the Saint-Antonin-sur-Bayon Maison Sainte-Victoire Visitor Center (page 141).

- Hike the iconic **Sentier Martel,** which plunges you into the heart of the Gorges du Verdon (page 178).

- Hike from **Monte-Carlo** to **La Turbie.** Starting at the Monaco train station, make your way up to the village of La Turbie and its famous Roman monument, the **Trophée des Alpes,** taking in the **Tête de Chien** (a rock named for its resemblance to the head of a dog) and incredible views of the coast (page 410).

- Hike from **Carnolès** to **Roquebrune village.** Following the coastal path, take in views of Monaco; Le Corbusier's modernist cabin, **Le Cabanon;** Eileen Gray's Villa E-1027; and a **10th-century château** (page 424).

Cycling

- Rent a bike and ascend **Mont Ventoux.** There's no shame in taking an e-bike, as the White Giant of Provence gets the knees of even professional cyclists quaking with fear (page 66).

- Take the 2-hour **Cavalon Velo Route** from Coustellet to Castellet, taking in some of the Luberon's most picturesque villages (page 158).

- Mountain biking in the rugged **Massif de l'Estérel** offers exhilarating views of the coastline and eastern end of the rust-red hills (page 364).

- Riding to the **Col de la Madone** above Menton is on most serious cyclists' bucket lists, and it's easy to rent a top-of-the-line racing bike in Menton, Nice, or Monaco (page 431).

Kayaking

- Rent a kayak and go with the flow down the **River Sorgue,** stopping for a refreshing swim in the summer (page 72).

- Hugging the coast in a kayak is a wonderful way to see **the Calanques,** just east of Marseille (page 219).

- Paddling across the **Haddock Strait** toward the **Île d'Or** off Le Dramont is a popular route for kayakers, inspired by a story from Hergé's Tintin series (page 363).

Camping

- View **Les Alpilles** from **Monplaisir campsite,** which is a short walk from the charming town of Saint-Rémy-de-Provence (page 106).

- Pitch a tent in the heart of the **Luberon Nature Park** at beautiful **Camping des Sources,** located just outside of Gordes (page 151).

- Sleep outside near the renowned Luberon village of **Lourmarin** at **Les Hautes Prairies** campground, which has tent sites and cabins (page 167).

- Just north of Fréjus, the **Domaine du Colombier** is an "open-air hotel" set among pines and olive trees with a water park, a fitness center, and plenty of restaurant options (page 356).

- The **Toison d'Or** campsite backs onto the Plage de Pampelonne in Ramatuelle and gives campers access to the wonderful white sand of the region's best beach, almost deserted in the early mornings and after dusk (page 385).

the Calanques

Wine-Tasting in Provence

Visiting a vineyard should be an intimate experience. Ideally, you'll meet the owner or a member of the family, or a staff member involved in the winemaking. You should be able to wander amid the vines and crouch down and crumble the soil between your fingers. Poking your head into the cellar where the wine is made is always fun. Usually, there's a pungent smell of fermentation. The best way to ensure a good visit is to **call in advance** and express an interest in the wines of the vineyard. If you do so, you'll be greeted with open arms and plentiful bottles to taste. The following is a suggested three-day itinerary to sample some of the best wines in Provence.

Day 1

Arrive in Avignon. Rent a car and immediately head 30 minutes north to **Châteauneuf-du-Pape.** The suburbs of Avignon give way to a sea of vines on either side of the Rhône. Upon arrival at Châteauneuf, head to the tasting rooms of **Beaurenard.** As well as a Châteauneuf-du-Pape, there's an excellent Rasteau full of spices and ripe fruit (€18/bottle). Have lunch at **La Mule du Pape,** a popular choice in the center of town.

In the afternoon, head into the nearby **Dentelles de Montmirail.** Stop at **Domaine Goubert** just outside Gigondas, where the wine is full of undertones of tobacco. Stay at **Hotel les Florets** outside Gigondas. Jump into the pool before an indulgent supper.

Day 2

Explore the village of **Gigondas** in the morning before driving 30 minutes to L'Isle-sur-la-Sorgue for lunch at the **Café Fleurs.** The restaurant has a carefully curated wine list that features the best local producers. After lunch, drive 30 minutes through the olive groves east toward the Luberon. Stop outside Bonnieux at vineyard **Château Canorgue,** where there are more than 10 different wines to taste.

Drop out of the Petit Luberon to the village of Lourmarin in the southern Luberon. Stay at **Le Moulin de Lourmarin** in the village center. Explore the village and then pop 2 km (1.2 mi) up the road for a tasting at **Domaine La Cavale,** which offers cellar tours and tastings of local wines, as well as great wines from around the world.

Day 3

Drive 45 minutes south to **Château La Coste,** just outside Aix-en-Provence. Earn your tasting by first taking the modern art trail through the vines. The walk is dotted with installations by famous artists. Afterward, enjoy lunch in one of the vineyard's several restaurants.

Then head south for another hour to **Domaine Tempier** outside Bandol. There's no fanfare or grand buildings, but this modest vineyard happens to make one of the best reds in Provence. The center of the domaine is a simple house surrounded by vines set at the end of a row of trees. The tasting room has a family feel, and the aged red is among the best you are ever likely to taste. Afterward, stay at the **Hotel du Golf** on the Plage Anse de Renécros in Bandol.

1: white wine of Provence **2:** Châteauneuf-du-Pape
3: Gigondas

Best Waterfront Dining

Food is a big part of the Southern France experience, and a silver platter of gleaming seafood alongside a glass of iced rosé and starched white table napkins is a vision of holiday extravagance.

Most beach restaurants are open from April until October. Ones actually on the sand often disappear completely in the autumn and are rebuilt in the spring. Others are boarded up and may reopen briefly for a week at Christmas, attracting regulars and winter tourists. These waterside restaurants are all worth planning a trip around:

MARSEILLE AND CASSIS

- In Marseille, escape from the crowds to the hidden Vallon des Auffes inlet and eat sensational bouillabaisse at **Chez Fonfon** (page 210).

- **Chez Gilbert** in Cassis is hard to beat for portside atmosphere and great seafood (page 219).

- **Le Château** between Marseille and Cassis in the Calanque Sormiou offers an idyllic setting (page 222).

THE VAR COAST

- **L'Espérance,** just outside Bandol, serves sublime food in a sublime setting (page 228).

- **O Petit Monde** on Plage Portissol in Sanary-sur-Mer offers great food accompanied by the sound of breaking waves on the nearby rocks (page 229).

NICE AND THE RIVIERA

- **Le Galet** is the pick of the beach restaurants on Nice's famous seafront promenade. It's a fun, family-friendly place for a beachside lunch, but it still feels pretty glamorous (page 258).

- Located at Plage Mala, one of the most undiscovered beach resorts on the Côte d'Azur, **Eden** restaurant-bar serves great seafood, burgers, and fries (page 281).

- Among the row of private beach clubs on Juan-les-Pins's seafront, diners eat with their feet in the sand at **Le Colombier** (page 307).

- In Théoule-sur-Mer, in between Cannes and Fréjus on the Baie de Cannes, **La Cabane du Pêcheur** is the epitome of seaside relaxation (page 344).

Best Cooking Schools and Classes

Cooking holidays are increasingly popular in Provence. You could build a whole week around cooking classes or just schedule half a day. Cookery classes are often combined with market tours, with the ingredients being purchased in the market. Here are some of the best classes available:

AIX, SAINT-RÉMY-DE-PROVENCE, AND MARSEILLE

- **Provence Gourmet** (www.provence-gourmet.fr) offers a friendly home cooking experience with local chef Gilles. Gilles was born and raised in Marseille and lived in the United States for four years. His courses often include vineyard visits and market visits in addition to the cooking class. Prices from €160.

- **L'Atelier des Chefs** (www.atelierdeschefs.fr), a cooking school right in the center of Aix, offers a variety of different classes and start times. Check the website for details.

THE RIVIERA

- **Les Petits Farcis** (www.petitsfarcis.com) in Nice is named after the local "stuffed vegetables" dish. Food writer and teacher Rosa Jackson and her team of English-speaking experts offer market tours, hands-on cooking classes, and lunch in the delightful surroundings of the old town. They also hold pastry- and macaroon-making workshops for groups of up to eight people in their stylish, air-conditioned studio. The market tour and cooking class costs €200 per person, and the pastry workshop is €95.

- **La Peetch** (www.lapeetch.com) is a residential cooking school in the holiday home of former TV host and French cookery icon Julia Child. Run by Makenna (who attended the same school as Julia), her husband, Chris, and a small team of foodie experts, the Courageous Cooking School occupies a rural setting in Châteauneuf-de-Grasse. The all-inclusive cottage accommodation sleeps eight for a six-night stay of recipe-free cooking and restaurant dining.

- **L'Amandier** (www.amandier.fr) restaurant in Mougins offers weekend cooking classes in French with Michelin-starred chef Denis Fétisson. The village also hosts an annual gastronomy festival, Les Etoiles de Mougins, every September where some of France's top chefs offer hundreds of cooking classes to the public.

Hidden Riviera

The Riviera is one of Europe's most popular holiday destinations, but in high summer the coast can feel overrun with tourists, the roads and ports are busy, the trains are crowded, and the restaurants are full, so it's nice to know that some places remain calm and relatively secluded even during July and August. These three locations are a little off the radar and ideal for those who want a calmer, more relaxed time, while still offering access to the lifestyle and activities of the Riviera. The pace here is a little slower—you may find yourself wanting to extend your trip by a few days or even a week amid this serene environment.

Menton
Day 1

After settling into your delightful room at the charming **Sous l'Olivier** B&B, have breakfast on the **promenade du Soleil** overlooking the Mediterranean, walk through **Marché des Halles,** and spend some time at the **Musée du Bastion,** getting familiar with the work of avant-garde French writer and artist Jean Cocteau. Wander through the **old town,** perhaps stepping into the magnificent baroque **Basilique Saint-Michel Archange,** and pay a visit to the many international residents of Menton's **cemetery.**

Spend late afternoon on the **Sablettes beach** or wandering through one of Menton's many **botanical gardens.**

Le Cannet
Day 2

Head west along the coast to **Le Cannet** either on the **A8** autoroute (about 1 hour) or on the slower **coastal roads.** Your base for the next two days will be **La Villa d'Emma,** a pleasant bed-and-breakfast with a pool. It's close enough to visit Cannes if you want to do some shopping or enjoy the nightlife.

Mougins

Spend some time on **rue Saint-Sauveur,** where there are some great restaurants, and enjoy a meal on the place Bellevue overlooking Cannes. Pass in front of the many murals dedicated to the artist Pierre Bonnard and visit the **Musée Pierre Bonnard** on Le Cannet's main boulevard. Follow the **Sur les Pas de Bonnard** route, a 2-hour walk tracing where Bonnard set up his easel around the town that takes you along the covered Siagne canal. Head to **Le Bistro des Anges** for a moonlit supper on the hillside overlooking Cannes.

Ramatuelle

Day 3

The beautiful village of **Mougins** is only 10 minutes' drive away on a hilltop toward Grasse. Spend the morning taking in the art galleries and wandering around the village before lunch. In the afternoon, visit both the **Centre de la Photographie de Mougins** and the **FAMM** (Femmes Artistes du Musée de Mougins), which has an exceptional collection of art from more than 80 women from all over the world. Dine at one of the restaurants on Mougins's main thoroughfare, or head back to Le Cannet for a meal at **Le Coin Gourmet.**

Ramatuelle
Day 4

Check out of the Villa d'Emma and head west once more, this time to Ramatuelle, an old village ingeniously built in the shape of a snail, its narrow lanes spiraling into each other. Spend the morning enjoying the shops, restaurants, and cafés before heading out for some wine-tasting in the many vineyards that dot the area surrounding the village, some of the oldest in France. Make your way to the famous **Plage de Pampelonne** in the late afternoon, after most of the visitors have left, and take advantage of the less crowded **beach clubs** for an indulgent dinner. You'll spend the night at one of **Toison d'Or's** luxury cabins, one of the lucky few that gets to skip the traffic back to Saint-Tropez and enjoy this special part of the coast when it quiets down.

Best Beaches

France's Mediterranean coastline offers a huge range of beaches, from the turquoise waters of Marseille's calanques to the craggy red coastline of the Estérel and the designer gravel at Plage du Larvotto in Monaco. Here is a small selection of some of the region's best beaches:

- **Plage de La Pointe Rouge, Marseille:** Go full bling and wear the most expensive designer shades you can afford on Marseille's hippest beach (page 205).

- **Plage du Bestouan, Cassis:** Escape from the portside hubbub to this secluded cove with a view over the Med toward Cap Canaille and Europe's highest sea cliff (page 217).

- **Plage Anse de Renécros, Bandol:** This moon-shaped sandy bay with a narrow entrance has lagoon-blue waters and is sheltered from the wind (page 227).

- **Plage des Marinières, Villefranche-sur-Mer:** Easy access from the train station and plenty of parking help make Villefranche's long curve of sand a great place for a beach day (page 270).

- **Le Grande Plage, Juan-les-Pins:** The gentle slope into the water makes this an ideal place for families to spend the day on the sand (page 306).

- **Plage de Pampelonne, Ramatuelle:** Expensive clubs welcome yacht parties, but there is space for everyone, including sections for naturist sunbathers and water sports enthusiasts (page 383).

Avignon and the Vaucluse

Readers of this guide who are planning a trip to Europe might find themselves with the same tricky choice that Pope Clement V faced back at the beginning of the 14th century when he was trying to pick a location for the papal court: Which city to choose, Rome or Avignon?

Perhaps it was a glass of the local red (now known as Châteauneuf-du-Pape) that nudged Clement toward Avignon. The blessings of the region must have been evident in every sip of the intoxicating, velvety wine whose high levels of alcohol have been known to induce a state of almost transcendental rapture in today's top tasters. The rest, as they say, is history. The immense, fortress-like Palais des Papes was thrown up in just 20 years and is now the main reason people visit Avignon.

Highlights

Look for ★ to find recommended sights, activities, dining, and lodging.

★ **Palais des Papes:** In the early 14th century, Pope Clement V opted for Avignon as a base for the papal court rather than Rome. The finery of this medieval Gothic palace has been revived with the magic of technology, and a tour of the building induces wonder, reverence, and a hint of trepidation (page 43).

★ **Pont Saint-Bénézet:** A reported miracle worked by a shepherd boy spurred construction of this iconic bridge completed in 1185 (page 45).

★ **Pont du Gard:** Pick your own superlative. Whatever word you alight on will barely do justice to this 275-m-long (902-ft-long), 49-m-high (160-ft-high) Roman aqueduct. For a close-up view of this ancient engineering masterpiece, rent a kayak (page 47).

★ **Wine-Tasting:** Châteauneuf-du-Pape is the blockbuster name, but also allow time to visit the vineyards of Gigondas, Vacqueyras, Rasteau,

and Beaumes-de-Venise. The countryside, as well as the wine, is a delight (page 56).

★ **Roman Theatre of Orange:** Seating 10,000 people, this amphitheater is unique thanks to its surviving stage wall. It comes alive every summer with a major opera festival (page 58).

★ **Roman Ruins of Vaison-la-Romaine:** Vaison has one of the most impressive and extensive collections of Roman ruins in Provence, including the remains of a residential quarter and a theater (page 60).

★ **Antiques-Hunting in L'Isle-sur-la-Sorgue:** Indulge in some of the best antiques shopping in the whole of France against a backdrop of churning waterwheels (page 69).

★ **Fontaine-de-Vaucluse:** The source of the River Sorgue is the largest karst spring in France. It's a deep cavern from which, at certain times of year, clear, cold water explodes in a bubbling torrent (page 70).

Avignon and the Vaucluse

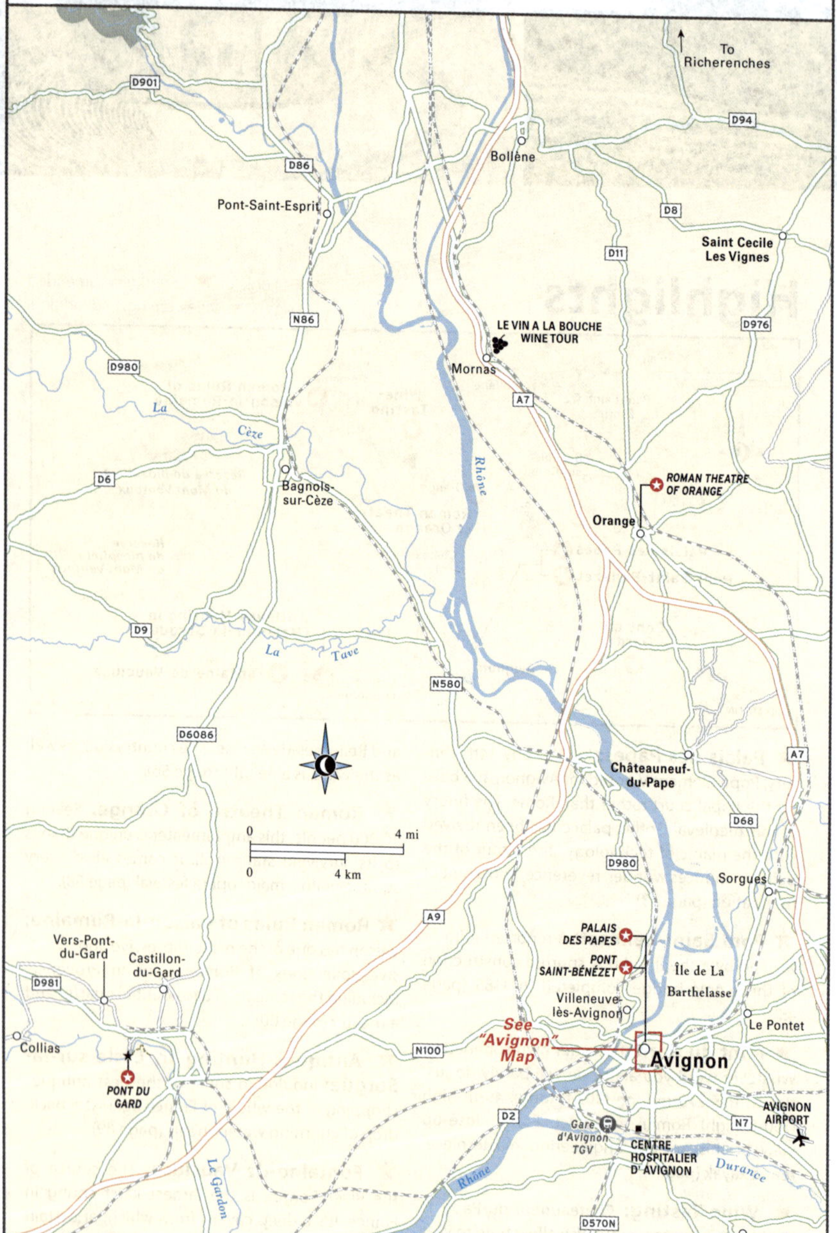

To Richerenches

D901

D94

D86

Bollène

Pont-Saint-Esprit

D8

D11

Saint Cecile Les Vignes

N86

LE VIN A LA BOUCHE WINE TOUR

D976

D980

Mornas

La

Cèze

A7

Rhône

D6

Bagnols-sur-Cèze

ROMAN THEATRE OF ORANGE

Orange

D9

La *Tave*

N580

A7

D6086

Châteauneuf-du-Pape

D68

| 0 | | 4 mi |
| 0 | | 4 km |

D980

Sorgues

A9

PALAIS DES PAPES

PONT SAINT-BÉNÉZET

Île de La Barthelasse

Vers-Pont-du-Gard

Castillon-du-Gard

Villeneuve-lès-Avignon

Le Pontet

D981

See "Avignon" Map

Collias

N100

Avignon

PONT DU GARD

AVIGNON AIRPORT

D2

Gare d'Avignon TGV

N7

CENTRE HOSPITALIER D'AVIGNON

Le Gardon

Rhône

Durance

D570N

Châteaurenard

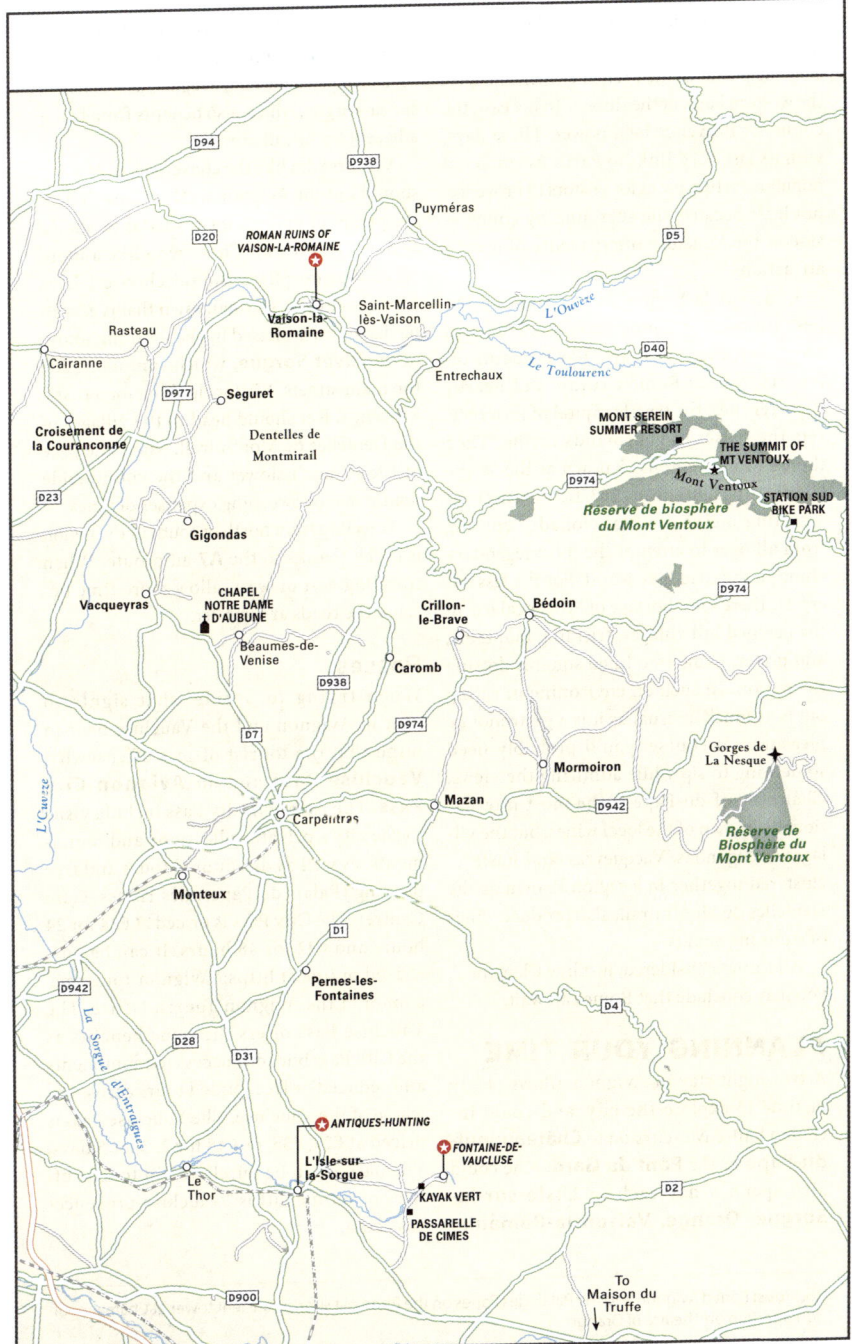

D94

D938

Puyméras

D20

ROMAN RUINS OF VAISON-LA-ROMAINE

D5

L'Ouvèze

Saint-Marcellin-lès-Vaison

Vaison-la-Romaine

Rasteau

Le Toulourenc

D40

Cairanne

Entrechaux

D977

Seguret

Croisement de la Couranconne

Dentelles de Montmirail

MONT SEREIN SUMMER RESORT

THE SUMMIT OF MT VENTOUX

D23

D974

Mont Ventoux

STATION SUD BIKE PARK

Réserve de biosphère du Mont Ventoux

Gigondas

Vacqueyras

CHAPEL NOTRE DAME D'AUBUNE

Crillon-le-Brave

Bédoin

D974

Beaumes-de-Venise

Caromb

D938

D974

D7

L'Ouvèze

Mormoiron

Gorges de La Nesque

Mazan

Carpentras

D942

Réserve de Biosphère du Mont Ventoux

Monteux

D1

Pernes-les-Fontaines

D942

D4

D28

D31

La Sorgue d'Entraigues

ANTIQUES-HUNTING

FONTAINE-DE-VAUCLUSE

L'Isle-sur-la-Sorgue

KAYAK VERT

Le Thor

D2

PASSARELLE DE CIMES

D900

To Maison du Truffe

Avignon has long been one of the principal gateways to Provence. Until the late 15th century the power of the French crown ended on the western bank of the Rhône. To the east, the counts of Provence held power. These days, with its fast TGV links to Paris, Avignon is a popular starting point for visitors to Provence, not least because the surrounding countryside of the Vaucluse offers plenty of nearby attractions.

L'Isle-sur-la-Sorgue boasts some of the best antiques shopping in France; the area around Orange and Vaison-la-Romaine offers impressive Roman ruins; and nearby Richerenches, the truffle capital of Provence, sets the pulses of gourmands racing. Then there's Mont Ventoux: Known as the White Giant, this 1,909-m (6,263-ft) mountain is a cycling mecca, with aficionados coming from all over to attempt the most legendary climb in the road bike world. For the less energetic there's no shortage of Provençal feel in the perched hill villages, trickling fountains, and pretty, plane tree-lined squares dotting the region. In such an environment, eating out is as much a visual as it is a gastronomic pleasure. Of course, you'll probably need something to sip while admiring the views. Châteauneuf-du-Pape is the most internationally known of the local wines, but the villages of Gigondas, Vacqueyras, and Rasteau, clustered together in a region known as the Dentelles de Montmirail, also produce wines of stunning quality.

All things considered, just like Clement V, you may conclude that Rome can wait.

PLANNING YOUR TIME

A two-night stay in Avignon allows plenty of time to explore the city, and could include a half-day excursion to **Châteauneuf-du-Pape** or the **Pont du Gard**. One could also spend a day each in **L'Isle-sur-la-Sorgue, Orange, Vaison-la-Romaine,** Mont Ventoux, and the **Dentelles de Montmirail.** Avignon can easily be used as a base to explore the entire region, as can L'Isle-sur-la-Sorgue, which also benefits from being adjacent to the Luberon.

Visitors who like the conveniences of a city should opt for Avignon as their base. There are plenty of options for restaurants, shops, and entertainment. Those who like a more relaxed atmosphere should choose L'Isle-sur-la-Sorgue. It's a small town that is easy to navigate and is blessed by the soothing sound of the **River Sorgue,** which runs alongside the main streets. Visitors looking for a rustic Provençal feel should head to the villages of the Dentelles de Montmirail, where the pace of life is much slower and the countryside dominated by sweeping expanses of vines.

Traveling from north or south in the region is quick thanks to the **A7** autoroute. When traveling east or west, allow more time because the roads are smaller.

Passes

When trying to decide what sights to visit in Avignon and the Vaucluse bear in mind that the tourist offices offer both a **Vaucluse Pass** and an **Avignon City Pass.** The Avignon City Pass include visits to the city's private collections and monuments, as well as discounted tours and free parking (Palais du Papes; Les Halles, Gare Centre). The City Pass is priced at €24 for 24 hours and €32 for 48 hours. It can be purchased online at https://avignon-tourisme. com or at the Avignon Tourist Office. The Vaucluse Pass offers the same benefits as the City Pass but adds access to monuments and reduced or free guided tours across the whole of the Vaucluse. The Vaucluse Pass is priced at €29, €35, or €48 for 2, 3, or 5 days. The passes can be purchased in tourist offices or online: https://vaucluse-provence-pass.com.

Previous: Pont d'Avignon and the Palais des Papes on the Rhône at sunrise; ancient tower at Châteauneuf-du-Pape; Roman Theatre of Orange.

Itinerary Ideas

THE BEST OF AVIGNON AND THE VAUCLUSE

Day One

1 Make the sumptuous **Maison sur la Sorgue** in L'Isle-sur-la-Sorgue your base to explore the region.

2 Head into nearby Avignon on the D901 (40-minute drive) and explore the city, visiting the magnificent **Palais des Papes** and the Pont d'Avignon.

3 Shop for lunch provisions among the colorful stalls at **Les Halles Food Market.**

4 Enjoy a picnic at the **Rocher des Doms** garden, overlooking the Place du Palais.

5 Leave Avignon and head north for 20 minutes on the D907 to **Châteauneuf-du-Pape** for an afternoon of wine-tasting.

6 Return on the D192 to L'Isle-sur-la-Sorgue for supper at **Le Vivier** restaurant on the banks of the River Sorgue.

Day Two

1 Rise early to visit the source of the River Sorgue, taking the 10-minute drive on the D25 to **Fontaine-de-Vaucluse.**

2 Canoe down the river with **Kayak Vert.** The descent starts just outside Fontaine-de-Vaucluse and takes 2 hours.

3 Have lunch 20 minutes away in Pernes-les-Fontaines (take the D25 and the D938) at **Au Fil du Temps.** Later, work off the calories by visiting as many of the town's 40 fountains as possible.

4 Return to L'Isle-sur-la-Sorgue on the D938 and explore the town's **antiques shops.**

5 Have supper in the excellent **Café Fleurs** in the center of town.

Day Three

1 Drive north and satisfy your inner historian by visiting the Roman ruins at **Vaison-la-Romaine.** Drive time is just under an hour via the D938.

2 Enjoy lunch in Gigondas, 20 minutes away just off the D977, and sample the excellent local wine on offer at **Le Bistrot de L'Oustalet.**

3 Drive to the summit of **Mont Ventoux** for the best view in Provence. Allow 30 minutes' drive time to reach Bédoin from Gigondas, and then another 30 minutes to get from there to the summit. Adrenaline junkies can hire a descent bike and meet their more timid friends back at base camp in Bédoin.

4 Return to L'Isle-sur-la-Sorgue (40 minutes on the D938) and round out the three days with supper in the atmospheric **Le Carré d'Herbes.**

Itinerary Ideas

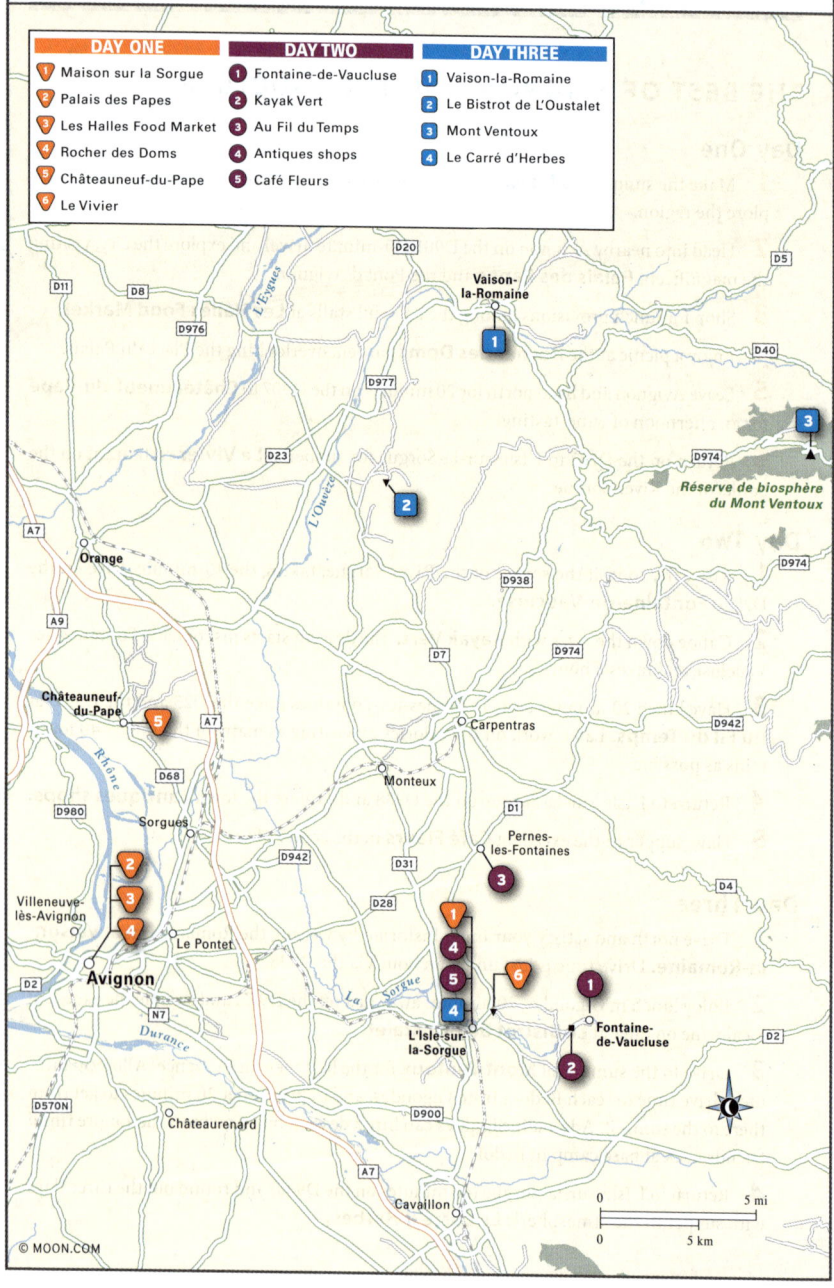

DAY ONE
1. Maison sur la Sorgue
2. Palais des Papes
3. Les Halles Food Market
4. Rocher des Doms
5. Châteauneuf-du-Pape
6. Le Vivier

DAY TWO
1. Fontaine-de-Vaucluse
2. Kayak Vert
3. Au Fil du Temps
4. Antiques shops
5. Café Fleurs

DAY THREE
1. Vaison-la-Romaine
2. Le Bistrot de L'Oustalet
3. Mont Ventoux
4. Le Carré d'Herbes

© MOON.COM

Avignon

To tourists Avignon will always be the City of the Popes, visited for the Palais des Papes and the romance of a kiss on the Pont d'Avignon. To oenophiles it is the capital of the Côtes du Rhône wine region, an area that produces some of the world's most renowned reds. For thespians Avignon is the home of one of Europe's top performing arts festivals. Visit in summer and the streets will be alive with theater troupes promoting their shows. Finally, it's a university city with a large student population sustaining plentiful bars and clubs. Yet the pressure of having so many different people to please means that away from the main streets and squares the city can feel a little run-down. Even so, there is a sense of innovation and creativity. Pop-up shops selling artisan designs are common, and the plentiful theaters host year-round events such as dance and yoga classes. Adding to the allure is a vibrant restaurant scene on and around the various church squares that are a legacy of papal rule.

ORIENTATION

The **historic center** of Avignon, where you will spend nearly all of your time, is surrounded by 4 km (2.5 mi) of walled ramparts. The bus station and central train station (for local trains, not TGV/bullet trains, which arrive farther out) are located opposite **Porte de la République,** just outside the walls. From there a broad avenue (cours Jean Jaurès and then rue de la République) runs south to north through the historic center, arriving at **Place de l'Horloge.** This square is home to the town hall and the 19th-century opera house, and is considered the heart of the city. Just to the north is the **Place du Palais,** the **Palais des Papes,** and the entrance to the **Pont Saint-Bénézet (Pont d'Avignon).** The main road running from west to east, branching off the bottom of Place de l'Horloge, is rue Carnot.

SIGHTS

TOP EXPERIENCE

★ Palais des Papes

Place du Palais; tel. 04 32 74 32 74; https://palais-des-papes.com; Jan.-Feb. and Nov.-Dec. daily 10am-5pm, Mar.-Nov. 9am-7pm; €12, ages 8-17 €6.50, under 8 free, City Pass

The Palais des Papes has long been one of Provence's must-see sights. The sheer size and scale of the Palais are impressive, and that's before you learn it was built in less than 20 years. Between 1335 and 1352, successive popes Benedict XII and Clement VI oversaw the construction. The Palais is a stark monolith of forbidding stone that also served as a fortress and inquisitional court. Little remains of the original furnishings, but holding up the Histopad tablet (provided with the entrance fee) allows visitors to travel back 700 years: Ghosts of cardinals sit in chairs, banquet tables are laden with roast chickens, and the faded frescoes on the walls are suddenly brought to life. The papal gardens have also recently been revived and are now open to the public. They can be included in the visit for an additional €2.50 or visited independently for €5.

Musée du Petit Palais

Palais des Archevêques, Place du Palais; tel. 04 90 86 44 58; www.petit-palais.org; Wed.-Mon. 10am-1pm and 2pm-6pm; free

Built between 1318 and 1320, the Petit Palais predates its big brother on the other side of the square by a decade or so. It was purchased by Pope Benedict XII in 1335 for use as the Episcopal Palace during construction of the Palais des Papes. It now houses a remarkable collection of 13th- to 15th-century paintings and sculptures.

Avignon

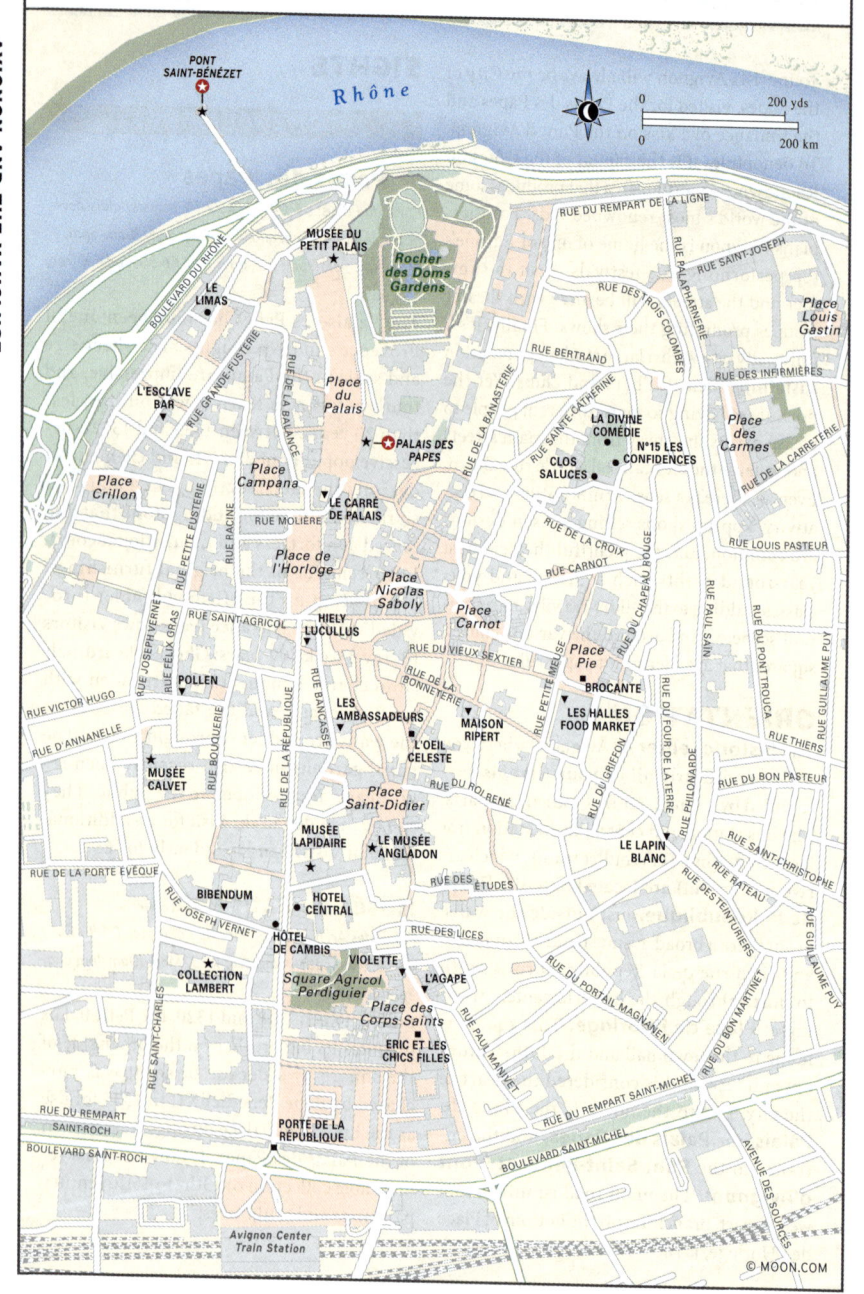

Rhône

PONT SAINT-BÉNÉZET

0   200 yds
0   200 km

RUE DE REMPART DE LA LIGNE

RUE SAINT-JOSEPH

RUE PALAPHARNERIE

RUE DES TROIS COLOMBES

Place Louis Gastin

MUSÉE DU PETIT PALAIS

Rocher des Doms Gardens

LE LIMAS

RUE BERTRAND

RUE DES INFIRMIÈRES

BOULEVARD DU RHÔNE

RUE GRANDE FUSTERIE

L'ESCLAVE BAR

RUE DE LA BALANCE

Place du Palais

RUE SAINTE-CATHERINE

LA DIVINE COMÉDIE

Place des Carmes

PALAIS DES PAPES

RUE DE LA BAMASTERIE

N°15 LES CONFIDENCES

RUE DE LA CARRETERIE

Place Campana

CLOS SALUCES

Place Crillon

LE CARRÉ DE PALAIS

RUE MOLIÈRE

RUE DE LA CROIX

RUE LOUIS PASTEUR

RUE PETITE FUSTERIE

RUE RACINE

Place de l'Horloge

RUE CARNOT

RUE DU CHAPEAU ROUGE

RUE PAUL SAIN

Place Nicolas Saboly

Place Carnot

RUE DU PONT TROUCA

RUE GUILLAUME PUY

RUE JOSEPH VERNET

RUE FÉLIX GRAS

RUE SAINT-AGRICOL

HIELY LUCULLUS

RUE DU VIEUX SEXTIER

Place Pie

POLLEN

RUE DE LA BONNETERIE

BROCANTE

RUE THIERS

RUE VICTOR HUGO

RUE BANCASSE

LES AMBASSADEURS

RUE PETITE MEUSE

LES HALLES FOOD MARKET

RUE DU FOUR DE LA TERRE

RUE D'ANNANELLE

RUE BOUQUERIE

MAISON RIPERT

RUE DU BON PASTEUR

RUE DE LA RÉPUBLIQUE

L'OEIL CELESTE

RUE DU GRIFFON

MUSÉE CALVET

RUE DU ROI RENÉ

Place Saint-Didier

LE LAPIN BLANC

RUE SAINT-CHRISTOPHE

MUSÉE LAPIDAIRE

LE MUSÉE ANGLADON

RUE DES RATEAU

RUE DE LA PORTE EVÊQUE

RUE DES ÉTUDES

RUE DES TEINTURIERS

RUE GUILLAUME PUY

BIBENDUM

RUE JOSEPH VERNET

HOTEL CENTRAL

RUE DES LICES

HÔTEL DE CAMBIS

VIOLETTE

RUE DU PORTAIL MAGNANEN

RUE DU BON MARTINET

COLLECTION LAMBERT

Square Agricol Perdiguier

L'AGAPE

RUE SAINT-CHARLES

Place des Corps Saints

RUE PAUL MANIVET

ERIC ET LES CHICS FILLES

RUE DU REMPART SAINT-ROCH

PORTE DE LA RÉPUBLIQUE

RUE DU REMPART SAINT-MICHEL

AVENUE DES SOURCES

BOULEVARD SAINT-ROCH

BOULEVARD SAINT-MICHEL

Avignon Center Train Station

© MOON.COM

Musée Calvet

65 rue Joseph Vernet; tel. 04 90 86 33 84; www.musée-calvet.org; Wed.-Mon. 10am-1pm and 2pm-6pm; free

The Musée Calvet contains a large and eclectic mix of paintings, sculptures, and archaeological finds, including an Egyptian collection. It's located in a spacious, airy 18th-century palace, and you don't have to pay a cent to enjoy it.

Musée Lapidaire

27 rue de la République; tel. 04 90 85 75 38; www.musée-lapidaire.org; Tues.-Sun. 10am-1pm and 2pm-6pm; free

Housed in an old Jesuit chapel on the main shopping street, this museum is a pleasant place to stumble upon. Luring you in from the doorway is an open view of the overflow archaeological collection of the Musée Calvet.

Collection Lambert

5 rue Violette; tel. 04 90 16 56 20; www.collectionlambert.fr; Sept.-June Wed.-Fri. 1pm-6pm, Sat.-Sun. 11am-6pm, July-Aug. daily 11am-7pm; €10, ages 12-25 €5, under 12 free, City Pass

Housed in two hôtel particuliers, the Collection Lambert is something of a haven from the bustling streets of the city. The collection is composed of 20th- and 21st-century contemporary art donated to the state by Yvon Lambert, including works by Claire Fontaine, Robert Ryman, and Bertrand Lavier.

Le Musée Angladon

5 rue Laboureur; tel. 04 90 82 29 03; https://angladon.com; Apr.-Oct. Tues.-Sun. 1pm-6pm, Nov.-Mar. Tues.-Sat. 1pm-6pm, closed Jan.; €8, ages 15-25 €3, ages 4-14 €1.50, City Pass

Picasso, Sisley, van Gogh, Cézanne, Degas: The list reads like a who's who of the art world. The paintings form part of the collection of couturier Jacques Doucet (1853-1929). Doucet's great-nephew Jean Angladon and Angladon's wife, Paulette Martin, established the museum as a place for the public to enjoy intimate experiences with some of the great works of modern art.

★ Pont Saint-Bénézet (Pont d'Avignon)

Pont d'Avignon, boulevard de la Ligne; tel. 04 32 74 32 74; www.avignon-pont.com; Jan.-Feb. daily 10am-6pm, Mar.-Nov. daily 9am-7pm, Nov.-Dec. daily 10am-5pm; €5, joint ticket with Palais des Papes €14.50, City Pass

In 1177 Saint-Bénézet, then a young boy, was out tending his sheep when he experienced a vision: a message from God calling him to build a bridge over the Rhône. People laughed at the divinely appointed bridge builder. They laughed no more when he placed the first stone, a boulder so massive that no man could lift it. Wealthy backers flocked to support the project, and a 22-arch, 900-m (2,952-ft) bridge was completed. God's blessing did not last forever. Parts of the bridge were swept away in successive floods and rebuilding work continued over the centuries, until in 1688 the city finally accepted the inevitable and let the bridge fall to ruin. Its fame today is as much a product of the children's song "Sur Le Pont d'Avignon" as it is of the legend of Saint-Bénézet. The song dates back to the 17th century, and the original version is more likely to have mentioned people dancing *under* the bridge—where the arches crossed the Île de la Barthelasse—than on it. The Pont d'Avignon is accessed via rue Ferruce. Stairs take you up the tower to the level of the bridge. A ramp and lift facilitate disabled access. An audio guide accompanies the visit, which lasts about an hour.

Villeneuve-lès-Avignon

www.villeneuvelezavignon.fr

A short taxi or bus ride (Line 11 from cours President Kennedy) takes you across the river to Villeneuve-lès-Avignon. Until 1790 Avignon itself belonged to either the papacy or the Duchy of Naples, and the French crown exercised its authority from the town of Villeneuve on the opposing bank. These days Villeneuve is a prosperous suburb of Avignon, offering excellent views back across the river. Place Jean Jaurès is a nice shady square for lunch.

★ Pont du Gard

A short hop over the Rhône from Avignon is the Pont du Gard Roman aqueduct. At nearly 50 m (164 ft) tall and 490 m (1,607 ft) long, this was the highest aqueduct of the Roman world, boasting 47 arches at the time of its construction. Its job was to carry water to the city of Nîmes 25 km (16 mi) away. The sheer scale of the endeavor provokes most people to pause in their stride, stop, and stare. The three-tier aqueduct was built with 50,000 tons of stone. A visit includes entry to the on-site museum, which delves into the science of how the Romans managed to construct such a behemoth. To drive to Pont du Gard from Avignon, take the N100, covering 26 km (16 mi) in 28 minutes.

GUIDED TOURS
Avignon Gourmet Tour

Avignon city center, meeting point and time confirmed on reservation; tel. 06 62 89 44 55; www.avignongourmetours.com; Tues.-Sat.; from €55

A guilt-free way to sample the best local produce is to take a walking tour. You'll enjoy tasting olive oil, cheeses, and plentiful Côtes du Rhône wine.

Noctambules d'Avignon

Avignon city center, meeting point and time confirmed on reservation; tel. 06 14 23 41 31; www.lesnoctambulesdavignon.com; info@lesnoctambulesdavignon; Apr.-late Sept. daily 9:30pm; from €27

For those wishing to avoid the hot summer sun, these tours are a great atmospheric option.

SPORTS AND RECREATION
Parks
Île de la Barthelasse

Located on a large island in the middle of the Rhône, this park is popular with cyclists and walkers. The views back toward Avignon are

beautiful. In the summer a **shuttle boat** runs every 15 minutes from boulevard du Quai de la Ligne, a short distance upriver from Pont Saint-Bénézet, across to the Île. Out of the main tourist season the hours are more irregular. The park can also be accessed from Pont Edouard Daladier.

The Rocher des Doms Garden

Overlooking the Place du Palais is this delightful garden. It's a good spot for views across the Rhône to Villeneuve-lès-Avignon. The garden, which has a small snack bar, is also ideal for a picnic in the shade.

Cycling
Rampart Circular

A 4-km (2.5-mi) cycle route has been created around the old city wall. The route also extends over the Pont Edouard Daladier and onto Île de la Barthelasse. The circular route hugs the Rhône before turning inland and heading through the administrative center of the city. The route can be picked up at any point along the city walls. Try starting at Place Crillon, just off the Place du Palais.

Velopop

Avignon stations include Place Pie, Hotel de Ville, Carmes, Port Saint-Lazare, Porte Thiers, Porte Lambert, Inurès; tel. 08 10 45 64 56; from €1.50

Avignon operates a bike rental scheme as part of its transport service. Sign up via the app (Velopop) or on the website (www.velopop.fr) and you will be sent a code to unlock a bike. After use, return the bike to any Velopop station.

FESTIVALS AND EVENTS
Avignon Festival and Avignonleoff

Various venues across town; www.festival-avignon.com; runs for three weeks from end of the first week in July; free to €50 depending on performance

Created in 1947, the Avignon Festival is one of the most important contemporary performing arts festivals in the world. Sixty shows

1: the imposing Palais des Papes **2:** Pont Saint-Bénézet **3:** Roman aqueduct at Pont du Gard **4:** view of Avignon from Île de la Barthelasse

are produced across 40 venues, attracting an audience of 155,000 people. The festival is so successful that it has spawned a more riotous and irreverent baby brother, Avignonleoff (www.avignonleoff.com), which puts on 1,000 shows across town. The websites for both the Avignon Festival and Avignonleoff are available in English and contain full program information.

SHOPPING
Brocante
Place Pie; Tues. and Thurs. 8am-5pm
Twice a week year-round, 30 brocante (secondhand) traders set up in the picturesque Place Pie. From antique mirrors to limited-edition books, there are always surprises, and it's always fun browsing.

Eric et les Chics Filles
7 Place des Corps Saints; tel. 04 90 82 54 67; https://ericetleschicsfilles.wordpress.com; Tues.-Sat. 10am-7pm
Nine different designers present their collections of clothing, knitwear, jewelry, bags, and textiles in a funky creative space.

L'Oeil Celeste
48 rue des Fourbisseurs; tel. 04 90 85 29 12; Tues. 2:30pm-6:30pm, Wed.-Sat. 10:15am-12:30pm and 2:30pm-6:30pm
Designer Patricia Gauthier has put together a collection of silver jewelry using stones, gemstones, and minerals from around the world.

FOOD
Regional Cuisine
★ L'Agape
21 Place Corps Saints; tel. 04 90 85 04 06; www. restaurant-agape-avignon.com; Tues. noon-1:15pm, Wed.-Sat. noon-1:15pm and 7:30pm-9:15pm; €25-80
A great location on a pretty square, seasonal good-value menus, a convivial atmosphere, and inventive cooking make this one of the most popular restaurants in the city.

Pollen
3 rue Joseph Vernet; tel. 04 86 34 93 74; http://pollen-restaurant.fr; Mon.-Tues. and Thurs.-Fri. noon-1:30pm and 7:30pm-9:30pm, Wed. noon-1:30pm; €40-120
The vibe at Pollen is cool, modern, and minimalist, with a clientele of serious foodies who love nothing better than chatting about what's on the plate in front of them. Reservations are made through the website only.

Hiely Lucullus
5 rue de la République; tel. 04 90 86 17 07; www. hiely-lucullus.com; Thurs.-Mon. noon-1pm and 7:30pm-8:30pm; €60-120
Food is a very serious business at Hiely Lucullus. The crisp table linens and polished cutlery impress before you even sit down. Presentation of dishes, such as saddle of rabbit stuffed with olives and served with a chicken jus, is immaculate.

Bistros
Le Lapin Blanc
101 rue de la Bonneterie; tel. 07 71 59 68 65; www. lelapinblanc.fr; Mon.-Tues. and Thurs.-Sat. noon-1:45pm and 7pm-9:45pm; €13-24
Bilingual Melanie greets you front of house while co-owner Amelie works magic in the kitchen at Le Lapin Blanc. An excellent-value seasonal menu is served on the terrace in the summer and in the cozy interior in the winter.

Bibendum
83 rue Joseph Vernet; tel. 04 90 91 78 39; www. bibendumavignon.fr; Tues.-Sat. noon-1:30pm and 7pm-9pm (Wed. lunch only); €17-32
Bibendum is a buzzy, glamorous venue that also houses a wine and cocktail bar. The presentation is slick and the food bistronomic, with a menu composed of modern twists on classics. Try the succulent roast baby chicken with a mustard and tarragon sauce.

1: Rocher des Doms garden **2:** Avignon Festival **3:** shopping at the brocante market **4:** lunch with a view of the Palais des Papes

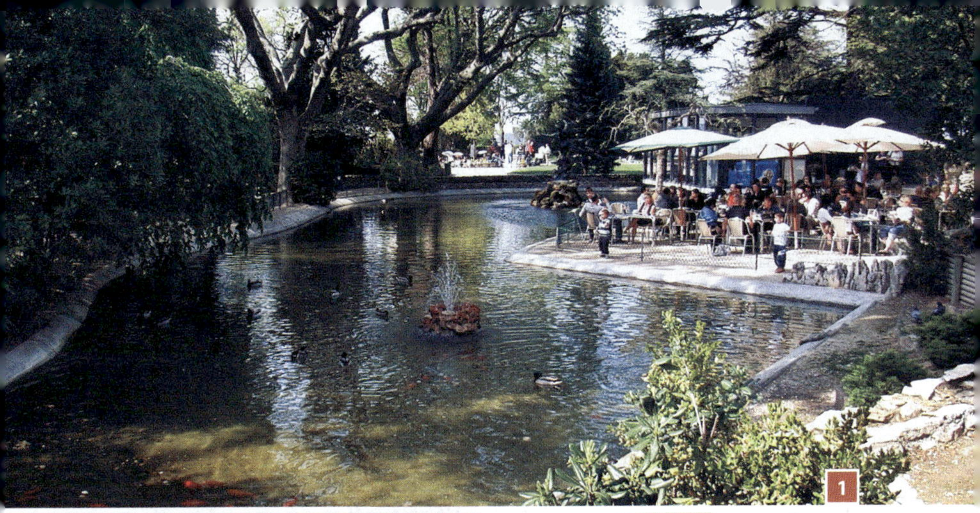
1

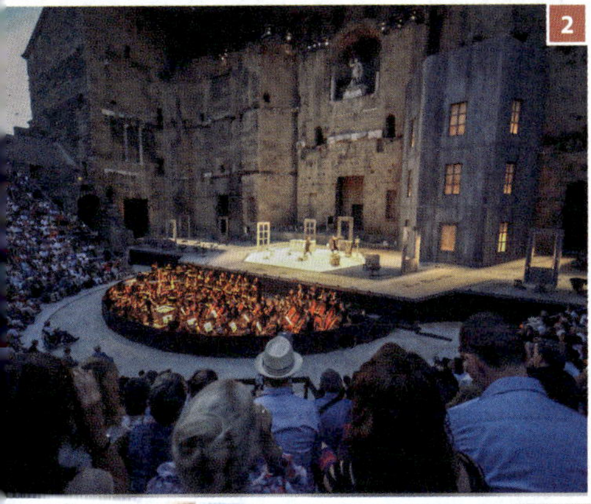
2

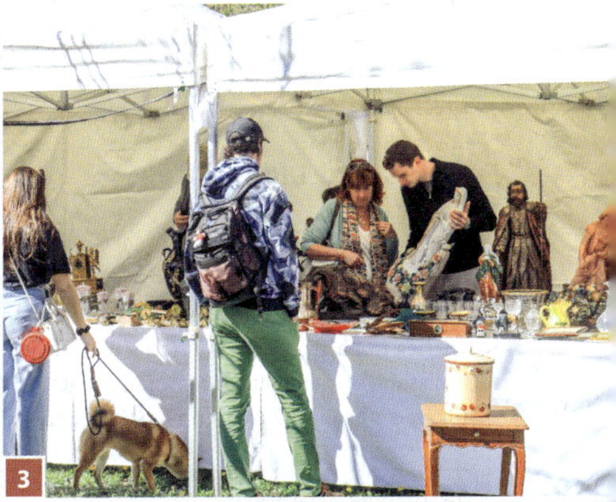
3

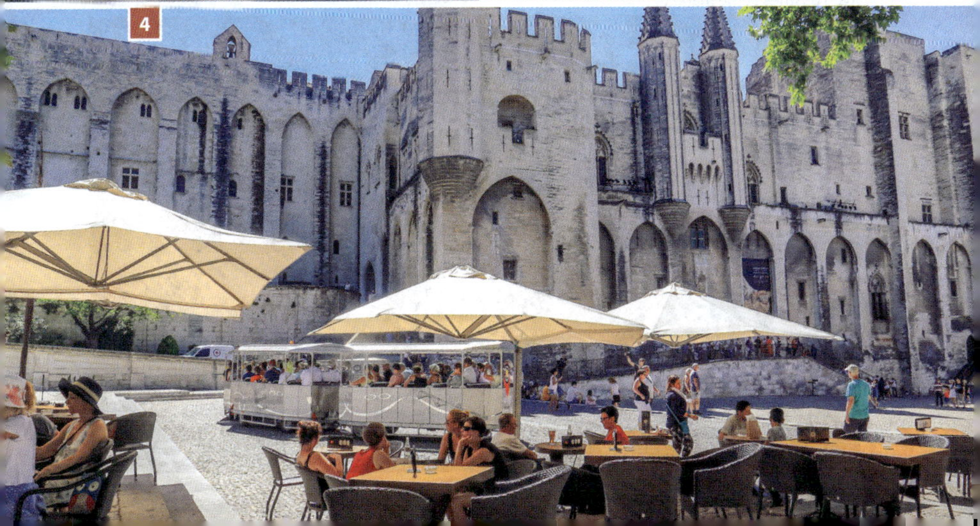
4

Maison Ripert

28 rue de la Bonneterie; tel. 04 90 27 37 97; www.
ripert.business.site; Mon.-Tues. and Thurs.-Sat. noon-
2pm and 7pm-10pm; €26-45

Set in an old patisserie dating back to 1820, the interior here is a delight, with wooden chairs, checked tablecloths, and gilded mirrors. The menu is fairly standard bistro fare with dishes such as lamb shank and duck breast.

Le Carré de Palais

1 Place du Palais; tel. 04 65 00 00 01; www.
carredupalais.com; daily noon-10pm; €29-46

Situated in the old Bank of France building, the Carré is a large bistro with a terrace facing the Place du Palais with a view of the Palais des Papes. Matching whatever you order from the seasonal menu with the perfect glass of Côtes du Rhône is the house specialty.

Cafés and Light Bites
Violette

30 Place Saint Corps; tel. 04 90 47 45 50; Tues.-Sat.
7:30am-7:30pm; from €3

This is a good place for a snack, from sandwiches to pain au chocolat (chocolate-filled croissant) to delicious cakes and pastries. It's located on one of Avignon's pretty church squares, and you can either eat in or take your delicacy to a nearby shady bench.

Markets
Les Halles Food Market

18 Place Pie; tel. 04 90 27 15 15; www.avignon-leshalles.
com; Tues.-Sat. 6am-2pm

Even if you have just eaten, a visit to Les Halles is a must. This covered market is like a window into the French soul. The pride that each of the market traders takes in the produce and the care with which it is arranged speak to the French people's enduring relationship with the land and their cuisine.

BARS AND NIGHTLIFE
L'Esclave Bar

12 rue des Limas; tel. 04 90 85 14 91; www.facebook.
com/esclavebar; Tues.-Sun. 11:55pm-7am; drinks from
€10

L'Esclave is Avignon's only city-center gay bar. Tuesday is karaoke night, which is particularly popular.

Les Ambassadeurs

27 rue Bancasse; tel. 04 90 86 31 55; www.
clublesambassadeurs.fr; Thurs.-Sun. 11:30pm-late; €10
admission Thurs., €20 Fri.-Sun., drinks from €10

Cozy booths surround the usually packed

fresh pastries at the market

dance floor of Les Ambassadeurs. Dress well to make it past the doorman.

Bibendum–Bar a Cocktail

83 rue Joseph Vernet; tel. 04 90 91 78 39; www. bibendumavignon.fr; Tues.-Sat. 6pm-12:30am; drinks from €12

Choose from an extensive cocktail list filled with classics and a few local specialties, and then sit back and relax in sumptuous surroundings listening to live piano music or DJ sets.

ACCOMMODATIONS
€50-100
Hotel Central

31 rue de la République; tel. 04 90 86 07 81; www.hotel-central-avignon.com; €89 d

Hotel Central offers 35 simply furnished rooms. The location is, as the name suggests, central, with Avignon's main shopping street just out the front door.

€100-200
Le Limas

51 rue de Limas; tel. 06 69 00 60 37; www.le-limas-avignon.com; €140 d

This four-bedroom bed-and-breakfast right in the historic heart of Avignon has a pretty roof terrace overlooking the Palais des Papes. In the summer months it's the perfect place for a croissant and coffee.

Hôtel de Cambis

89 rue Joseph Vernet; tel. 04 90 14 62 73; www. hoteldecambis.com; €150 d

A centrally located boutique hotel with an emphasis on the region's wines, the Cambis is an excellent choice for oenophiles.

Nº15 Les Confidences

15 rue Saluces; tel. 06 11 52 51 38; www.15avignon.fr; €160 d

This charming bed-and-breakfast is near the Palais des Papes and has a verdant garden and small pool.

Maison Orsini

21 rue Montée de la Tour, Villeneuve-lès-Avignon; tel. 06 82 27 65 94; www.maisonorsini.com; €170 d

So often in Provence, it is better to be on the outside looking in. The view from the terrace of the Maison Orsini in Villeneuve over the Rhône to the Palais des Papes makes the stay.

Over €200
★ Clos Saluces

11 rue Saluces; tel. 06 72 75 49 37; www.leclossaluces. fr; €240 d

It took three years to renovate this beautiful 19th-century town house close to the Palais des Papes. The interior decor is a rich mixture of vintage furnishings and objects from the 1950s to the 1970s. There's a shady garden to relax in.

La Divine Comédie

16 Impasse Jean Pierre Gras; tel. 06 77 06 85 40; www. la-divine-comedie.com; €400 d

Boasting the largest private garden in Avignon, with a 15-m (49-ft) swimming pool and wellness center, this five-suite guesthouse is the latest addition to Avignon's collection of luxury retreats. The inside is furnished with gorgeous antiques.

INFORMATION AND SERVICES
Tourist Office

41 cours Jean Jaurès; tel. 04 32 74 32 74; www.avignon-tourisme.com, Mon. Sat. 9am-6pm, Sun. 10am-5pm

Avignon's central tourist office has English-speaking staff who can help with anything from finding accommodations to getting tickets for events.

Centre Hospitalier d' Avignon

305 rue Raoul Follereau; tel. 04 32 75 33 33; www. ch-avignon.fr

Located to the south of the city near the TGV station, the hospital has a 24-hour emergency service.

Post Office

2 rue Petite Meuse; tel. 04 32 75 33 339; Tues.-Fri. 9am-noon and 1:30pm-4pm, Sat. 9am-noon

On the corner of the attractive Place Pie in the center of Avignon, the post office offers the standard services.

GETTING THERE

Air

Marseille Provence Airport is the main hub for Avignon and is 45 minutes' drive away. **Avignon Airport** (335 avenue Clément Ader, Avignon-Montfavet; www.avignon. aeroport.fr) has in the past offered seasonal flights with UK low-cost carriers, but at the time of writing there were no routes available.

Car

The city sits at the heart of the motorway network. It is located on the A7 and A9 autoroutes and the national N7 and N100 routes. Traveling from **Paris** to Avignon by car takes just under 7 hours on the A6 and A7 autoroutes. From **Arles,** take the D2; the journey is 45 km (28 mi) and takes about 50 minutes. From **Aix-en-Provence,** take the A7 for 88 km (54 mi); the trip takes 70 minutes. From **Marseille,** take the A7, covering 105 km (65 mi) in 70 minutes.

The most convenient parking for access to the historic center is at the **Palais des Papes,** located on the north side of the city on the banks of the Rhône. There are also parking lots outside the city center at **Île Piot** and **Parking Les Italiens** that are free of charge. A park-and-ride bus takes you to the city center.

Train

The **TGV station** is on avenue de la Gare in the Courtine district southwest of the city center. It takes about 2 hours 40 minutes to travel to Avignon from Paris. It takes 1 hour to get there from Lyon and 30 minutes from Marseille.

The price of a TGV train varies widely depending on how far ahead you book and when you travel. Complicating matters further, there is a new low-cost TGV service called Ouigo that offers a no-frills service between the cities. If you travel in the week in the middle of the day, it is possible to pick up a train from Avignon to Paris for as little as €25. Travel in peak times and the same journey could cost you more than €200. There are new direct TGV connections between Avignon and Barcelona, Madrid, Frankfurt, Geneva, and Amsterdam.

A train connection from the TGV station to the city center runs 35 times a day; the journey time is 5 minutes on a TER regional train. The city-center train station on boulevard Saint-Roch just outside the town walls offers connections to local towns such as Carpentras.

GETTING AROUND

Most people will find that the easiest way to get around Avignon is to walk. Many of the roads are pedestrian-friendly and the distances between sights are relatively short.

Electric shuttle buses connect the out-of-town parking lots on Île Piot and Parking Les Italiens to the city center. Within the city **La Baladine,** seven-seat electric vehicles, travel through the city center. A ride costs €0.50; you simply need to wave your hand for the vehicle to stop. **Velopop cycles** are available for rent at 17 cycle stations across Avignon.

The Vaucluse

Cheaper, quieter, and more authentic than the tourist hot spots, the northern Vaucluse is an area of Provence that is often overlooked by visitors. Yet at Orange and Vaison-la-Romaine are Roman ruins that can compete with those in Arles, and dotted throughout the Dentelles de Montmirail are picturesque villages that rival those in the Luberon or Les Alpilles. Added to this are three unique reasons to visit: First, L'Isle-sur-la-Sorgue is a town unlike any other in Provence. Built at the branching of the Sorgue river, a network of canals and bridges link its pretty streets and give this town its nickname of "the Venice of Provence." The sight of the waterwheels (there are still 20 turning throughout the town) is alone worth a visit, but most people come for the antiques, from chandeliers to leather club armchairs to smaller items such as kitchen tools; browsing the stores is an utter pleasure.

Second, there's Mont Ventoux, which dominates the skyline of Provence. It's visible from the Luberon, Arles, Les Alpilles, and throughout the northern Vaucluse, peeking between trees and jutting majestically above other lesser peaks. There is no better way to understand the geography of Provence than by standing on its summit How you get to the top is up to you, but note the ascent once famously killed a professional cyclist in the Tour de France, so if in doubt, drive.

Finally, although the vignerons of Bandol on the Var coast might quibble, the southern Rhône is the finest wine-producing region in Provence. Be careful though, and check the alcohol content listed on the label, because these days some wines contain as much as 15.5 percent alcohol, which is practically the level of a fortified wine. The effect can be not so much to loosen your tongue as to unravel it and lasso it around a nearby tree.

ORIENTATION

The Vaucluse stretches from its capital, Avignon, in the west, north along the banks of the Rhône, and includes the wine-producing village of **Châteauneuf-du-Pape** and the town of **Orange,** 30 km (19 mi) away. Orange is famed for its Roman theater. To the northeast of Orange is another celebrated Roman site, **Vaison-la-Romaine,** and south of Vaison is the small mountain chain the **Dentelles de Montmirail.** Here you will find well-known wine-producing villages such as Gigondas, Vacqueyras, Rasteau, and Beaumes-de-Venise. Immediately to the east of the Dentelles is the unmissable highest mountain in Provence: **Mont Ventoux.** Head south from Mont Ventoux and you'll encounter the ever-churning waterwheels of the renowned antiques center of **L'Isle-sur-la-Sorgue.** The Vaucluse extends east from here and into the Luberon, an area covered later in this guide.

TOP EXPERIENCE

CHÂTEAUNEUF-DU-PAPE

The construction of Châteauneuf's eponymous château was started under Pope John XII in 1317. Today only the ruins of its walls still overlook the village. However, the wine industry that the Avignon popes nurtured survives. Look carefully at the fields of vines and you will see round stones surrounding the bases of the vines. These were left behind after the retreat of an ancient sea. Today they soak up the heat of the sun during the day and act like radiators at night, keeping the soil around the vines warm, speeding the maturation of the grapes and maintaining moisture in the soil. All of this helps to produce sumptuous red wines. Upon arrival in Châteauneuf-du-Pape, you'll see that the outskirts of the village are dominated by the tasting rooms of

1

2

3

the major vineyards. Once you have had a few nips of wine, the village is worth exploring. Head up the hill toward the château, and then duck off the main drag to find picturesque and sleepy side streets.

Wineries
Maison Brotte
avenue Saint-Pierre de Luxembourg; tel. 04 90 83 59 44; www.museeduvinbotte.com; daily 9am-11:45am and 2pm-5:45pm; tasting of 3 wines free, longer experiences €4-17
On the outskirts of Châteauneuf-du-Pape, this domaine offers the most informative overall visit. A small on-site museum explains the unique terroir of Châteauneuf-du-Pape. Reserve your tasting on the website.

Beaurenard
10 avenue Saint-Pierre de Luxembourg; tel. 04 90 83 71 79; www.beaurenard.fr; Mon.-Fri. 10am-noon and 1:30pm-6pm; tasting free
This renowned producer has a tasting room and wine cellar near the entrance to town. If you want to take a tour of the cellar it helps to phone in advance, but you can taste the estate's wines without a reservation. In addition to Châteauneuf-du-Pape, there's an excellent Rasteau full of spices and ripe fruit available for €18 a bottle.

Andre Mathieu
3bis route de Courthézon; tel. 04 90 83 72 09; www.domaine-andre-mathieu.com; Apr.-Sept. Mon.-Fri. 9am-noon and 2pm-6pm, Sat. 10am-noon and 2pm-6pm, Oct.-Mar. Mon.-Fri. 9am-noon and 2pm-5:30pm, Sat. by reservation; tasting free
Visiting Domaine Mattieu is an intimate experience: You might meet the winemaker and you'll see the land where the vines are growing. Prices for a bottle of Châteauneuf start around €20.

★ Vinadea
8 rue Maréchal Foch; tel. 04 90 83 79 60; www.vinadea.com and https://wineactivities.net; daily 10am-1pm and 2pm-6:30pm; tasting free
This shop is recommended by locals as the best place to buy and taste wine. More than 200 different wines are on sale from Châteauneuf and surrounding appellations, and different bottles are opened every day for free tastings. There is also Vinotheque, an associated experience center where different tasting experiences can be reserved (€25-45). See the website for details.

Tours
Wine Prestige Tour
Pickup and drop-off available for addresses within 10 km (6 mi) of Châteauneuf-du-Pape; tel. 06 83 53 39 79; https://wineprestigetour.com; daily; €90 half day, €140 full day
Pierre Fernandez was born in Châteauneuf-du-Pape. His grandmother was one of the first female winemakers in the appellation, and he's a qualified sommelier. In other words, what he doesn't know about the wines of the region isn't worth knowing. Tours are friendly and informative, and Pierre's connections open doors (and bottles) that would otherwise be closed to the public.

Avignon Wine Tour
Pickup and drop-off available for addresses within Avignon; tel. 06 28 05 33 84; www.avignon-wine-tour.com; weekly tours include Châteauneuf-du-Pape on Wed.; €130/day
Francois Marcou leads wine tours throughout Provence. As you would expect from a former hotelier and restaurateur, he's charming and very knowledgeable about wine. The tour includes tastings at four vineyards and a lunch stop.

Festivals
Verraison Festival
Throughout the village; www.poptourisme.fr/sortir/agenda-chateauneuf-du-pape; first weekend in Aug.
Châteauneuf holds a huge party to celebrate the ripening of the grapes. There's usually a

---

1: The pebbles in Châteauneuf-du-Pape soak up the heat and give the wine its unique flavor.
2: ruins of Châteauneuf-du-Pape's château
3: countryside around Châteauneuf-du-Pape

☆ Wine-Tasting

Terroir is one of the most frequently used words in the French wine vocabulary. It's also the word that is most frequently misunderstood by enthusiastic amateur tasters. Literally translated it means "the land." But when a professional says they can taste the terroir they mean not just the soil type, but also the altitude, the surrounding geology, the weather, and even the ethos of the winemaker. And nowhere in Provence is this concept more easily understood than on a hot summer's day in the southern Rhône. On days like this you can smell the grapes fermenting on the vine, and even perhaps fry an egg on the surface of one of the large, rounded stones that litter the old floodplain in places. To truly taste wines from the southern Rhône, you have to experience the terroir for yourself by heading out into the vines for a tasting. The red wines are the stars of the show, but the whites are a better value and are pleasingly long in the mouth. You might like to visit:

red wine of the southern Rhône

DOMAINE GOUBERT

235 chemin de Jardinieres, Gigondas; tel. 04 90 65 86 38; www.lesgoubert.fr; Mon.-Fri. 9am-noon and 2pm-6pm

The domaine produces fantastic Gigondas, full of ripe fruit flavors, spices, and an undertone of tobacco.

CLOS DE CAVEAU

1560 chemin de Caveau, Vacqueyras; tel. 04 90 65 85 33; www.closdecaveau.com; summer daily 10am-noon and 2pm-6pm, Nov.-Mar. Mon.-Fri. 10am-noon and 2pm-6pm

The signature Vacqueyras red tastes of ripe fruit and plums with a touch of mint.

ANDRE MATHIEU

3bis route de Courthézon; tel. 04 90 83 72 09; www.domaine-andre-mathieu.com; Apr.-Sept. Mon.-Fri. 9am-noon and 2pm-6pm, Sat. 10am-noon and 2pm-6pm, Oct.-Mar. Mon.-Fri. 9am-noon and 2pm-5:30pm, Sat. by reservation; free tasting

Many Châteauneuf producers have moved their tasting rooms into town, but not Andre Mathieu, where you can still visit the vineyard.

medieval banquet and a "papal" procession. In a village where it's never hard to find a drink, it becomes impossible to move without having a glass thrust into your hands.

Food
La Mule du Pape
2 rue de la République; tel. 04 90 83 79 22; Tues.-Wed. 10am-7pm, Fri.-Mon. 10am-10pm; €23

The restaurant takes its name from a famous Alphonse Daudet short story. Grab a copy and read all about the pope's vindicative donkey in between courses of the excellent-value and beautifully presented three-course menu.

Le Comptoir de la Mere Germaine
Place Jean Moulin; tel. 04 28 69 00 60; http://lameregermaine.chateauneufdupape.fr; Thurs.-Mon. noon-1:45pm and 7pm-9:30pm; €22-28

A welcome addition to the Châteauneuf restaurant scene, the Comptoir has a gorgeous terrace overlooking the vines. The food is modern, as is the decor. It's all very slick and well produced. Each dish has an accompanying wine recommendation.

Le Verger des Papes
rue Montée du Château; tel. 04 90 83 50 40; Tues.-Sat. noon-1:30pm and 7:30pm-8:30pm; €18-39
Nestled in the heights of Châteauneuf-du-Pape, Le Verger des Papes is a village stalwart. The menu is well thought out, with something for everyone. There's even an opportunity for a wine-tasting after lunch in the adjoining shop.

Accommodations
Hotel Sommelier
2268 route de Roquemaure D17 Grange Neuve; tel. 09 70 35 60 29; www.hotel-charme-vaucluse.com; €155 d
Three km (2 mi) outside Châteauneuf, with its own pool and restaurant, this hotel is a good place from which to explore the surrounding vineyards. The are 16 rooms, of which 2 are large family suites. Most rooms overlook the swimming pool, which nestles beneath the picturesque ivy-covered walls of the L-shaped hotel building.

★ Château Fines Roches
1901 route de Sorgues; tel. 04 90 83 70 23; https:// chateaufinesroches.com; €199 d
This hotel is set in a picturesque 19th-century château that is surrounded by vines. Views from the turret room out across the countryside are magnificent. There's an onsite restaurant with a terrace looking out to Châteauneuf and Mont Ventoux.

Hotel La Mere Germaine
3 rue Commandant le Maitre; tel. 04 90 22 78 34; www.lameregermaine-chateauneufdupape.fr; €250 d
The Mere Germaine is a village-center institution offering rooms, a Michelin-starred restaurant, and a brasserie. The rooms eschew the usual subdued Provençal color palette in favor of bold, bright colors. They offer sweeping views over the surrounding countryside.

Getting There and Around
The village of Châteauneuf-du-Pape is best explored on foot, but you will also want to see the surrounding area. If you do not have a car, consider renting a bike to get out and about in the vineyards. Bicycles are available for rent from **Sun-e-bike** (rue des Consuls, Châteauneuf-du-Pape; tel. 07 58 14 33 97; www.location-velo-provence.com).

Car
From **Avignon** to Châteauneuf-du-Pape, take the D907; the trip is 17 km (10.5 mi) and takes 23 minutes. From **Orange,** take the D68 for 10 km (6 mi); the drive takes 16 minutes. If you're coming from **Vaison-la-Romaine,** take the D977; the journey is 35 km (22 mi) and takes 40 minutes; and from **L'Isle-sur-la-Sorgue,** take the D16 31 km (19 mi) and you will arrive in 43 minutes.

Bus
From Avignon, **Zou line 902** (https://zou. maregionsud.fr) runs to Sorgues; this part of the journey can also be made by train from the Avignon central station. From Sorgue, a connecting bus (line 923) takes you to Châteauneuf-du-Pape. The journey, including the connection, takes just over 1 hour and costs €4. Using the train as a substitute where necessary, there are five services a day. **Line 23** also runs five times a day between Châteauneuf-du-Pape and Orange; the trip takes 25 minutes and costs €2.

Bike
A green cycle route—the **ViaRhona RV17**—links Châteauneuf-du-Pape with Orange and Avignon. Details can be found online at www.provence-cycling.co.uk; navigate to "Experiences" and "A Slow Holiday Along the Via Rhona."

ORANGE
Few motorists who whiz past Orange on the way to the coast ever think of stopping. The enlightened ones get to experience a historic

town with a medieval center and a couple of major Roman monuments. The historic center is ringed by the busy N7 and D976 roads. The Roman Arc de Triomphe is located on the N7 just as the road approaches the town from the north. The Roman Theatre (Théâtre Antique) is at the southern end of the historic center.

Sights

★ Roman Theatre of Orange (Théâtre Antique)

rue Madeleine Roch; https://theatre-antique.com; Jan.-Feb. and Nov.-Dec. daily 9:30am-4:30pm, Mar. and Oct. daily 9:30am-5:30pm, Apr.-May and Sept. daily 9am-6pm, June-Aug. daily 9am-7pm; €11.50

Academic study of Roman theater continues to explore why performances took place, who performed, and who watched. Central to the understanding of the art form is the theater in Orange. Built around 1 BCE, it is the only Roman theater in the world whose stage wall is still standing. The wall measures 103 m (338 ft) long and 36 m (118 ft) high.

Spaced along both the interior and exterior facades are entrance and exit points for the actors. Above the principal door is a statue of the Emperor Augustus. A visit to the site includes an excellent audio guide that adds

historical color, plus the *Ghosts of the Theatre* performance, a multimedia presentation depicting the different ways the theater has been used through the ages—the latest of which is the Odysee Sonore, a spectacular sound and image show that runs when night falls from May to September. Check the website for performances, which are irregular.

Arc de Triomphe

Av l'Arc de Triomphe; open 24/7

This arch, built during the reign of Augustus (27 BCE to 14 CE), is found in a small park where the N7 road splits as it enters Orange. Engravings depict naval battles, spoils of war, and Roman battles against the Gauls and Germanic tribes.

Festivals and Events

Chorégies d'Orange

Théâtre Antique, rue Madeleine Roche; www.choregies. fr; early July-early Aug.

Dating back to 1860, the chorégies are thought to make up the oldest surviving music festival in France. The format has changed over the years and now concentrates exclusively on opera, with around six performances taking place over a one-month period.

the Roman Theatre of Orange

Truffles: The Black Diamond of Provence

Provence is home to one of the most expensive gastronomic delicacies in the world: the *tuber melanosporum*, or black truffle. Depending on harvesting conditions, the price for 1 kilo (2.2 lb) of truffle can near €1,000, hence the affectionate nickname "black diamond."

Part of the fun of eating truffles, and the justification for the price, is the associated mystique. The fact that truffles are so hard to find means they are all the more prized. As a fungus, they can only grow from trees (usually oak) infected with a specific virus. In the soil underneath such trees, a thin thread attaches the truffle to its host tree. Truffles are also very choosy about exactly when they grow. In general, truffle season is December to February. For a bumper harvest, rain in spring needs to be followed by a dry summer and a cold winter. Then there's the lunar cycle: Some hunters insist on a full moon before they go out foraging.

truffles

To find truffles an expert dog is a must, ideally suckled from birth on a mother who has had truffle oil rubbed on her nipples.

All this effort is worthwhile because truffles taste majestic. Grated over some scrambled eggs, stirred into a pasta or risotto, or even mixed with a little butter in a baked potato, the truffle transforms everyday food into a banquet.

WHERE TO FIND TRUFFLES

- **Richerenches:** Thanks to a combination of soil and climate, the small village of Richerenches in the northern Vaucluse is the center of Provençal truffle production. Every Saturday from mid-November to the end of March, Richerenches hosts **a truffle market,** luring restaurant buyers from across France.

- **Maison du Truffe** (Place de l'Horloge, Menerbes; www.vin-truffe-luberon.com): Groups can organize truffle-hunting demonstrations and truffle-themed cooking classes at this restaurant and activity center in Menerbes in the Luberon. Or stop by the shop for truffle-themed gifts.

- **La Truffe du Ventoux** (634, chemin du Traversier, Monteux; tel. 04 90 66 82 21): The Jaumard family, truffle farmers and traders based at the foot of Mont Ventoux, offer outings into their truffle grove to hunt truffles accompanied by their dogs, followed by a tasting meal, during a winter weekend.

Food
Le Parvis
55 cours Pourtoules; tel. 04 90 34 82 00; https://leparvisorange.com; Wed.-Sat. noon-1:45pm and 7:30pm-9pm, Sun. noon-1:45pm; €22-48

This restaurant is building a reputation for beautiful, tasty plates of food. Tourists and locals enjoy dishes like filet mignon of pork with sweet potato mash, curried zucchini, and cider vinegar.

Au Petit Patio
58 cours Aristide-Briand; tel. 04 90 29 69 27; Mon.-Tues. and Fri.-Sat. noon-1:30pm and 7pm-9:30pm, Wed.-Thurs. noon-1:30pm; lunch €24, dinner €38-48

Au Petit Patio offers fun bistro cooking and an attractive outside terrace. The menu is based on Provençal favorites, with a slight bias toward fish. Popular with locals, the atmosphere is convivial, and it's sensible to reserve, particularly in the evenings in season.

Accommodations

L'Herbier

8 Place aux Herbes; tel. 04 90 34 09 23; www.
lherbierdorange.com; €85 d

This is an excellent-value two-star hotel in
the historic center of Orange. It's located on a
pretty square and has a pleasant terrace, surrounded by a stone wall, where breakfast is
served.

Le Grand Hotel D'Orange

8 Place Langes; tel. 04 90 11 40 40; www.
grandhotelorange.com; €140 d

Just up the road from the Théâtre Antique,
the Grand Hotel has 40 contemporary rooms,
a small pool, a garden, and a popular bistro.

Au Vin Chambre

15 avenue Frederic Mistral; tel. 04 88 84 12 19; www.
auvinchambre.com; €150 d

This boutique hotel has a large garden with a
pool, table tennis, and a pétanque court. There
are four spacious and beautifully decorated
rooms and an excellent restaurant on-site. It's
a short walk to the center of Orange.

Getting There and Around

Most people will find that the easiest way to
get around Orange is to walk. Many of the
roads are pedestrian-friendly and the central
area can be crossed in just over 10 minutes
on foot.

Car

To reach Orange from **Avignon** take the A7.
The journey is 34 km (21 mi) and takes 35
minutes. From **Vaison-la-Romaine,** take the
D977; the journey is 29 km (18 mi) and takes
35 minutes. From **L'Isle-sur-la-Sorgue,** take
the A7 for 50 km (31 mi); the trip will take
about 45 minutes.

Train

The Orange **train station** (www.sncf-
connect.com) is located on rue Pierre Semard.
There are 23 trains a day running between
Orange and Avignon. The journey time is
around 20 minutes and costs €6.60.

Bus

Zou line 904 (https://zou.maregionsud.fr)
from Vaison-la-Romaine runs hourly, and the
bus departs from the train station; the trip
takes an hour and costs €2. **Line 922** from
Châteauneuf-du-Pape runs six times a day
and departs from La République; it takes 25
minutes and costs €2. **Line 902** to Avignon
runs twice an hour and departs from rue
Pourtoules, taking 1 hour and costing €2.

Bike

A cycle route—the **ViaRhona RV17**—links
Châteauneuf-du-Pape with Orange and
Avignon. Details can be found online at
www.provence-cycling.co.uk; navigate to
"Experiences" and "A Slow Holiday Along
the Via Rhona."

VAISON-LA-ROMAINE

Vaison is split in two by the Ouvèze river. On
the hill to the south is the medieval town.
Here the streets are narrow and cobbled, and
the houses are made of stone. Every now and
then there's also a shady square with a plane
tree and a fountain. The medieval town tends
to be quieter than the new town, which is on
the opposite side of the river where the land
is flat, and the town has grown out from the
main tourist sight, the Roman remains that
give Vaison its name.

Sights

★ Roman Ruins

rue Burrus; www.provenceromaine.com; Nov.-Dec. and
Feb. daily 10am-noon and 2pm-5pm, Mar. and Oct.
daily 10am-12:30pm and 2pm-5:30pm, Apr.-Sept. daily
9:30am-6pm; €9, ages 10-17 €4

The main Roman remains are a few minutes'
walk from the bridge over the Ouvèze river
on either side of rue Burrus. The two different
Roman neighborhoods, named Puymin and
Vilasse, are visible from the road. Paying the
entrance fee allows you to wander amid the
ruins of Roman streets and examine up close
the floor plans of houses, including bathing
rooms and the remains of mosaic tiles. On
the Puymin site there is also a museum that

holds decorative articles and tools discovered during the dig, as well as a Roman theater, which is testament to the wealth of Vaison at the time.

Cathedral Notre-Dame de Nazareth and Cloisters

rue Alphonse Daudet; www.provenceromaine.com; Oct.-May daily 10am-5pm, June-Sept. daily 9am-6pm; free

This cathedral was built on the site of a former Roman temple, and you can still see the Roman columns supporting the exterior wall. It almost looks like the ancient temple wants to erupt from the earth and displace the Christian church. Inside, the cathedral is simply decorated and the most aesthetically pleasing part of a visit is seeing the cloisters, which surround a small garden at the northern end of the cathedral.

Festivals and Events
Dance Festival

Théâtre Antique; www.vaison-danses.com; late June-late July; €10-48

Performances here range from ballet to modern dance, and even hip-hop. The old Roman theater provides a memorable backdrop to the shows.

Food
The Girocedre

4 rue de Portulet, Puyméras; tel. 04 90 46 50 67; www legirocedre.fr; reservations legirocedre@club-internet. fr; Wed.-Sat. noon-1:15pm and 7pm-8:30pm, Sun. noon-1:30pm; €32

Just north of Vaison, this restaurant is particularly popular on summer evenings thanks to an outside dining area festooned with lights. Twice a week the chef goes to Marseille to pick up the catch of the day from a childhood friend who fishes off the coast.

Le Moulin a Huile

1 quai Maréchal Foch; tel. 04 90 36 04 56; www. lemoulinahuile84.fr; Thurs.-Mon. 12:15pm-1:15pm and 7:15pm-9pm; €38-55

In a lovely setting on the Ouvèze river, this gastronomic institution is a good place to stop and enjoy intricately prepared dishes.

★ Bistro Panoramique

rue Gaston Gevaudan; tel. 04 90 41 72 90; www. maisonsduo.com; Tues.-Sat. 12:15pm-1:30pm and 7:15pm-9:30pm; €35-70

This is an address for serious foodies. The dishes are almost too beautiful to eat, but paradoxically you long to devour them the moment they arrive. There's also a tapas bar and hotel rooms.

Accommodations
Auberge d'Anais

132 chemin de Anais, Le Peyréras, route de Saint-Marcellin, Entrechaux; tel. 04 90 36 20 06; www. aubergeanais.com; €100 d

Ten km (6 mi) south of Vaison-a-Romaine, this small seven-bedroom hotel is surrounded by vines and olive trees. There's a large pool and a good restaurant. The overall ambience is one of pared-down simplicity, with the beauty of the location being the prime reason to stay.

Le Beffroi

2 rue de l'Évêché; tel. 04 90 36 04 71; www.le-beffroi. com; €150 d

Located in medieval Haute Vaison, Le Beffroi offers plenty of period charm with oak-beamed ceilings and traditional terre cuite (baked earth) tiled floors. The hotel is split between two buildings. There's a pool in the garden.

Getting There and Around

Vaison is a small town best explored on foot.

Car

To reach Vaison from **Avignon,** take the A7 and D977. The trip is 52 km (32 mi) and takes 54 minutes. From **Orange,** take the D23 and D977, covering 30 km (19 mi) in 33 minutes; from **L'Isle-sur-la-Sorgue,** take the D977, covering 55 km (34 mi) in 50 minutes.

1

2

Bus

This region is not very well served by public transport, but **Zou line 904** (https://zou. maregionsud.fr) runs between Orange and Vaison-la-Romaine 10 times a day, passing through the Dentelles villages of Seguret and Sablet, and costing €2. The bus stop is on avenue des Choralies in Vaison-la-Romaine. The stop in Orange is the train station on rue Pierre Semard.

DENTELLES DE MONTMIRAIL

Paradise for wine lovers and hikers, the Dentelles are a range of limestone rocks folded upward during the Jurassic age. Their name stems from a resemblance to the seemingly haphazard collection of pins on a lace-making board. On the southwestern slopes of the Dentelles are the renowned red wine-producing villages of Gigondas, Vacqueyras, and Beaumes-de-Venise. Across the Ouvèze river are the winemaking villages of Rasteau and Cairanne.

Sights
Château and Hospice Gigondas
Montée du Château and Places des Vignerons, Gigondas; accessible all day

Gigondas is such a lovely village that its Roman name was Jocanditus, meaning "joy." Following the **Montée du Château,** a marked path takes you past dry stone terraces planted with historic grape varieties and Mediterranean plants to the ruins of the château at the top of the hill. Also of interest is the old hospice, now used to house art exhibitions.

Chapel Notre-Dame d'Aubune
chemin de Notre-Dame d'Aubune, Beaumes-de-Venise; tel. 04 90 62 94 39; Thurs. 10am-noon

Even if you don't happen to be in Beaumes on a Thursday morning when the chapel opens its doors, it's worth taking the road adjoining the Canal de Carpentras to take a quick look from the outside. The chapel was built in the 12th century and is one of the finest examples of Provençal Romanesque art.

Wineries
Domaine Goubert
235 chemin de Jardinieres, Gigondas; tel. 04 90 65 86 38; www.lesgoubert.fr; Mon.-Fri. 9am-noon and 2pm-6pm

The oldest parts of this vineyard date back to the 17th century, making the building one of the first to be constructed outside the village walls. The domaine produces fantastic Gigondas, full of ripe fruit flavors, spices, and an undertone of tobacco.

Clos de Caveau
1560 chemin de Caveau, Vacqueyras; tel. 04 90 65 85 33; www.closdecaveau.com; Mon.-Sat. 10am-6pm

The domaine is a welcoming place to visit, with a 1.4-km (0.9-mi) wine trail that winds through the different varietals of grapes and a tasting room with a panoramic view. The signature Vacqueyras red tastes of ripe fruit and plums with a touch of mint.

Rouge Bleu
Le Petit Alcyon, La Bouillon, Sainte-Cécile-les-Vignes; tel. 07 61 00 47 92; www.rouge-bleu.com

Co-owner Caroline Jones spent her early career in the wine industry in her native Australia. She is dedicated to making natural wines, almost completely cutting out chemical use in the fields and during the vinification process. The resulting wines are wonderfully expressive of the land. Email contact@rouge-bleu.com to reserve a tasting.

Le Vin a la Bouche Wine Tour
33 Grande Rue, Mornas; tel. 04 90 46 90 80; www.levinalabouche.com; €140/day (minimum 2 people)

Celine Viany attended France's most prestigious wine university before becoming a sommelier and working in some of Provence's most exclusive restaurants. She now runs informative wine tours specializing in the Côtes du Rhône region. Her professional contacts

1: Roman tile work in Vaison-la-Romaine
2: vineyards along the Montée du Château path

are excellent, which means that clients get to enjoy bottles that casual tasters never experience.

Sports and Recreation
Le Rocher du Midi
Distance: *12 km (7.5 mi) round-trip*
Time: *4.5 hours*
Information and Maps: *Tourist Office, 5 rue du Portail, Gigondas; tel. 04 90 65 85 46; www.ventouxprovence.fr*
Trailhead: *13 Pl. Gabrielle Andéol, Gigondas*
The hike follows the routes des Dentelles toward the Rocher du Midi. The Rocher (rock) is a natural viewpoint and picnic spot that offers a lovely panorama of the Dentelles, the Cévennes, and Mont Ventoux. The path is marked and relatively flat, with a total climb of only 200 m (656 ft).

Dentelles Cycle Route
Beaumes-de-Venise Tourist Office, 140 Place du Marche, Beaumes-de-Venise; tel. 04 90 62 94 39; www.ventouxprovence.fr
Departing from Beaumes-de-Venise, follow the cycle route in the direction of Lafare. The circular route covers 21 km (13 mi) and takes approximately 2 hours. There are extensive views of the Dentelles, the terraced vineyards, and the hill villages of Lafare and La Roque Alric.

Benoit Igoulen Bike Hire
Beaumes-de-Venise; tel. 06 14 11 18 15; www.igoulen-location-velo.fr; daily 8am-8pm; from €18/half day
This well-established shop in the center of Beaumes rents road, mountain, and electric bikes. The shop also provides advice on routes.

Festivals
Gigondas sur Table
Gigondas; tel. 04 90 37 79 60; www.gigondas-vin.com; promotion@gigondas-vin.com; mid-July (usually July 16); €50/person
The evening (for which reservations are required) includes tastings of wines from all the Gigondas domaines, complemented by food

prepared by the top local chefs. Fine wine and great food—it couldn't be more French!

Vacqueyras Fete du Vin
Vacqueyras; www.fetedesvins-vacqueyras.fr; mid-July; €72 meal, €15 tastings
The festival involves two days of wine-tasting in the streets of Vacqueyras with a procession of vignerons, bands, and a holy mass, all topped off by a grand gourmet meal. As always in this wine-producing area, remember the wines are not only delicious, but also high in alcohol. Consume in moderation.

Food
Le Bistrot de L'Oustalet
rue de Rouvis, Gigondas; tel. 04 90 37 66 64; www.loustalet-gigondas.com; Fri.-Tues. noon-1:30pm and 7:15pm-8:30pm; €16
Wine and food are viewed as inseparable in this gastronomic temple in the heart of Gigondas. The bistro is the more relaxed baby brother of the adjoining Michelin-starred restaurant, but still offers a selection from the wine list of 100 different producers. The focus is on fresh seasonal dishes.

Le Mesclun
rue des Poternes, Seguret; tel. 04 90 46 93 43; July-Aug. Thurs.-Tues. 12:15pm-1:30pm and 7:15pm-8:30pm, Sept.-June Tues. 12:15pm-1:30pm, Thurs.-Sat. 12:15pm-1:30pm and 7:15pm-8:30pm, Sun. 12:15pm-1:30pm; €22-25
Benefiting from a lovely shady terrace and a setting high in the village with views out over the vines, Le Mesclun is brimming with Provençal atmosphere. The presentation of the food is immaculate, and the menu a showpiece for the flavors of the south.

Côteaux et Forchettes
3340 route de Carpentras, Croisement de la Couranconne, Cairanne; tel. 04 90 66 35 99; www.côteauxetforchettes.com; Fri.-Wed. noon-2pm and 7pm-9pm; €28-58
The terrace of the Côteaux et Forchettes is so close to the vines that in the summer you can smell the sugar rising in the grapes as you eat.

Inventive internationally influenced cooking pulls in a crowd of locals and in-the-know tourists.

Accommodations
Le Bastide Bleue
route de Sablet, Seguret; tel. 04 90 46 83 43; www. bastidebleue.com; €90 d

From its shady courtyard restaurant serving local food and wine, to the stone walls and blue shutters of the building, the Bastide exudes Provençal charm. It's relatively inexpensive, and all seven rooms have en suite bathrooms and Wi-Fi. There's also a pool to enjoy, and two of the rooms are larger and can accommodate families. The nearby village of Seguret is one of the most attractive in the area.

Hotel Les Florets
1243 route des Florets, Gigondas; tel. 04 90 65 85 01; www.hotel-lesflorets.com; €120 d

Located in the countryside just outside Gigondas, this small hotel has a picturesque pool and restaurant with an excellent reputation. Rooms all have en suite bathrooms and Wi-Fi. Although not overly luxurious, Les Florets feels like a hidden haven, a place where an afternoon can easily slip by as you lounge by the pool with a good book before enjoying an indulgent evening meal.

Villa Sainte Anne
345 route de Sablet, Gigondas; tel. 04 65 89 00 20; www.oenotourisme.pierre-amadieu.com; chambres@ pierre-amadieu.com; €220 d

This charming boutique hotel has five tastefully furnished rooms and is set in the heart of a vineyard. There is a swimming pool and sauna.

Getting There and Around
Car
The D977 and the D938 are the main roads that encircle the Dentelles, taking you to Vaison-la-Romaine as you head north and the town of Carpentras as you travel south. Vaison-la-Romaine is 55 km (34 mi) from Avignon and 30 km (19 mi) from **Orange.** Carpentras is 28 km (17 mi) from Avignon and 26 km (16 mi) from Orange.

Bus
The region is not very well served by public transport. **Zou line 904** (https://zou. maregionsud.fr) runs between Orange (departing from the train station on rue Pierre Semard) and Vaison-la-Romaine (departing from avenue des Choralies) 10 times a day, passing through the Dentelles villages of Seguret and Sablet and costing €2.

MONT VENTOUX AND VICINITY
Some people, when they see a mountain, have an immediate urge to climb to the top. Others are happy to enjoy the scenic view from afar. Whichever camp you fall into, Mont Ventoux provides a majestic backdrop to a holiday in this area. Bédoin, the main village at the base of the mountain, is an attractive place that bustles with activity thanks to the daily influx of cyclists preparing to take on the climb to the summit. The Monday market is one of the largest in the Vaucluse. High up near the peak of Mont Ventoux, the Mont Serein resort offers hiking, pony riding, and plenty of children's activities. In the winter Mont Serein becomes one of the mountain's two ski resorts. Back down at the foot of the mountain, the little-known Gorges de la Nesque is a scenic excursion for a day.

Sights
The Summit of Mont Ventoux
Col des Tempetes, Mont Ventoux

It's wise to check the weather forecast before setting off for the summit of Mont Ventoux. Motorists (and brave, extremely fit cyclists) depart from either Malaucene or Bédoin, taking the D974 in both cases. From December to the end of February the road is sometimes closed due to snow. Mont Ventoux is some 1,909 m (6,263 ft) high and geologically, if not geographically, is part of the same mountain chain as the Alps. Besides snow, the other meteorological threat is the wind. At the summit, a gust can knock people off their bikes or feet; the last part of the road is called the Col de Tempetes, which translates as Canyon of the Storms. The good news is that once you reach the summit you can justifiably claim to have seen the best panorama in Provence.

Gorges de la Nesque

D942, Villes-sur-Auzon

The baby sibling of the Gorges du Verdon is accessed by taking the D942 between Villes-sur-Auzon and Castelleras. The road is narrow and the gorge cut by the River Nesque is precipitous enough to set your heart pumping. In places the drop is 400 m (1,312 ft), and it feels impossible to reconcile the geological feature with the tiny trickle of the river. Indeed, halfway along the gorge, the Nesque, almost as if it were ashamed of itself, disappears underground to reappear farther south at Fontaine-de-Vaucluse.

Sports and Recreation
Bédoin Location Cycle Hire

20 route de Malaucene, Bédoin; tel. 04 90 65 94 53; www.bedoin-location.fr; Wed.-Mon. 9am-1pm and 2pm-6pm; from €35/half day

Bikes and advice for all riding skill levels are available here, including electric bikes to ease your way to the summit, and descent bikes for those who want the buzz of coming down without the hassle of going up. If Mont Ventoux, the "beast of Provence," is too ambitious for you, the shop proposes plenty of other routes.

Station Sud Bike Park

Chalet Reynard, route de Mont Ventoux, Bédoin; tel. 04 90 65 63 95; opening time depends on weather, call for information; €14/day

At 1,500 m (4,921 ft), Station Sud, one of Mont Ventoux's winter ski resorts, is transformed during the summer into a bike park. Cyclists use the ski lifts to climb the mountain, then descend cross-country trails of varying difficulty.

Mont Serein Summer Resort

Station du Mont Serein, Beaumont du Ventoux; tel. 04 90 63 42 02; www.stationdumontserein.com; daily in summer except in the case of high winds; price varies

Escape the extreme heat of lowland Provence and enjoy walking and horse-riding trails. There are also trampolines and a downhill go-kart descent to keep the kids occupied.

Gorges de la Nesque canyon pass

The Tour de France and Mont Ventoux

cycling to the top of Mont Ventoux

The Tour de France is the annual cycle race that the French nation obsesses over every July. For three weeks, every detail of every stage is scrutinized. Experts agree that the tour is won or lost on the climbs, and there's none more fearsome than Mont Ventoux. The White Giant of Provence was first added to the rota of Tour stages in 1951, and now appears on the list every four years or so. The battle to reach its summit first has pushed countless riders to exhaustion and revealed the depths of their determination and courage. Nobody suffered more than Englishman Tom Simpson, who fell from his bike in the scree near the summit and died of exhaustion during the 1967 race. A memorial now marks the spot.

Tour spectators arrive as much as a few days before the passage of the cyclists. Superfans camp by the roadside to secure the best view. On the actual day, excitement mounts as police outriders secure the route, followed by a caravan of trucks decorated by tour sponsors that distribute free gifts to the crowd. After that, the cyclists pass in an instant of sporting perfection, apparently untroubled by the preternatural effort required to propel a bike at speed over immense distances. In just one or two seconds it's over, but it's worth it, because the Tour is as important psychologically to France as any of its national monuments. Check the Tour website (www.letour.fr) for details.

There are plenty of places to satisfy your inner professional cyclist around Mont Ventoux. Here are a few:

- **The Summit of Mont Ventoux:** Whether you get there by the power of your own legs or with the assistance of an e-bike, the view from the top is amazing (page 65).

- **Gorges de la Nesque:** The D942 between Villes-sur-Auzon and Castelleras is a favorite with cyclists, with thrilling views of the gorge (page 66).

- **Station Sud Bike Park:** A ski resort turned cross-country bike park for adrenaline junkies (page 66).

The GR4 walking trail passes the station on its way to the summit. The walk from the station to the top takes 1.5 hours.

A Ventoux Rando

Bédoin; tel. 06 10 33 55 12; www.aventoux-rando.com; weekly in tourist season

You're in safe hands with guide Cedric Demangeon, a part-time firefighter and mountain rescue worker. The sunrise walk begins around 11pm, under the light of the moon and headlamps. There are many planned breaks throughout the night, including a rest stop in an old sheepfold; an hour before sunrise, you'll ascend the summit to a magnificent view. The sunset walk departs around 7pm. Over the next 2 hours, the color gently fades from the sun, and the shadows lengthen on the slopes of the mountain. Deer can usually be seen foraging on the bare scree near the summit. Finally around 9pm, watch from the summit as the light is turned off across Provence.

Markets

Bédoin Monday Market

avenue Barral des Baux, Bédoin

The largest market around Mont Ventoux takes place in Bédoin every Monday. There's a nice mix of food and arts and crafts items, including olive wood chopping boards and locally made jewelry.

Wineries

Château Pesquie

1365bis route de Flassan, Mormoiron; tel. 04 90 61 94 08; www.châteaupesquie.com; daily 10am-noon and 2pm-6pm

A regal line of centuries-old plane trees lines the road leading up to this impressive château. Running right up to the foot of the balustraded terrace are formal gardens overlooked by Mont Ventoux. There's a vineyard walk and the possibility of a picnic lunch provided by the vineyard (book in advance).

Food

Le 6 à Table

6 Place Nationale, Caromb; tel. 04 90 62 37 91; www.pascal-poulain.com; Tues.-Sat. noon-1:30pm and 7pm-9pm; €22-28

Set opposite the Saint Maurice church, Le 6 à Table is a traditional French brasserie offering great food at reasonable prices. It's well worth a stop.

La Terrace du Chalet Reynard

route du Mont Ventoux, Bédoin; tel. 04 90 60 48 25; www.chalet-reynard.fr; Wed.-Sun. noon-2pm; €20-30

Hikers and bikers can sample specialties such as truffle omelet and melted Mont d'Or cheese impregnated with truffles. It's mountain food for mountain people. The ground-floor café is open 9am-6pm daily and serves more basic dishes.

La Colombe

3890 route du Mont Ventoux, Bédoin; tel. 04 90 65 61 20; www.la-colombe.fr; Thurs.-Mon. noon-1pm and 7pm-9:30pm, Wed. 7pm-9pm; €48-72

This charming small restaurant just outside Bédoin offers delicate food prepared from local seasonal products. Expect pork from Mont Ventoux and plenty of truffles in season. Depending on whether you arrive before or after an ascent of the White Giant, there is a spacious terrace for you to relax or collapse.

Accommodations

Château Mazan

rue Napoléon, Mazan; tel. 04 90 69 62.61; www.chateaudemazan.com; €200 d

Birthplace of the Marquis de Sade's father and uncle, this château was built around 1720. It has been converted into an atmospheric 30-bedroom hotel that is chock-full of antiques. In the summer there's a romantic restaurant on the château's terrace. All rooms have Wi-Fi, air-conditioning, and en suite bathrooms.

★ Hotel Crillon-le-Brave

Place de l'Eglise, Crillon-le-Brave; tel. 04 90 65 61 61;
www.crillonlebrave.com; €650 d

Crillon-le-Brave is a perched village just outside Bédoin. It's now dominated by a luxury hotel that occupies eight different houses in the small village. Each room is individually decorated with antique furniture, and many boast large picture windows. There are two restaurants, a pool with a view, and countless hidden corners to escape to.

Getting There and Around
Car

To reach Bédoin from **Avignon,** take the D942 and the D974, covering 40 km (25 mi) in 45 minutes. From **L'Isle-sur-la-Sorgue,** take the D163; the journey is 31 km (19 mi) and takes 40 minutes. From **Orange,** take the D974 for 37 km (23 mi); it will take 45 minutes. From **Vaison-la-Romaine,** take the D938; the trip is 22 km (13.5 mi) and takes 30 minutes.

Bus

Transcove line L (www.transcove.com) runs eight times a day between Carpentras, Mazan, Crillon-le-Brave, and Bédoin. The bus takes 45 minutes to reach Bédoin and costs around €4.

L'ISLE-SUR-LA-SORGUE AND FONTAINE-DE-VAUCLUSE

Built on an island created by the branching of the Sorgue river, L'Isle-sur-la-Sorgue is a vibrant market town that draws visitors from miles around thanks to its fame as an antiques center. Markets on Thursday (Provençal) and Sunday (antiques and Provençal goods) are extremely popular. Bridges and waterwheels are hidden away in unexpected places, and enjoying a coffee on a riverside terrace allows you to watch the spectacle of ducks paddling furiously to avoid being swept downstream. For more aquatic experiences, the nearby village of Fontaine-de-Vaucluse is located at the bubbling source of the Sorgue, and the village of Pernes-les-Fontaine is best known for its 41 public fountains.

Sights
Eglise Notre Dame des Anges

Place de la Liberté, L'Isle-sur-la-Sorgue; https://
islesurlasorguetourisme.com; May-June and Sept.
Mon.-Fri. 10am-noon and 3pm-6pm, July-Aug. Mon.-
Fri. 10am-6pm, Oct.-Mar. Mon.-Fri. 10am-noon and
3pm-5pm

The boutique-filled streets of L'Isle-sur-la-Sorgue wind away from the river toward a large, open central square that is dominated by the Eglise Notre Dame des Anges. Built in the 12th century, the church was extended and renovated in the Italian style during the 18th century. The exterior, which opens onto Place de la Liberté, is relatively plain, but inside the scale is impressive. The stalls and chapels are particularly fine examples of the baroque style with plentiful use of vibrant colors, such as rich blues and decadent golds.

★ Antiques-Hunting

The main antiques shops in L'Isle-sur-la-Sorgue line the southern bank of the river, and a pleasant hour can easily be spent hopping from one to another. Note that most antiques shops are closed Tuesday-Thursday.

Le Village des Antiquaires de la Gare

2 avenue de l'Égalité, L'Isle-sur-la Sorgue; tel 04 90 38
04 57; www.levillagedesantiquairesdelagare.com; May-
Oct. Thurs. 10am-5pm, Fri.-Mon. 10am-7pm, Nov.-Mar.
Fri.-Mon. 10am-6pm

Probably the best of several antiques villages (collections of small antiques shops in the same building) to be found around the town, this complex has a nice, relaxed atmosphere with the traders sipping on wine and gossiping while waiting for the next sale. Leather club sofas, chandeliers, and mirrors are among the eclectic items on sale.

Ramis Antiquites

Le quai de la Gare, 4 avenue Julien Guigue, L'Isle-sur-la-Sorgue; tel. 06 09 33 43 68; www.ramisantiquites.com; Fri. 10am-6pm and Sat.-Sun. 10am-7pm

Tables, paintings, sculptures, and other exceptional works of art are for sale over the two floors of this showroom set in a historic town house. The collection is at the higher end of the price range available in the town.

Hotel Dongier

15 Espl. Robert Vasse, L'Isle-sur-la-Sorgue; tel. 04 90 21 50 24; Thurs.-Mon. 10am-7pm

Grouped together in this 18th-century town house is a wonderful collection of antiques. At times it feels more like a museum than a shop, because entire walls are filled with oil paintings and some of the furniture dates back to the 16th century.

La Cours de Créateurs

11 rue Carnot, L'Isle-sur-la-Sorgue; tel. 06 35 41 15 31; Tues.-Sun. 10am-7pm

There's a nice mix of brocante, pop art, vintage items, and contemporary artisan goods in this cooperative shop set in an old convent. Prices are more reasonable than in some of the more upscale shops.

Candy and Cloud

9 rue Carnot, L'Isle-sur-la-Sorgue; www.candyandcloud. com; Tues.-Sun. 9am-7pm

A collection of funky objects from home clearances and brocante markets fill this colorful shop. There are also some contemporary home decor wares mixed in.

L'Ile aux Brocantes

7 avenue des 4 Otages, L'Isle-sur-la-Sorgue; tel. 06 20 10 58 15; Fri. 2pm-7pm and Sat.-Mon. 10am-7pm

There can be few more pleasant places to shop for antiques. Cross over a little bridge and enter a shady square ringed with the stands of 40 or so dealers. If it all gets too much, there's an on-site restaurant to relax and consider the pros and cons of a purchase. The website is an excellent resource to identify potential purchases in advance of a visit.

Weekly Market

quai Jean Jaurès, Place de la Liberté and throughout town, L'Isle-sur-la-Sorgue; Thurs. and Sun.

L'Isle-sur-la-Sorgue has two weekly markets. On Thursday morning the market is smaller and concentrates mainly on food. On Sunday, the town comes to a stop with a combined food and antiques market. It's great fun because you never know quite what you might find, from glorious antique furniture to 3.5-m-tall (12-ft-tall) mock space rockets.

Bi-Annual Antique Fair

L'Isle-sur-la-Sorgue; Easter weekend and Aug. 15

Twice a year the number of antiques dealers in L'Isle-sur-la-Sorgue swells from 300 to 700 when 400 traveling dealers join the resident sellers, creating one of the largest antiques fairs in the world. Almost every inch of pavement is filled with traders selling anything from small old-fashioned coffee grinding machines to tables, beds, and chests of drawers. Part of the fun is haggling, which is best done with a smile on your face and a readiness to walk away if the price is not right.

Hiking
★ Fontaine-de-Vaucluse

Distance: *2 km (1.2 mi) round-trip*
Time: *65 minutes*
Information and Maps: *www.alltrails.com/fr/ explore/trail/france/vaucluse/gouffre-de-fontaine-de-vaucluse*
Trailhead: *rue Jean Moulin, Fontaine-de-Vaucluse*

You can hear the roar from the village of Fontaine-de-Vaucluse, 1 km (0.6 mi) away from the source. An easy 30-minute walk up a level, rocky path with a slight incline at the end takes you to the spring. The experience varies dramatically depending on flow levels; they can be checked in advance on the tourist office website (https://islesurlasorguetourisme. com). Either water cascades from a rocky hole in the ground, plunging down the valley, or you can peer down through a gap in the rocks

1: L'Isle-sur-la-Sorgue **2:** antiques shop in L'Isle-sur-la-Sorgue **3:** spring above Fontaine-de-Vaucluse **4:** Fontaine-de-Vaucluse

into a still, mirror-like pool. The spring can discharge up to 700,000,000 cubic meters of water a year, making it the fifth-largest karst spring in the world.

The Fountain Walk of Pernes-les-Fontaines

Distance: *3.2 km (2 mi) round-trip*
Time: *1 hour*
Information and Maps: *Tourist Office, Place Gabriel Moutte, Pernes-les-Fontaines; tel. 04 90.61 31 04*
Trailhead: *Tourist Office*

In the 18th century Pernes was transformed by the discovery of the Saint-Roch source. Suddenly it was possible to bring water to every corner of the town. Four commemorative fountains were built to celebrate: Cormoran, Souchet, Reboul, and Hopital. As the saying goes, there's no such thing as too much of a good thing, and so the industrious people of Pernes kept on investing in fountains. Today there are 41 public fountains and an estimated 60 private ones. Don't expect to be wowed by great torrents of cascading water; the fountains of Pernes are largely simple affairs, but it's fun to pick up a map for the circular walk, which starts and finishes at the tourist office, and tick off the fountains as you go. For lunch before or after the walk, try **Restaurant Au Fil du Temps** (51 Place Louis Giraud; tel. 04 90 30 09 48).

Sorgue Waterwheel Walk

Distance: *2 km (1.2 mi) round-trip*
Time: *1 hour*
Information and Maps: *Tourist Office; Place de la Liberté, L'Isle-sur-La-Sorgue; tel. 04 90 38 04 78*
Trailhead: *Notre Dame des Anges, Place de L'Eglise, L'Isle-sur-la-Sorgue*

Up until the 16th century, the traditional trades of L'Isle-sur-la-Sorgue (wool and silk weaving) were done by hand. New waterwheel technology gradually transformed the work, providing power to factories. Around 60 waterwheels were in use by the 19th century. Today nearly 20 remain dotted around the town, slowly churning the water. Invariably,

they are coated by heavy moss, which softens their appearance and brings greenery to the center of town. Pick up a map from the tourist office to follow the Parcours des Roues.

Other Sports and Recreation

Kayak Vert

avenue Robert Garcin, Fontaine-de-Vaucluse; tel. 04 82 29 42 42; www.canoevert-vaucluse.com; June-Aug., descents every 30 minutes 10am-5pm, out-of-season groups by appointment; adults from €20, ages 12-18 €15, ages 6-11 €10

Kayaking down the emerald-green, icy-cold waters of the Sorgue is a great escape from the summer heat. The 2-hour route is picturesque as you float through wooded countryside past bathers and picnickers. You can nose into the back gardens of houses on the river and even jump out for a swim in various places. It is advisable to have an adult in the boat with children because the water can flow quickly and there are frequent overhanging branches.

Passerelle de Cimes

route de Cavaillon, Lagnes; tel. 04 90 38 56 87; https:// accrobranche-vaucluse.com; check online reservation system for available times; €13-21

This activity is like an obstacle course in the trees. Catering to all levels, and starting with children as young as four, Passerelle de Cimes is one of the best accrobranching centers in the region. The advanced courses really are only for the brave.

Golf Saumane

1141 route de Fontaine-de-Vaucluse, Saumane-de-Vaucluse; tel. 04 90 20 20 65; www.golfdesaumane.fr; greens fee €85

Just along the road from L'Isle-sur-la-Sorgue is this picturesque parkland golf course. As you might expect on a course close to the Sorgue, there are plenty of water hazards.

Food

★ Le Carré d'Herbes

13 avenue des Quatre Otages, L'Isle-sur-la-Sorgue; tel. 04 90 38 23 97; www.lecarredherbes.eu; Wed.-Sat.

noon-2:30pm and 7pm-9pm, Sun. noon-2:30pm; €20-24

Locals rave about the lunchtime three-course menu and they book well in advance to secure a table for Sunday lunch (the day of the main market). Lamb with sun-dried tomatoes and eggplant cannelloni is a specialty.

Solelh

30 avenue des Compagnons de la Libération, L'Isle-sur-la-Sorgue; tel. 04 90 89 01 42; www.solelh-restaurant. fr; Thurs.-Sun. noon-1:30pm and 7pm-9:30pm, Mon. noon-2pm; €24

Solelh is a popular brasserie nestled among the antiques shops. Either compose a meal from the selection of tapas or pick from a small range of larger dishes like braised beef and polenta. The food is big on flavor.

Olive et Raisins

20 All du Juin, L'Isle-sur-la-Sorgue; tel. 04 90 21 17 36; www.olive-et-raisin.com; daily 9am-10pm; €18-34 for two people

This little deli on the banks of the Sorgue has a small restaurant and offers takeaway picnic hampers. Food is served on sharing plates, and there's an almost infinite choice of nibbles: saucisson (dry, cured sausage), cured hams, foie gras, pickled artichokes, cheeses, and more. Tables are limited, so book in advance to secure a seat.

Café Fleurs

9 rue Théodore Aubanel, L'Isle-sur-la-Sorgue, tel. 04 90 20 66 94; www.cafefleurs.com; Wed.-Sun. noon-1:45pm and 7:15pm-9:15pm; €26-33

L'Isle-sur-la-Sorgue is rarely quiet. The restaurant tables along the river quickly fill with crowds. To escape the throngs and enjoy refined dining in a quiet courtyard, try the Café Fleurs. The beef filet with grated summer truffle is perfectly accompanied by a red Côtes du Rhône.

Le Vivier

800 cours Fernande Peyre, L'Isle-sur-la-Sorgue; tel. 04 90 38 52 80; www.levivier-restaurant.com; Wed.-Fri.

and Sun. 12:30pm-1:30pm and 7:30pm-9:30pm, Sat. 7:30pm-9:30pm; €30-60

Le Vivier offers inventive high-quality food on the banks of the River Sorgue 1 km (0.6 mi) or so from the center of town. There's a large terrace overlooking the river, and three-course menus range €30-60. Fish is a specialty; try the perch with wild rice, seaweed, and clam broth. For a Michelin-star restaurant, the atmosphere is noticeably relaxed and not too fussy. In the summer season reservations are advisable.

Summer Floating Market

First Sun. in Aug.; from 7:30am

Once upon a time the fishers who lived along the Sorgue river used to pole their narrow low-bottomed boats into L'Isle-sur-la-Sorgue to sell their fish. In the Provençal language the boats were called *nego chin*, the less-than-optimistic translation of which is "dog drowner." This tradition has been revived, although in this one-off event it is more likely to be melons, olives, and wine rather than fish on sale. Making a purchase is not easy: First there are the crowds to negotiate and then there is the large drop down to the boats. Overeager shoppers have been known to tumble into the river! Due to these hazards the floating market is more of a spectacle than a serious opportunity to shop. Arrive early to avoid the crowds and see the boats pulled up by the banks, filled with their goods.

Accommodations

Hotel les Névons

205 chemin des Névons, L'Isle-sur-la-Sorgue; tel. 04 90 20 72 00; www.hotel-les-nevons.com; €150 d

This large hotel offers exceptional value just a few minutes' walk from the center of the town. The decor is basic, but there's secure parking (which is something of a luxury in busy L'Isle-sur-la-Sorgue) and a rooftop pool. It's not the place for those who love Provençal character, but the Névons remains a practical, nononsense choice.

Mas de Cure Bourse

120 chemin de la Serre, L'Isle-sur-la-Sorgue; tel. 04 90 38 16 58; www.masdecurebourse.com; €200 d

This 17th-century post house has plenty of charm. Located a 5-minute drive from the town, the hotel is nestled in verdant parkland planted with lavender and roses. There's a pool, private parking, and an on-site restaurant.

★ Maison sur la Sorgue

6 rue Rose Goudard, L'Isle-sur-la-Sorgue; tel. 06 87 32 58 68; www.lamaisonsurlasorgue.com; €400 d

As you would expect in L'Isle-sur-la-Sorgue, this guesthouse is overflowing with antiques and collectibles from around the world. The four rooms are exceptionally spacious and beautifully designed. The location in the center of town is perfect for a refined and cosseted stay.

Getting There and Around

L'Isle-sur-la-Sorgue is best explored on foot.

Car

To reach L'Isle-sur-la-Sorgue from **Avignon** take the D942, traveling 30 km (19 mi) in 38 minutes. From **Orange,** take the A7, covering 46 km (28 mi) in 33 minutes. From **Vaison-la-Romaine,** take the D938; the journey is 48 km (30 mi) and takes 57 minutes.

Train

L'Isle-sur-la-Sorgue train station (www.sncf-connect.com) is located just off avenue de l'Egalité. Trains run 10 times a day from Avignon's central station on boulevard Saint-Roch; singles cost €5.20, and the journey takes just over half an hour.

Bus

Zou line 906 (https://zou.maregionsud.fr) from Avignon runs 10 times a day from the train station on boulevard Saint-Roch; it takes 44 minutes and costs €2.80, arriving on cours Emile Zola in L'Isle-sur-la-Sorgue.

Arles and Les Alpilles

Arles is a town blessed by culture. It is overflow-ing with Roman monuments, immortalized in paintings such as *The Night Café* by Vincent van Gogh, and it boasts world-class museums and art galleries. It's full of fountains and squares, chic boutiques, and famous chefs, all of which combine to make it a quintessentially Provençal destination. Yet Arles is also the gateway to the Camargue and the home of bullfighting in the region. A wild Camarguaise spirit underlies everything, and it frequently escapes during the city's many unmissable festivals and fairs.

Driving out of Arles, you cross into the Camargue wetland delta. In low-lying fields, black bulls lower their horns and paw the ground. Rare migratory birds arrow overhead. Mosquitos swarm and the mistral wind

Highlights

Look for ★ to find recommended sights, activities, dining, and lodging.

★ **Les Arènes:** The cries of the gladiators echo through the ages thanks to the superbly preserved architecture of this 2nd-century Roman arena (page 84).

★ **Fondation Van Gogh:** Changing collections celebrate van Gogh's artistic legacy, and there is always an original or two on display (page 87).

★ **Luma Arles:** A multimillion-euro redevelopment of an old railway repair yard into a space for artistic exchange has changed the face of Arles forever; the Frank Gehry skyscraper is impossible to miss (page 89).

★ **Parc Ornithologique du Pont de Gau:** Get up close with the flamingos at this bird park with viewing blinds and marked trails (page 96).

★ **Camargue Safari:** Whether you ride in the back of a jeep or on a horse, an official safari is the best way to see the Camargue wildlife (page 97).

★ **Glanum:** Visualize the everyday life of a Roman citizen in the ruins of this fortified town outside Saint-Rémy-de-Provence. Particularly impressive are the mausoleum and the triumphal arch (page 101).

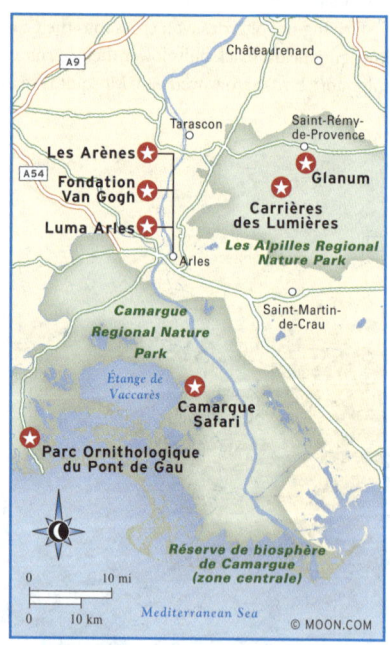

★ **Carrières des Lumières:** This old bauxite mine is now used for high-quality large-scale art projections set to music (page 107).

Arles and Les Alpilles

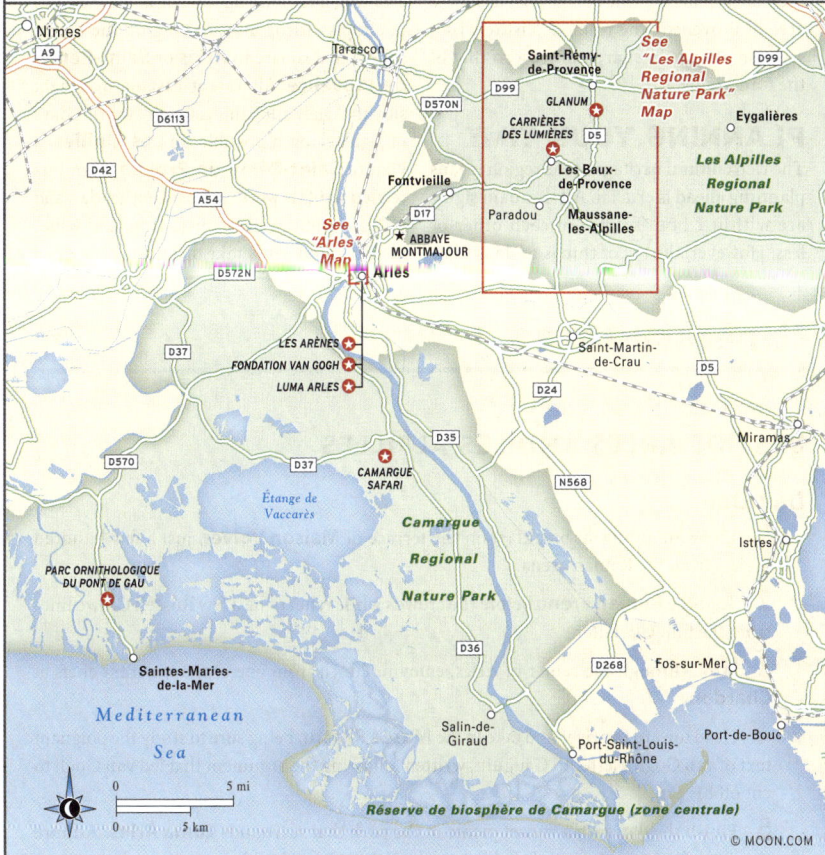

© MOON.COM

howls. Stop at one of the many roadside restaurants and you'll often discover a guitarist playing flamenco music. Wine flows, couples dance, and there are feats of eating and drinking that make a party anywhere else in Provence look like a dull workplace affair. The flamboyant culture of the wild, windswept, waterlogged nature reserve makes a statement: This is the Camargue. Don't be fooled by the pink flamingos and the picturesque white horses: Life is hard here, and we dance today because tomorrow is unknown.

The Rhône crossing near Port-Saint-Louis-du-Rhône in the Camargue toward Les Alpilles is made via a small ferry pulled by a chain. The journey takes 5 minutes, during which passengers are transported from one Provençal universe to another. Les Alpilles is epitomized by its main town, Saint-Rémy-de-Provence. It is a gorgeous place, all cobbled streets, plane trees, and chic cafés sheltering under the jagged peaks of nearby low, rocky mountains. The surrounding countryside is

Previous: Arles; Glanum; Arles's Roman arena.

full of olive groves, vineyards, and wealthy, attractive villages. As a result the area vies with the Luberon for the title of the most desirable area of Provence. The refined, leisurely lifestyle on offer is in sharp contrast to that of the Camargue.

PLANNING YOUR TIME

The rich cultural heritage of the region means planning ahead is crucial. All the main sights are within 1 hour's drive of each other or less. However, the list of things to do is long.

Visiting all the Roman sights alone could easily take a week. Identify in advance what interests you, and don't underestimate the appeal of simply sitting and soaking up the atmosphere in a square in **Arles** or **Saint-Rémy-de-Provence.** Allocate at least one overnight stay in Arles, allowing yourself two full days of sightseeing if possible. In **Les Alpilles,** a visit to **Saint-Rémy-de-Provence** and its related sights is worth at least another day, and village-hopping, olive oil-tasting, and wine-tasting could easily take up another.

Itinerary Ideas

BEST OF ARLES AND LES ALPILLES

Day One

1 Enjoy an indulgent breakfast on the terrace of **Maison Volver,** just a few hundred meters from the Roman arena.

2 Tour the **Roman arena** before the crowds while rehearsing a few Russell Crowe lines from the film *Gladiator.*

3 Stroll through the center of Arles, enjoying lunch from one of the itinerant chefs at **Chardon.**

4 Head toward the river and take in the **Musée Réattu,** being sure to study the poignant text of van Gogh's letter to Gauguin, written following the argument that led van Gogh to cut off his ear.

5 Finish the day with a sunset picnic in the park that surrounds **Luma Arles.** Admire the play of light on the metal cladding of the Frank Gehry building.

Day Two

1 From Arles, drive north on the D570 and check in 35 minutes later at the **Mas de Carassins** boutique hotel.

2 Take a 5-minute walk into **Saint-Rémy-de-Provence** to explore the picturesque streets of the town.

3 For a very local experience, have lunch at **Bar Tabac les Comptoirs du Gigot** on the bustling ring road.

4 Spend what remains of the afternoon visiting the Roman ruins of **Glanum,** just up the road from the hotel.

5 Return to **Mas de Carassins** for supper by the pool, gazing at Les Alpilles as the light fades from the rocks.

Arles

For a long time—some would say too long—Arles has traded on its past. It's the go-to city for Roman sites: the arena, the theater, the baths, the graveyard, the underground crypts, and the finds collected in the city's archaeological museum. It's also the center of the Provençal heritage movement started by poet Frederic Mistral at the turn of the 20th century. Arles has its own queen, who oversees parades of hundreds of women in traditional costume. And for a little intrigue, Arles is the city where van Gogh inexplicably cut off his ear and gave it to a prostitute.

Yet if you look closely, there's something missing that no amount of gladiator reenactments or bullfights can fix: A city that does little but celebrate the past slowly dies. Every summer for the past 50 years, a dash of much-needed modernism has been provided by the Rencontres d'Arles International Photo Festival, which in 2023 attracted just under 150,000 visitors. Yet, outside the tourist season, the streets of Arles still fall silent.

The Fondation Van Gogh, which opened in 2014, was a step in a new direction. The new center is forward- rather than backward-looking in that it explores the contemporary artistic response to van Gogh's work, rather than rehashing the history of the artist's life. Now there are two more new, challenging projects shaping the future of the city. First, there's Luma Arles, which comprises a multimillion-euro collaborative center for artists, a park, and an exhibition and performance venue. Its landmark Frank Gehry building is a skyscraper so avant-garde in design it would raise eyebrows in Manhattan. And right next door to Luma Arles, France's national photography school occupies glitzy new premises designed by Marc Barani.

Watching change as it happens, observing an entire city transform itself, is something to behold. If Arles gets it right, it will have everything: an unrivaled heritage coupled with the vibrancy of youthful creation and ideas. There's never been a more exciting time to visit.

ORIENTATION

Arles is located on the east bank of the **Rhône,** nestled into a bend in the river. The historic center, which contains most of the main sights, is small and largely pedestrian, and it can be crossed on foot in just over 15 minutes. Most people start their tour of the city from the **Boulevard des Lices,** where there is parking and a pedestrian square containing the main tourist office. Across the Boulevard des Lices from the tourist office is **Rue Jean Jaures,** which takes you to the large **Place de la République,** which in turn is bordered by the lively **Place du Forum.** Also across the Boulevard des Lices from the tourist office is the **Jardin d'Eté park.** Walk through the park, climb a small hill and some steps, and you will pass the **Roman Théâtre Antique,** followed by **Les Arènes** (the Roman arena). Continuing toward the river you'll encounter other major sights, including the **Fondation Van Gogh, Thermes de Constantin** (Roman baths), and **Musée Réattu.**

Outside the center, the main sights are **Luma Arles,** the **Alyscamps,** and the **Musée Departemental Arles Antique.** Luma Arles and Les Alyscamps are adjacent to each other. They are within walking distance of the tourist office and easy to find, just past the junction of Boulevard des Lices and Boulevard Emile Combes. The Musée Departemental Arles Antique is located farther out of town, near the main N113 river crossing.

SIGHTS

The most cost-effective way of discovering the sights of Arles is to buy either a Pass Avantage or a Pass Liberte. The **Pass Avantage** (€19, reduced €16) is valid for six months and

Itinerary Ideas

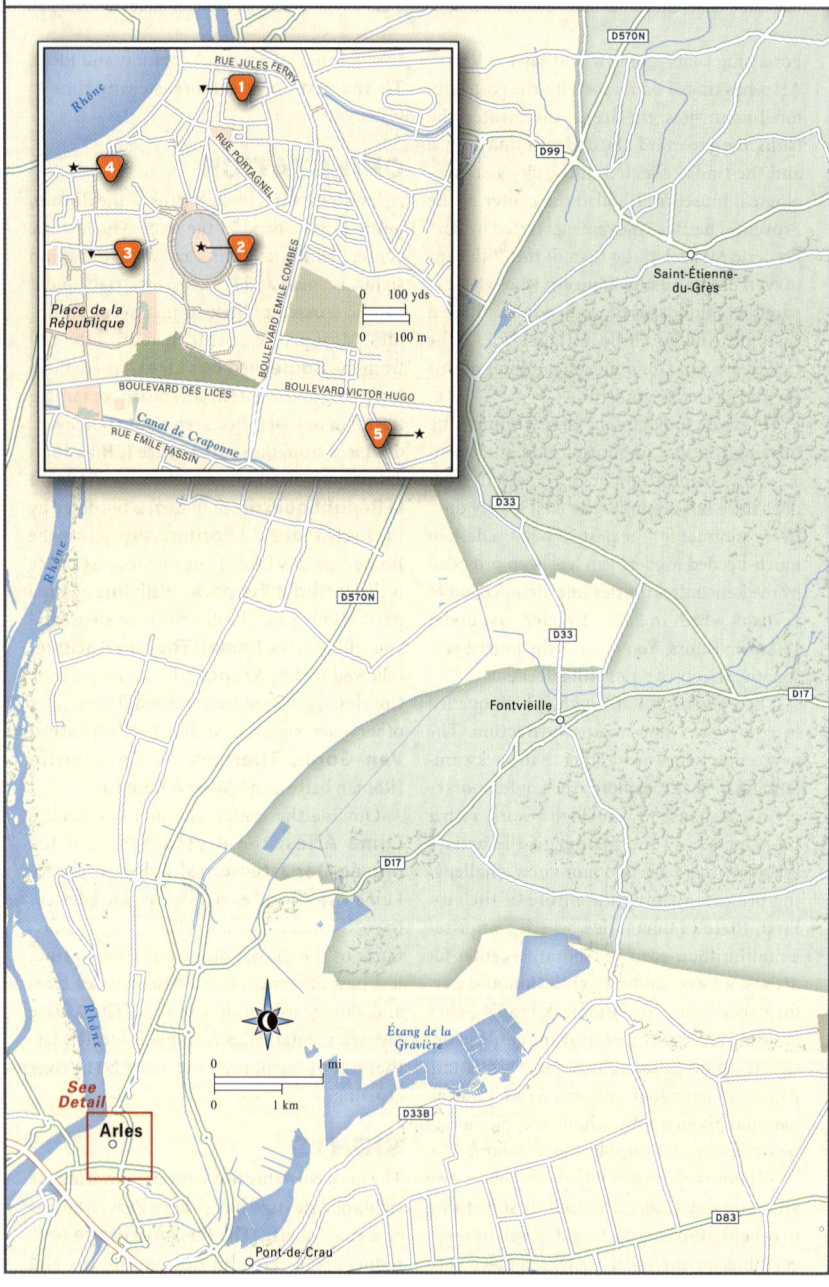

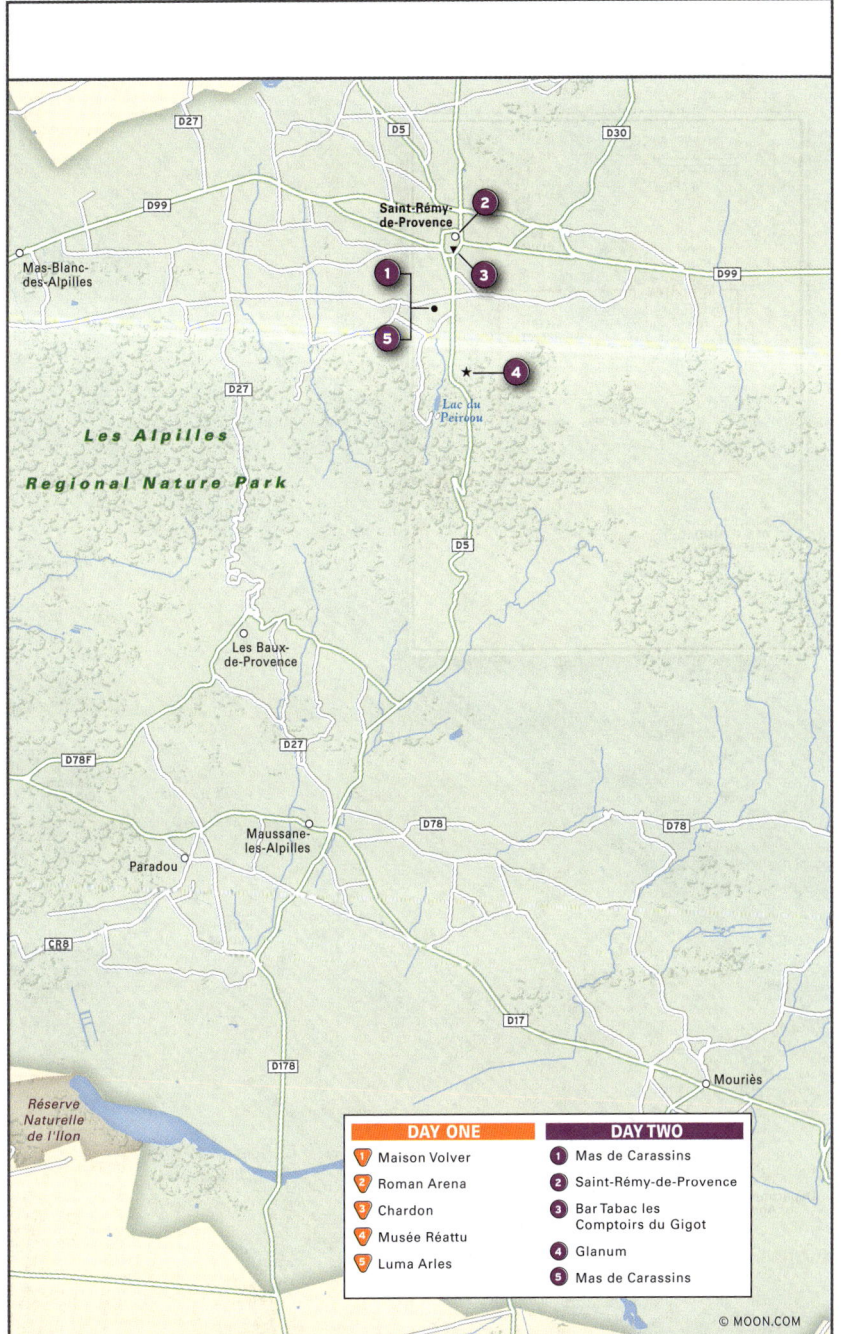

DAY ONE
1 Maison Volver
2 Roman Arena
3 Chardon
4 Musée Réattu
5 Luma Arles

DAY TWO
1 Mas de Carassins
2 Saint-Rémy-de-Provence
3 Bar Tabac les Comptoirs du Gigot
4 Glanum
5 Mas de Carassins

© MOON.COM

Arles

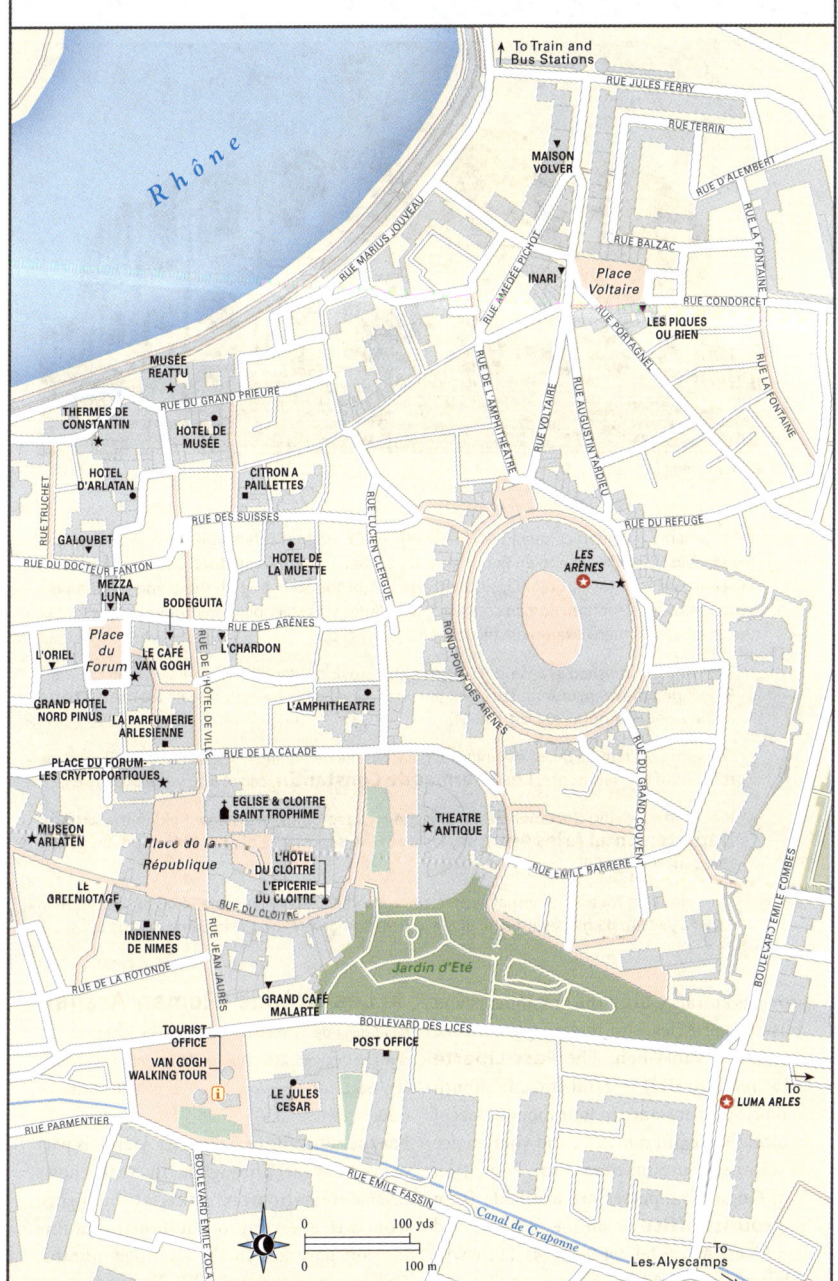

Rhône

To Train and
Bus Stations

RUE JULES FERRY

RUE TERRIN

RUE D'ALEMBERT

MAISON
VOLVER

RUE BALZAC

RUE LA FONTAINE

INARI

*Place
Voltaire*

RUE CONDORCET

LES PIQUES
OU RIEN

RUE PORTAGNEL

RUE LA FONTAINE

RUE MARIUS JOUVEAU

RUE AMÉDÉE PICHOT

MUSÉE
REATTU

RUE DU GRAND PRIEURE

THERMES DE
CONSTANTIN

HOTEL DE
MUSÉE

RUE DE L'AMPHITHEATRE

RUE VOLTAIRE

RUE AUGUSTIN-MIDIEU

RUE DU REFUGE

HOTEL
D'ARLATAN

CITRON A
PAILLETTES

RUE DES SUISSES

RUE TRUCHET

GALOUBET

RUE DU DOCTEUR FANTON

HOTEL DE
LA MUETTE

RUE LUCIEN CLERGUE

LES
ARÈNES

MEZZA
LUNA

BODEGUITA

RUE DES ARÈNES

*Place
du
Forum*

LE CAFÉ
VAN GOGH

L'CHARDON

L'ORIEL

ROND-POINT DES ARENES

GRAND HOTEL
NORD PINUS

LA PARFUMERIE
ARLESIENNE

L'AMPHITHEATRE

RUE DE L'HOTEL DE VILLE

RUE DE LA CALADE

RUE DU GRAND COUVENT

PLACE DU FORUM-
LES CRYPTOPORTIQUES

MUSEON
ARLATEN

*Place de la
République*

EGLISE & CLOITRE
SAINT TROPHIME

THEATRE
ANTIQUE

L'HOTEL
DU CLOITRE

L'EPICERIE
DU CLOITRE

RUE EMILE BARRERE

LE
GREENIOTAGE

RUE DU CLOITRE

INDIENNES
DE NIMES

RUE JEAN JAURES

RUE DE LA ROTONDE

Jardin d'Eté

GRAND CAFÉ
MALARTE

BOULEVARD DES LICES

BOULEVARD EMILE COMBES

TOURIST
OFFICE

POST OFFICE

VAN GOGH
WALKING TOUR

To
LUMA ARLES

LE JULES
CESAR

RUE PARMENTIER

BOULEVARD EMILE ZOLA

RUE EMILE FASSIN

Canal de Craponne

0   100 yds

0   100 m

To
Les Alyscamps

Roman Arles

Roman arena in Arles

Arles became a Roman colony in the 1st century BCE when it sided with the Romans against Greek-held Marseille. The town famously built Caesar 12 ships in 30 days, and in gratitude he raised it to the status of Colonia, settling veterans of the Sixth Legion there and beginning a program of construction. Roman construction in Arles was still going on during the twilight of the empire, with many examples still dotting the city today.

- With Arles established as an important port and provincial center, a theater was built around 1 BCE during the reign of Augustus, and this was followed by the 20th-largest **Roman arena** in the world in around 90 CE (page 84).

- Emperor Constantine built a magnificent palace at the beginning of the 4th century CE, and the remains of the bathhouse, **Les Thermes de Constantin,** can still be visited (page 85).

- Plenty of surviving documentation, and the archaeological finds now held in the **Musée Departemental Arles Antique,** testify to the importance of Arles to the Romans during this period (page 89).

- Those looking for a quick visual appreciation of Arles's significance need go no farther than **Les Alyscamps** graveyard, one of the largest Roman graveyards in Europe (page 89).

provides entrance to all the monuments and museums of Arles with the exception of the Fondation Van Gogh. The **Pass Liberte** (€15, reduced €13) is valid for one month and provides entrance to four monuments of choice, the Réattu museum, and your choice of another museum except the Fondation Van Gogh. The passes are available from the **tourist office** or online (9 boulevard des Lices, Arles; tel. 04 90 18 41 20; www. arlestourisme.com).

★ Les Arènes (Roman Arena)

Rond-point des Arènes; tel. 04 90 49 38 20; https:// arenes-arles.com; Mar., Apr., and Oct. daily 9am-6pm, May-Sept. daily 9am-7pm, Nov.-Feb. daily 10:30am-4:30pm; €9

For the residents of Arles, Les Arènes is part of everyday life. Throughout the year it hosts diverse events from pop concerts to bullfights, from a rice festival to traditional-costume fashion parades. In the summer months, Camarguaise (nonlethal) bull running can be seen three nights a week. There are also frequent gladiatorial reenactment shows. It's a working part of the city and, as such, is liable to be taken for granted.

To a visitor, it's a marvel. The 20th-largest Roman arena in the world, it was built in 90 CE. Modeled on the Colosseum in Rome, it could accommodate more than 20,000 spectators for a gladiatorial contest. It has 120 arches and is 136 m (446 ft) in length. Visit just after opening time, even in the summer, and the arena will be almost deserted. The silence is easily filled by the imagination: the cries of the gladiators, the roars of the lions, the jeers and applause of the crowd. There are few places in the world where it is possible to channel the spirit of ancient Rome in such a fashion.

Yet climb to the top of one of the two towers added on either end of the arena in medieval times, and you can see modernity in the form of the Luma Arles building glinting in the distance. After the fall of the Roman Empire in the 5th century CE, the arena enjoyed a second life as a walled protective citadel. Houses were built inside—even a town square and two churches. In the 19th century a decision was made to demolish this residential neighborhood and return the arena to its Roman splendor.

Roman Theater (Théâtre Antique)

rue de la Calade; tel. 04 90 49 59 05; www. arlestourisme.com; Mar., Apr., Oct. daily 9am-6pm, May-Sept. daily 9am-7pm, Nov.-Feb. daily 10:30am-4:30pm; €9

While the Roman masses packed the arena and roared at the blood being spilled, the more refined classes filled the theater a couple of minutes' walk away. There they watched Roman or Greek comedies, tragedies, mimes, and pantomimes. The theater in its pomp had three parts: the cavea (the semicircular seating area for spectators), the stage and set, and an enclosing rear wall. Today, the best-preserved area is the cavea. Some remains of the set area do still stand, particularly two columns,

nicknamed the Two Widows, which must have once framed the stage. The theater was built at the end of the 1st century BC during the reign of Augustus. It was one of the first stone theaters. Nearly 3,000 years later, in the summer months the theater is still used for concerts and shows. It's possible to get a good feel for the monument by peering in from the outside, but paying the entrance fee plus the extra few euros for the audio guide provides a much more immersive experience.

Place du Forum—Les Cryptoportiques

Place de la République, Hotel de Ville; tel. 04 90 49 59 05; www.arlestourisme.com; Mar., Apr., and Oct. daily 9am-6pm, May-Sept. daily 9am-7pm, Nov.-Feb. daily 10:30am-4:30pm; €5

The Forum was the economic and political center of Roman life in Arles. Little is known about the Forum buildings, but the foundations in the form of the cryptoportiques remain. Accessed from the town hall, these underground galleries supported the square and buildings above. However, they are dark and often damp. Unless you happen to be a historian of structural engineering, they are the least interesting of Arles's Roman sites.

Thermes de Constantin

rue Dominique Maïsto; tel. 04 90 49 59 05; www. arlestourisme.com; Mar., Apr., and Oct. daily 9am-6pm, May-Sept. daily 9am-7pm, Nov.-Feb. daily 10:30am-4:30pm; €5

The Thermes de Constantin (Roman baths) formed part of Emperor Constantine the Great's (272 to 337 CE) palace in Arles. They are estimated to have covered some 3,715 sq m (40,000 sq ft) and included rooms for undressing and taking cold and hot baths. There was also a warm air room and a dry resting room. The baths were once a social center where people met, gossiped, conducted business, and hatched political plots. Today only the remains of the warm air room, the hot baths, and the underfloor heating system survive. Even so, the scale is impressive and

it's possible to picture the place in its operational heyday.

★ Fondation Van Gogh

35 rue du Dr Fanton; tel. 04 90 93 08 08; www. fondation-vincentvangogh-arles.org; daily 10am-6pm; €10, ages 12-18 €4

Van Gogh's dream was to create an artistic commune in his yellow house in Arles. His famous fight in late 1888 with Paul Gauguin put pay to this idea. Gauguin was staying with van Gogh at the time, and when Gauguin threatened to leave van Gogh severed his own ear with a razor blade. The art world never forgot this incident or van Gogh's dream of an artistic community. In the 1980s a foundation was created to give new life to the idea of an international center of artistic exchange in Arles. Artists including Francis Bacon donated works, but it was not until the intervention of philanthropist Luc Hoffmann in 2010 that the Fondation Van Gogh finally achieved momentum, opening in a physical space four years later. The building usually hosts a few van Gogh originals and temporary collections chosen to illustrate van Gogh's artistic legacy. The intent is to stimulate thought through art that references the work of van Gogh. Unless you're an art historian, the "conversation" between the modernist works on display and van Gogh's legacy can at times be opaque. Even so, the Fondation is a stimulating place to visit. An old hôtel particulier (town house), the building was extensively remodeled to host the center. Particularly striking is the entrance gate, commissioned from the artist Bertrand Lavier, which re-creates the shaky Vincent signature, surrounded by dashes of yellow and green paint.

Espace Van Gogh

Place Félix Rey; daily 7am-7pm; free

Names can be misleading. From the name of this sight, you would be forgiven for expecting to find a space filled with van Gogh's work.

Instead, visitors see a pleasant courtyard filled with flowers and surrounded by shops selling postcards and prints of van Gogh's work. Constructed in the 16th century, the Espace Van Gogh housed Arles's main hospital until the 20th century. It's notorious as the place where van Gogh was committed after the infamous episode of cutting off his left ear in December 1888. During his time here, van Gogh painted the hospital. The courtyard has been restored to resemble his most famous work here, titled *Le Jardin de L'Hotel de Dieu*. For those who enjoy matching the scene in an artwork to contemporary reality the Espace merits a visit, but with all the shops the atmosphere can feel a little too commercial.

Eglise and Cloitre Saint-Trophime

Place de la République; tel. 04 90 18 41 20; www. arlestourisme.com; Apr. and Oct. daily 9am-6pm, May-Sept. daily 9am-7pm, Nov.-Feb. daily 10:30am-6pm; free

Ducking into the Eglise and Cloitre Saint-Trophime is a good way to escape the oven-like Place de la République in the heat of summer. Named after the first bishop of Arles, construction of the church began in the early 12th century. The particularly impressive Romanesque entrance was added around 1180. Some of the material used for the arched doorway was salvaged from the city's Théâtre Antique. The scene depicted as you enter is the Last Judgment, with the damned heading to hell in chains and the saved floating up to heaven accompanied by angels. Inside, the church is much larger than might be expected. The interior is largely notable for Romanesque sculpture, as well as late Roman sarcophagi and baroque tapestries.

Entrance to the church is free, but to see the celebrated cloister there is a fee (€6, reduced €5). Accessed from a separate, slightly hidden entrance, the Cloitre Saint-Trophime was added to the southeastern corner of the church in the same period as the entrance arch. It's a remarkable example of Romanesque architecture with arches supported by pillars

1: the Roman arena 2: Roman Theater 3: Fondation Van Gogh 4: spring flowers in the Espace Van Gogh

Loving Vincent

The official story of Vincent van Gogh's time in Provence is well known. In February 1888 van Gogh arrived in Arles from Paris, drawn by the clear light and bright colors. He dreamed of living in an artist's commune, and in May he rented the yellow house that was to become the subject of one of his pictures. In October fellow artist Paul Gauguin came to live with him. In December they fought and van Gogh slashed off his own ear. He was then admitted to a hospital in Arles. The people of Arles petitioned to have van Gogh thrown out of town, and he moved to Saint-Rémy-de-Provence where he entered a mental hospital. There he spent a year before moving to Auvers-sur-Oise outside Paris, where he shot himself. During this time in Provence, van Gogh painted 350 works.

Vincent van Gogh self-portrait

Mere words cannot convey the pain and torment that van Gogh's gift for painting brought to his life. His obsession with fulfilling his talent despite a lack of contemporary appreciation, his fight against epilepsy and mental illness, and the enduring beauty of his work all contributed to his story, which continues to resound through the ages. In 2016 local author Bernadette Murphy published the *New York Times* best-seller *Van Gogh's Ear* after new evidence was uncovered about van Gogh's fight with Gauguin. The year 2017 saw the release of the Oscar-nominated animated feature *Loving Vincent,* calling into question the accepted version of van Gogh's death. Then in 2018, *At Eternity's Gate,* a feature film starring Willem Dafoe as van Gogh, attempted to capture the light and shade of the painter's time in Provence.

Though the people of Arles once kicked van Gogh out of town, they are proud of him now. The artist's legacy can be seen all over the area:

- **Fondation Van Gogh:** Founded in the spirit of the artist's desire to create an artistic community, this old hôtel particulier usually has a few van Gogh originals on display (page 87).

- **Espace Van Gogh:** Now filled with shops selling van Gogh prints and postcards, this 16th-century building once housed Arles's main hospital, where van Gogh was committed after cutting off his ear in 1888 (page 87).

- **Le Café Van Gogh** (11 Place du Forum; www.restaurant-cafe-van-gogh.com): This café, though not necessarily a culinary highlight of Arles, is featured in *Café Terrace at Night.*

- **Les Alyscamps:** Van Gogh and Gauguin enjoyed painting this ancient Roman graveyard (page 89).

- **Abbaye Montmajour:** Van Gogh visited this massive monastery more than 50 times, drawing the imposing structure from different angles (page 90).

- **Van Gogh Walking Tour:** Follow signposts throughout Arles marking the spots where 10 seminal paintings were completed (page 90).

- **Monastere Saint-Paul de Mausole:** Van Gogh had himself voluntarily interned at this mental hospital in Saint-Rémy-de-Provence (page 101).

- **The Rock with Two Holes:** The artist painted this rock, accessible by a pleasant circular walk through vineyards and the rocky Alpilles (page 103).

surrounding a small garden. Themes from the Old and New Testaments are depicted in the carvings in the stone.

Museon Arlaten

31 rue de la République; tel. 04 13 31 59 99; www. museonarlaten.fr; daily 9:30am-6pm; €8

This museum is a must for anyone interested in Provençal society. The collection of local costumes and memorabilia was begun by the poet Frederic Mistral in the 19th century, but the museum ran out of money at the beginning of the 21st century and has only just reopened. The reimagining of the space is a spectacular success, with modern architecture and multimedia displays blending seamlessly into one of the most ancient buildings in Arles. The remains of part of the Roman Forum can even be seen in the courtyard of the museum.

Musée Réattu

10 rue du Grand Prieuré; tel. 04 90 49 37 58; www. museereattu.arles.fr; Mar.-Oct. Tues.-Sun. 10am-6pm, Nov.-Feb. 10am-5pm; €8

Located on the wind-whipped apex of a bend in the Rhône, this former priory of the Knights of Malta, completed at the end of the 15th century, now houses Arles's art museum. And what a museum it is, with a collection ranging from 18th-century works by its former owner Jacques Réattu to a contemporary sonic art installation. The museum holds intriguing pieces of art history, such as a poignant letter from van Gogh to Gauguin, written in January 1889, that touches on the breakup of their artistic relationship. There are also 57 sketches by Picasso donated to the museum two years before the artist's death, an impressive sculpture gallery, and, as you would expect in Arles, an extensive collection of photography.

Musée Departemental Arles Antique (Departmental Museum of Ancient Arles)

Presqu'île du Cirque Romain; tel. 04 13 31 51 03; www. arles-antique.cg13.fr; Wed.-Mon. 9:30am-6pm; €8,

reduced €5; Hortus Garden Apr.-Sept. Wed.-Mon. 10am-7pm, Oct.-Mar. Wed.-Mon. 10am-5:30pm; free

From a distance the big, blue modern box on the bank of the Rhône looks like a branch of Swedish furniture giant IKEA. But rather than flat-pack furniture, this blue-clad structure houses an extensive collection of archaeological artifacts found in the Arles region. Sculptures, mosaics, sarcophagi, numerous objects of everyday life, and even a 31-m (101-ft) Roman barge are laid out in the spacious interior. The main theme is the Romanization of Gaul—in other words, how the Romans, came, saw, conquered, and then made the place feel like home. The arena may be the go-to monument in Arles, but to really understand the city and its history, a visit to the archaeological museum is essential. Outside the museum is a Roman-inspired garden.

Les Alyscamps

avenue des Alyscamps; tel. 04 90 18 41 20; www. arlestourisme.com; Mar., Apr., and Oct. daily 9am-6pm, May-Sept. daily 9am-7pm, Nov.-Feb. daily 10:30am-4:30pm; €4.50

Les Alyscamps was one of the most important Roman burial grounds in Europe. Its significance endured long after the empire ended, with wealthy Christian families sending funeral barges down the Rhône to have their loved ones buried in this supposedly blessed spot. The name is the Provençal Occitane derivation of Elysian Fields, a Greek perception of the afterlife where the chosen were admitted to join the gods. Looting over the years has robbed the burial ground of some of its more spectacular monuments. Today only one main row of graves remains, leading to the Saint-Honorat 12th-century Romanesque church. Van Gogh and Gauguin enjoyed painting together here, but a 2,000-year-old graveyard is not everybody's idea of a fun outing.

★ Luma Arles

45 chemin des Minimes; tel. 04 90 47 76 17; www.luma-arles.org; Sept.-June Wed.-Mon. 10am-6pm, July-Aug. daily 10am-7:30pm; reserve online, free

Love it or hate it, the Luma Arles project is

having a transformative effect on the city of Arles. Funded by philanthropist Maja Hoffmann, the site occupies an old railway repair yard. The hangars and workshops have been imaginatively transformed into exhibition centers for art shows and resident dance and theater companies. The Frank Gehry building, which naysayers point to as a modernist catastrophe inflicted on the city, is a co-working space for resident artists and an integral part of any visit to Luma Arles. The jarring juxtaposition of the glinting silver tower, all jagged angles and boxes, with the symmetrical perfection of the Roman arena is impossible to ignore, which perhaps was Gehry's intention. The surrounding park has added welcome green space close to the city center. It is open daily from 7am-8:30pm.

Abbaye Montmajour

route de Fontvieille, Arles; tel. 04 90 54 64 17; www. abbaye-montmajour.fr; daily 10am-5pm (summer until 6:30pm); €6, under 18 free

Built between the 10th and 18th centuries, the Abbaye was until the late Middle Ages surrounded by marshes and only accessible by boat. Today there's plenty to see, most notably the rock cemetery, the chapel of the holy cross, and the ruined Maurist Abbaye. By the beginning of the 20th century Montmajour had been split up and was owned by 20 different people. Parts were even used as a sheep barn. It was gradually taken back into state ownership and is now one of fewer than 100 classed national monuments.

WALKING TOURS
Van Gogh Walking Tour

9 boulevard des Lices; tel. 04 90 18 41 20; www. arlestourisme.com; free with downloadable map

Van Gogh arrived in Arles in February 1888 and in just over a year produced more than 300 paintings. *The Night Café* and *Yellow House* are just two of the works that changed the art world forever. Signposts on the walk mark where these and other seminal paintings were completed. There are 10 spots in all,

and completing the circuit, depending on how long you stop and stare, takes around 3 hours.

ENTERTAINMENT AND EVENTS
Festivals
Feria d'Arles

Les Arènes; tel. 04 90 18 41 20; www.arlestourisme. com; Easter weekend

The opening of the bullfighting season is celebrated with one hell of a party in Arles. There's music, dancing, lots of drink, and crowds spilling from one bar to another. Around 500,000 people cram into Arles, and street bands (peñas) roam the city followed by streams of sangria-swilling bullfighting fans. The arena hosts daily bullfights.

Fête des Gardians

Les Arènes; tel. 04 90 18 41 20; www.confrerie-des-gardians.com; May 1; bull run €10

Some 500-plus years ago, the Camargue Gardians formed one of France's first confreries (brotherhoods). Today there are around 200 Gardians working on the manades (ranches) in the Camargue. Their work includes upkeep of the land and enclosures, castrating the bulls and horses, and participating in the summer course camarguaise. On May 1 the Gardians ride into Arles, celebrate a holy Mass, have a long lunch, and then stage a Course Camarguaise bull run in the arena. The proceeds of the event support Gardians who are injured during their work or become ill.

Course Satin

Manade Mogador Petit Route de Tarascon; tel. 04 88 96 47 00; www.festivarles.com; second Sun. in June; free

Dating back to 1589, this bareback horse race is as fiercely contested as ever. Only purebred Camarguaise horses from the best stud farms are eligible to enter. Watching the race is an exhilarating experience; winning, rather than

1: the Frank Gehry building at Luma Arles
2: a Camargue Gardian herding horses **3:** Abbaye Montmajour

1

2

3

the safety of the riders, is all that matters. The victor leaves with a gold-embroidered satin scarf, which he keeps for a year. If the same rider wins three years in a row, he keeps the scarf forever. The Course Satin caps a full day of horse races and horsemanship displays.

L'Abrivado des Bernacles

Departs from Mas des Bernacles in the Camargue, arrives at Les Arènes in Arles; tel. 04 90 18 41 20; www. festivarles.com; first Mon. in July

Taking place on the same day as one of the most important Course Camarguaise bull runs (Le Concorde d'Or), the abrivado is a display of horsemanship by the Camargue Gardians. Departing from 15 km (9 mi) outside Arles, they form a phalanx of horses to drive bulls into the center of the city. As the distance to the city shrinks, so the crowd swells.

Fête du Costume

boulevard des Lices and Théâtre Antique; tel. 04 90 18 41 20; www.festivarles.com; first Sun. in July and the Fri. before

Created by the poet Frederic Mistral, the Fête du Costume was instigated to preserve Provençal dress and traditions. In 1903 just 16 girls turned up for the first Arlesian costume festival. However, after this inauspicious start, the festival quickly took off, and with hundreds participating, it soon moved venues to the Roman theater. These days the festival has two parts. On the Friday night the pegoulado takes place: Crowds gather in traditional dress for a torch-lit procession that ends with a Farandole (Provençal dance) in the arena. On Sunday there is a dress parade in front of the queen of Arles in the Roman theater. Every three years a newly elected queen is presented to at the costume festival.

Arles Photography Festival

Exhibitions throughout the city; tel. 04 90 96 76 06; www.rencontres-arles.com; July-Sept., hours depend on venue; €32 day pass, €39 unlimited pass

In the summer months Arles hosts a major photography exhibition. Thought-provoking images are displayed throughout the city. Major venues host ticketed shows by well-known photographers, but even without a pass you can partake in the atmosphere of the festival. Photos decorate windows and street corners and animate a visit to the city. The opening week is particularly busy with various VIP and press events. The festival attracts an estimated 100,000 people to the city.

Feria du Riz

boulevard des Lices and the Arena; tel. 04 90 18 41 20; www.festivarles.com; second weekend in Sept.; free

Thanks to pump stations on the Petit Rhône and associated irrigation channels, the Camargue region produces France's entire output of rice. The harvest usually begins after this festival. Floats decorated with sheaves of rice and all sorts of other paraphernalia pass throught the streets of Arles. There's traditional dress, dance, bodegas spilling onto the streets, and of course bullfighting, in the form of a Spanish corrida.

The Arts
Le Corridor

3 rue de la Roquette; tel. 04 90 43 63 62; https:// lecorridor-artcontemporain.com; Mon.-Sat. 3pm-7pm; free

Annick and Michel Ray opened Le Corridor in 2016. Passionate about contemporary art, they have been collectors for 25 years and have converted part of their town house into a gallery to provide a space for artists to display and sell their work.

SHOPPING
La Main qui Pense

15 rue Tour de Fabre; tel. 04 90 18 24 58; www. cecilecayrol.com; Mon.-Fri. 10am-noon and 2pm-6pm

Cecile Cayrol eschews the colorful palette that characterizes many Provençal ceramics. Instead, form and design take precedence, and her workshop contains collections of bowls, cups, water jugs, and plates presented in subdued grays and reds.

La Parfumerie Arlesienne

5 rue de la Palais; tel. 04 90 97 02 07; www.la-parfumerie-arlesienne.com; Mon.-Sat. 10am-7pm

Inspired by Arles and the surrounding Camargue countryside, this boutique perfumerie produces a small range of scents, candles, and soaps. Bestsellers include the floral Cloud of Flamingos scent.

Citron A Paillettes

9 rue Réattu; tel. 04 90 93 54 31; www.facebook.com/citronapaillettes; Tues.-Sat. 10am-7pm

This small boutique specializes in original clothes and accessories. The clothes are either one-off creations or part of a limited line.

Indiennes de Nimes

14 Place de la République; tel. 04 90 18 21 52; www.indiennesdenimes.fr; Mon.-Sat. 10am-12:30pm and 2:30pm-7pm

Channel your inner cowboy or cowgirl at this clothing store selling traditional clothing for Camargue Gardians. From boots to hats and everything in between, the store can transform you into a Gardian in under 20 minutes.

FOOD

Arles has a good selection of restaurants. The city gets very busy during the tourist season from June to the middle of September, and the restaurants tend to be small with around 30 covers. To make sure you get a good table, indeed any table, it's advisable to make a reservation at the restaurants listed below. While touring the sights of Arles, look for these restaurants, poke your head in the door, and if you like what you see, reserve a table for lunch or supper. There are, of course, plenty of less memorable restaurants where you can get a table at the last moment.

Regional Cuisine

★ L'Epicerie du Cloitre

18 rue de Cloitre; tel. 04 65 88 33 01; www.lecloitre.com; daily noon-2:30pm (until 5:30pm Sat.) and 7pm-9:30pm; €7-17

This small, atmospheric courtyard restaurant bubbles with conversation and serves tapas made from high-quality local ingredients. Seafood dominates, with squid and sardines both on the small menu. Vegetables are sourced from the renowned potager of the Michelin two-starred Le Chastagnette restaurant in the Camargue.

Les Piques ou Rien

53 rue Condorcet; tel. 04 86 32 29 44; https://les-piques-ou-rien-restaurant-arles.eatbu.com; Wed.-Sat. noon-1:30pm and 7:30pm-9:30pm, Sun. noon-1:30pm, Tues. 7:30pm-9:30pm; €29

For an inventive concept restaurant, try Les Piques ou Rien. Here, after telling the chef about any allergies, diners simply wait to be served. What comes out of the kitchen is a surprise. There's usually a cold tapas starter, followed by a selection of hot, tasty dishes and five melt-in-the-mouth desserts.

Galoubet

18 rue du Docteur Fanton; tel. 04 90 93 18 11; daily noon-1:30pm and 7pm-9:30pm; €35

Rain or shine, this is a popular restaurant. Inside there's a cozy bistro atmosphere with vintage furnishings; outside there's a vine-covered terrace. The cooking is an inventive take on local classics featuring seasonal ingredients. Try the delicious medallions of pork with figs.

Le Greeniotage

7 rue de Carmes; tel. 04 90 91 07 69; www.rabanel.com; Wed.-Sun. noon-2pm and 7:30pm-9:30pm; €39-49

The sister bistro of the gastronomic Greensronome, the Greeniotage is the brainchild of celebrity chef Jean Luc Rabanel. The menu is short, with only four main courses on offer. Usually they are classic dishes such as roasted cod and camargue rice or long-cooked lamb shank. The dining room has a nice, cozy atmosphere.

L'Oriel

6 rue du Forum; tel. 04 90 49 49 21; www.restaurantoriel.com; Wed.-Sat. noon-2:30pm and 7:30pm-10:30pm, Tues. 7:30pm-10:30pm; €36-64

Classical French cooking in a luxurious

dining room is the winning formula of this newcomer to the Arles restaurant scene. The menu describes dishes with an unusual economy of vocabulary; "Pork, potatoes, mushrooms, apricot" hardly has you salivating. However, chef Quentin Lepilliet is as big on flavor as he is short on words.

International
Chardon

37 rue des Arènes; tel. 09 72 86 72 04; www. hellochardon.com; Fri.-Mon. noon-1:30pm and 7:30pm-9:15pm, Thurs. 7:30pm-9:15pm; €39

Chardon is an ever-changing delight of a restaurant. From one season to the next, it's impossible to know who will be cooking. Leading chefs from across the world are invited to take up residency, enjoy the atmosphere of Arles, and put something special on the plates of the locals. What's served depends on the nationality and influences of the particular chef.

Inari

16 Place Voltaire; tel. 09 82 27 28 33; www.inari-arles. com; Thurs.-Sun. 12:30pm-2pm and 7:30pm-9:30pm, Wed. 7:30pm-9:30pm; €35-55

Set in a 13th-century chapel, this restaurant is the hottest spot in Arles right now. Celine Pham has hit the sweet spot of the local palates with her asiastique fusion cusine. The mackerel marinated in Riesling vinegar and served with Japanese cucumber and chili is a must. Make sure you book in advance.

Cafés and Light Bites
Maison Volver

8 rue de la Cavaliere; tel. 04 90 96 05 88; www. maisonvolver.com; daily 8am-9pm

This is a nice spot for a leisurely breakfast just a couple hundred meters (650 ft) from the Roman arena. There is a large, comfortable terrace, and the owners take great pride in welcoming clients. Service continues all day with more substantial dishes for lunch and supper.

Mezza Luna

1 Place du Forum; daily 9am-10pm; €10-18

The bustling Place du Forum is filled with cafés and restaurants. For a quick bite, Mezza Luna is a good pit stop. On the menu are salads, vegetable tarts, and bruschettas. The Place du Forum in general is a good place to find a table if you can't get into the restaurants listed here.

Markets
Saturday and Wednesday Market

boulevard des Lices and boulevard Emiles Combe; Sat. and Wed. 8am-12:30pm

In Arles's twice-weekly market, there are clothes, linens, and gift stalls, but food dominates. In particular, look for butchers selling the locally reared bull meat, which has a richer, denser texture than normal beef, and specialty rices from the Camargue. Often these rices require lengthy cooking times (up to 25 minutes) and are specifically intended for salads. Clams fresh from the Camargue are also on sale. The Saturday market is larger and more popular.

BARS AND NIGHTLIFE
Grand Café Malarte

2 boulevard des Lices; tel. 04 90 54 56 74; www.grand-cafe-malarte-restaurant-arles.com; daily 8am-11pm; drinks from €3

It's hard to miss the Grand Café; its large terrace spills out onto Arles's main street, the boulevard des Lices. There's a brasserie restaurant, but if you are not hungry or have already eaten it's a good place to stop and soak up the Arlesian atmosphere. Inside there are nice touches like a wall-to-wall glass fridge filled with rosé.

Cargo de Nuit

7 avenue Sadi Carnot; tel. 04 90 49 55 99; www. cargodenuit.com; check website for events; pricing and hours depend on event

Part nightclub, part concert venue, the Cargo de Nuit is now more than 20 years old and still going strong. If there's a big-name DJ in town or a well-known artist, then chances are they'll be playing here. There's a wine and cocktail bar and restaurant on-site.

Bodeguita

49 rue des Arènes and 16 rue d'Hotel de Ville; tel. 04 90 96 68 59; www.bodeguita.fr; Wed.-Sat. 12:15pm-2pm and 7:30pm-10:30pm; tapas from €5

This authentic, all-singing, all-dancing tapas bar offers cocktails, shooters, imported Spanish charcuterie, and classic tapas such a patatas bravas (fried potatoes served with tomato sauce). It's so popular that there are now two sites open, and during weekends each does two sittings.

ACCOMMODATIONS
€50-100
Hotel de la Muette

15 rue des Suisses; tel. 04 90 96 15 39; www.hotel-muette.com; €99 d

There are 18 recently refurbished rooms in this city-center hotel. Old stone walls and beams create a warm ambience. It's a perfect budget choice for families, with triples, quadruples, and even a quintuple room, all with en suite bathrooms. The owners are big environmentalists and as a result of their efforts the hotel has a Green Key eco label.

L'Amphitheatre

5-7 rue Diderot; tel. 04 90 93 98 69; www.hotelamphitheatre.fr; €99 d

Tucked away down a side street, just minutes from the arena, this old hôtel particulier is full of character. There's a good choice of standard rooms, all with en suite bathrooms, air-conditioning, and Wi-Fi, and a small outside garden. The room decor is in need of updating, and parking is 300 m (984 ft) away on the boulevard des Lices.

€100-200
Hotel de Musée

11 rue Grand Prieuré; tel. 04 90 93 88 88; https://hoteldumusee.net; €110 d

The large interior courtyard is the main attraction of this budget city-center hotel near the Roman baths and the Musée Réattu art museum. It's a relaxing, shady place to sit and enjoy breakfast or immerse yourself in a book. The rooms are basic with old-fashioned decor, but they all have air-conditioning. Private parking is available for €20 per night.

★ L'Hotel du Cloitre

18 rue du Cloitre; tel. 04 88 09 10 00; www.hotelducloitre.com; €139 d

Seemingly effortlessly cool, L'Hotel du Cloitre has a young, vibrant vibe, with an on-site tapas restaurant, roof terrace, and funky rooms. Whimsical pieces of furniture, attentive service, and the charm of the old building work together to put a broad smile on the face of each guest. All rooms have ceiling or floor fans rather than air-conditioning. The public parking can be booked in advance for €11 per night.

Hotel L'Arlatan

26 rue Sauvage; tel. 04 90 97 20 29; www.hotel-arlatan.fr; €155 d

A spectacular renovation of Arles's most renowned hôtel particulier has transformed this historic building a few streets from the Roman baths into the destination hotel for Arles. Everywhere you look there are works of art; the scale of the endeavor and the attention to detail make the Arlatan a unique place to stay. Private parking is €18 per night.

★ Grand Hotel Nord Pinus

Place du Forum; tel. 04 90 93 44 44; www.nord-pinus.com; €190 d

The Nord Pinus is an Arlesian institution on the bustling Place du Forum that exudes the spirit and history of the city. Fashion designer Christian Lacroix describes it best: "To me when I was a child it was the temple of every summer holiday of high society and of course bull fighting. Starting from the Place du Forum, but never leaving it, it was like a voyage of adventure toward Paris, Spain, the world. Images of … white and gold capes spring to mind, of Cocteau and Picasso in black capes." All rooms have air-conditioning, Wi-Fi, and en suite bathrooms. Parking can be reserved in advance for €25 per day.

Camargue Regional Nature Park

Invest some time in the Camargue and it seduces like few other areas of Provence. The inhabitants of this wetland delta, formed by the Rhône as it meets the sea, are fiercely proud of their way of life: rearing bulls for the Course Camarguaise and corridas; farming the land for rice; producing wine and some of the finest table salt in France. Away from the main roads, out in the heart of the countryside, flamboyances of flamingos fly low overhead, their bodies improbably long and pink, illuminated by the setting sun. Gardian cowboys come whooping on horseback around corners, driving bulls before them. Humans, horses, and herds change direction as one, driven by centuries-old instinct.

In the bars and restaurants, the fact that life is hard here seems to have generated a reckless freedom of spirit, a desire to dance and party that is more Spanish than French. It all seems a million miles from the manicured streets of Saint-Rémy-de-Provence. For nature lovers, particularly bird-watchers, a trip to the Camargue and its nature reserves is a must.

SIGHTS

Eglise Notre-Dame-de-la-Mer

2 place de L'Eglise, Saintes-Maries-de-la-Mer; tel. 04 90 97 80 25; www.sanctuaire-des-saintesmaries.fr
It's no coincidence that Notre-Dame-de-la-Mer resembles a fortress. Built between the 8th and 12th centuries, its secondary purpose was as a refuge from pirate attacks on the port. It's free to enter and visitors can descend into the crypt where the statue of the gypsy patron saint, Black Sarah, stands lit by flickering candles. If you look above the choir pews, you will see the Chapel of Saint-Michel where the relics of Mary Salomon, Mary Jacobi, and Black Sarah are held. During the Roma pilgrimage held to celebrate the mythical landing of Jesus's partner Mary Magdalene in Provence, these statues are lowered from the chapel and carried out to sea. The barge that carries the statues of the Marys and Sarah is also on display in the church.

Outside the church are steps leading onto the roof. For a couple of euros you can climb up and enjoy an unparalleled view of Saintes-Maries-de-la-Mer, the Mediterranean, and the Camargue.

Musée de la Camargue

Pont de Rousty, Arles; tel. 04 90 97 10 82; www.arlestourisme.com; Apr.-Sept. daily 9am-5:30pm, Sept.-Mar. daily 10am-5pm; €7
Housed in an old sheep barn, the Camargue Museum depicts daily life in the Camargue from the 19th century to the present day. Focusing on key activities such as agriculture, livestock breeding, hunting, fishing, and annual celebrations and traditions, the museum traces the evolution of the Camargue into its modern form, explaining how the economic activities that developed over the 20th century—hydraulic works, viticulture, rice growing, and sea salt production—secured the future of the region. A visit to the museum also includes a 3.5-km (2.1-mi) walk through the grounds of the Mas de Pont de Rousty, where daily life in the Camargue is illustrated through various information points and observatories.

★ Parc Ornithologique du Pont de Gau

RD 570 Lieu dit Pont de Gau, Arles; tel. 04 90 97 87 62; www.parcornithologique.com; daily 9am-6pm; €8, ages 4-12 €6
The ornithological park is spread out over 60 hectares (148 acres) and includes more than 7 km (4 mi) of nature trails that offer unrivaled proximity to the Camargue birdlife. A map of suggested walks is provided on arrival. These vary in length, but on average the marked circular trails take 1.5 hours to walk.

The management of the park has subtly altered the watercourses and natural environment to facilitate sightings. By bringing nature closer to visitors, the park aims to help people appreciate it better and support conservation efforts. The birdlife varies with the time of year. From March to May many species are migrating north. Depending on the water levels in the Camargue, they will stop to rest and feed. Some species end their migration in the reserve, with large colonies of herons and egrets nesting in the trees. During the winter resident species include ducks, geese, cranes, birds of prey, and a number of rarer species, such as the tiny penduline tit. Winter is also the best time to watch flamingos; the population shows off colorful new plumage during courtship displays.

Guided visits in French are available and can be arranged by filling in an inquiry form on the website.

RECREATION AND TOURS

wild horses in Camargue Regional Nature Park

★ Safaris Camargues Alpilles
pickup point 1 rue Emile Fassin, Arles, or 17 rue de la République, Saintes-Maries-de-la-Mer; tel. 04 90 93 60 31; www.camargue.com; €52

The Camargue covers 930 sq km (359 sq mi), one-third of which is marshland or inland water. Unmarked, often private dirt tracks riddled with potholes crisscross the countryside between the tarmac roads. In short, you need to know where you are going to get the best out of a visit. Various companies offer educational safaris that take you in an open-topped Land Rover into the undiscovered Camargue for a couple of hours or a whole day. You'll enjoy wildlife-spotting (bulls, horses, and flamingos) as well as an excellent introduction to the culture of the Camargue, from rice growing to bull herding. It is simply the best way to gain insight into the Camargue in a short amount of time. In half a day, you will see more wildlife and learn more about the culture of the Camargue than you would on your own in a week.

Bateau Camargue
Port Gardian; tel. 06 17 95 81 96; www.bateau-camargue.com; €16 adults, €9 children

A good way of getting up close with the bulls and wild horses of the Camargue is to take a boat trip up the Petit Rhône. Departing from Saintes-Maries, the excursion lasts 1.5 hours, taking visitors up to the boundaries of the Pont de Gau ornithological park.

Les Arnelles Equestrian Tourism
RD570; tel. 06 03 89 23 79; www.arnellescamargue.com; €50/2 hours

Les Arnelles offers a wide variety of horseback rides, from a short ride out in the countryside to a whole-day safari. One of the most popular is the sunset ride. To avoid disappointment it's important to describe your skill level in advance. All levels are catered for, but the rides and groups need to be tailored to ability.

Over €200
Le Jules Cesar
9 boulevard des Lices; tel. 04 90 52 52 52; www.hotel-julescesar.fr; €215 d

A former Carmelite monastery that's been revamped by star interior decorator M Lacroix, this central hotel is full of color and moody artistic pieces. Just moments from the Roman arena and theater, the hotel offers a bar, a good restaurant with a terrace overlooking the boulevard des Lices, a swimming pool, and a spa. All rooms have air-conditioning, Wi-Fi, and en suite bathrooms. Private parking is available for €20 per night.

L'Hôtel Particulier
4 rue de la Monnaie; tel. 04 90 52 51 40; www.hotel-particulier.com; €315 d

This small luxury haven in the middle of Arles seduces with its pristine white interiors and black-and-white photos. Every detail has been thought through and every comfort provided, including a small swimming pool and spa treatments. All rooms have air-conditioning, Wi-Fi, and en suite bathrooms. The hotel has its own private garage.

INFORMATION AND SERVICES
Tourist Office
Esplanade Charles de Gaulle, boulevard des Lices 13200; tel. 04 90 18 41 20; www.arlestourisme.com; Mon.-Fri. 9am-12:45pm and 2pm-4:45pm, Sat. 9am-1:15pm and 2:30pm-4:30pm, Sun. 9am-12:45pm and 2pm-4:45pm

Arles's central tourist office has English-speaking staff who can help with anything from finding accommodation to getting tickets for events. The office is stocked with helpful maps and brochures that cover not just Arles but also the Camargue.

Hospitals and Pharmacies
For medical services, the main **pharmacy** is on 4 rue Jean Jaures (tel. 04 90 96 16 08). The closest hospital is **Joseph Imbert** (Quartier Fourchon; tel. 04 90 49 29 29).

Post Office
5 boulevard des Lices; Mon.-Fri. 8:30am-noon and 2pm-7pm, Sat. 8:30am-midday

The post office offers the usual range of services for sending letters and parcels within France and internationally. It is located half a kilometer south of the Roman theater.

GETTING THERE
Car
Arles is easily accessible by car, although the surrounding roads can get congested, particularly at the beginning and end of the day. To reach Arles from **Marseille,** take the A7 and then the A54 to Arles. It takes about an hour to travel the 91-km (56-mi) distance.

From **Aix-en-Provence,** Arles is about 75 km (47 mi) on the A54. The drive takes about an hour. From **Avignon,** take the D2 for 46 km (28 mi); it will take about 45 minutes. From **Saint-Rémy-de-Provence,** take the D570N. It takes about 30 minutes to cover the 30 km (19 mi) between Saint-Rémy and Arles. From **Saintes-Maries-de-la-Mer,** take the D570; it will take you about 35 minutes to cover 37 km (23 mi).

Parking near the center of the city can be problematic. The largest **parking lot** is on boulevard des Lices near the tourist office.

Train
The **train station** is located a short walk from the center, next to the bus station, on avenue Paulin Talabot. Service runs from the following towns and cities.

From Marseille, direct trains run approximately every half hour, and the journey from Marseille Saint-Charles Station takes about an hour. Tickets are about €15 one-way. From Avignon, direct trains run from the central station every half hour or so and take 20 minutes to reach Arles. Tickets cost €9 one-way. Consult www.sncf-voyageurs.com for up-to-date timetables.

Bus
The **bus station** is located a short walk from the center, next to the train station on Avenue

Paulin Talabot. Note there is no direct bus from Aix-en-Provence to Arles.

From Avignon, in the summer **Zou line 707** (https://zou.maregionsud.fr) costs €4 and takes about 90 minutes. From Saint-Rémy-de-Provence, **Zou line 704** costs €3 and takes about 60 minutes. Buses run frequently.

GETTING AROUND

No matter how you arrive in Arles, the best way to see it is on foot. All the sights listed in this chapter are within 15 minutes' walk of each other and much of the historic city is pedestrianized. Roads that are open to traffic tend to be narrow and one-way, and they're best avoided. The only sight you will need to drive to is the Arles Archaeological Museum.

Taxis can be booked through your hotel or **Arles Taxi Services** (tel. 04 28 31 41 06; www.arles-taxis-services.com).

Les Alpilles Regional Nature Park

The Alpilles Regional Nature Park is filled with fields of olive trees and vines interspersed with jagged rocky outcrops and charming, typically Provençal villages such as Maussane-les-Alpilles and Eygalières. There are plentiful cycling, hiking, and horseback riding routes taking you into the hills where wild herbs sprout from the bases of rocks and pine trees shimmer in the blazing sun. In Saint-Rémy-de-Provence, there are the ruins of the Roman town of Glanum to visit and van Gogh's artistic legacy from his yearlong sojourn in the town to enjoy. Nearby Les Baux-de-Provence, a village perched precariously on a rocky outcrop, is in danger of becoming overrun by tourists, but it's still well worth a visit for its medieval château and one-of-a-kind art projections in the bauxite quarry. Less well known, Fontvieille is a sleepier place but one that still boasts a wealth of sights, such as the 10th-century Montmajour Benedictine Monastery.

ORIENTATION

The Alpilles (literally little Alps) sit just east of **Arles.** The name of the area derives from the limestone massif that runs east to west between the Durance and Rhône rivers. It is 25 km (16 mi) long and approximately 8 km (5 mi) wide. To the north of the massif sits bustling, picturesque **Saint-Rémy-de-Provence** on the D99 road between Cavaillon and **Beaucaire.** It's a natural base from which to explore the Alpilles. Still on the northern side of the massif is the village of **Eygalières.** Its main street is a picture-postcard image of Provence, and the village is another good base for exploring the area. From Saint-Rémy the D5 winds through rocky crags to the southern side of Les Alpilles and the villages of **Maussane-les-Alpilles** and **Les Baux.** Les Baux is surrounded by luxury hotels, but the medieval village itself suffers from overtourism, making Maussane a better place to stay.

SAINT-RÉMY-DE-PROVENCE

The town of Saint-Rémy-de-Provence dominates Les Alpilles. Its shops, restaurants, and bourgeoise atmosphere make it feel like a mini Aix-en-Provence. Days can disappear in the embrace of its small squares, bubbling fountains, and narrow cobbled streets lined with art galleries and boutiques. The pedestrianized old town is encircled by a ring road. Thankfully the traffic is slow moving and there are plentiful pedestrian crossings. Shade from centuries-old plane trees shelters café terrace after café terrace, making the ring road the main social and people-watching hub of Saint-Rémy. All the sights listed below participate in the Saint-Rémy Pass entry system. If you pay one entry at full price, then entry

Les Alpilles Regional Nature Park

© MOON.COM

to all the other sites is at a reduced rate. Ask for the pass when you pay if this offer interests you. The pass is free.

Sights

Monastere Saint-Paul de Mausole

chemin Saint-Paul; tel. 04 90 92 77 00; www.saintpauldemausole.fr; Apr.-Sept. daily 9:30am-7pm, Oct.-Mar. daily 10:15am-5:15pm; €7

In May 1889 Vincent van Gogh had himself voluntarily interned in this mental institute. His ambitious plans for an artist's commune in the yellow house in Arles had come to nothing, and after a fight with fellow artist Paul Gauguin he lost his sanity and cut off his ear, wandering through the streets in a deranged state. After this incident the people of Arles drove him from the town.

Van Gogh found refuge in Saint-Rémy, where he came to accept the madness induced by his epileptic fits and resumed painting. The tour includes a visit to the room where van Gogh slept and painted, which has been restored to an approximation of its condition at the time; a walk through the gardens where large panels illustrate the views that van Gogh painted; the "Roman Cloisters," considered a masterpiece of 12th-century Provençal architecture; and the baths, where van Gogh was treated for his condition by repeated immersion in cold water. The audio guide is well worth the extra couple of euros, giving a detailed history of van Gogh's life, particularly concentrating on his work during the year he spent in Saint-Rémy. Part of the building is still a working mental health institution, so visitors are asked to be respectful and quiet at all times.

Musée Estrine

8 rue Lucien Estrine; tel. 04 90 92 34 72; www.musee-estrine.fr; June-Sept. Tues.-Sat. 10am-6pm, reduced hours out of season, closed Dec.-Feb.; €7, under 12 free

Built in 1749, this elegant town house was the residence of the representatives of the prince of Monaco. It is now a museum that concentrates on 20th- and 21st-century art. Rather poignantly, its collections are dedicated to van Gogh's desire "that living artists are not unjustly unknown." There's a space set aside to explain van Gogh's artistic legacy, including a short film about the artist. However, there are no permanent van Gogh originals. Nonpermanent exhibitions change regularly. Check the website for details.

Musée des Alpilles

1 Place Favier; tel. 04 90 92 68 24; www.mairie-saintremydeprovence.com; May-Sept. Tues.-Sun. 10am-6pm, Oct.-Apr. Wed.-Sat. 1pm-5:30pm; €5, children free

Small provincial museums can be a bit of disappointment; they have no budget for the sort of glitzy multimedia displays employed by their urban counterparts, so they can seem to be stranded in the past. The Alpilles museum is an exception. Located in the Hotel Mistral de Montdragon mansion house, the architectural quality of the building, and particularly the inner courtyard, sets the tone for the visit. The purpose of the museum is to help visitors understand the current landscape, natural and human, in this part of Provence. Accordingly, exhibits cover topics ranging from the historic cultivation of poppy seeds to bullfighting. It's a well-thought-out introduction to the region.

★ Glanum

route des Baux-de-Provence; tel. 04 90 92 23 79; www.site-glanum.fr; Apr.-Sept. daily 9:30am-6pm, Oct.-Mar. Tues.-Sun. 10am-5pm; €8

If you only have a passing interest in the Romans, you can soak up a bit of culture and see a couple of the most impressive Glanum monuments for free: a triumphal arch and the Julii Mausoleum. The triumphal arch was built around 14 CE as a symbol of Roman power and authority. Study the sculptures carefully and you will see lots of Gauls in chains being lorded over by their imperial masters. The mausoleum is one of the most impressively preserved burial monuments from the Roman era, dating to around 40 BCE. The upper part consists of a circular chapel, the middle section has four arches

topped by gorgon's heads, and the square base is carved with legendary scenes of battle.

Those with an appetite for more ruins can cross the road and pay the entrance fee to discover the remains of Glanum. The town was founded on the site of a sacred spring by a Celto-Ligurian tribe between the 2nd and 4th centuries BCE. The tribe had regular contact with the Greek settlement at Marseille, and some of the remaining structures show a Greek influence. In 49 BCE when Julius Caesar conquered Marseille and the Romanization of Gaul began, Glanum prospered. Located on the Via Domitia Roman road that linked Italy and Spain, the city even minted its own coin. As they enter, visitors can see scale models of the town in its different eras. The main ruins include the sacred well, the Hellenic fountain, the First Forum, the Second Forum, the Baths, and the Market Square.

Hotel de Sade

1 rue du Parage; tel. 04 90 92 64 04; www.hotel-de-sade.fr; June 16-Sept. 17 Tues.-Sun. 9:30am-1pm and 2pm-6pm; €4, under 18 free

This ancient building, with parts date back to the 5th century, hosts the collection of archaeological finds from the nearby Roman Glanum site. It's one for lovers of antiquity, with plenty of crumbling statues and gruesome pieces such as a stone lintel that was used to display the severed heads of enemies. There are also sculptures of Greco-Roman gods such as Dionysos-Bacchus. In addition, there are often temporary art exhibitions. As its name suggests the building was once the town house of the Sade family. It belonged to the grandfather of the famous marquis.

Festivals and Events
Fête de la Transhumance

Saint-Rémy; www.saint-remy-de-provence.com; seventh Sun. after Easter

For those not in the know, it can come as

quite a shock to suddenly discover 5,000 or so sheep, along with some donkeys and goats, cantering gaily through the center of Saint-Rémy on Whit Sunday (the seventh Sunday after Easter). The fête commemorates the transhumance, a 10-day hike to the Alps to find fresh green pastures for Provence's sheep. These days the sheep are transported in the back of trucks, but that doesn't stop the fun part: letting animals take over the town for a few hours.

Summer Feria

Saint-Rémy; around Aug. 15

This four-day festival in the middle of summer combines music (brass bands in the day and techno at night), bull running, and traditional dress processions. Wine and pastis fortify the young men of the town, who try to snatch ribbons from the horns of the bulls as they pass through the streets on the way to the arena.

The Arts
Galerie du Pharos

36 rue Lafayette; tel. 04 90 92 57 90; www.lapeinture.com; Wed.-Sun. 10:30am-12:30pm and 2pm-7pm, Tues. 2pm-7pm

The gallery displays permanent collections by established local artists. The works range from pop art to landscapes.

Galerie Lacaux

14 rue Jaume Roux; tel. 04 90 26 91 33; www.galerie-lacaux.com; Tues.-Sun. 10.30am-1pm and 2:30pm 7pm

Denis Lacaux specializes in surrealist fantasy paintings that are full of vibrant colors. The gallery also contains permanent collections of sculpture and works by other artists.

Hiking
The Alpilles Ridge and the Rock with Two Holes

Distance: *8 km (5 mi) round-trip*
Time: *2.5 hours*
Information and Maps: *www.alltrails.com/fr/randonnee/france/bouches-du-rhone/rocher-des-deux-trous-depuis-saint-remy-de-provence*

1: Fête de la Transhumance **2:** the Rock with Two Holes **3:** the ancient ruins of Glanum

Trailhead: *Glanum car park*

This is a pleasant circular walk that takes you through vineyards and into the rocky Alpilles. Starting from the Glanum parking lot a few kilometers to the south of Saint-Rémy, the dirt path takes you past the mental institution where van Gogh was interned and out into the hills. There are no refreshments along the route, but you can stop at the Rock with Two Holes, which was painted by van Gogh during his time in Saint-Rémy, and admire the view out across the countryside. The return loop takes you back through the vineyard to Glanum. Much of the route is on marked walking trails, but even so there are some steep points where the rocky scree can get slippery after wet weather. Only intermittent shade is provided by pine trees, so take plenty of water and wear a hat.

Lac du Pieroou

Distance: *2 km (1.2 mi) round-trip*
Time: *30 minutes*
Information and Maps: *www.alltrails.com/trail/ france/bouches-du-rhone/circuit-du-lac-du-peirou- glanum*
Trailhead: *avenue Antoine de la Salle*

For a pleasant stroll and a picnic, this man-made lake is hard to beat. Surrounded by trees and rocks, it's a peaceful place to visit just a few kilometers from Saint-Rémy. The circular lake walk is 2 km (1.2 mi) and takes 30 minutes.

Cycling

Les Alpilles is an extremely enjoyable place for cycling.

Saint-Rémy-Eygalières-Maussane Loop

This circular route allows you to take in some of the best scenery of Les Alpilles. Leave Saint-Rémy on the chemin la Croix des Vertus/Ancienne Voie Aurelia toward Eygalières. This is a flat road that meanders through the outskirts of the town, passing the **Domaine Milan** vineyard, a natural

winery that is well worth a stop for a tasting. From Eygalières the D24 and then the D78 roads toward Maussane-les-Alpilles are still relatively flat, and the countryside is a mix of pine forest, vines, and olive trees. Maussane makes a nice place for lunch before you tackle the windy climb back into the hills on the D5, which descends back to the starting point in Saint-Rémy. The entire route is about 45 km (28 mi) and takes about 2.5 hours. In the summer the roads can be busy, so it is important to be aware of traffic and cycle single file.

Sun-e-bike

2 rue Camille Pelletan; tel. 04 32 62 08 39; www.sun-e-bike.com; daily 9am-6pm; €40/day

Sun-e-bike has a wide choice of bikes and suggests eight different circuits you can follow to explore Les Alpilles.

Shopping
Market

Place de la Mairie and Place de la République; every Wed.

Saint-Rémy hosts one of the biggest and best markets in Provence. You can find the usual mixture of fresh fruit and vegetables, snacks that pair perfectly with a glass of wine, and plenty of gifty products, including jewelry and textiles. There is a smaller mainly food market on Saturday.

Libellule

10 rue Jaume Roux; tel. 04 90 21 19 89; https:// libellule13210.fr; Tues.-Sat. 10am-1pm and 3pm-7pm

The shop is full of cushions, throws, and other soft furnishings to help you get that Provençal look at home. Expert interior decorating advice is available if needed.

Le Comptoir des Alpilles

2 Place Jules Pelissier; tel. 04 90 94 86 76; www. facebook.com/ComptoirdesAlpilles; daily 10am-7pm

This homeware shop specializes in colorful cushions, tablecloths, and tableware. It's hard to leave without a little something to brighten up a room back home.

Le Savoir Faire des Alpilles

1 boulevard Marceau; tel. 04 90 94. 53 52; www.le-savoir-faire-des-alpilles.business.site; daily 10am-7pm

Located on the site of the old Saint-Rémy forge, the shop features the creations of a collection of local producers and artisans, including jewelry, furniture, shoes, and paintings.

Joel Durand

3 boulevard Victor Hugo; tel. 04 90 92 38 25; Tues.-Sat. 9:30am-12:30pm and 2:30pm-7pm, Sun. 10am-1pm

There's nothing like a chocolate fountain in the window to lure in shoppers. Joel Durand is one of France's top chocolatiers, and delicious boxes of melt-in-the-mouth chocolate creations are the specialty of this shop.

Food

Olga

19 boulevard Victor Hugo; tel. 04 90 92 34 49; Tues.-Sat. noon-2:30pm and 7pm-10pm; €24

This restaurant on the bustling exterior ring road serves classic French cuisine prepared under the watchful eye of talented young chef Jeremy Scalia. The wine list is worth exploring, and the staff are helpful in matching wine with food. It's a small restaurant, and reservations are advisable even outside the main tourist season.

Bistrot les Pieds Dans l'Eau

22 boulevard Victor Hugo; tel. 04 90 90 74 49, www.bistrot-les-pieds-dans-l-eau.business.site; Mon.-Fri. 10am-4pm and 6pm-1am, Sat.-Sun. 10am-12:30am; €24

Just off the circular road, this bistrot has a large terrace centered on a fountain. The chairs are low and comfortable, the atmosphere relaxed and convivial. It's a place for a cocktail or aperitif as well as a meal. There are plates of cold meats and cheeses served on long wooden boards, salads, and larger mains such as rib of beef.

Bar Tabac les Comptoirs du Gigot

21 boulevard Victor Hugo; tel. 04 90 92 02 17; Thurs.-Tues. noon-2:30pm and 7pm-10:30pm, Wed. 7pm-10:30pm; €25

The menu at Bar Tabac les Comptoirs du Gigot is limited to say the least, with just a few mains. There's one dish, though, that is far and away the most popular: roast leg of lamb served with very garlicky gratin dauphinoise potatoes. It's delicious, and waiters tour the tables replenishing empty plates. This restaurant is big on local atmosphere and low on vegetarians, although there are a few salads on offer.

L'Auberge

12 boulevard Mirabeau; tel. 04 90 92 15 33; www.aubergesaintremy.com; Tues.-Sat. noon-1:30pm and 7pm-9:30pm

Sample the cooking of Fanny Rey, the 2017 Michelin Female Chef of the Year. Rey's cuisine is powerful and unashamedly modern, using, for example, algae instead of oil in the cooking of some of her dishes. Reserving ahead is imperative.

Accommodations

Hotel Gounod

18 Place de la République; tel. 04 90 92 06 14; www.hotel-gounod.com; €150 d

The Gounod has a central location with comfortable rooms and a large courtyard set around a swimming pool. The decor is slightly quirky, with black and white predominating and the repeated motif of musical notes. Rooms all have en suite bathrooms and Wi-Fi. A small spa offers beauty treatments. Public parking is available directly opposite the hotel.

★ Mas des Carassins

1 chemin Gaulois; tel. 04 90 92 15 48; www.masdescarassins.com; €200 d

Away from the summer crowds but still within walking distance of the center, the Mas des Carassins cultivates a traditional Provençal atmosphere. The old stone house has bedrooms with quilted bedspreads and a mature garden full of trickling fountains. The pool has views over the countryside toward Les Alpilles. The hotel has its own private parking.

La Maison du Village

10 rue du Huit Mai 1945; tel. 04 32 60 68 20; www. lamaisonduvillage.com; €300 d

A village house was converted into this cozy boutique hotel with four suites. Each has a bedroom, sitting room, and private bathroom. There's a peaceful and shady interior courtyard for relaxing. The nearest public parking is 200 m (656 ft) away.

Domaine de Chalamon

291 chemin Chalamon; tel. 04 87 83 10 10; www. lesdomainesdefontenilles.com; €300 d

The Chalamon is the latest uber-chic outpost of the Fontenille group, and like its sister hotels it majors in interior design. Every element of this careful restoration has been planned to perfection. The shady formal garden is a joy to kick back in while listening to the sound of the cicadas.

Monplaisir

chemin Monplaisir; tel. 04 90 92 22 70; www.camping-monplaisir.fr

A short walk from the charming town of Saint-Rémy-de-Provence, this campsite has it all: professional service, a swimming pool, a restaurant, and plenty of shady pitches. It's simply a great place to stay to see Les Alpilles.

Getting There and Around

Once you have arrived at Saint-Rémy-de-Provence, the center of the town is pedestrianized and best visited on foot.

Car

Saint-Rémy-de-Provence is easily accessible by car. Upon arrival you will be channeled onto a ring road that loops around the historic center.

To reach Saint-Rémy from **Marseille,** take the A7 and then the D99; the drive takes 1 hour and 10 minutes, covering 89 km (55 mi); from **Aix-en-Provence,** take the A8, A7, and the D99; the drive takes 1 hour and 5 minutes, covering 73 km (45 mi). From **Avignon,** take the D35 and then the D5; the drive takes 30 minutes, covering 24 km (15 mi). If you're

coming from **Arles** take the D570N; the 30-km (19-mi) journey will take 30 minutes. To get to Saint-Rémy-de-Provence from **Saintes-Maries-de-la-Mer,** take the D570; the journey is 65 km (40 mi) and takes 1 hour 5 minutes.

Parking is difficult in the peak summer months, and you can expect to have to walk for up to 10 minutes to get to the center of town. Out of the main season it is relatively easy to find a place to park.

Bus

The **bus station** (24 boulevard Marceau) is adjacent to the historic center. From Arles **Zou line 704** (https://zou.maregionsud.fr) costs €3 and takes about 60 minutes, and in the summer **Zou line 707** from Avignon costs €4 and takes about 90 minutes. Buses run frequently.

LES BAUX-DE-PROVENCE

Les Baux-de-Provence is perhaps the only village in Provence to suffer from the curse of overtourism. From a distance the silhouette of the ancient ramparts, a single flag fluttering in the wind, is a majestic sight. Up close the narrow streets of the village are overrun on a daily basis with coach parties. In season, parking anywhere near the hilltop sight is increasingly difficult. Shops and cafés sell touristy products to the hordes. Yet, if you are willing to put up with masses, or you can visit out of season, the sights make the aforementioned inconveniences seem incidental. From April to September, you can buy the €18 Pass Baux, which entitles you to entry to the Château des Baux and Carrières des Lumières, with half-price entry to the Musée Yves Brayer.

Sights
Château des Baux

Grande Rue; tel. 04 90 49 20 02; www.chateau-baux-provence.com; June-Sept. daily 9am-7pm, Oct.-Apr. daily 9am-5:30pm; €8.50

Built in the 11th century by the lords of Les

Baux and variously demolished and rebuilt over the following centuries, the fortress is a wonderfully evocative place to visit. In the summer months there are daily displays of medieval arts and crafts, and more excitingly, for the children at least, siege reenactments using full-scale siege engines. The views over the vines and olive groves of the surrounding countryside are unsurpassed. A tour visits different parts of the château, including watchtowers, chapels, pigeon houses, caves, and a windmill.

★ Carrières des Lumières

route de Maillane; tel. 04 90 49 20 03; www.carrieres-lumieres.com; Jan., Mar., and Nov.-Dec. daily 10am-6pm, Apr.-May, June, and Sept.-Oct. daily 9:30am-7pm, July-Aug. daily 9:30am-7:30pm; €12.50, ages 7-17 €10.50, family ticket €40, under 7 free

The old bauxite mines, from which Les Baux-de-Provence takes its name, have been converted into a venue for multimedia art displays. Each year, a new show is conceived around famous painters and their works. Paintings are projected to fill all the walls of the old mine and classical music plays in the background, guiding the mood of the experience from melancholy to cheerful.

The extensive caves and their immense walls can be disorienting at first, but soon the environment starts to feel like a cocoon in which to experience art in a different way. The most arresting parts of the show occur when a three-dimensional universe is created by the paintings, which the watching crowd walks through and explores. The ability of the creative directors to manipulate characters and images within the paintings and seemingly move them toward and away from the viewer makes viewers feel like part of the art. Ultimately, the combination of paintings and music, and the unique viewing experience, is very moving, and more than worth the price of admission. A new show debuts every January and runs throughout the year. The experience lasts 30 minutes.

Musée Yves Brayer

Place François Hérain; tel. 04 90 54 36 99; www.yvesbrayer.com; Apr.-Sept. 10am-12:30pm and 2pm-6:30pm, Oct.-mid-Mar. Wed.-Mon. 1pm-7:30pm, closed Jan.-early Mar.; €8, under 18 free

Provence is so renowned for its star painters that it's easy to overlook lesser-known artists. Yves Brayer was born in 1907 and worked in the age of Picasso. Unlike Picasso, his name is not associated with any particular artistic school and many will not have heard of him before their visit to this museum. Indeed, visitors are often drawn to the museum by temporary exhibitions by more well-known artists on the 3rd floor; recent shows have included works by Marc Chagall and Paul Signac. However, discovering a new artist is a delightful experience. The range, quality, and detail of Brayer's work surprise many visitors and often make this small museum a highlight of a trip to Les Baux.

Santon Museum

Place Louis Jou; tel. 04 90 54 34 03; www.lesbauxdeprovence.com; Mon.-Fri. 9:30am-5pm, Sat.-Sun. 10am-5:30pm; free

One consequence of the French Revolution and the establishment of a republic was a crackdown on religion. Churches were defaced and some even destroyed. Traditional nativity services with life-size donkeys and locals playing the parts of the wise men were abandoned. In their place people celebrated the nativity in their own homes with carved wooden figures: santons. The anti-religious zeal soon dissipated, but the Provençal people's love of the santon never disappeared. This museum includes extensive collections of santons by the most celebrated Provençal craftsmen, as well as collections of nativity figures from around the world.

Golf
Domaine de Manville Golf Course

Les Baux-de-Provence; tel. 04 90 54 40 20; www.domainedemanville.fr; greens fee €103

Centered on a luxury hotel, this challenging 18-hole course has scenic views of Les Baux.

1

2

3

4

Food
Une Table au Soleil
rue du Trencat, cours des porcelet; tel. 06 38 68 24 92; www.instagram.com/loustaudipastre; €18-25

Eating well in the historic center of Les Baux is difficult. So many tourists pass through that quality is not as high as elsewhere. Une Table au Soleil is the exception. There's a shady courtyard, welcoming staff, and good food ranging from lobster to simple pasta dishes. Reserving in advance is advisable.

L'Ousteau de Baumaniere
D27 Les Baux-de-Provence; tel. 04 90 54 33 07; www.baumaniere.com; €130-330

This is one of the best, if not the best, restaurants in Provence. In a stunning setting underneath the rocks of Les Baux, the garden vibrates with cicadas and exudes the scent of wild herbs and flower blossoms. The food is delicate and complex, and the prices are accordingly high. Reserving in advance is necessary.

Accommodations
Le Mas d'Aigret
D27A, Les Baux-de-Provence; tel. 04 90 54 20 20; www.masdaigret.com; €205 d

A friendly family hotel in the shadow of Château des Baux, Le Mas d'Aigret has good views over the fields of olives and vines below. Comfortable rooms with en suite bathrooms and Wi-Fi, a good restaurant, and the opportunity to visit Les Baux before and after the crowds make this a popular choice.

Mas d'Oulivie
route d'Arcole; tel. 04 90 54 35 78; www.masdeloulivie.com; €250 d

Located on the olive tree-filled plain beneath Les Baux, this hotel has a fragrant garden and a picturesque pool. The rooms are comfortably furnished with en suite bathrooms and Wi-Fi, and there is a good choice of sizes. The on-site restaurant is only open at lunchtime.

Getting There and Around
Car
The road network around Les Baux is good, even in high season. However, when you arrive, **parking** is extremely limited. Arrive early or visit in winter to avoid a 15-minute uphill walk to reach the town from your parking spot.

To reach Les Baux from **Saint-Rémy-de-Provence,** take the D5 and D27, covering 10 km (6 mi) in 20 minutes.

Bus
In the summer **Zou line 707** (https://zou.maregionsud.fr) from Avignon to Saint-Rémy is extended to Arles, stopping at Les Baux. It runs five times a day and the fare is about €3 for a single journey. Journey time from Avignon is 1 hour.

MAUSSANE-LES-ALPILLES AND LE PARADOU
Maussane-les-Alpilles and Le Paradou are two separate villages, located no more than 1 km (0.6 mi) apart. Maussane-les-Alpilles is prosperous and picturesque. On a daily basis it draws in well-heeled people from the villa-rich surrounding countryside for a morning coffee and a spot of food shopping. Le Paradou is smaller and is famed for the restaurant the Bistrot du Paradou, which has attracted its share of celebrities over the years. There's not a lot to see in either village, but both ooze Provençal charm.

Sights
Place de L'Eglise
D17, Maussane-les-Alpilles

Maussane-les-Alpilles's Place de l'Eglise central square is the social center of the village. It's filled with café terraces shaded by plane trees. Construction of the **Sainte-Croix church,** which overlooks the square, began in 1750. The relaxed atmosphere of the place

1: a biker overlooks Les Baux-de-Provence
2: market in Saint-Rémy-de-Provence
3: an art show at Carrières des Lumières
4: Maussane-les-Alpilles

is enhanced by the Four Seasons fountain. Built in 1860, it is unusually large for a village of Maussane's size and it bubbles away, providing the soundtrack for long Provençal lunches.

Moulin Jean-Marie Cornille

rue Charloun Rieu, Maussane-les-Alpilles; tel. 04 90 54 32 37; www.moulin-cornille.com; Mon.-Sat. 10:30am-11:45am and 1:30pm-3:30pm; €4, under 10 free

The process of harvesting the olives and then extracting the oil is explained in an exhibition on the 1st floor of this mill. Afterward there are various oils to taste, ranging from green, thin, and slightly tart offerings to my preferred style: peppery, smooth, and viscously golden. A large boutique (Mon.-Sat. 9:30am-12:30pm and 2:30pm-6:30pm) sells the mill's olive oil as well as other local products.

Golf
Golf de Servanes

Domaine de Servanes, Mouries; tel. 04 90 47 59 95; http://golfservanes.com; greens fee €94

This mature golf course winds through the pretty scenery of Les Alpilles. Don't get too distracted by the surroundings: There are plenty of water hazards.

Horseback Riding
Le Petit Roman Horse Riding

route du Ferigoulas, D5, Maussane-les-Alpilles; tel. 06 98 16 02 01; www.randochevalalpilles.com; from €45

The stable offers horseback rides into the Alpilles countryside, including longer rides that become gourmet escapades with tastings at local producers.

HD Bike

102 avenue de la Vallée des Baux, Maussane-les-Alpilles; tel. 06 43 31 76 98; www.hdbike.fr; May-Sept. Mon.-Sat. 9am-12:30pm and 2pm-7pm, Sun. 9am-12:30pm; from €25/half day

The countryside around Maussane is perfect for cycling. HD Bike can suggest plenty of different routes to hop between the local villages.

Festivals and Events
Village Festival

Maussane-les-Alpilles, Aug. 13-15

This two-day festival includes music, dancing, a communal meal in the street, boule competitions, fireworks, and, of course, a bull run.

Shopping
Market

Place Henri Giraud, Maussane-les-Alpilles; Thurs.

A typical Provençal market takes place every Thursday in Maussane. There's a good mixture of food, clothes, textiles, and gifts available.

Food
Café de la Fontaine

70 avenue de la Vallée des Baux, Maussane-les-Alpilles; tel. 04 90 54 30 15; daily 8am-11pm, kitchen noon-3pm and 7pm-10:30pm; €16-32

The heartbeat of the village, Café de la Fontaine has tables out on the square, next to the Four Seasons Fountain. Inside there's a funky, very French dining room and a bar backed by a large picture of a topless woman waving the tricolor. The café takes great pride in its food and has its own meat-aging fridge and chicken rotisserie. In summer it is advisable to reserve.

Bistrot du Paradou

57 avenue de la Vallée des Baux, Le Paradou; tel. 04 90 54 32 70; Tues.-Sat. 11:30am-2pm and 8pm-10pm; €49 including wine

This bistro is one of the most renowned restaurants in Provence. People come for the atmosphere as much as the food. Photos of celebrity clients hang from the walls, and the place feels a bit like a club where everybody knows one another. The menu changes daily and there's little choice; you just eat what the chef proposes. Thankfully, it's excellent. Reservations are always advisable.

Maison Drouot

18 impasse Michel Durand, Maussane-les-Alpilles; tel. 06 61 07 38 54; http://maisondrouot.com; Tues.-Sat. 7:30pm-10:30pm; €63

How to Taste Olive Oil

Les Alpilles is filled with olive groves and produces some of the finest olive oil in the world. It's possible to pop in and taste at many of the local mills. The flavor of an oil is determined by a number of factors, including the varietal of olive, ripeness at harvest, climate, soil type, crop maintenance, and the milling process. Oil made with unripe (green) olives contains peppery flavors, usually described as grassy. Riper olives display softer flavors that are usually described as buttery.

To taste an olive oil, first pour a small amount into a wine glass, then swirl the oil to free the aroma. Inhale deeply from the top of the glass to get an idea of the fruitiness or other characteristics of the oil. At this point, the process is still rather like tasting wine, but here's the difference: With wine you sip then gargle/swish, but with oil you slurp. Pretend you are a naughty child and make as much noise as possible. Slurping mixes the oil with air and spreads it throughout your mouth. And unlike with wine, you get to swallow. Note the burning sensation in your throat.

tasting olive oil from a wine glass

Once you can taste, there's a whole new vocabulary to master. The IOC (that's the International Olive Council, not the International Olympic Committee) suggests a host of descriptive terms:

- Apple/Green Apple: indicative of certain olive varietals

- Almond: nutty (fresh, not oxidized)

- Artichoke: green flavor

- Astringent: puckering sensation in the mouth created by tannins, often associated with bitter, robust oils

And that's just the "A"s!

WHERE TO TASTE

- **Château Estoublon Estate Circuit** (Route de Maussane; tel. 04 90 54 64 00; www.estoublon.com): Enjoy a walk on the grounds of this château, then stop in for a tasting of their various olive oils.

- **Moulin Jean-Marie Cornille** (page 110): This olive oil mill gives free tours and tastings Tuesdays and Thursdays in the summer. Tours cover the process of harvesting the olives and then extracting the oil. There are various different oils to taste, ranging from green, thin, slightly tart offerings to peppery, smooth, and viscously golden.

This upmarket restaurant is housed in a converted village house and offers outdoor service around a pool in the summer and a cozy modern dining room for the winter. Reservations are always advisable. Lamb from Les Alpilles is usually on the menu. The rooms above the restaurant are available to rent.

Accommodations
Castillon des Baux

10 bis avenue de la Vallée des Baux, Maussane-les-Alpilles; tel. 04 90 54 31 93; www.castillondesbaux.com; €159 d

Perched on a small hill on the outskirts of Maussane, this hotel is a good budget choice.

There's a good-sized pool and easy access to the restaurants of Maussane and the surrounding sights.

Hameau des Baux

285 chemin de Bourgeac, Le Paradou; tel. 04 90 54 10 30; www.hameaudesbaux.com; €400 d

A couple of kilometers outside of Maussane, and set amid olive trees, rocks, and pines, this luxury hideout offers 21 welcoming rooms and suites. There's a pool and a gastronomic restaurant, and a cute food truck serves snacks in the summer months.

L'Hotel Particulier

13 rue de l'Escampadoul, Maussane-les-Alpilles; tel. 04 90 52 51 40; www.les-maisons.hotel-particulier.com; €400 d

The Hotel Particulier is Maussane's newest boutique hotel. It's a beautifully renovated combination of two ancient houses on the outskirts of Maussane. The vibe is one of peaceful pampering.

Getting There and Around
Car

To reach Maussane and Le Paradou from **Arles,** take the D83 and the D27, covering 24 km (15 mi) in 28 minutes. From **Saint-Rémy,** take the D5, covering 10 km (6 mi) in 15 minutes. **Parking** in Maussane is usually not a problem. If you can't find a spot in the center of the village, try the large parking lot a couple of minutes' walk away on avenue des Alpilles.

Aix-en-Provence, the Luberon, and the Gorges du Verdon

Most cities have their detractors. It's human nature. Give people the slightest opportunity to complain and they will find something to complain about. Yet, in 15 years of living just north of Aix-en-Provence, I have never heard a bad word said about the place. Mention Aix (pronounced "X") and people smile, almost involuntarily. The city seems to have an inherent positive energy. It makes people happy.

There are obvious reasons to visit Aix: Cézanne's artistic legacy, the world-class Musée Granet, picturesque daily markets, eclectic shops, and numerous excellent restaurants. What is less apparent from the pages of most guidebooks and what surprises people about Aix is how time slips pleasantly away in the embrace of her streets. For a

Highlights

Look for ★ to find recommended sights, activities, dining, and lodging.

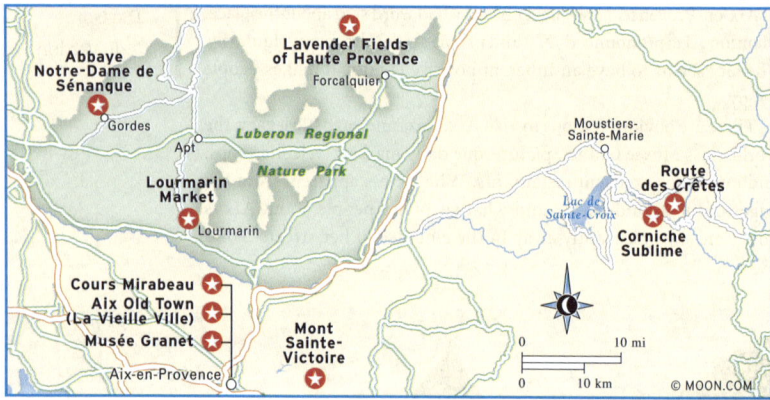

★ **Cours Mirabeau:** The year-round heart of Aix-en-Provence, the Cours is shady in the summer and a sun trap in the winter—the epitome of café society at its southern best (page 122).

★ **Aix Old Town (La Vieille Ville):** With squares, fountains, narrow streets, eclectic shops, and plentiful cafés and restaurants, this area is simply a joy to wander around (page 122).

★ **Musée Granet:** The Granet Museum hosts an extensive collection of modern art, including works by Picasso, Matisse, Monet, and Klee, as well as 10 Cézanne originals (page 126).

★ **Mont Sainte-Victoire:** Much-painted by Cézanne, this mountain to the east of Aix offers great hiking and/or a driving tour, taking in chic Le Tholonet and Vauvenargues, where Picasso is buried (page 141).

★ **Abbaye Notre-Dame de Sénanque:** The Cistercians chose this remote and beautiful spot to found an abbey in 1148. Today it still has a community of monks and is best known for the picture-postcard view when the lavender is in bloom (page 148).

★ **Lourmarin Market:** Stalls selling artisanal products and Provençal food specialties stretch through the streets in a theatrical show of color (page 165).

★ **Lavender Fields of Haute Provence:** Picturesque Provençal villages and the fields of surrounding lavender compete for attention. It's best to visit at the end of June and in early July to catch the lavender in full bloom (page 171).

★ **Route des Crêtes and Corniche Sublime:** Take a drive along either of these sinuous cliffside roads to appreciate the true majesty of the Gorges du Verdon (pages 177 and 182).

city there is very little traffic noise, as both the old town and most of the main street, the Cours Mirabeau, is pedestrianized. In the Mazarin Quarter, filled with its hôtel particuliers, people stroll happily on the roads, rarely disturbed by cars but lulled instead by the sound of trickling fountains. This is the unnoticed song of the city.

To the north, the Luberon, with its soft-folded mountains filled with pungent pine trees and sweet-smelling wild herbs, can seem too good to be true. It's almost like someone planned it as a theme park for tourists. Nearly every village has a castle, and sweeping views are commonplace. The palette of colors—red ochers, blazing purple lavender, and brittle green olive—packs an unrivaled visual punch. Hiding amid this riot for the senses are some of the best boutique hotels in the world. Nearly all the food and wine is local. To the east, the Luberon blends into Haute Provence where, in the summer, hill villages appear to float dreamlike on seas of lavender.

Farther east, the landscape in unrecognizable. The Gorges du Verdon (Verdon Gorge) is a 700-m-high (2,300-ft-high), 25-km-long (16-mi-long) chasm cut into limestone cliffs. It is a natural wonder, a one-off, an area that resembles nowhere else in France. Yes, lavender still clings to high plateaus, but the turquoise waters of Lac de Sainte-Croix are almost otherworldly. Rather than sit in cafés admiring the view, tourists seek adrenaline, flinging themselves on bungee ropes from bridges, or shooting in special wetsuits down canyons in the rock. The most sedentary activity is hiring a paddleboat.

PLANNING YOUR TIME

Aix-en-Provence is a small city. Although historic sites and museums are plentiful, most people will feel that two days is long enough to explore the center of Aix. Another day could easily be added to explore the outskirts of the city. If you plan to stay longer than this, use Aix as a base to head into the surrounding countryside (**Pays d'Aix-en-Provence**); perhaps schedule a couple of wine-tastings and consider hiking and biking options. In the evening you can return to the city to enjoy the excellent choice of restaurants, the myriad cultural events, and the city vibe.

A car is almost essential to explore the Luberon. Public transport is very irregular. A determined and committed sightseer (with a car) can get a feel for the Luberon Regional Nature Park in just one day. The villages are rarely more than 10 km (6 mi) apart, and so to hop from one to another is relatively easy. An additional day would be needed for a whistle-stop tour of Haute Provence and the Montagne de Lure. For those planning a more relaxed trip, the ideal way to visit would be to pick a base in one of the Luberon villages for two or three nights. The days would easily slip by as you visit many of the villages and their associated sights. To see the lavender fields (in bloom between late June and mid-July), an additional night in Haute Provence would be advisable.

Adventure and water sports enthusiasts and keen hikers could easily spend a week or more in the Gorges du Verdon. However, if you plan to visit just to get a feel for the remarkable landscape, then one or two nights will suffice. Villages in the Gorges, although bustling in the summer, have a remote, slightly Alpine feel. Out of season, they fall very quiet.

Previous: Abbaye Notre-Dame de Sénanque; Fontaine de la Rotonde; shopping for vegetables and flowers at one of Aix-en-Provence's daily markets.

Aix-en-Provence, the Luberon, and the Gorges du Verdon

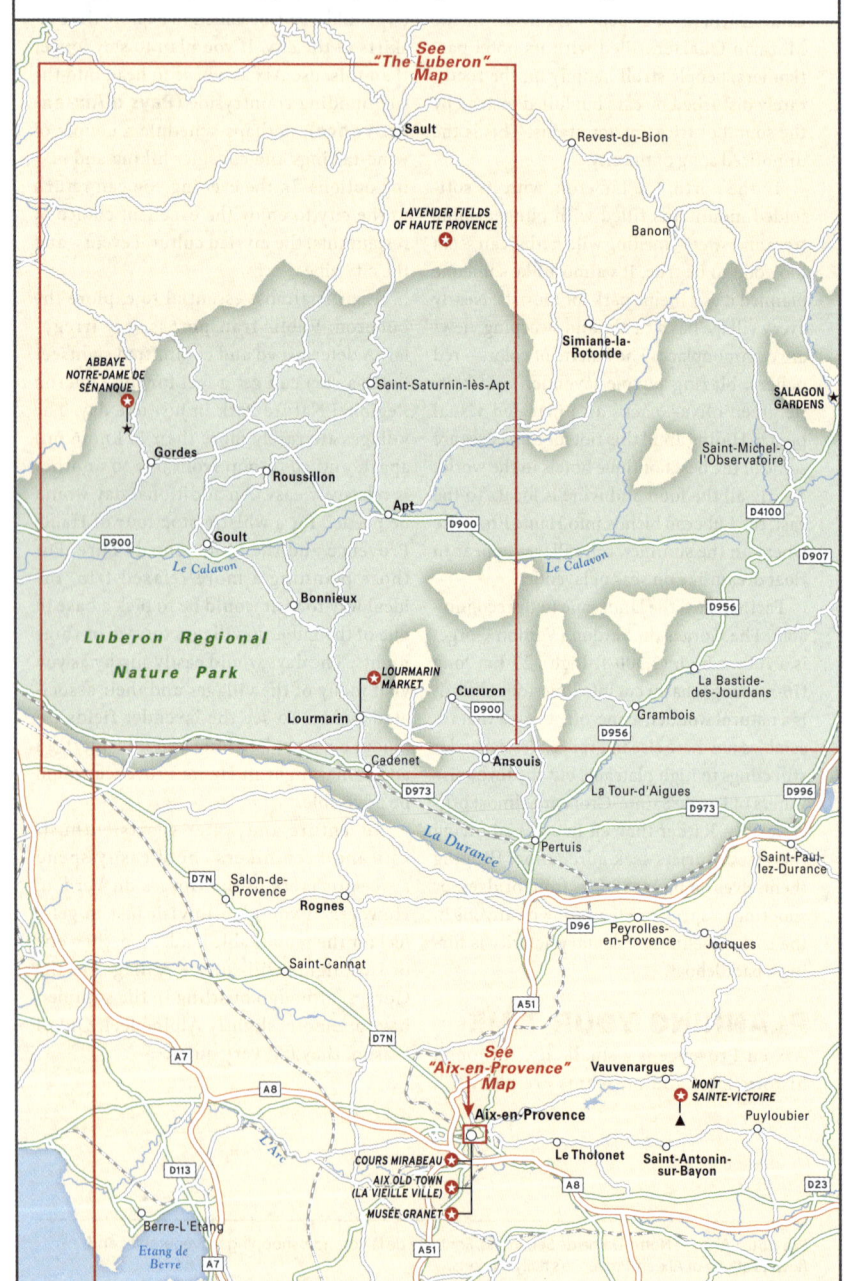

See "The Luberon" Map

Sault

Revest-du-Bion

LAVENDER FIELDS OF HAUTE PROVENCE

Banon

Simiane-la-Rotonde

Saint-Saturnin-lès-Apt

SALAGON GARDENS ★

ABBAYE NOTRE-DAME DE SÉNANQUE ★

Saint-Michel-l'Observatoire

Gordes

Roussillon

Apt

D900

Le Calavon

Le Calavon

D4100

D907

D900

Goult

Bonnieux

D956

Luberon Regional Nature Park

LOURMARIN MARKET

Cucuron

La Bastide-des-Jourdans

Lourmarin

D900

Grambois

Cadenet

Ansouis

D956

D973

La Tour-d'Aigues

D973

D996

La Durance

Pertuis

Saint-Paul-lez-Durance

D7N

Salon-de-Provence

Rognes

D96

Peyrolles-en-Provence

Jouques

Saint-Cannat

A51

D7N

See "Aix-en-Provence" Map

Vauvenargues

MONT SAINTE-VICTOIRE

A7

Aix-en-Provence

Puyloubier

A8

COURS MIRABEAU

Le Tholonet

Saint-Antonin-sur-Bayon

D23

D113

AIX OLD TOWN (LA VIEILLE VILLE)

MUSÉE GRANET

A8

Berre-L'Etang

Etang de Berre

A7

A51

L'Arc

AIX-EN-PROVENCE, THE LUBERON, AND THE GORGES DU VERDON

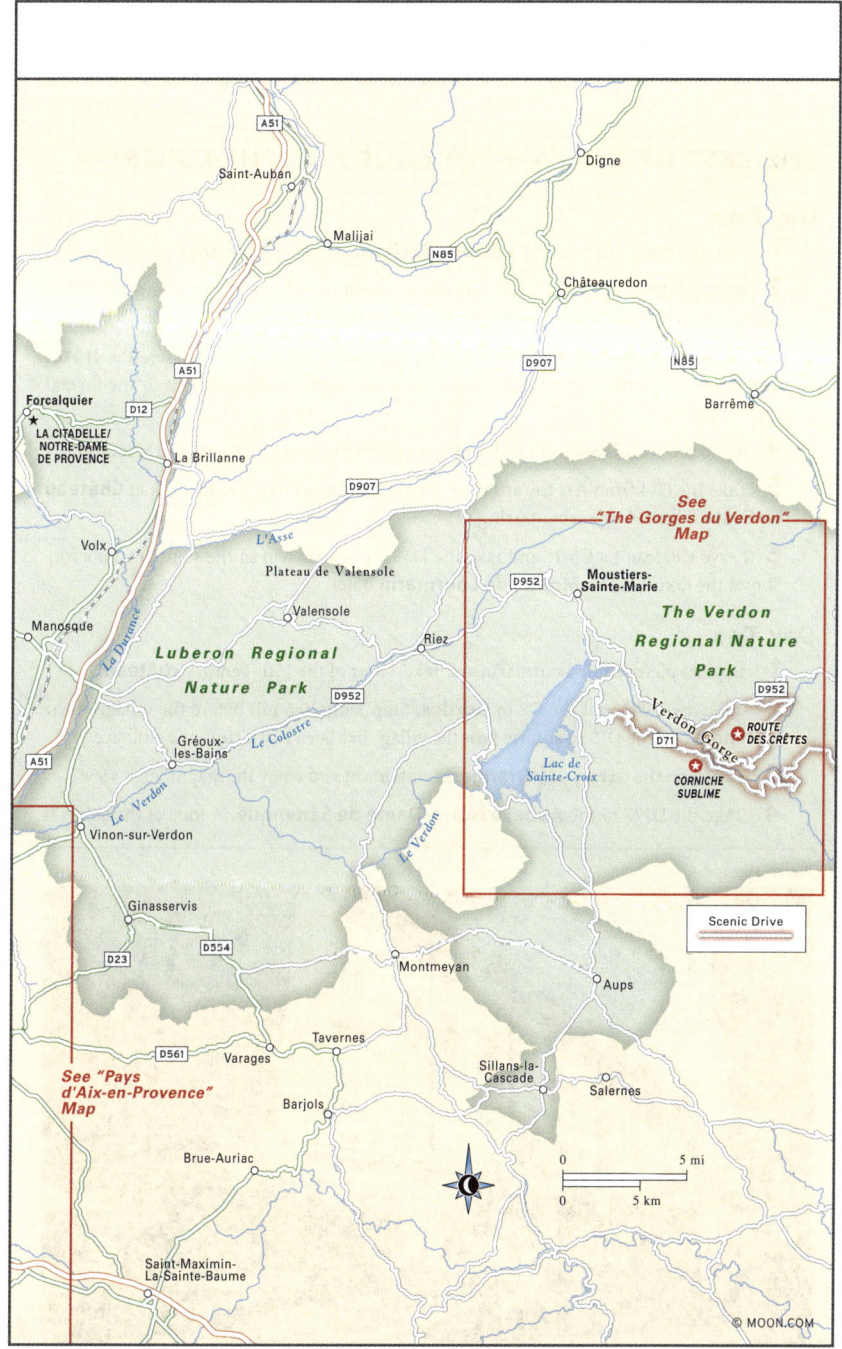

© MOON.COM

Itinerary Ideas

THE BEST OF AIX-EN-PROVENCE AND THE LUBERON

Day One

1 Pick up a breakfast treat at **Pâtisserie Béchard** at the bottom of the Cours Mirabeau.

2 Walk up the Cours Mirabeau and take any right into the old town. Browse the boutiques of the old town before heading for the vibrant **daily food market** in Place Richelme.

3 Explore the grandeur of one of Aix's most celebrated town houses, the **Hôtel Caumont.** Tour the latest art exhibition and stop for a prelunch drink in the formal gardens.

4 Avoid the crowds and eat inside at **Drole d'Endroit,** a hidden gem of a restaurant.

5 Take the D14 from Aix toward the Luberon. Stop and enjoy the art walk at **Château La Coste** followed by a wine-tasting.

6 Leave Château La Coste and take the D561 to Lourmarin in the southern Luberon. Stay at the luxurious **Le Moulin de Lourmarin** hotel.

Day Two

1 Explore picturesque Lourmarin and take a tour of the 15th-century **château.**

2 Take the D943 and the D2 to **Gordes.** Stop 1 km (0.6 mi) before the village at the **viewpoint** on the D15 to admire how the village has been hewn from the cliff face.

3 Lunch on the terrace of **L'Orangerie** restaurant and enjoy the magnificent view.

4 Take the D177 to the **Abbaye Notre-Dame de Sénanque.** A tour of the abbey is

Aix-en-Provence

particularly spectacular at the end of June or beginning of July when the lavender is in bloom.

5   Take the D177 and then the D103 toward Menerbes, and then the D3 toward the hill-top village of **Bonnieux.** The D3 road, which runs along a vine-filled valley, is one of the most picturesque in the Luberon.

6   Return on the D943 for another night in Lourmarin. Have supper at the **Café L'Ormeau,** which often has live music.

ONE DAY IN THE GORGES DU VERDON
Wake up in the the picturesque village of Moustiers-Sainte-Marie.

1   In the morning, get the worst of the vertigo over and done with by driving the circular **Route des Crêtes,** departing from the village of La Palud-sur-Verdon. Spot some vultures as you peer 700 m (2,300 ft) down to the Verdon river below.

2   For lunch, picnic at **Pont du Galetas** at the Moustiers-Sainte-Marie end of Lac de Sainte-Croix.

3   In the afternoon hire a kayak from **Base Nautique l'Etoile** to view the gorge, looking from the river upward.

4   Finish the day with a 5-km (3-mi) loop walk from Aiguines to the **Col d'Illoire viewpoint** to watch the sunset.

5   After all that activity, reward yourself with a blowout meal at **La Bastide de Moustiers.**

Sainte-Croix-du-Verdon

Itinerary Ideas

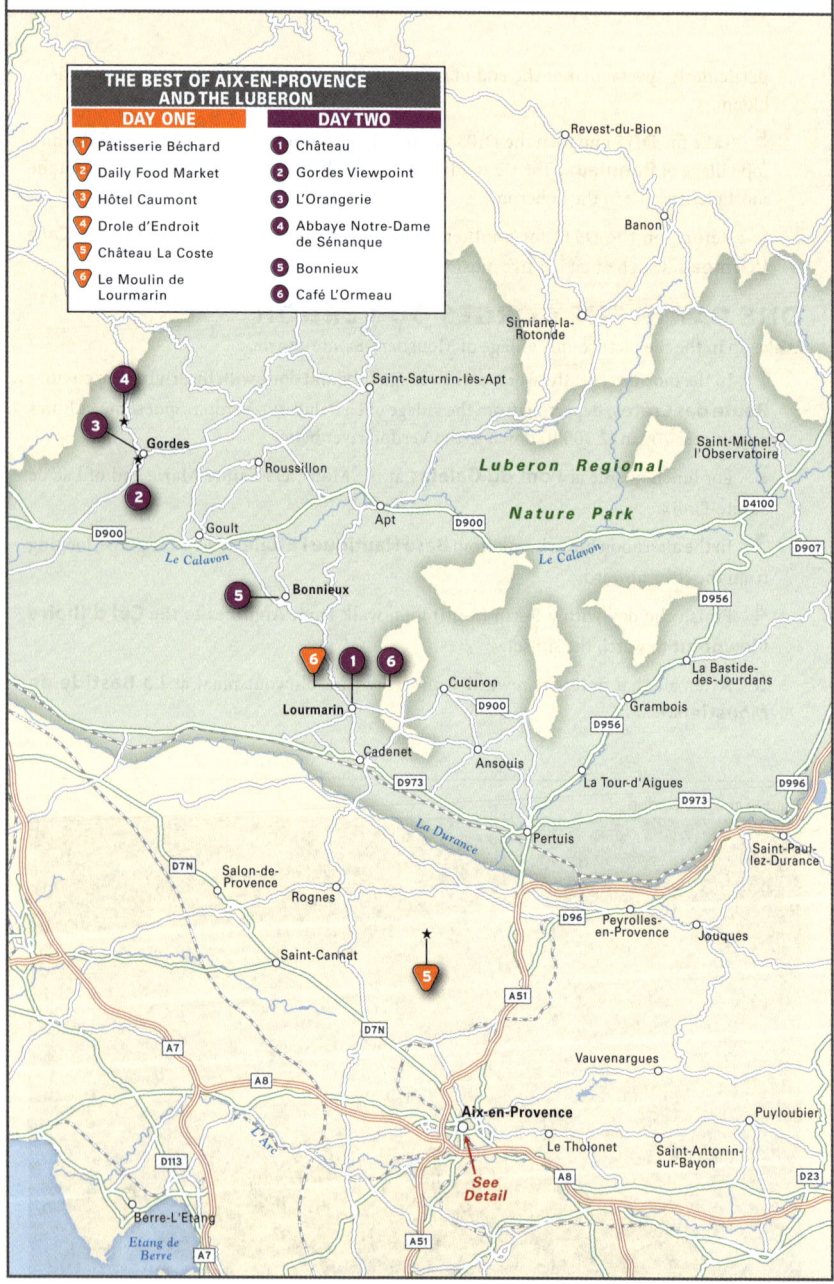

THE BEST OF AIX-EN-PROVENCE AND THE LUBERON

DAY ONE
1. Pâtisserie Béchard
2. Daily Food Market
3. Hôtel Caumont
4. Drole d'Endroit
5. Château La Coste
6. Le Moulin de Lourmarin

DAY TWO
1. Château
2. Gordes Viewpoint
3. L'Orangerie
4. Abbaye Notre-Dame de Sénanque
5. Bonnieux
6. Café L'Ormeau

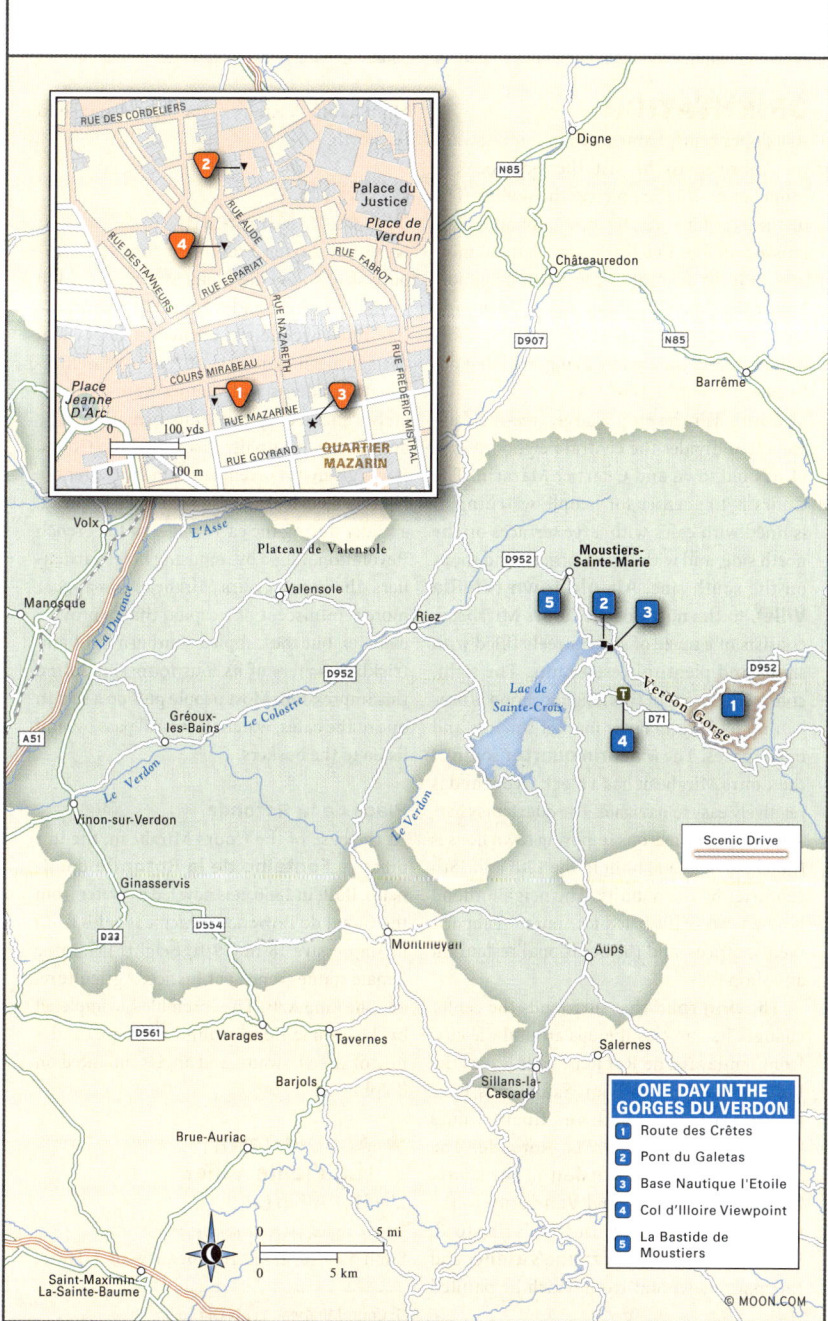

ONE DAY IN THE GORGES DU VERDON

1. Route des Crêtes
2. Pont du Galetas
3. Base Nautique l'Etoile
4. Col d'Illoire Viewpoint
5. La Bastide de Moustiers

© MOON.COM

Aix-en-Provence

ORIENTATION

Aix slopes gently from north to south toward the Arc river. To the east, the rocky peak of Mont Sainte-Victoire pierces the skyline. The historic center is relatively small and can be crossed on foot in a little more than 20 minutes. It is divided into three main neighborhoods: the Cours Mirabeau, Aix old town, and Quartier Mazarin. These central neighborhoods are surrounded by a ring road that delineates the outskirts of Aix.

Cours Mirabeau is a large, tree-lined avenue that divides the central neighborhoods of Aix old town and Quartier Mazarin. The Cours is the center for people-watching; it is lined with cafés with large terraces on the north side, and with banks, offices, and shops on the south side. **Aix old town (Vieille Ville),** to the north of the Cours Mirabeau, consists of a maze of small streets filled with shops and plentiful restaurants. The tight, confusing layout of this neighborhood is broken up by several large market squares and open spaces. The **Mazarin quarter** south of the Cours Mirabeau has a block layout and is relatively easy to navigate. The quarter is characterized by numerous imposing town houses (hôtel particuliers) built in the 17th and 18th centuries by the nobility. The neighborhood is now a mix of historic sites, museums, apartments, offices, and the occasional restaurant and shop.

The **ring road** that surrounds the center changes its name as it loops around the city, from boulevard de Roi René as it leaves the Rotonde fountain, to Cours Saint Louis, boulevard Aristride Briand, and finally Cours Sextius as it returns to La Rotonde. The city's main parks—**Jourdan** to the south, **Rambot** to the east, and **Vendôme** to the northwest—are located just off this ring road, in the **Aix outskirts. Cézanne's atelier** and the main viewpoint from which he painted

Mont Sainte-Victoire are a short distance to the north.

SIGHTS
★ Cours Mirabeau

No visit to Aix-en-Provence is complete without a stroll along the Cours Mirabeau. At 440 m (1,443 ft) long and 48 m (147 ft) wide, the tree-lined avenue was created in 1649 when the southern ramparts of Aix were demolished. Residents and visitors were immediately charmed by the transformation. The avenue was originally just called the Cours. It was given its current name in 1876 after the Honoré Gabriel Riqueti, Count of Mirabeau, a leader during the early stages of the French Revolution. Lined by imposing hôtel particuliers, the grand Cours Mirabeau can appear more reminiscent of a capital than a provincial city, but the dappled southern light and trickling waters of its four fountains soften this impression. Most people pull up a seat in one of the cafés, watch the world pass by, and listen to the buskers.

Place de la Rotonde

At the base of the Cours Mirabeau, the impressive **Fontaine de la Rotonde** dominates. Built in 1860, it is now fed by water from the Canal de Provence, which cascades from its impressive 18-m (59-ft) height. The three female statues represent Justice, Agriculture, and the Fine Arts. The ensemble is completed by 12 bronze lions around the base, and a mix of sirens, swans, and angels mounted on dolphins.

★ Aix Old Town (La Vieille Ville)
L'Hôtel Albertas

Place Albertas; www.aixenprovence.fr

Such is the arresting beauty of L'Hôtel

1: Cours Mirabeau 2: Aix Old Town

Aix-en-Provence

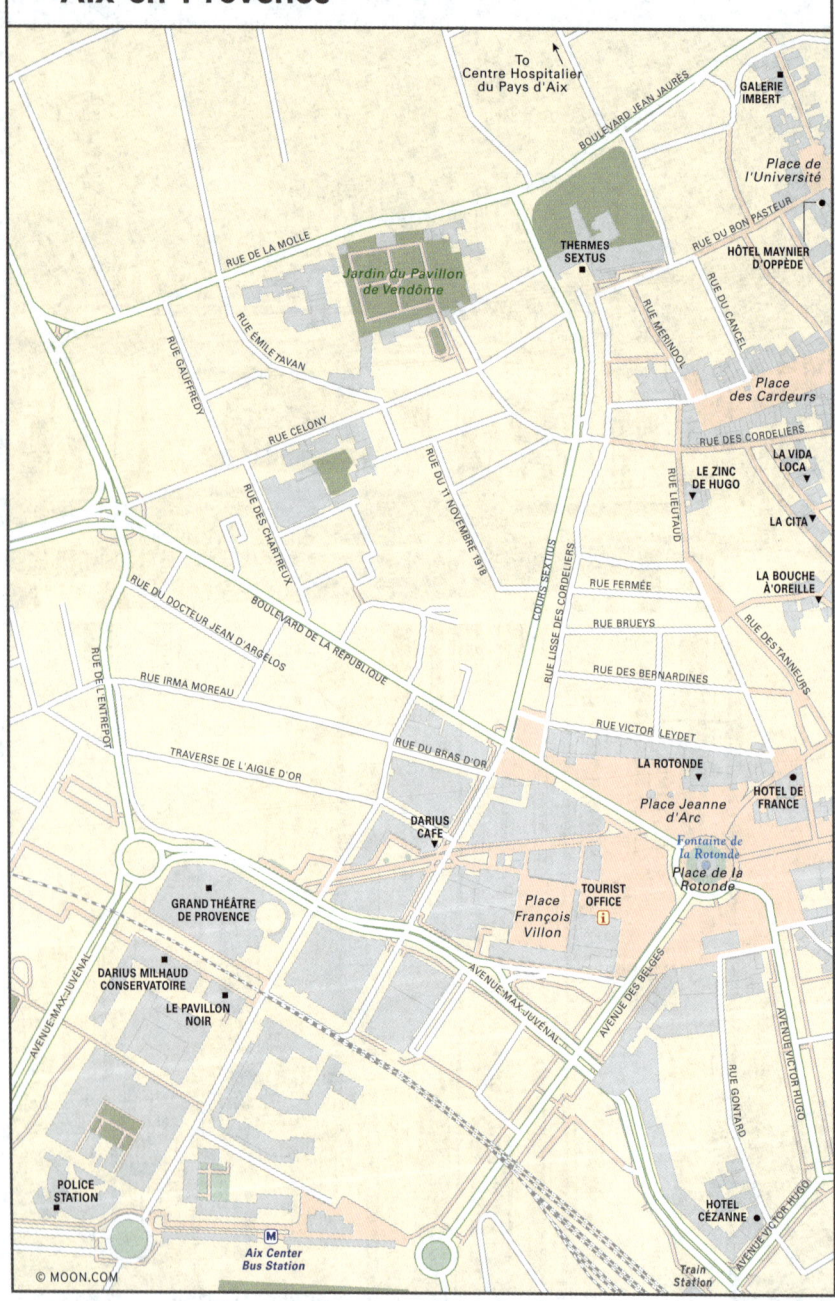

To
Centre Hospitalier
du Pays d'Aix

GALERIE
IMBERT

*Place de
l'Université*

BOULEVARD JEAN JAURÈS

RUE DE LA MOLLE

*Jardin du Pavillon
de Vendôme*

THERMES
SEXTUS

RUE DU BON PASTEUR

HÔTEL MAYNIER
D'OPPÈDE

RUE ÉMILE TAVAN

RUE MERINDOL

RUE DU CANCEL

*Place
des Cardeurs*

RUE GAUFFREDY

RUE CELONY

RUE DES CORDELIERS

LE ZINC
DE HUGO

LA VIDA
LOCA

RUE DU 11 NOVEMBRE 1918

RUE DES CHARTREUX

RUE LIEUTAUD

LA CITA

COURS SEXTIUS

RUE LISSE DES CORDELIERS

RUE FERMÉE

LA BOUCHE
À OREILLE

RUE DU DOCTEUR JEAN D'ARGELOS

BOULEVARD DE LA RÉPUBLIQUE

RUE BRUEYS

RUE DES TANNEURS

RUE DE L'ENTREPÔT

RUE IRMA MOREAU

RUE DES BERNARDINES

RUE VICTOR LEYDET

TRAVERSE DE L'AIGLE D'OR

RUE DU BRAS D'OR

LA ROTONDE

HOTEL DE
FRANCE

*Place Jeanne
d'Arc*

DARIUS
CAFÉ

*Fontaine de
la Rotonde*

*Place de la
Rotonde*

GRAND THÉÂTRE
DE PROVENCE

*Place
François
Villon*

TOURIST
OFFICE

AVENUE MAX-JUVÉNAL

DARIUS MILHAUD
CONSERVATOIRE

LE PAVILLON
NOIR

AVENUE MAX-JUVÉNAL

AVENUE DES BELGES

AVENUE VICTOR HUGO

RUE GONTARD

POLICE
STATION

HOTEL
CÉZANNE

AVENUE VICTOR HUGO

Ⓜ
Aix Center
Bus Station

*Train
Station*

© MOON.COM

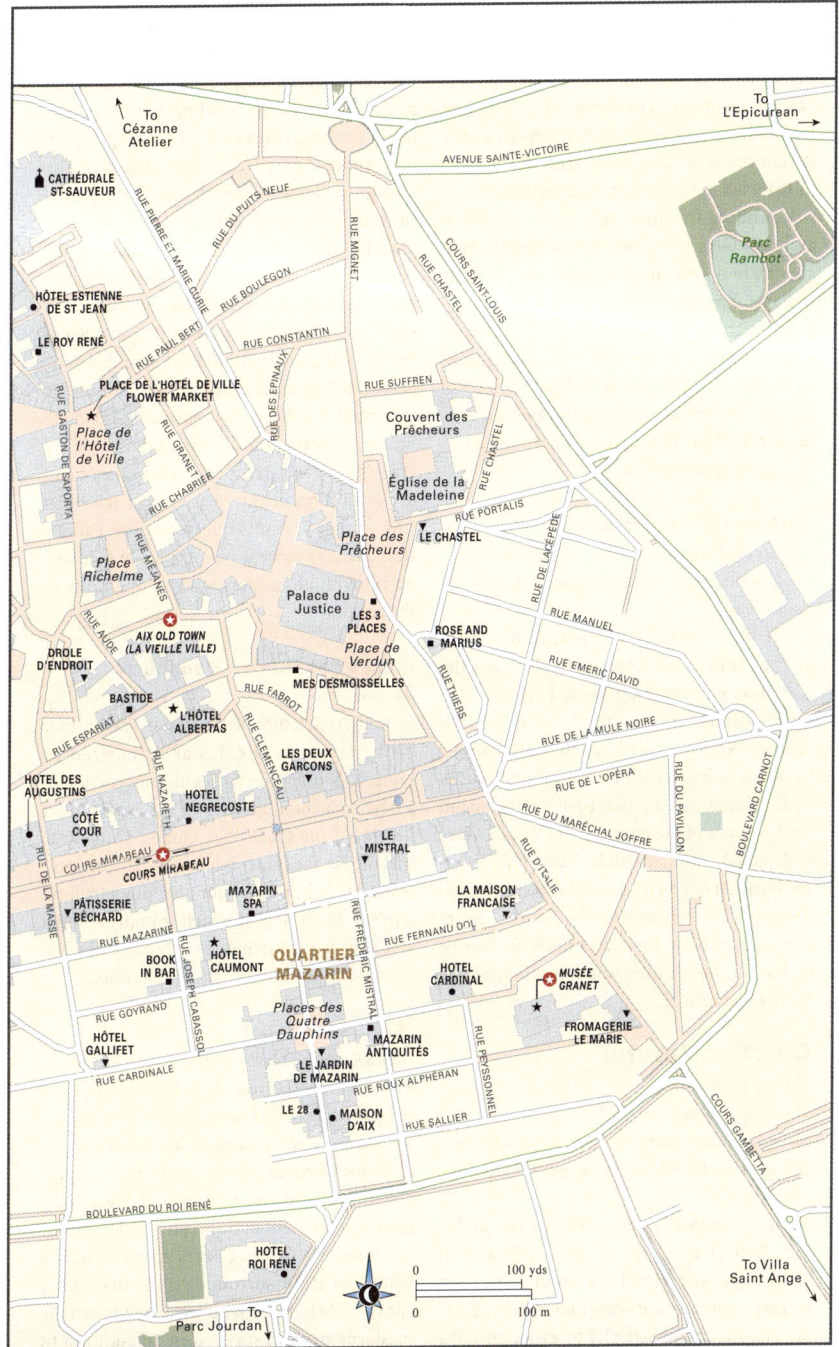

Albertas that people can't help themselves: They stop, they stare, they take multiple photographs, and then they leave shaking their heads in wonder. The hotel takes its name from a family of parliamentarians from the Italian town of Alba. Its current form dates to 1745, when the family acquired the houses surrounding the hotel and, with the help of architect George Vallon, demolished them to create a look reminiscent of the royal squares in Paris. The architecture of the facade is regency with a nod to the baroque. The central fountain was added in 1912 and was designed by students of the Aix art school. The hotel is now divided into offices and private residences, and so, although Casanova once visited, the public now can't.

Cathédrale Saint-Sauveur

34 Place des Martyrs-de-la-Résistance; tel. 04 42 23 47 40; www.aixenprovencetourism.com; daily 8am-7:30pm; free

Cathédrale Saint-Sauveur is said to be built on the site of an old Roman temple to Apollo. Christian construction started around 500, and continued intermittently until the 18th century. Art lovers are lured to Saint-Sauveur by the chance to see the *Burning Bush Triptych* by the Renaissance painter Nicolas Froment. In the center of the painting the Virgin Mary is depicted with her child, sitting on a burning bush. At her feet, Moses is shown, amazed by the vision before him. On either side are portraits of King René and his wife, Jeanne de Laval. The painting was commissioned by King René in 1475.

Quartier Mazarin
Hôtel Caumont

3 rue Joseph Cabassol; tel. 04 42 20 70 01; www.caumont-centredart.com; May-Sept. daily 10am-7pm, Oct.-Apr. daily 10am-6pm; €14.50, ages 7-17 €10, under 7 free

Built between 1715 and 1742 by the architect Robert de Cotte for the Marquess of Cabannes, the Hôtel Caumont has been largely restored to its original state. It presents a unique opportunity to experience the grandeur and luxury in which the French nobility lived in the 18th century.

A visit begins with the exploration of the restored part of the hôtel particulier, starting in the large courtyard, where ample space was provided for the carriages of guests. The facade, in classical French style with plentiful horizontal and vertical lines, was completed by both Aixoise and Parisian artisans, creating a pleasing mix of styles. Inside, the entrance hall wows with its intricate high staircase, considered a showstopper even at the time of construction. Climbing the staircase, the bedroom of the young Pauline de Caumont has been re-created, its intimate nature reflecting the fashion of the Louis XV era. The second part of the visit takes in an art exhibition in the east wing of the hotel (the former servants quarters). The exhibition changes twice a year, typically focusing on individual painters with strong links to the region.

Places des Quatre-Dauphins

www.aixenprovence.fr

In 1646 Archbishop Richelieu of Aix, brother of the famed cardinal, was authorized by Louis XIV of France to begin construction of a new residential quarter of Aix. At its heart he placed the Places des Quatre-Dauphins. Installed at the center is a fountain by sculptor Jean Claude Rambot. Four dolphins surround a pyramid, on top of which initially stood a statue of Saint Michel, later replaced by the Maltese Cross, and finally a pine cone.

★ Musée Granet

Place Saint-Jean de Malte; tel. 04 42 52 88 32; www.museegranet-aixenprovence.fr; June 15-Oct. 29 Tues.-Sun. 10am-6pm, Oct. 29-June 15 Tues.-Sun. noon-6pm; peak season €8, out of season €6.50, under 18 free

Cézanne's artistic legacy can be truly appreciated at the Musée Granet and the associated Planque collection. The collection at the Granet anchors Cézanne's work in the history of art by surrounding the 10 original canvases held there with work by his artistic predecessors and successors. In addition to

Les Hôtel Particuliers

Place d'Albertas, lined by hôtel particuliers

Don't let the name fool you: Hôtel particuliers are town houses—not hotels—built largely during the 17th and 18th centuries and were the main residences of nobles. Thanks to its parliament and university Aix attracted the upper classes of French society in unparalleled numbers. At the beginning of the 17th century, it is estimated 13 percent of the city's population was composed of nobility. Of course they all needed somewhere to live, and so began a construction boom of lavish town houses. Today the concentration of hôtel particuliers on the Cours Mirabeau lends the road its unique charm, even if many are now used for mundane purposes such as supermarkets and insurance brokers.

Of particular note is number 20, Hotel Forbin. In 1807, Napoleon's sister Pauline Borghese received guests from the old French nobility in this building, taking the advice of her brother that in the "good town of Aix … people and especially the fair sex do not like the classes to mix." This theme was picked up by the writer Émile Zola, who in the latter half of the 19th century would sit in the Deux Garcons café on the Cours Mirabeau and observe Aix society. The fictional town of Plassans that Zola created in his *Les Rougon-Macquart* novel series is loosely based on Aix, where the nobility strolled up the south side of the Cours backed by a quarter filled with hôtel particuliers, and the people frequented the cafés on the north side. The two sections of society never met.

HÔTEL PARTICULIERS WORTH VISITING

- **Hôtel Estienne de Saint-Jean,** 17 rue Gaston de Saporta, used to be called the "musée du vieil Aix." It houses a collection of objects (furniture, clothes, santons, and pottery) that together tell the story of the history of Aix-en-Provence. It is open Wednesday-Monday 10am-12:30pm and 1:30pm-5pm.

- **Hôtel Caumont Art Center,** 3 rue Joseph Cabassol, is a restored hôtel particulier and the best place to gain an appreciation of the lifestyle of the nobility in the 18th century (page 126).

- **Hôtel Maynier d'Oppède,** 21 rue Gaston de Saporta, now houses an institute of French studies for foreign students. There are also summer concerts in the courtyard. There are no specific opening hours but you can usually poke your head in and look around.

- **Hôtel Gallifet,** 52 rue Cardinale, is an art center, café, and venue for jazz concerts.

the permanent collections of paintings spanning the 14th to the 20th centuries, there is an impressive sculpture gallery and a display of archaeological finds from the pre-Roman Celto-Ligurian Entremont settlement located just to the north of Aix. The Planque collection, housed in La Chapelle des Penitents Blancs, a short 2-minute walk from the main museum, showcases 300 paintings lent to the Granet Museum by the Fondation Jean Planque. A visit to the collection is included in the entry price to the main museum, and it is well worth the small detour to see works by Monet, van Gogh, Cézanne, Picasso, de Stael, and Klee.

Aix Outskirts

Cézanne Atelier

9 avenue Paul Cézanne; tel. 04 42 21 06 53; www. cezanne-en-provence.com; Apr.-May daily 9:30am-12:30pm and 2pm-6pm, June-Sept. daily 9:30am-6pm, Oct.-Mar. Tues.-Sat. 9:30am-12:30pm and 2pm-5pm; €6.50, ages 13-25 €3.50, under 13 free

In 1886 Cézanne's father died, freeing the painter from financial reliance. The family house at Jas de Bouffan was sold in 1899 and Cézanne purchased a site in the hills above Aix, in the Lauves neighborhood. Here he began construction of a villa with an atelier in which he could work. Today the 1st-floor atelier has been restored to mirror its condition during the years before Cézanne died. Many of the original objects remain. For example, the olive jar sitting on the table in the middle of the room appeared in Cézanne's paintings no less than 22 times. Visiting the atelier should be an intimate experience, a chance to sit and contemplate the life of the painter. However, 70,000 people climb the steps to the studio every year, with around 15 people admitted at a time. The best way to feel the spirit of the painter is to pay the extra €3 for an audio guide and sit on one of the benches, slowly soaking up the atmosphere as you listen to the narrative. It takes approximately 20 minutes to walk from the center of Aix-en-Provence and up avenue Cézanne to the atelier. Alternatively, take bus line 5 or line 10 from the Rotonde Poste stop next to the Apple store.

Terrain des Peintres

49 avenue Paul Cézanne; www.cezanne-en-provence. com

A short distance uphill from the Cézanne Atelier is the "Land of the Painters" viewpoint. It was from here that Cézanne painted the obsession of his artistic life: Mont Sainte-Victoire. In all, Cézanne painted the mountain 87 times, with the most famous paintings created at this, the Sainte Marguerite viewpoint. Reproductions of some of these works are available to view on-site, enabling visitors to make the link between the landscape before them and Cézanne's paintings. The walk from the atelier to the Land of the Painters takes about 15 minutes.

Les Carrières de Bibémus

3090 chemin de Bibémus; tel. 04 42 16 11 61; www. cezanne-en-provence.com; open for 1-hour tours only Apr.-Oct. Mon.-Sat., 10am tours in French with printed info in English; €19, under 7 free

Cézanne's ability to cross between artistic styles is illustrated by the work he did at the Bibémus quarry west of Aix-en-Provence. In his renderings of the rocks at the quarry, Cézanne laid the foundation for the cubist movement of the early 20th century. Cubist painters, such as Picasso, often present multiple views of the same subject at the same time. The beginning of this technique can be discerned in the blurring of perspectives in Cézanne's paintings at Bibémus, such as *The Red Rock* (1797).

The sight is only open to prebooked tours on selected days. See the website for days and reservations. The tours depart from the parking de Bibémus.

Le Jas de Bouffan

17 route de Galice; tel. 04 42 16 11 61; www.cezanne-en-provence.com; at the time of writing this sight was closed for repairs, see website for updates

Le Jas de Bouffan was the home of the Cézanne family. Between 1859 and 1899,

Cézanne's Provence

Objects from Cézanne's paintings are dotted throughout his atelier.

Paul Cézanne was born in Aix-en-Provence in 1839 and died there in 1906. After an early unsuccessful career as a banker, he left Aix for Paris to begin training as a painter. His subsequent work changed the way painters interpret the world around them, but like many outliers, he was scorned by the contemporary establishment. Rejected by the École des Beaux-Arts in Paris, he worked in the Académie Suisse where he encountered painters from the emerging impressionist school. Even for impressionists, themselves no strangers to causing outrage, Cézanne's work was a bridge too far. Claude Monet dismissively said that Cézanne "paints with his trowel."

Classifying Cézanne's work is difficult; calling it Postimpressionism probably gets the closest. He painted landscapes, still lifes, and nudes using a range of different techniques, notably favoring revolutionary color rather than a linear perspective.

Cézanne's early paintings were rejected from prestigious Parisian shows, and every time he left the capital to return to Provence, laughter echoed in his wake. Lack of success forced him into a double life. He met his wife-to-be, Hortense Fiquet, in Paris, where she modeled for him. They kept her existence secret from his father for fear he would lose his allowance (and with it his ability to paint the other love of his life, Provence).

Cézanne died a painter's death, working on a canvas in a rainstorm and catching pneumonia as a result. It was only after his death that Cézanne's work became widely recognized. Picasso famously called him "the father of us all," referring to how the cubist school built upon Cézanne's paintings of Bibémus quarry.

FOLLOWING CÉZANNE'S FOOTSTEPS

- A visit to Cézanne's **atelier** (page 128) or his family home, the **Jas de Bouffan** (page 128), helps foster more understanding of his life and work.

- His particular obsession was **Mont Sainte-Victoire,** which he painted 87 times from various observation points (page 141).

- Other favored places to paint included the **quarry at Bibémus** (page 128) just outside Aix and the fishing port of **L'Estaque** just outside Marseille.

Cézanne completed 39 oil paintings and 17 watercolors here, inspired by the garden, the line of chestnut trees, and the surrounding woods. Major renovation work is under way, designed to make Le Jas de Bouffan the main center for Cézanne-related tourism and study in Aix.

Fondation Victor Vasarely

1 avenue Marcel Pagnol; tel. 04 42 20 01 09; www. fondationvasarely.org; daily 10am-6pm; €15, ages 10-15 €8, ages 5-9 €4, under 5 free

Arriving at the Fondation Victor Vasarely is like stepping onto the set of a science-fiction movie. First there is the building itself, a concoction of giant black-and-white squares that resemble one face of a half-finished Rubik's cube. Step inside and everywhere there are walls of color, geometric shapes, and odd angles. It's truly an alien world, a place that challenges the senses, constantly fooling the eye by playing with perceptions of space.

Vasarely is a unique figure in the history of 20th-century art, the creator of an entirely new movement: optical art. While painting at Gordes, an hour to the north of Aix, Vasarely had a revelation: In the intense summer light, he noticed a contradictory perspective to the linear one he had been using. "Never can the eye identify to what a given shadow or strip of wall belongs, solids and voids merge into one another, forms and backgrounds alternate," he said of Gordes. "Thus identifiable things are transmuted into abstractions."

Camp des Milles

40 chemin de la Badesse; tel. 04 42 39 17 11; www. campdesmilles.org; daily 10am-7pm; €9.50, under 9 free

Less than 10 km (6 mi) from the leafy shade and joyously bubbling fountains of the Cours Mirabeau, in the Provençal landscape that was a crucible of 20th-century artistic creativity, lies, almost inconceivably, the site of a former concentration camp. Between 1940 and 1942 Camp des Milles was a detention center for "undesirables" and eventually became the base from which more than 2,000 Jews were deported to Auschwitz. The building (an old tile factory) and the rooms within stand as a memorial to the people who passed through the camp. The exhibits in the museum attempt to explain how "free" Vichy, France, agreed to deliver 10,000 Jews to the Germans, including children under 16.

SPORTS AND RECREATION

For most people, recreation in Aix consists of finding a shady spot on a café terrace, ordering an espresso, and watching the world pass by. However, for those with itchy feet there are popular walking tours offered by the tourist office. It is also still possible to take the thermal waters that first attracted the Romans to this area, but modern spas proliferate and are a better choice for beauty treatments and pampering.

Guided Tours
Footsteps of Cézanne Walk

Tourist office, 300 avenue Giuseppe Verdi; www. aixenprovencetourism.com; tel. 04 42 16 11 61; Thurs. 10am-noon; €12

This tour is available by joining the weekly guided group, by picking up a map at the tourist office, or by downloading the Sur Les Pas de Cézanne app. Starting next to the statue of Cézanne outside the tourist office, the walk takes you up into the Mazarin Quarter before crossing the Cours Mirabeau and looping around the old town. There are embossed "C"s set into the paving stones to help you find the way. The trail includes Cézanne's school, various apartments he lived in, the café he frequented, and the church where he was married and where his funeral was held.

Visit Aix Old Town

Tourist office, 300 avenue Giuseppe Verdi; tel. 04 42 16 11 61; www.aixenprovencetourism.com; Fri.-Sat. 10am-noon; from €12

The tours on Friday and Saturday morning are slightly different but both are full of fascinating historical anecdotes. They are a great way to explore the old town for the first time, allowing you to get your bearings and discover the best hidden squares.

Parks
Parc Jourdan

Main entrance rue Anatole France; tel. 04 42 16 11 61; www.aixenprovence.fr; June-Aug. daily 9am-8pm, out of season usually closes around sunset

The Aixoise have been enjoying spending their leisure time in this park since the 1930s. It's one of the largest parks in the city and has plenty of shady, grassy areas. There's a good children's play area and a bouldrome for fans of Provence's favorite sport, pétanque. During the summer months the park hosts large exhibitions and concerts.

Jardin du Pavillon de Vendôme

Entrance 32 rue Celony or rue de la Molle; www. aixenprovence.fr; June-Aug. daily 9am-8pm, out of season usually closes around sunset

A real hidden treasure of Aix-en-Provence, these gardens comprise 9,000 sq m (2.2 acres) and are designed in a formal French style. Bushes, plants, and trees are all trimmed with geometric precision to be as aesthetically pleasing as possible. The Pavillon overlooking the garden houses a fine-art museum that has a permanent collection of 17th- and 18th-century portraits and hosts regular modern art exhibitions.

Spas
Mazarine Spa

27 rue Mazarine; tel. 04 42 50 27 27; www.srmazarine. fr; Mon.-Sat., reserve treatments in advance

Probably the pick of the many spas in Aix-en-Provence, this spa is located on the genteel rue Mazarine. Clients enter through a boutique selling luxury candles and cosmetic products. There's a small indoor swimming pool, a hammam, and the full range of massages and treatments.

Thermes Sextius

55 avenue des Thermes; tel. 04 42 23 81 82; www. thermes-sextius.com; Mon.-Sat. 8am-8pm; from €100

The architecture may be modern, but the thermal waters here have been used as a cure since Roman times. The range of spa treatments available at the Sextius spa relies heavily on the in-house source for the thermal water, filtered by Mont Sainte-Victoire, which bubbles up beneath Aix at a perfect 33°C (91.4°F). Full pricing information can be found on the spa's website.

the Pavillon de Vendôme

ENTERTAINMENT AND EVENTS

Entertainment is all around you in Aix-en-Provence. Street performers with tubs of soapy liquid blow bubbles the size of small children, entire groups of buskers jazz up the atmosphere on the Cours Mirabeau, and the ubiquitous human street statues doff their caps in return for a coin or two.

Aix is also a world capital for opera and classical music. From Easter until the end of summer, concerts and performances draw top stars and large audiences to venues across the city.

The Arts

Grand Théâtre de Provence

380 avenue Max Juvénal; tel. 04 42 91 69 70; www. lestheatres.net; hours vary; from €10

Designed by Milanese architect Vittorio Gregotti and inaugurated in 2007, the Grand Théâtre is one of a trio of modernist cultural buildings. Together they form a distinct contemporary district of Aix. The stones chosen for the construction of the building were selected in a variety of shades to mirror the way light plays on the rocky faces of Mont Sainte-Victoire. With 1,370 seats, the Grand Théâtre can host all but the largest events, and there's a lovely roof terrace where the public can look over Aix.

Le Pavillon Noir

530 avenue Wolfgang Amadeus Mozart; tel. 04 42 93 48 14; www.preljocaj.org; hours vary; from €25

Le Pavillon Noir has been the home of the Preljocaj dance company since 2006. When the company is not in residence, other dance companies from around the world are invited to perform. The building is architecturally arresting, designed as a cube by Rudy Ricciotti; there are four rehearsal rooms and one stage.

Darius Milhaud Conservatoire

380 avenue Wolfgang Amadeus Mozart; tel. 04 88 71 84 20; www.aixenprovence.fr; open for performances

Part music school, part performance center, the Conservatoire has 62 classrooms, 4 dance studios, and a 500-seat auditorium. The arresting design using sheet-metal folds is by architect Kengo Kuma. The Conservatoire plays a full role in the two major Aix music festivals: the Festival de Paques and the Festival d'Aix. Of particular interest are music masterclasses from well-known musicians.

Festivals
Festival de Paques

Various locations, including Grand Théâtre de Provence, 380 avenue Max Juvénal; tel. 08 20 13 20 13; www. festivalpaques.com; two weeks around Easter; €25-105

This classical music festival runs for two weeks over the Easter period. It attracts some of the best orchestras in the world, including in recent years the London Symphony Orchestra. Expect performances of works by the big names of the classical music world: Bach, Vivaldi, Brahms, and Mozart, as well as some lesser-known composers.

Le Sma'art

Parc Jourdan; tel. 04 42 49 97 52; www.salonsmart-aix. com; mid-May; €10, under 10 free

Never has a stroll in the park been so much fun. Thousands of paintings, mostly by young up-and-coming artists, are displayed in Parc Jourdan.

Festival d'Aix-en-Provence

Various locations; tel. 08 20 92 29 23; www.festival-aix. com; July; from €20

Whether you are an afficionado or an amateur, Aix-en-Provence is the place to experience opera. The festival tends to feature five major operas and a host of subsidiary events, including performances of everything from classical music to flamenco guitar. Now over 70 years old, the Festival d'Aix even has an offshoot event, Aix-in-June, offering free plays, masterclasses, and classical music performances across the city and out into the countryside, taking place in venues such as Silvacane Abbey.

SHOPPING

Aix-en-Provence is often called the Paris of the South. The reputation comes partly from the attention to detail with which the residents dress. Big brands such as Hermes, Longchamps, and Agnes B are on display, and there are plenty of jewelers where locals can source that finishing touch of bling. For on-trend purchases, there's Zadig et Voltaire, Kooples, and American Vintage. The main shopping area is the old town, where there are plenty of owner-operated boutiques selling original collections mixed in with national brands. Les Allées Provençal is a modern development on the opposite side of La Rotonde fountain from the Cours Mirabeau. It is largely filled by international chain stores such as H&M and Zara.

Beauty
Bastide
14 rue Espariat; tel. 04 84 47 00 29; https://bastide. com; Mon.-Thurs. 10am-7pm, Fri.-Sat. 10am-7:30pm, Sun. 11am-6pm

Bastide specializes in simple, clean, natural skin and hair-care products made from Provençal ingredients. Examples include lavender honey body wash, summer fig hand cream, and rose and olive tree eau de toilette. The shop, located in the heart of the old town, is a joy to visit.

Rose and Marius
3 rue Thiers; tel. 09 82 59 35 35; www.roseetmarius. com; Tues.-Sat. 10:30am-1pm and 2pm-7pm

Rose and Marius sells fragranced candles in beautiful Limoges porcelain holders, with intricate colored geometric patterns inspired by the tiled floor of the family bastide where the owner grew up. Look for the unique rosé eau de toilette. By appointment, the shop also offers smell workshops to help you recognize scents and find the perfect fragrances for you and your house.

Clothing
Mes Desmoisselles
12 rue Marius Reynaud; tel. 04 42 29 67 46; www. mesdemoisellesparis.com; Tues.-Sat. 10am-7:30pm

Mes Desmoisselles is a Parisian fashion house now open in Aix. It specializes in the chic, stylish womenswear that makes French women seem so effortlessly graceful.

Specialty Food
Le Roy René
11 rue Gaston de Saporta; tel. 04 42 26 67 86; www. calisson.com; daily 10am-7pm

Calissons are small almond-shaped treats, not much larger than a mint candy, made from a paste of preserved melons and almonds and topped with icing. They have been produced in Aix since the 15th century. These days Le Roy René is the self-declared king of the calisson business. While small bakeries turn out hand-crafted sweets, this large enterprise ships Aix's specialty worldwide. At its out-of-town factory (5380 route d'Avignon) there's even a calisson museum.

Fromagerie Le Marie
30 rue d'Italie; tel. 04 88 41 58 27; www.fromagerie-lemarie.com; Tues.-Fri. 9:30am-1pm and 3:30pm-7pm, Sat. 9am-7pm

The cheeses here are simply sensational. Don't be afraid to ask the helpful staff for a recommendation. It's a local favorite so be prepared to queue.

Bookstores
Book in Bar
4 rue Joseph Cabassol; tel. 04 42 26 60 07; www. bookinbar.com; Mon.-Sat. 9am-7pm

Book in Bar is so much more than a bookshop. It's the center for Anglophones in Aix-en-Provence. There are book clubs, book signings, and talks by authors. Like all good bookshops, it has a slightly cluttered feel, as if there are too many books to fit within its walls. If browsing gets tiring, coffee and cake are served from a hole-in-the-wall kitchen, and you can relax on sofas and various tables and chairs.

Markets

Scarcely a day passes in Aix without a market of some form or another.

Cours Mirabeau

Cours Mirabeau; Tues. and Thurs. 8:30am-1pm

On Tuesday and Thursday the Cours Mirabeau is lined with a clothes market. There's a good range of items including coats, shoes, gloves, underwear, and fashion wear, and prices are much lower than in the Aix boutiques. The backdrop, of course, is gorgeous: the long view down to the tumbling waters of La Rotonde, the imperious rows of hôtel particuliers, and the watchful plane trees. It's hard to imagine a more scenic place to shop.

Les 3 Places

place des Prêcheurs, place de Verdun, place de la Madeleine (squares intersect with rue Portalis); Tues., Thurs., and Sat. 8:30am-1pm

There really is something for everyone in this market, from rare books and brocante to fruit, vegetables, and spit-roasted chickens, to artisan-made leather wallets, belts, and handbags. It's a joy to lose yourself amid the stalls and see what you emerge with at the end.

Place de l'Hotel de Ville Flower Market

Intersects with rue Vauvenargues; daily except first Sun. of the month, 8:30am-12:30pm

Even if you have no interest in buying flowers, this market is a delight to see. Seasonal flowers and plants are laid out around an 18th-century fountain, capped with a column salvaged from the ruins of the Palace of the Counts of Provence.

Place Richelme Food Market

place Richelme; daily 8:30am-1pm

This market has a real local feel. It's where the Aixoise go to pick up their groceries. There's loads of fruit and vegetables as well as a good fish stall.

Brocante Art and Antiques

Galerie Imbert

7 rue Jacques de la Roque; tel. 09 72 58 37 30; https:// galerieimbert.com; Tues.-Sat. 10:30am-1pm and 3pm-6:30pm

Galerie Imbert opened nearly 25 years ago near the cathedral in Aix. Landscapes and still life paintings line the walls of the narrow space. The gallery represents around 20 local artists, and prices range from the hundreds to the thousands of euros.

Mazarin Antiquités

8 rue Frédéric Mistral; tel. 06 21 21 82 76; Tues.-Sat. 10am-1pm and 3pm-6:30pm

Here you'll find a haphazard treasure trove of items from the Provençal past. Finds might include chandeliers, grandfather clocks, coffee grinders, plates, and paintings. Don't be afraid to negotiate on price.

FOOD

The best way to eat out in Aix is to start thinking about your ideal meal early in the day. As a rule of thumb, when you eat outside on a terrace in one of the more touristy areas, you have to sacrifice a little on value for your money. Weather is crucial: The midday sun in summer turns some restaurant terraces into furnaces, but for the rest of the year the same terraces are pleasant. If you go to one of the many more local restaurants hidden in side streets, you may not be able to sit outside, but you'll pay less and often eat better.

Cours Mirabeau

Pâtisserie Béchard

12 Cours Mirabeau; tel. 04 42 26 06 78; www. maisonbechard.fr; Tues.-Sat. 8am-7pm; from €2.50

Beloved by the Aixoise for generations thanks to its sweet concoctions, this pastry shop also offers plentiful savory snacks. Try the feuilleté à la saucisse, a hot dog encased in crumbling pastry. Depending on your view, the serving system is either charming or antiquated: You line up for one person to take your order. This is then written on a slip of paper. You then

Calissons

Calissons are Aix's signature sweet treat. Small and almond-shaped, they're made from a paste of preserved melons and almonds, topped with icing. Some chefs add a dash of orange blossom, some lemon zest or even a little vanilla, but tinkering more than this is frowned upon.

Calissons have been delivering little moments of pleasure since the 15th century. They were famously served at the marriage of King Roi René to his second wife, Jeanne de Laval, in Aix. Rumor has it they owe their name to a hug (calin) bestowed at this event. True or not, they are deliciously indulgent and the perfect gift to take home for friends and relatives.

Here's a simple recipe for a moment of bliss during a visit to Aix-en-Provence:

calissons, the signature pastry of Aix

- Stop at favorite local bakery **Pâtisserie Béchard** on the Cours Mirabeau and treat yourself to a box of calissons (page 134).

- With your box of calissons in hand, cross the road and settle into a table with a view at the iconic **Deux Garcons** café (page 135). Depending on the time of day and your mood, order any one of the following: a glass of champagne, a glass of sweet wine, a coffee, or a tea. Take a sip of your drink and then a bite of calisson, and then repeat until a broad smile inevitably spreads across your face.

queue again to pay before finally being given your goodies. Fortunately, the wait is worth it.

La Rotonde

2A place Jeanne d'Arc; tel. 04 42 91 61 70; www. larotonde-aix.com; daily 7:30am-1:30am; €21-35

La Rotonde is the best place to sit and enjoy the cascading waters of its namesake fountain. The atmosphere is vibrant and young. The food—a mixture of Provençal (pied pacquets), Italian (pizzas), and American (burgers)—is good but a little overpriced.

Côté Cour

19 Cours Mirabeau; tel. 04 42 93 12 51; www. restaurantcotecour.fr; Tues.-Sat. noon-2pm and 7:30pm-10pm; €28-36

Probably the best restaurant on the Cours Mirabeau is Côté Cour. The restaurant is nestled in a courtyard behind one of the street's grand old buildings. There's a retractable roof and pleasing modern decor. The service

is slick and the food excellent. Try the pungent truffle risotto around Christmastime.

★ Les Deux Garcons

53 Cours Mirabeau; 04 42 26 00 51; www. lesdeuxgarcons.fr; €30-45

The most famous café brasserie in Aix burned down in 2019. It remained closed at the time of writing, with a reopening date scheduled for 2025. The restoration promises to be worth the wait, with the aim of returning the historic building to its former glory in the days when Cézanne and Zola were patrons.

Aix Old Town (La Vieille Ville)
Drole d'Endroit

1 rue Annonerie Vieille; tel. 04 42 38 95 54; http:// droledesite.fr; Tues.-Sat. noon-2pm and 7pm-9pm, Wed. 7pm-10pm; €15-22

Drole d'Endroit is a hard place to find, which makes good sense, as the name means "funny location." It's tucked down a side alley, and thousands of people must walk past every week not knowing it's there. The menu changes every week and attracts a loyal following of in-the-know locals.

La Bouche à Oreille

1 rue Aumone Vieille; tel. 09 72 89 19 19; www. le-bouche-a-oreille-aix-en-provence.com; Tues.-Sat. noon-2pm and 7:30pm-10pm; €26

Tucked away in a quiet little square, Le Bouche à Oreille is a rustic brasserie with a small menu chalked up on a blackboard. Inside the decor is simple but charming, and outside there's a small, shady terrace. The salted pig's knuckle is delicious.

Le Chastel

18 place des Prêcheurs; tel. 04 42 29 70 64; www. lechastel.com; Tues.-Sat. noon-2:30pm and 7pm-10pm; €20-34

Le Chastel has a small, intimate feel, and is always crowded. It offers excellent French food plus a few Italian classics, such as spaghetti vongole (with clams) and veal Milanaise (breaded and fried), thrown in for good measure.

Le Zinc de Hugo

22 rue Lieutaud; tel. 04 42 27 69 69; https://zinc-hugo. com; Tues.-Sat. noon-2:30pm and 7pm-10:30pm; €20-36

Meat lovers should look no further than Zinc de Hugo, a cozy brasserie with an open-fire grill. Meat is sourced from across the world. Generous portions, traditional decor, an excellent wine list, and a convivial atmosphere make the Zinc enduringly popular.

Les Vieilles Canailles

7 rue Isolette; tel. 04 42 91 41 75; www.vieilles-canailles. fr; Tues.-Sat. noon-2pm and 7:30pm-9:45pm; €25-35

It's hard to find a more atmospheric dining experience in Aix than Les Vieilles Canailles. Over the sound system Serge Gainsbourg croons away and bottles of wine, personally chosen by owner Pierre Hochart, line the walls. Dishes might include Corsican charcuterie and a roast rack of lamb en croute (in a pastry crust) with purple mustard.

Quartier Mazarin
Hôtel Gallifet

52 rue Cardinale; tel. 09 53 84 37 61; http:// hoteldegallifet.com; June-Sept. Thurs.-Sat. noon-11pm, Sun. and Tues.-Wed. noon-6pm; €24

This is a great summer pop-up restaurant-café in the garden of a hôtel particulier. At lunch there's a small menu with dishes such as breaded chicken and salmon filet. After 3pm the restaurant transforms into a tea and coffee shop. Then in the evening a more gastronomic menu is offered. The garden is a lovely, shady place to sit, with the hôtel particulier forming a beautiful backdrop. There are occasional jazz concerts and art exhibitions.

Le Jardin de Mazarin

15 rue du 4 Septembre; tel. 04 66 42 05 31; www. jardinmazarin.com; Mon.-Sat. noon-3pm and 7:30pm-10:30pm; from €25

Le Jardin de Mazarin is a refined, discreet restaurant, and the perfect bolt-hole for the residents of this expensive quarter of Aix. Hidden behind an unassuming facade are a warmly furnished dining room and a long terrace opening onto a small garden.

La Maison Francaise

2 rue Fernand Dol; tel. 04 42 12 62 09; www. lamaisonfrancaise.com; Tues.-Sat. noon-11:30pm; €25-40

This is a great place to eat away from the crowds. There's plenty of space between tables and a lounge area for an apero before eating. The food is modern French.

Aix Outskirts
Darius Café

115 avenue Giuseppe Verdi; tel. 04 42 27 98 97; Mon.-Sat. 6am-midnight; from €18

This very popular café in Les Allées Provençal shopping area, just outside the center of Aix, is a 2-minute walk from the tourist office. There's a big terrace and an atmospheric interior that bubbles with conversation. This is a good place to stop for a break and mix with the locals.

BARS AND NIGHTLIFE

The city is one of the campuses of Aix Marseille university, which, with approximately 74,000 students, is the largest university in the French-speaking world. Cafés and bars are filled during the day with youngsters trying to look both hip and studious at the same time. In the evening they drop the pretense of studiousness and head for Irish bars and tapas restaurants. Given its student population, the club scene in the city is surprisingly understated.

La Cita

16 rue Félibre Gaut; tel. 04 86 31 52 43; http://lacita.fr; Mon.-Sat. 7pm-midnight; beer or sangria €3.50/glass or €12/liter, tapas from €5.50

This traditional Spanish tapas bar gets rowdier and rowdier as the evening progresses. Jugs rather than glasses of alcohol are the norm, and the tapas are an essential way to soak up the booze.

La Vida Loca

11 rue Félibre Gaut; tel. 06 37 67 04 45; Tues.-Sat. 2pm-2am; drinks from €4

Just along the street from La Cita is La Vida Loca. As the name suggests, this is not the place to go for a quiet drink and intimate conversation, but to dance with students and locals. Expect to have to shout to be heard.

Le Mistral

3 rue Frédéric Mistral; tel. 06 20 38 50 25; http://mistralclub.fr; Tues.-Sat. 11:55pm-6am

The grande dame of Aix nightclubs, Le Mistral is 65 years old and still drawing the crowds. A great central location, big-name DJs, and a revamped interior by designer Gianni Fasciani make Le Mistral the place to see and be seen.

ACCOMMODATIONS

The center of Aix has a good choice of hotels, but room prices are expensive compared to the rest of France and space can be limited. Hotels with a more spacious feel can be found in the Aix outskirts.

Cours Mirabeau
Hotel Negrecoste

33 Cours Mirabeau; tel. 44 22 77 42 22; https://hotel-negre-coste.com; €170 d

Hotel Negrecoste has a central location halfway up the Cours Mirabeau with a large terrace bar and a restaurant. Rooms are modern and nicely designed. A Nuxe spa offers some much-needed pampering after a day's sightseeing. Private parking is available for €20 per day.

Aix Old Town (La Vieille Ville)
Hotel de France

63 rue Espariat; tel. 04 42 27 90 15; https://hoteldefrance-aixenprovence.com; €120 d

Hotel de France is located on the bustling pedestrian rue Espariat, which runs parallel to the Cours Mirabeau. The breakfast room is large, airy, and a charming mix of old and new. Bedrooms are decorated in a clean, minimalist style; rooms with windows onto the rue Espariat may experience some nighttime noise.

★ Hotel des Augustins

3 rue de la Masse; tel. 04 42 27 28 59; https://hotel-augustins.com; €120 d

For a sense of history, try the Hotel des Augustins, a converted chapel with links to the 12th-century Augustins Convent, which once stood in the area. In particular, the reception area has a wonderful ecclesiastical feel. Some rooms still have a touch of historical character, with the occasional exposed stone arch and views of the bell tower of the Augustins church.

Quartier Mazarin
Hotel Cardinal

24 rue Cardinale; tel. 04 42 38 32 30; www.hotel-cardinal-aix.com; €90 d

Hotel Cardinal is a good budget option right next door to the Musée Granet. The decor is a little dated with an overuse of chintzy florals and old-fashioned paintings, but the rooms are comfortable. Suites with kitchenettes and small terraces are available for those planning a longer stay in Aix.

Le 28

28 rue du 4 Septembre; tel. 07 83 15 75 92; www. hotelparticulier-le28.com; €300 d

Le 28 is a close neighbor of Maison d'Aix but it's a touch more flamboyant in its interior decor. This is another hôtel particulier converted into a boutique hotel. The hotel offers access to the pool and spa at the nearby Spa Mazarin. There are seven rooms in total, so early booking is recommended.

★ Maison d'Aix

25 rue du 4 Septembre; tel. 04 42 53 78 95; www. lamaisondaix.com; €420 d

Maison d'Aix plays on the reputation of its former owner Henriette Reboule, a "priestess of love" who purchased the 18th-century hôtel particulier in 1903. Guests now sleep in the rooms where Henriette once entertained her lovers and there is still an underlying erotic feel to the place.

Aix Outskirts

Hotel Cézanne

40 avenue Victor Hugo; tel. 04 42 91 11 11; www. boutiquehotelcezanne.com; €200 d

Hotel Cézanne has 57 rooms and is a popular choice with families, thanks to plenty of interconnecting rooms. Although this property is not central, the walk to the Cours Mirabeau takes only a couple of minutes.

Hotel Roi René

24 boulevard du Roi René; tel. 04 42 37 61 00; www. accorhotels.com; €200 d

Hotel Roi René is part of the Accor Sofitel chain, and what it lacks in individual charm it makes up for in convenience. The lobby is large and perfect for a relaxing drink or coffee, and there's a good restaurant and a sizeable open-air pool. It's 2 minutes' walk from

the center and backs onto the open spaces of Parc Jourdan.

Villa Saint Ange

7 Traverse Saint Pierre; tel. 04 42 95 10 10; www. villasaintange.com; €450 d

Villa Saint Ange is a luxury bolt-hole just a couple of minutes' walk from the city center. There's a wonderful pool with an associated spa and fitness center. The rooms are elegantly decorated with sumptuous furnishings.

INFORMATION AND SERVICES
Tourist Information

Aix-en-Provence has a large, well-staffed **tourist office** (300 avenue Giuseppe Verdi; tel. 04 42 16 11 61; www. aixenprovencetourism.com). It is the departure point for many walking tours and the best place to pick up maps and information. It is located at the bottom of the Cours Mirabeau, at the entrance to Les Allées Provençal shopping center.

The tourist office offers a **City Pass,** which includes visits to 25 cultural sites and a further 30 for a reduced price, plus access to Aix-en-Bus and English-speaking guided tours of the city. The adult prices are €29 for 24 hours, €39 for 48 hours, and €49 for 72 hours. There are reduced prices for children under 13; those under 3 years are free.

Book in Bar (4 rue Joseph Cabassol; www.bookinbar.com) bookshop and coffee shop is an excellent place to get in touch with the local expat network. It has a notice board full of postings for events and clubs for English-speakers.

Post Offices

There are numerous post offices in Aix-en-Provence. The main office near La Rotonde fountain is on 2 rue Lapierre. It is open Monday-Friday 9am-6:30pm and Saturday 9am-12:30pm. More centrally, in the old town there is a post office in Place de l'Hôtel de Ville. It is open Monday-Friday 9am-noon and 1pm-6pm and Saturday 9am-noon.

Health and Safety

The **police station** is located at Hotel de Police, avenue de l'Europe (tel. 04 42 93 97 00). To phone the police dial 17 or European emergency number 112.

For health care, the **Centre Hospitalier du Pays d'Aix** (avenue des Tamaris, Aix-en-Provence; tel. 04 42 33 50 00) offers emergency services, including an excellent children's emergency department. There are numerous pharmacies in Aix-en-Provence. They all display a prominent green cross jutting out into the street. The pharmacy on the Cours Mirabeau is located at 17 bis Cours Mirabeau, and it is open Monday-Saturday 8am-8pm.

Communications

Many, but by no means all, cafés offer free Wi-Fi. Ask before sitting down. Les Allées Provençale shopping center (95 avenue Giuseppe Verdi) offers free Wi-Fi.

GETTING THERE
Air
Marseille Provence Airport

Marignane

Marseille Provence Airport is 32 km (20 mi) from Aix and offers 152 regular flight routes. The drive to Aix-en-Provence takes about 30 minutes. Traffic in the morning and evening is bad, so allow an hour at these times of day. The airport has a dedicated low-cost flight terminal that offers budget flights with carriers such as **Ryan Air** and **EasyJet** to many European capitals. A **shuttle bus** runs every half hour from the airport to the center of Aix, also stopping at the Aix **TGV train station.** The journey takes 1 hour and costs €7.

Car

At the heart of the autoroute network of Provence, Aix-en-Provence is easy to get to and it makes an ideal base for exploring the rest of the region.

From **Marseille Provence Airport,** take the D9 for 28 km (17 mi), about 29 minutes. From **Avignon,** take the A7 for 88 km (55 mi), about 60 minutes. From **Arles,** take the

A54 for 77 km (48 mi), about 1 hour. From **Cannes/Nice,** take the A8 for 175 km (109 mi), about 2 hours.

Train
Aix-en-Provence TGV Station

9 route Départementale, Aix-en-Provence; www.sncf-connect.com

The TGV station is located a short distance outside the city center. It offers high-speed connections to major European destinations (Paris 3 hours, Barcelona 4 hours). A **shuttle bus** runs every half hour to the center of town, taking 30 minutes and costing €4. All TGV train tickets need to be booked in advance, either online or from a ticket office at a train station.

Local Trains

Local trains, including ones from Marseille, stop at the **central train station** on rue Gustave Desplaces.

Bus
Aix Center Bus Station

boulevard Victor Coq; tel. 08 09 40 04 15; www.lametropolemobilite.fr

The bus station, Aix Center, is located 500 m (0.3 mi) from the town center. Buses run every 30 minutes to Marseille airport and the TGV station (taking 60 minutes and 30 minutes, respectively) and frequently to Marseille, Avignon, and Arles. There is an extensive network of local destinations.

GETTING AROUND

Once visitors have arrived in the center of Aix-en-Provence, there is little need for public transportation. The old town is largely pedestrian. It takes a little over 15 minutes on foot to cross from one side of the old town to the other. Probably because of this, there is no pickup and drop-off cycle scheme in Aix, and in central Aix cycling is not a popular method of transportation. For the less mobile, the Diabline electric vehicles are a good way of getting around. The bus network is not heavily used by tourists.

Bus and Diablines

The local bus network is provided by **Aix-en-Bus** (www.aixenbus.fr).

The historic center of Aix-en-Provence is served by small electric vehicles called **Diablines.** They follow three routes and have access to the otherwise pedestrianized old town area. Tickets cost €0.60 and are purchased directly from the driver. There are no specified stops; a Diabline will stop when hailed.

Taxi/Uber

The main **taxi stand** is at the bottom of the Cours Mirabeau near La Rotonde fountain. Taxis can also be booked through **Taxi Radio Aixoise** (tel. 04 42 27 71 11) or by using the **Uber** app or asking at your hotel. Licensed taxi journeys are metered. Note that taxis are not needed to travel around Aix. It is much easier to walk, although visitors may consider taking one to Cézanne's atelier, which is about a 20-minute walk uphill from the center of town.

PAYS D'AIX-EN-PROVENCE

To the east of Aix is some of the most beautiful countryside in Provence. Mont Sainte-Victoire is, of course, the star of the show, drawing the eye with the ever-changing interplay of light and rock that so entranced Cézanne. Hiking, biking, and wine-tasting are the main activities, although in July and August many people choose to just sit by a pool and listen to the chorus of cicadas.

Orientation

Le Tholonet and **Saint-Antonin-sur-Bayon** are the main bases for exploring **Mont Sainte-Victoire** from the south, and **Vauvenargues** is the main base from the north. These villages can be easily reached by car from the center of Aix. The **D17** runs through Le Tholonet (15 minutes from the center of Aix) and Saint-Antonin-sur-Bayon

(20 minutes from the center of Aix). The **D10** from the center of Aix takes you to Vauvenargues in 30 minutes. It's possible to drive the entire circuit of Mont Victoire: The D17 and the D10 are linked just to the east of the village of Puyloubier by the D23.

Other key points of orientation around Aix are the town of **Rognes,** 30 minutes north of Aix, which hosts a famous piano festival in the summer and is the home of a much-loved truffle market before Christmas. And the **Cistercian Silvacane Abbey** is just north of Rognes outside the village of La Roque-d'Anthéron. Also of interest, **Château La Coste** vineyard and art center is to the south of Rognes on the way back toward Aix, near the village of Le-Puy-Sainte-Réparade. South of Aix on the way toward Marseille is the town of **Aubagne,** which is nestled west of the **Sainte-Baume Massif,** a rock formation that divides the Var and Bouches-du-Rhone départements. Aubagne is also home to the French Foreign Legion Museum.

Sights

Silvacane Abbey

RD561, La Roque-d'Anthéron; tel. 04 42 50 41 69; www. abbaye-silvacane.com; June-Sept. daily 10am-12:45pm and 2pm-6pm, Oct.-Mar. Tues.-Sun. 10am-12:45pm and 2pm-4:45pm, Apr.-May Tues.-Sun. 10am-12:45pm and 2pm-5:15pm; €8, under age 12 free

One of three Cistercian abbeys in Provence, Silvacane was built in the 12th century. The abbey's bare Gothic- and Roman-influenced architecture still conveys a sense of the life of devotion led by the monks. The Cistercians—known as the "white monks" because their robes were left undyed—divided their days between devotion, manual work, and intellectual improvement. The visit takes in the abbey's various rooms, such as the salle des monies where the monks copied manuscripts and the salle capitulaire for hearing confessions, as well as the gardens and the vivier (fishpond), which supplied the monks with food.

Pays d'Aix-en-Provence

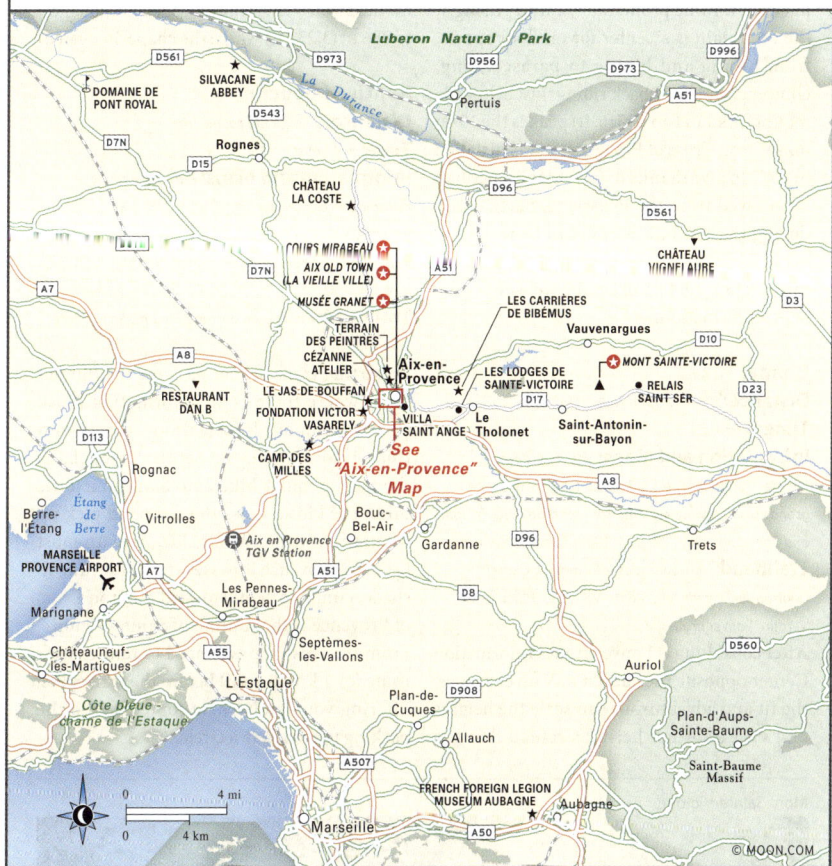

French Foreign Legion
Museum Aubagne

chemin de la Thuilière, Aubagne; tel. 04 42 18 10 96; https://musee.legion-etrangere.com; Tues.-Sun. 10am-noon and 2pm-6pm; free

Since 1831 many of the most forlorn, love-sick young men in the world have joined the French Foreign Legion to forget their troubles and spill their blood for France. At least that's how joining has often been portrayed in film and literature. A more accurate history of the Legion can be en-joyed at its museum, whose purpose is to remember and honor the legionnaires of

the past and to integrate and educate new recruits.

Hiking
★ Mont Sainte-Victoire

Pays d'Aix-en-Provence; tel. 04 42 26 67 37; www. grandsitesaintevictoire.com

Mont Sainte-Victoire dominates the landscape of the Pays d'Aix-en-Provence. The garrigue, or scrubland, which to the untrained eye ap-pears uniform, contains numerous species of plants, including rosemary, thyme, gorse, and wildflowers, all of which explode into color in spring, draping the bare rock in pinks, whites, and yellows. Pine, oak, olive, almond, mulberry, and cypress trees complete a landscape in which birdlife proliferates. Unsurprisingly, the mountain is a center for outdoor sports, from hiking and biking to parascending, climbing, and horseback riding. Just below the mountain's 1,011-m (3,316-ft) peak is the 19-m (62-ft) Iron Cross of Provence. It was installed in 1877 to give thanks to God that France had been saved from the Prussians. Particularly during the summer, access can be restricted because of the fire risk. Always check with the information centers or on the website for the latest information.

Chapel Hike

Distance: *12 km (7.5 mi) round-trip*
Time: *4 hours*
Information and Maps: *Vauvenargues Information Center; tel. 04 42 26 67 37; www. grandsitesaintevictoire.com; Apr.-Sept. daily 10am-1pm and 2pm-6pm, Oct. Sat.-Sun. 10am-1pm and 2pm-6pm*
Trailhead: *Vauvenargues Information Center (Maison du Grande Site Sainte-Victoire), Place de Verdun, Vauvenargues*

After consulting at Vauvenargues Information Center opposite the church in Vauvenargues, the fit and adventurous can scale the heights and visit a former hermit's refuge that has been converted into a small chapel. The route takes you from the Vauvenargues valley on a marked trail (GR9, white and red markers) up to 986 m (3,234 ft) where the chapel is located.

Sentier Imoucha

Distance: *9 km (6 mi) round-trip*
Time: *3-4 hours*
Information and Maps: *Barrage Bimont Information Center; tel. 06 70 74 97 78; www. grandsitesaintevictoire.com; Apr.-May and Sept.-Nov. 9am-noon and 2pm-6pm, June-Sept. 9:30am-12:30pm and 3pm-6pm*
Trailhead: *Parking du barrage Bimont, Saint-Marc Jaumegarde*

Departing from the car park for the Bimont dam, where there is an information kiosk (Barrage Bimont Information Center) with route information, the Sentier Imoucha is the most popular hike onto Sainte-Victoire. From the kiosk, cross the dam and pick up the trail posts on the far side. As with most hikes in Provence it passes from sunshine to shade, climbing gently toward the Iron Cross of Provence at the summit of Sainte-Victoire. From the cross you will need to retrace your route for 1 km (0.6 mi) before the trail splits, offering you an alternate route back to the parking to complete a circle.

Mont Sainte-Victoire

Summit Sentiers

Distance: *11-15 km (7-9 mi) round-trip*

Time: *4 hours*

Information and Maps: *Saint-Antonin-sur-Bayon Maison Sainte-Victoire Visitor Center; tel. 04 13 31 94 70; www.departement13.fr; Tues.-Sun. 9am-12:30pm and 1:30pm-6pm*

Trailhead: *Saint-Antonin-sur-Bayon Maison Sainte-Victoire Visitor Center parking lot, 17 chemin départemental, Saint-Antonin-sur-Bayon*

From the visitor center parking lot, follow the signs for the Mont Sainte-Victoire sentiers. It's possible to follow four different routes, marked as yellow, red, green, and black, to the Iron Cross at the top of the mountain. Yellow is the hardest and is occasionally closed because of falling rocks. Always ask for route information at the visitor center before starting out. Allow at least 4 hours for the climb and descent. For those who do not wish to reach the summit, the brown sentier offers great views of Mont Sainte-Victoire, taking you east 8 km (5 mi) to Puyloubier. It is an easy hike, but it's not circular. Inquire at the visitor center about guided walks.

Saint-Ser Chapel Hike/Botanic Walk

Distance: *1-3 km (0.6-1.8 mi) round-trip*

Time: *1-1.25 hours*

Information and Maps: *https://travel.aixenprovencetourism.com*

Trailhead: *Parking Saint-Ser, D13 avenue Cézanne, 1.5 km (0.9 mi) east of Puyloubier*

For an easier climb, head to the parking Saint-Ser in Puyloubier and follow the trail shown by the red markers to the Saint-Ser Chapel. From the chapel, there is an excellent view of the vineyards of Puyloubier, the valley of the River Arc (which runs into Aix), and the Sainte-Baume Massif that marks the boundary between the Bouches-du-Rhone departement and the Var.

Another option from the same parking area is to follow a botanic walk, which takes around 1 hour, covers 1 km (0.6 mi), and introduces you to the main species of plants on the mountain.

Cycling

On a spring or autumn day there are few more scenic bike routes than looping around Mont Sainte-Victoire, stopping at vineyards for a wine-tasting or at one of the many restaurants for lunch. The road, which you will have to share with cars, is gentle and passes from sun to shade, and there are plenty of places to stop and admire the view. Follow the D17 on the southern side and the D10 on the northern side. The road that links north and south is the D23.

Station Bees

Le Tholonet, parking Paysager des Infernets, 3333 route Cézanne; tel. 07 81 64 47 34; www.stationsbees.com; Mar.-Oct. daily 9am-1pm and 2pm-6pm, check website out of season; from €25

Station Bees rents e-bikes for access to Mont Sainte-Victoire.

Golf
Domaine de Pont Royal

Mallemort; tel. 04 90 57 40 79; www.golf-pontroyal.com; greens fee €75-98

Domaine de Pont Royal is one of the best golf courses in Provence, designed by the late Severiano Ballesteros. Views from certain points on the course are spectacular.

Wineries
Château La Coste

2750 route de la Cride, Le-Puy-Sainte-Réparade; tel. 04 42 61 92 92; https://chateau-la-coste.com; daily 10am-7pm; cave visit €20, art walk €15, tasting free

Arriving at Château La Coste is visually shocking. First there's the winery, an elongated silver dome glinting in the sun that looks more suited to housing fighter jets than making wine. Next there's the sculpture of a *Lord of the Rings*-size spider, created by Louis Bourgeois, skating on a thin film of water. In the distance, an elliptical silver pod rotates, appearing as if it has just landed from outer space. The château is a modern art and architecture destination as much as a vineyard. There are frequent evening concerts during the summer at the Frank Gehry-designed

Music Pavilion. After contemplating the often-challenging works you can taste the domaine's wines, take a full tour of the winemaking facilities, and enjoy a meal in one of the estate's four restaurants.

Château Vignelaure

route de Jouques, Rians; tel. 04 94 37 21 10; www.
vignelaure.com; Mar.-Oct. and Dec. daily 10am-5pm,
Jan.-Feb. and Nov. Mon.-Sat. 10am-5pm; visit to cave
and art gallery €10, tasting free

The famed American wine critic Robert Parker calls Château Vignelaure "one of the showpiece properties not only of Provence, but of France." Cabernet sauvignon from the vines of Château La Lagune in Bordeaux was planted here in 1966, making this the first winery to use this Bordeaux grape varietal in Provence. There are now five subterranean levels to age the wine, each holding its own secrets, including an art gallery filled with photos by Henri Cartier Bresson. The garden of the château is home to numerous sculptures.

Festivals and Events

Rognes Truffle Market

Cours Saint Etienne, Rognes; www.festivites-rognes.fr;
Sun. before Christmas

Since 1988, year after year, the Rognes Truffle Market has grown in size. A variety of Christmas culinary treats are on sale, but the star of the show is undoubtedly the truffle. The deep, earthy smell of the black diamond of Provence hangs in the cold winter air. The traders are always ready with cooking instructions, suggesting, for example, a last-minute shaving of truffle over the traditional Christmas bird of Provence, the capon.

La Roque-d'Anthéron International Piano Festival

Château de Florans, La Roque-d'Anthéron; tel. 04 42
50 51 15; www.festival-piano.com; mid-July-mid-Aug.;
tickets €30-65

Lovers of classical music will already have heard of La Roque-d'Anthéron International Piano Festival. Nearly 80,000 tickets are sold every year. Most recitals take place in the enchanting outdoor arena on the grounds of the Château de Florans or in the nearby Cistercian abbey of Silvacane. In recent years the festival has spread its wings, staging concerts as far afield as the Théâtre Antique in Arles.

Food

★ Restaurant Saint Esteve

2250 route Cézanne, Le Tholonet; www.
leslodgessaintevictoire.com; tel. 04 42 27 10 14; Mon.-

Frank Gehry-designed Music Pavilion at Château La Coste

Sat. noon-2pm and 7:30pm-10pm, Sun. lunch only; mains €69-89

There's an additional presence at every table in the Restaurant Saint Esteve, a dining partner who says little but surveys all. This inescapable "third person" is Mont Sainte-Victoire. So perfect is the view of the mountain that it almost feels like you are dining inside a painter's canvas. The food more than matches the environment with classical French dishes such as guinea fowl with girolle mushrooms executed to perfection.

★ Restaurant Dan B

1 rue Frédéric Mistral, Ventabren; tel. 04 42 28 79 33; www.danb.fr; Thurs.-Sun. noon-1:15pm and 7:45pm-9:15pm, Mon.-Wed. 7:45pm-9:15pm; €81-149

Architecture and fine dining meet in perfect harmony in Restaurant Dan B. A small stone Provençal terrace with a view over the surrounding countryside has been transformed into a modern temple to gastronomy. An entire glass wall slides away, opening the restaurant to the countryside.

Accommodations
★ Relais Saint-Ser

avenue Cézanne, Puyloubier, Pays d'Aix-en-Provence; tel. 04 42 66 37 26; www.relaisdesaintser.com; €125 d

The Relais is a recently renovated old farmhouse just outside of Puyloubier. In the clear skies above, paragliders swirl on thermals before swooping to their landing point adjacent to the hotel. Best of all is the sensational panorama from the terrace. Fields of vines and olive trees sweep away toward the Massif de Sainte-Baume. It's a view to revel in. There's an excellent restaurant and a pool with the same scenic view.

Hotel Sainte-Victoire

Vauvenargues, Pays d'Aix-en-Provence; tel. 04 42 54 01 01; www.hotelsaintevictoire.com; €240 d

The Sainte-Victoire is a modern luxury hotel just outside Vauvenargues. There are lovely views of the surrounding countryside and Mont Sainte-Victoire from the pool. The hotel's excellent restaurant

persuades many to sleep over rather than head home.

Les Lodges de Sainte-Victoire

2250 route Cézanne, Le Tholonet, Pays d'Aix-en-Provence; tel. 04 42 50 51 15; www.leslodgessaintevictoire.com; €360 d

Staying at Les Lodges de Sainte-Victoire is one of the most relaxing ways to see Aix and the surrounding countryside. The location is wonderful: a hotel set in a mature park with views of Mont Sainte-Victoire. Centuries-old plane trees shade the terrace, and there are two restaurants to choose from. Hiking and biking are on the doorstep. The city is a 5-minute drive away.

Getting There and Around

Once you have arrived, the villages listed in this section are best explored **on foot.** Walking between villages and sights, though, is not advisable; the distances tend to be long and the roads busy. The exception is when following a marked sentier, such as the brown sentier between Saint-Antonin-sur-Bayon and Puyloubier to the south of Mont Sainte-Victoire.

Car

From the center of **Aix,** the D7 takes you out toward Le Tholonet and the southern face of Mont Sainte-Victoire, and the D10 leads to Vauvenargues and the northern face of the mountain. The two roads meet at the village of Pourrières where the D23 takes you around to the opposite face of the mountain. Jouques and the River Durance can be reached by taking the D11 from near Vauvenargues. These roads are small and picturesque, and in places they wind up and down hills.

To head north from Aix toward **Silvacane Abbey, Rognes,** and **Château La Coste,** take the D14. Again, this is a small, picturesque road. The main autoroute north is the A7.

The A52 motorway takes you south out of Aix toward Aubagne and the **French Foreign Legion Museum.**

AIX-EN-PROVENCE, THE LUBERON, AND THE GORGES DU VERDON

AIX-EN-PROVENCE

Bus

Bus is the only way to explore the Pays d'Aix-en-Provence without a car. Buses run from the **central bus station** on boulevard Victor Coq in Aix (tel. 08 09 40 04 15). The main routes you may want to use are **line 140,** to Vauvenargues, approximately one bus every hour 7am-8pm; **line 110,** Le Tholonet, approximately one bus every hour 7am-7pm; and **No. 250,** Rognes and Le Roque-d'Antheron, approximately one bus every hour 6am-9pm. You can find precise times and maps for these routes online at www.lametropolemobilite.fr.

The Luberon

Without exception, all the villages in the Luberon in this guidebook merit a lengthy visit. Many visitors will never find time to see sights such as the Pont Julien Roman bridge let alone the Fort de Buoux, marooned as it is in the depths of the Grand Luberon. Instead, they will wander contentedly along cobbled village streets lined with art galleries and small boutiques. Sightseeing here often consists of no more than a leisurely lunch and a look at the view over an old stone wall.

Everyone, of course, has a favorite village. Some are drawn to Lourmarin, where Peter Mayle, the late author of *A Year in Provence* lived, others to Bonnieux and its views of Mont Ventoux. The fun of visiting the Luberon is to explore and fall in love with one of the villages yourself. Those who can tear themselves away from the pleasures of village-hopping will find biking and walking trails crisscrossing the mountains.

ORIENTATION

The **Luberon Regional Nature Park** is located some 40 km (25 mi) east of Avignon and 40 km (25 mi) north of Aix-en-Provence, right in the center of Provence. The Luberon is easy to explore, and the most well-known villages are all within a 30-minute drive of one another, from swanky **Gordes** and ocher-tinged **Roussillon** to the bustling market town of **Apt,** to charming, picturesque villages such as **Bonnieux** and **Lourmarin.** To the east are the famous **lavender fields of Haute Provence.**

GORDES

Nicknamed "the Lighthouse of the Luberon" thanks to the way the central tower watches over the plain below, Gordes is easy to fall in love with. From its Renaissance château to the steep, narrow, cobbled streets (calades) that wind between the houses, the village has an unparalleled seductive allure.

Sights
Gordes Viewpoint

D15; www.gordes-village.com; always accessible; free

As you approach Gordes on the route de Cavaillon (D15), look out for a small slip road (chemin de Bel Air) about 1 km (0.6 mi) before the town. This is the best place to stop if you want to take photos of the village. Walk around 15 m (49 ft) up the road and you'll come to a large rock, upon which inevitably a few tourists will be standing with selfie sticks. The reason for all the fuss is the iconic view of one of Provence's most famous villages. The houses seem to defy gravity by clinging to the precipitous cliffside, and it's easy to understand why the village was so successful in repelling invaders over the centuries.

Château de Gordes

Place du Château; tel. 04 90 72 02 68; www.chateaudegordes.com; June 14-Nov. 13 daily 10am-1pm and 2pm-6pm, Nov. 14-June 13 Wed.-Sun. 9:30am-12:30pm and 1:30pm-5pm; €6

Dating back to 1031, the Château of Gordes dominates the Luberon skyline. In that time it's been used as a prison, a storehouse, and a garrison. These days, it welcomes tourists

The Luberon

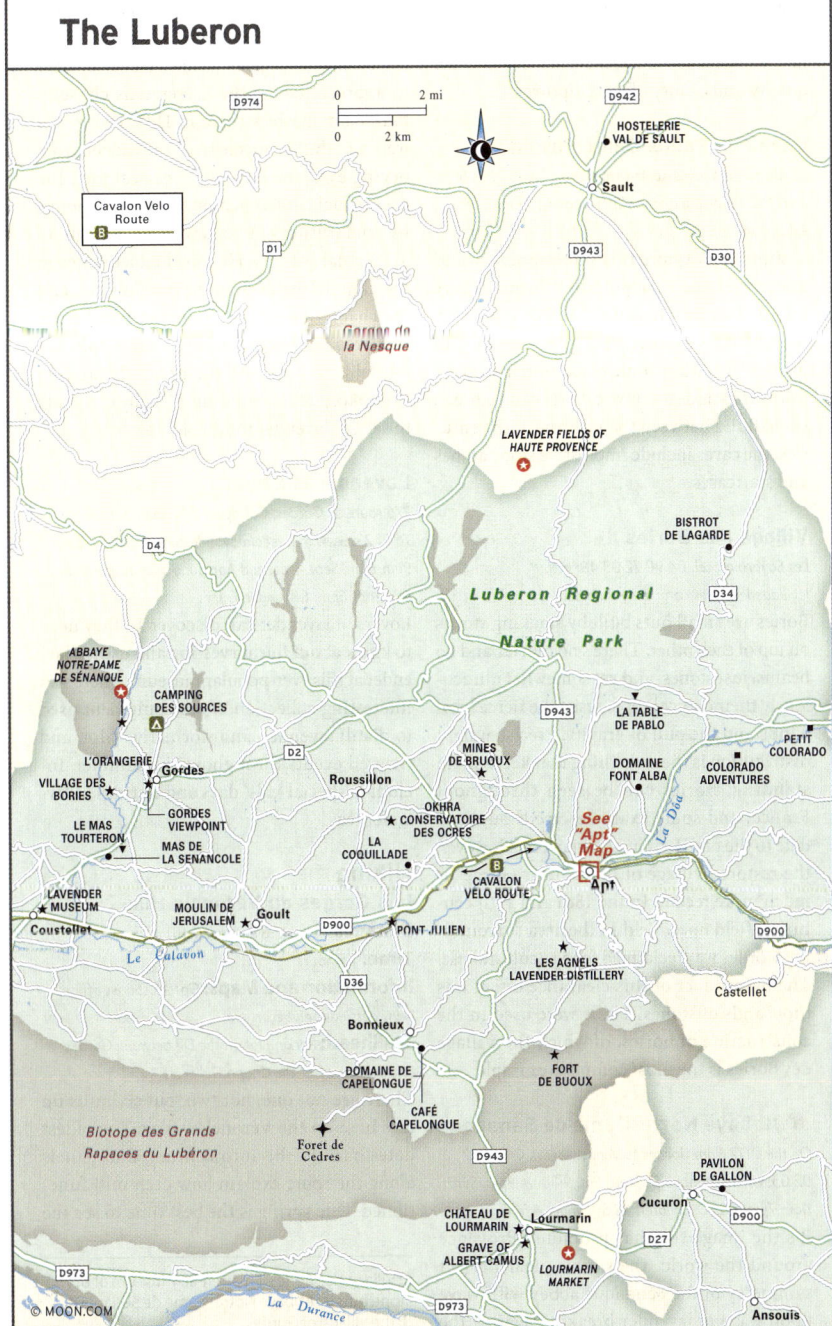

0   2 mi

0   2 km

Cavalon Velo Route

D974
D1
D942

HOSTELERIE
VAL DE SAULT

Sault

D943   **D30**

Gorges de
la Nesque

LAVENDER FIELDS OF
HAUTE PROVENCE

D4

BISTROT
DE LAGARDE

Luberon Regional

Nature Park

D34

ABBAYE
NOTRE-DAME
DE SÉNANQUE

CAMPING
DES SOURCES

LA TABLE
DE PABLO

D943

L'ORANGERIE

D2

Gordes

MINES
DE BRUOUX

DOMAINE
FONT ALBA

COLORADO
ADVENTURES

PETIT
COLORADO

VILLAGE DES
BORIES

Roussillon

OKHRA
CONSERVATOIRE
DES OCRES

LE MAS
TOURTERON

GORDES
VIEWPOINT

LA
COQUILLADE

See
"Apt"
Map

MAS DE
LA SENANCOLE

LAVENDER
MUSEUM

MOULIN DE
JERUSALEM

Goult

CAVALON
VÉLO ROUTE

Apt

Coustellet

Le Calavon

D900

PONT JULIEN

D900

LES AGNELS
LAVENDER DISTILLERY

Castellet

D36

Bonnieux

DOMAINE DE
CAPELONGUE

FORT
DE BUOUX

Biotope des Grands
Rapaces du Lubéron

CAFÉ
CAPELONGUE

Foret de
Cedres

PAVILON
DE GALLON

D943

Cucuron

D973

CHÂTEAU DE
LOURMARIN

Lourmarin

D27   **D900**

GRAVE OF
ALBERT CAMUS

LOURMARIN
MARKET

© MOON.COM

La Durance

D973

Ansouis

to art exhibitions within its historic walls. Be sure to take a look at the Renaissance chimney, which is remarkable for its size and the quality and beauty of its sculptures.

Caves du Palais Saint-Firmin

Gordes; http://caves-saint-firmin.com; Apr. 8-Nov. 5 daily 10:30am-1pm and 2:30pm-6pm; €6, ages 5-15 €4.50, under 5 free

Visitors can explore this subterranean world underneath the Palace of Saint-Firmin. There are 50 caves that descend 20 m (65 ft) beneath the palace. Together they give insight into parts of life that took place underground; features include an olive oil press and a bread oven. Build between the 11th and 18th centuries, the caves include impressive stone arches and staircases.

Village des Bories

Les Savornins; tel. 04 90 72 03 48; http:// levillagedesbories.com; daily 9am-sunset

Bories are small huts built by stacking stones on top of each other. There's no mortar and no beams, just stones, and yet somewhat miraculously their conical roofs resist the fierce local winds and the pull of gravity. Precise weight distribution is the only thing that keeps them standing. Bories can be seen throughout France, and some examples are thought to date to the pre-Roman Ligurian era. However, the restored village of Bories near Gordes is much more recent. In the 18th and 19th centuries, field upon field in the area surrounding Gordes was reclaimed for agricultural use. The by-product of this clearance effort was thousands of stones. These were used in the construction of bories, of which the Village des Bories is the most complete example.

★ Abbaye Notre-Dame de Sénanque

On the D177 from Gordes to Venasque; tel. 04 90 72 02 05; www.senanque.fr; Mon.-Sat. 9:30am-7pm, Sun. 11am-7pm; €8.50, children €3.50, under 6 free

It's the image that has promoted Provence around the world: the simple, near-perfect symmetry of the Sénanque abbey with wave upon wave of lavender breaking against the stone walls. The abbey shelters in an isolated valley and is approached by a narrow, winding road, which gives visitors plenty of time to appreciate why the Cistercians chose to found a monastery there in 1148. Fed by the waters of the Senancole river and hidden from prying eyes, the location is perfect for a life of spiritual contemplation. Today the monks welcome visitors. A visit, which is brought to 3D virtual life by a Histopad tablet, takes in the church, the dormitory, the cloisters, and the chapter house. Lavender essence, honey, and even liquors produced by the monks are available to buy, with the proceeds funding the upkeep of the building. Visit by early July to see the lavender in full bloom.

Lavender Museum

276 route de Gordes, Cabrieres d'Avignon; tel. 04 90 76 91 23; www.museedelalavande.com; July-Aug. daily 9am-6pm, Sept.-Dec. and Apr.-June daily 10am-noon and 2pm-6pm; €8, under 15 free

Lovers of lavender can discover all they need to know about the harvesting and uses of lavender at this ever-popular museum. There's an interesting collection of old equipment used to distill lavender, an informative film, and themed events throughout the summer, including special kids' days and tastings of lavender teas.

Hiking
Les Gorges du Véroncle Hike

Distance: *4 km (2.5 mi) round-trip*
Time: *2.5 hours*
Information and Maps: *Gordes tourist office; www.destinationluberon.com*
Trailhead: *Parking lot on the D2 between Gordes and Joucas, after the Chapo Gallery*

There are not one, not two, but six mills on this hike up the Véroncle valley. The oldest dates back to the 16th century, and panels along the route explain how each mill functioned. Late spring is the best time to see the

---

1: Village des Bories, an open-air museum of dry stone huts 2: Abbaye Notre-Dame de Sénanque 3: the village of Gordes

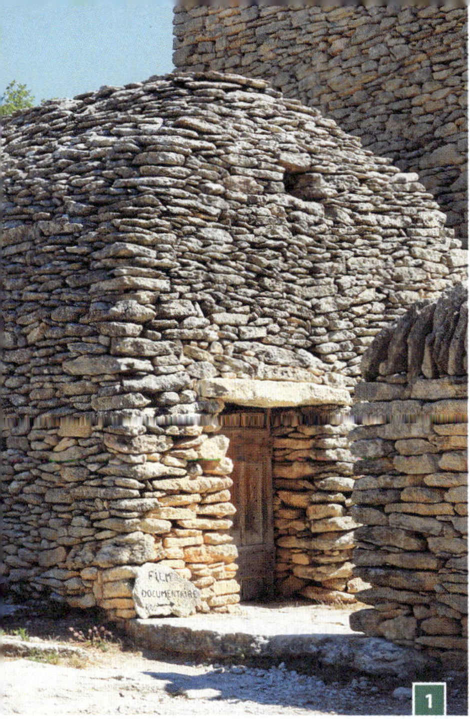

1

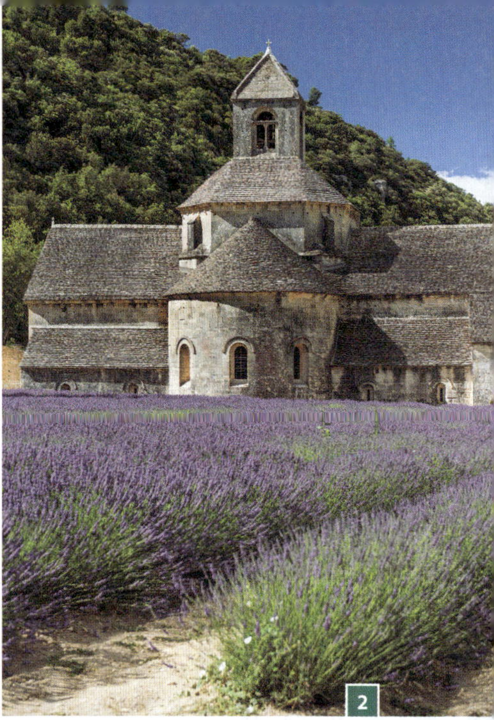

2

3

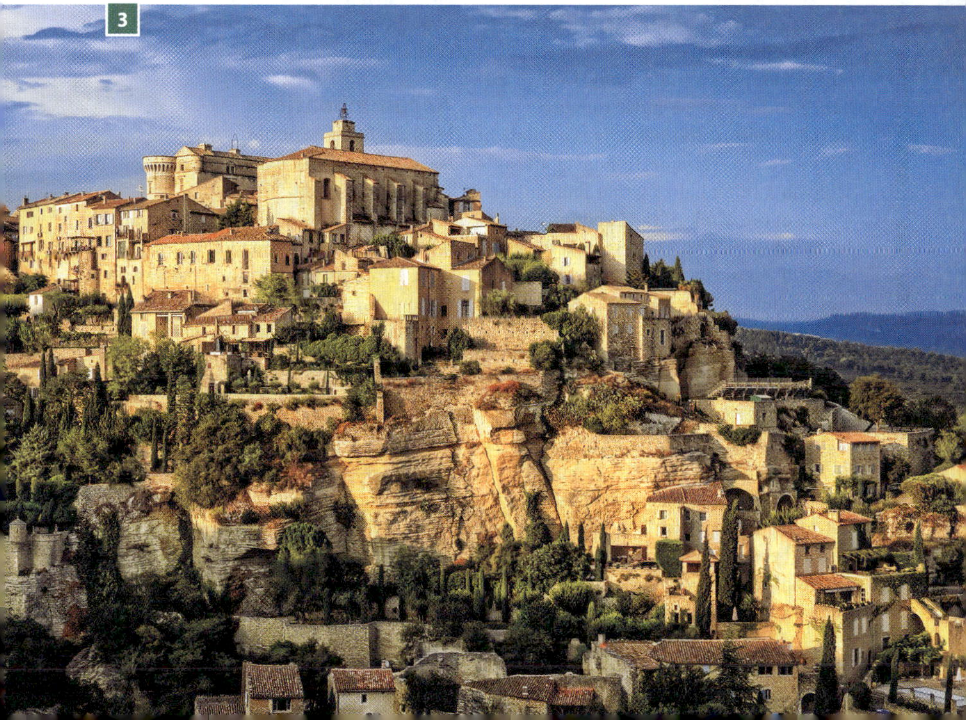

Véroncle river in full flow together with its picturesque waterfall. At other times of year the water can dry up. In places the walk can be slippery and the gorge narrow, so wear good walking shoes. The walk is 4 km (2.5 mi) there and back, with a 430-m (1,410-ft) incline. Look for the yellow signs.

Festivals and Events
Festival Les Soirées d'Été

Théâtre des Terrasses; tel. 04 90 72 65 05; www. festival-gordes.com; first week in Aug.; from €35 per performance

The performances are varied at the Festival Les Soirées d'Été (including jazz, piano, and acoustic guitar), but one thing never changes: The intimate 500-seat theater overlooking the Luberon amplifies the effect of the music, sweeping the audience away on an unforgettable emotional journey just as the stars blink into light overhead.

Food
La Trinquette

rue des Trapacelles; tel. 04 90 72 11 62; www.facebook. com/LaTrinquette; Wed.-Mon. 5pm-9:30pm; €16-30

Honeymooners should book well in advance to secure the most romantic table in the Luberon, and possibly the whole of Provence. La Trinquette boasts a single table on a Juliet balcony that juts out onto a vista of the whole of the Luberon. For unlucky couples who can't secure the table with the view, the excellent food is ample consolation.

L'Orangerie

Hotel Bastide du Gordes, 61 rue de la Combes Airelles; tel. 04 90 72 12 12; Mon.-Sat. 12:15pm-2:15pm and 7pm-9pm, Sun. brunch 11:45am-2:30pm; lunch €36-69, dinner from €130

A meal at L'Orangerie is not cheap, but it's definitely very pretty. The views from the terrace here are some of most beautiful in the Luberon. As you would expect, service is attentive and the food excellent.

Le Mas Tourteron

chemin Sainte Blaise; tel. 04 90 72 00 16; www. mastourteron.com; Mar. 2-Oct. 29 Thurs.-Fri. and Mon. 7:15pm-9:15pm, Sat.-Sun. 12:30pm-2pm and 7:15pm-9:15pm; €65

This no-nonsense family restaurant celebrates the strong and vibrant flavors of Provençal cuisine. Expect hearty fare with plenty of garlic, tomatoes, and eggplant, served on a beautiful, sweet-scented wisteria-covered terrace. The restaurant is located a couple of kilometers outside Gordes in a bucolic environment.

Coustellet Farmers Market

Place du Marché, Maubec; www. marchepaysandecoustellet.com; Wed. 5pm-7:30pm, Sun. 7:30am-12:30pm

Indisputably, the best fresh produce market in the Luberon takes place every Sunday morning at the Coustellet crossroads just a couple of kilometers from Gordes. The Wednesday evening affair is a smaller, more intimate version of the main event, a mere dress rehearsal for Sunday, when every farmer worth his bulb of garlic pitches up and sells his best fresh vegetables.

Accommodations
Mas de la Senancole

Hameau Les Imberts; tel. 04 90 76 76 55; www.mas-de-la-senancole.com; €180 d

At first glance this hotel set on the plain beneath Gordes seems nondescript. However, walk through the entrance hall and you come upon a ravishing swimming pool set in a beautiful mature garden. There's a good choice of room sizes, and some have little terraces that open onto the garden.

Mas de Romarins

2213 route de Sénanque; tel. 04 90 72 12 13; www. masromarins.com; €250 d

Set on a cliff top opposite the village, this is probably the hotel with the nicest view of Gordes. The rooms are simple but comfortable and there's a nice pool and restaurant.

The Wines of the Luberon

The statistics: Some 3,410 hectares (8,430 acres) of vines enjoy 130 days of sunshine, hot days, and cold nights, as well as the drying effect of the mistral wind and a clay limestone soil that is ideal for viticulture. The main grape varieties are Syrah and grenache for the reds, and rosé and grenache blanc and vermentino for the whites.

As wine has morphed from an agricultural industry into a tourist activity, upmarket tasting rooms have sprung up over the region, offering tours, tastings, and courses. The other trend has been a shift toward organic production and a continued concentration on rosé, which remains the region's best seller (89 percent of the wine produced is rosé). The quality of the wine produced has definitely improved as well, but with this improvement prices have risen. For a decent bottle of rosé, prices now hover around €10. Whites and reds can easily reach €15.

grapes in a vineyard

THE FIVE-STAR TREATMENT

- **Aureto winery** (www.aureto.fr) on the road between Apt and Cavaillon.

- **Domaine de La Fontenille** (www.domainedefontenille.com) in the southern Luberon, just outside Lourmarin.

- The extravagant tasting room of **Domaine La Cavale** (www.domaine-lacavale.com) just outside Lourmarin.

SMALLER DOMAINES AND ORGANIC PRODUCTION

- **Château Canorgue** (http://châteaulacanorgue.com) in Bonnieux formed the dreamy backdrop to the Ridley Scott and Russell Crowe adaptation of Peter Mayle's novel *A Good Year*.

- **Domaine de La Verrière** (www.domainedelaverriere.fr) outside Goult makes an excellent viognier for less than €10.

- One of the region's rare natural wine producers, **Les Tulles Bleus** (www.lestuilesbleues.fr), is just outside Cucuron.

Camping des Sources

route de Murs, Gordes; tel. 04 90 72 12 48; www. campingdessources.com

Just outside of Gordes in the heart of the Luberon Natural Park, this campsite has pitches, chalets, and mobile homes. There's a pool and restaurant, and in the summer, live music. It's a beautiful place to stay.

Getting There and Around

Once in Gordes, the château and the Saint-Firmin caves are in the center of the village and can be visited **on foot.** For the other sights listed you will need a car.

Car

Gordes, Roussillon, and **Goult** are roughly 10 km (6 mi) apart. Together, they form a triangle with Gordes at the most northerly point. To reach Gordes by car, take the D2 from **Saint-Saturnin-les-Apt;** the D15 from **Murs;** the D56 from **Goult;** and the

D177 from **Venasque.** Once you arrive, park in the village **parking lot,** which is a short 2-minute walk from the center. There is a €5 parking charge.

Traveling from **Avignon,** the village is reached by taking the A7 south and then the D900. The journey of 50 km (31 mi) takes around 50 minutes.

Bus

From Apt, there are four buses a day to Gordes. **Line 17** departs from outside the tourist office in **Apt** (788 avenue Victor Hugo), arriving at the square in front of the château in Gordes. The journey takes 50 minutes and costs €2.60 (www.apt.fr/Les-transports.html).

GOULT

Goult is noticeably quieter than neighboring Roussillon and Gordes. While no village in the Luberon can be described as undiscovered, Goult is at least understated. Year-round the village feels calm, and in summer it's good a good place to keep in mind if you want to escape from the bustle. The pedestrian streets encourage a slow, joyful meandering, and all paths eventually lead to the windmill that crowns the village.

Sights
Moulin de Jerusalem

Goult; tel. 04 90 72 38 58; www.goult.fr/patrimoine; daily sunrise-sunset; free

No one is quite sure how this windmill, which dates back to the end of the 17th century, was given its name. One theory is that it was named after the nobles from the Goult area who headed off to the Crusades centuries earlier. Although you can't actually see inside the windmill, it's easy to visualize how the mistral wind (which frequently nears 100 kph/62 mph) could power the milling process. The view of the Luberon from adjacent to the windmill is one of the best and worth the climb up through the picturesque village streets.

Conservatoire Terrasses de Culture

Goult; tel. 04 90 72 38 58; www.goult.fr/patrimoine; daily sunrise-sunset; free

Next to the Jerusalem windmill are signs indicating the way to the Conservatoire Terrasses de Culture. Maintained as a link to the region's agricultural past, these ancient terraces demonstrate how man tamed the harsh Provençal landscape through the use of abundantly available stone. Dating the terraces is difficult. The Romans used terracing to cultivate olive trees and vines, and some of the terraces might date back a couple of thousand years. However, it is likely that they fell in and out of use over the centuries as demand for land changed.

Food and Accommodations
Café de la Poste

rue de la Republique; daily 7am-10pm; €15-25

This café was made famous by Peter Mayle in his best-selling memoir *A Year in Provence.* The tables are now filled with tourists rather than Mayle's celebrated paysanne (local farmers). The food is simple but good, and the ambience convivial. The terrace is the unofficial social heart of the village.

Le Carillon

avenue du Luberon; tel. 04 90 72 15 09; www. restaurant-goult.com; Thurs.-Mon. 12:15pm-2pm and 7:15pm-8:30pm; €25-44

On the main square in Goult, the Carillon is a favorite with villagers. The menu is eclectic, with local ingredients complemented by the occasional more exotic dish such as slow-cooked cod in a ginger and coconut vinaigrette, served with sweet potatoes and caramelized banana chips.

La Bartavelle

29 rue Cheval Blanc; tel. 04 90 72 33 72; https:// labartavellegoult.com; Tues.-Sat. 7:30pm-8:30pm, reservations required; €52

La Bartavelle has been a runaway success of a

1: Moulin de Jerusalem in Goult **2:** colorful village houses in Roussillon

1

2

restaurant, and you need to book a couple of days in advance to be sure of getting a table. The setting, in the back streets of Goult, is idyllic, and the food resolutely seasonal and local. Visit in spring for delectable fresh asparagus dressed with parmesan shavings.

Notre-Dame des Lumières

Lumières; tel. 04 90 72 22 18; www.notredamedelumieres.com; €144 d

There are no hotels in Goult. However, a couple of minutes' drive from the village in the hamlet of Lumières, this former convent is a good base for exploring the Luberon villages. There's a large pool overlooked by a restaurant.

Getting There and Around

Once in Goult, the village is almost entirely pedestrianized.

Car

Goult is located off the D900, which runs between Apt and Coustellet. There is a large **parking** area opposite Café de la Poste.

Traveling from **Avignon,** the village is reached by taking the A7 south and then the D900. The journey of 50 km (31 mi) takes around 50 minutes.

Bus

Transvaucluse bus **line 15** (www.apt.fr/Les-transports.html) runs seven times a day between Avignon and Apt, stopping at the **Lumières hamlet** beneath Goult. Allow at least 30 minutes to walk uphill to the village.

ROUSSILLON

The red-orange cliffs of Roussillon are a sensational sight. A great shard of color plunges away from the villages into a deep canyon, which visitors can explore by taking the Sentier des Ocres. The village itself is a pleasing collection of vibrant-colored houses. The streets are pedestrianized and filled with art galleries and shops that cater largely to tourists. The best time to experience Roussillon is either at daybreak or sunset.

Sights
Okhra Conservatoire des Ocres

D104; tel. 04 90 05 66 69; http://okhra.com; Jan. by appointment, Apr.-June daily 10am-1pm and 2pm-5pm, July-Aug. daily 10am-7pm, Sept.-Oct. daily 10am-6pm, Nov.-Dec. daily 2pm-5pm; €9.50, under 18 €5.50, under 10 free

A cooperative society dedicated to preserving and sharing know-how relating to the use of color could not have chosen a better base than Provence. A visit to this old ocher factory is just the start of a journey into the world of color. Year-round presentations and workshops offer opportunities to delve into a creative artistic world.

Sentier des Ocres

Parking des Ocres; tel. 04 90 05 60 25; http://otroussillon.pagesperso-orange.fr/sentier.html; summer daily 9am-1 hour before sunset, out of season 10am-1 hour before sunset, hours may fluctuate with number of visitors; €2.50, under 15 €1.50, under 10 free

This walk, which starts adjacent to the main village car park (Parking des Ocres), is a pleasant one that winds through the cliffs and rocky crags, which at certain times of day glow a vibrant orange, calling to mind Mars as imagined in movies. Allow 30 or 60 minutes to complete the visit, depending on whether you choose the 2-km (1.2-mi) or 4-km (2.5-mi) route.

Hot-Air Ballooning

Simply the most sensational way to enjoy the Luberon is by hot-air balloon, floating silently from perched village to perched village, rising and falling with the birds. Flights depart at dawn, entering the park at the moment when nature rather than man dominates. In summer the land below looks like an impressionist painting, with deep swathes of purple competing with vibrant splashes of yellow, as the balloon drifts over lavender and sunflower fields. In spring, expanses of red poppies carpet the land, and white blossoms from almond and cherry trees blow in the air. In autumn a rich gold and red coat dresses the fields

Ocher

the Petit Colorado ocher mine near Apt

When science can't explain a phenomenon, myth and religion step in. For centuries, the inhabitants of Apt, Gargas, Roussillon, and Rustrel couldn't understand why the landscape around them glowed almost blood red, and so they invented fanciful stories. There's the one about the forlorn lover casting herself off the cliffs and another about the Archangel Gabriel slaying a battalion of fallen angels and their blood staining the ground red.

Buy a local a pastis and he or she will probably indulge you with a different tall tale. Of course, you may get lucky and hit upon an amateur geologist who'll explain that more than 100 million years ago Provence was covered by a sea. The retreat of the water left sandy deposits and clay, both of which had absorbed iron from the seawater. Over the millennia, they combined to produce ocher, the unique color of which is derived from the high concentration of iron. (Personally, I prefer the biblical version with angels swarming in combat over the Luberon hills.)

In any event, extraction of ocher began in 1780 in Roussillon, and by 1925 there were 22 mines in the area, sending the finished pigment in more than 20 different shades worldwide for use in paints and industrial products. The introduction of synthetic coloring agents after the Second World War led to a drastic decline in the industry. However, some ingenious and very Gallic uses were still found for ocher and its industrial legacy. Until mushroom cultivation in the otherwise abandoned mines of Bruoux was discontinued, it was possible to find an ocher-inspired menu of goat's cheese (with an ocher rind) and Bruoux Mine mushrooms on toast. Local rumor has it that an overindulgent diner's stomach once glowed in the dark.

WHERE TO SEE OCHER

· **Okhra Conservatoire des Ocres:** A society dedicated to sharing knowledge about using and preserving color, including ocher (page 154).

· **Sentier des Ocres:** Beware of ocher stains on this fascinating walk through an ocher-dominated landscape (page 154).

· **Restaurant David:** Eat dinner with a view of Roussillon's ocher cliffs (page 156).

· **Mines de Bruoux:** Not for the claustrophobic, a visit to this former ocher mine is a testament to what man can create without machines (page 158).

· **Petit Colorado:** Two signposted walks take you through the site of an old open-air ocher mine (page 158).

of vines as the Luberon experiences a fall to rival New England.

Montgolfiere Luberon Vol-Terre

1066 routes de Gaillanes; tel. 06 03 54 10 92; www. montgolfiere-luberon.com; flights by reservation; from €250/person

This company offers hot-air balloon rides out of Roussillon. Guests help to prepare the balloon for flight. The average altitude during the trip is 1,000 m (3,280 ft), and the flight time is 1 hour. Group sizes vary between 4 and 15 people. After the ride, enjoy a glass of champagne.

Food and Accommodations
Café les Couleurs

Place de la Mairie; tel. 04 90 06 12 83; daily 10am-10pm; €18-28

Right in the heart of Roussillon, this café is a convenient place to stop. On the menu are well-prepared French bistro classics, such as coq au vin (chicken in wine) with fresh tagliatelle. The service can be a little brusque in the high season, as the staff struggle to keep up with the influx of tourists.

Restaurant David (Omma)

Place de la Poste; tel. 04 90 05 60 13; www. leclosdelaglycine.fr; Thurs.-Sat. 12:15pm-1:30pm and 7:15pm-8:45pm (last reservation), Sun. lunch only; €25-34

Nearly every table in Restaurant David's elegant, light-filled dining room has a captivating view of the ocher cliffs of Roussillon. Dishes might include guinea fowl served with a crunchy honey-and-saffron polenta, or monkfish with a ravioli of artichoke and chorizo. It's the best restaurant for a serious meal in Roussillon.

Clos de la Glycine (Omma)

Place de la Poste; tel. 04 90 05 60 13; www. leclosdelaglycine.fr; €220 d

Clos de la Glycine is a charming village-center hotel. Rooms are well furnished and comfortable. If available, opt for a room overlooking the cliffs.

La Coquillade

Hameau Le Perrotet, Gargas; tel. 04 90 74 71 71; www. coquillade.fr; €600

Five minutes away by car from Roussillon, this luxury hotel has been created out of an old Provençal hamlet. Most of the rooms are located in their own individual stone buildings with accompanying terraces. Bike hire, three restaurants, a pool, a vineyard, extensive views, and a spa add to the reasons to stay, though the prices are high.

Getting There and Around
Car

Roussillon is reached by taking the D104 from **Goult,** the D169 from **Joucas,** or the D227 from **Saint-Saturnin-les-Apt. Parking** can be difficult in high season. Be prepared to walk for up to 10 minutes from the parking lot to the village. The village is pedestrianized.

Traveling from **Avignon,** the village is reached by taking the A7 south and then the D900. The journey of 50 km (31 mi) takes around 50 minutes.

Bus

Transvaucluse line 17 (www.apt.fr/Lestransports.html) departs four times a day from Apt to Roussillon. The journey takes 25 minutes and costs €2.60.

APT

The town of Apt is not as instantly appealing as many of the neighboring villages, but it is pleasant enough, with a long pedestrianized shopping street and plenty of cafés with large terraces shaded by plane trees. It comes alive on Saturday mornings for one of the most vibrant, and least touristy, of the region's markets.

As a base to explore the whole region, it's geographically ideal, as it is adjacent to the main roads heading in every direction. It is also home to the region's main tourist office (788 avenue Victor Hugo) as well as the Maison du Parc information center. A few kilometers outside of town near Rustrel, the major sight is the Colorado Provençal, where

Apt

open-air ocher mining has created a luminous red-orange landscape filled with seemingly alien rock formations.

Sights
La Maison du Parc

60 Place Jean Jaures; www.parcduluberon.fr; Mon.-Fri. 9am-noon and 2pm-5:30pm; free

All sorts of great information related to the Luberon is available at the Maison du Parc, including an introduction to the region's most famous bird, the Egyptian vulture, and an explanation of the region's geology, which is closely linked to the retreat of the

Aptian Sea from the area 120 million years ago.

Cathedral Sainte-Anne

187 rue des Marchands; tel. 06 88 47 23 42; daily 8:30am-6pm

The domed roof of the Sainte-Anne Cathedral stands out from afar, its golden cross glinting in the sunshine. Like Saint-Sauveur Cathedral in Aix, the building is a mix of styles. The earliest evidence of construction dates back to the 5th century, when a church was built on the old Roman Forum. A side nave was added in 1200, and the chapel of Sainte-Anne in 1660.

Perhaps the most exceptional feature is the 14th-century stained glass in the apse depicting Mary with the infant Jesus and Anne.

Les Agnels Lavender Distillery

route de Buoux; tel. 04 90 04 77 00; www.lesagnels. com; Apr.-Sept. Mon.-Fri. tours in English at 3pm; distillery visit €7, lavender field and distillery visit €15 (June-early Aug. only)

Probably the best lavender-themed experience in the region, Les Agnels distinguishes itself by being a working lavender distillery, allowing visitors to observe the process of extracting the essential oil from the flowers. The tour of the distillery is preceded by a short film on the history of lavender in the region, and there is a shop where you can purchase all things lavender, including soap, honey, bath salts, and the essential oil itself.

Mines de Bruoux

1434 route de Croagnes, Gargas; tel. 04 90 06 22 59; www.minesdebruoux.fr; Mar.-Nov. daily 10am-6pm, all visits guided and by reservation only; €9.50, ages 6-17 €7.50, under 6 free

The Bruoux mine is the place to appreciate the industrial scale of ocher mining in the area. It's almost impossible to imagine how man could have created this immense underground network of caves. In all there are 40 km (25 mi) of hollowed-out galleries. Men worked four hours a day, six days a week, using only pickaxes to carve the route. So intensive was the work that one man would typically change pickaxe three or four times a day. In places the ceilings are 15 m (49 ft) high and reminiscent of an underground cathedral.

Sports and Recreation
Petit Colorado

D22, Rustrel; tel. 04 90 75 04 87; https:// coloradoprovencal.fr; daily, hours vary according to seasonal daylight, reserve in advance on website, closed Jan.; parking €6-8

Taking a walk in the Colorado Provençal can be disorienting. From the olive green of the Luberon, you enter a landscape of vibrant reds and oranges. The two signposted walks take you through the site of an old open-air ocher mine. The first walk is 2 km (1.2 mi) and takes 40 minutes, and the second walk is 4 km (2.5 mi) and takes 1.75 hours. There are strange rock formations aplenty, reminiscent (on a much smaller scale) of Colorado in the United States.

Cavalon Velo Route

From Coustellet to Castellet; www.veloloisirprovence. com

This ride is billed by the tourist office as the ultimate family cycling outing: no cars, no gradient. Even if you are not a cycling enthusiast, the appeal is obvious. The path runs for some 37 km (23 mi) from Coustellet to the foot of Castellet, along the route of an old railway track. If you pick up the route at Apt, there is a choice of heading east into wilder and less inhabited countryside, or west toward Coustellet. There is a 1-km (0.6-mi) section on the way to Coustellet that runs alongside a small road. You can park and pick up the route in Apt adjacent to the tourist office at 788 avenue Victor Hugo.

Colorado Adventures

Quartier Nôtres Dames des Anges, Rustrel; tel. 06 78 26 68 91; www.coloradoadventures.com; July-Aug. 9am-7pm, last entrance 5pm, out of season check website for hours

For those who are tired of lazing by the swimming pool, Colorado Adventures offers the opportunity to spend a few hours swinging from branches and hanging from ropes. There's a range of color-coded routes through the treetops, with security assured by a metal guide rope, harness, and climbing clips.

Shopping
Atelier Buisson Kessler

140 rue de la Republique; tel. 04 90 04 89 61; www. atelierbuisson-kessler.com; Tues.-Sat. 10am-6pm

This is a charming family-run ceramics shop inspired by the natural palette of the Luberon. Plates and bowls are multicolored and often dotted with wild patterns. The ceramics are guaranteed to liven up any table setting, and they make great mementos of a holiday in Provence.

Marcel Richaud

112 quai de la Liberté; tel. 04 90 74 43 50; Tues.-Sat. 9:45am-12:15pm and 2:30pm-7pm

Apt is a world leader in the production of candied fruit (fruit confit). The process of preserving fruit by extracting the water and injecting a sugar solution that covers and conserves it was discovered in the late Middle Ages, and the town hasn't looked back since. This well-established shop sells everything from small bags to nibble on to beautifully arranged displays on trays.

Saturday Morning Market

Place Gabriel Péri; tel. 04 90 74 00 13; www.apt.fr/ Les-marchA-c-s.html; Sat. 8am-12:30pm

Apt market is busy, bustling, and authentic. Excellent fresh fruit and vegetables draw the locals from miles around. In the winter, the earthy scent of truffles drifts in the cold, frosty air.

Food
Au Platane

16 Place Jules Ferry; tel. 04 90 04 74 36; www. restaurant-apt.fr; Mon.-Sat. noon-2pm and 7:30pm-9:30pm; €25

Au Platane is a traditional French restaurant with a lovely terrace set beneath a plane tree. It is definitely one for the summer months. The menu is small, but caters to all.

La Table de Pablo

Les Petits Clements, Villars; tel. 04 90 75 45 18; www. latabledepablo.com; Thurs.-Tues. noon-1:30pm and 7:30pm-9pm, Sat. and Thurs. dinner only; €38-55

La Table de Pablo nestles a few kilometers into the countryside outside Apt. There's a pleasant, shady terrace bordered by greenery, and summer dining is accompanied by a chorus of cicadas. The food is excellent, with a seasonal menu nicely balanced between meat and fish; accompanying sauces are rich and intense.

Bistrot de Lagarde

RD34 Lagarde d'Apt, Lagarde d'Apt; tel. 04 90 74 57 23; http://lebistrotdelagarde.free.fr; Wed.-Sun. 12:30pm-1:30pm and 7:30pm-9:30pm; €50

For years, this bistro hidden amid fields of lavender outside Apt has been a word-of-mouth success. People think nothing of traveling an hour or so to taste the inventive cuisine of Lloyd Tropeano. Dishes include a medley of heirloom tomatoes presented in various mouthwatering ways.

Accommodations
Le Couvent

36 rue Louis Rousset; tel. 04 90 04 55 36; www. loucouvent.com; €110 d

As the name suggests, this bed-and-breakfast occupies the building of an old convent in Apt. Arched ceilings and ornate rugs create a warm ambience. There's a lovely pool set in a walled courtyard. All in all, it's the perfect antidote to dull chain hotels.

Domaine Font Alba

chemin des Coulets; tel. 04 90 06 12 83; www. chateaufontalba.com; €125 d

Vineyard hotels in France can be a little over the top. If they've got a few vines and a château, the owners too often think they can charge a fortune. Domaine Font Alba's ethos is the opposite. There are five rooms, each named after a grape variety, and one suite in the converted old sheepfold.

Getting There and Around

Apt is a small town, and the easiest way to get around is on foot. Walking from one side of the town to the other takes no more than 15 minutes.

Car

Apt is an important transport hub. It can be reached in under 20 minutes by car on the following roads: the D3 from **Bonnieux,** the D22 from **Rustrel,** the D48 from **Saignon,** the D900 from **Coustellet,** the D113 from **Buoux,** and the D943 from **Saint-Saturnin-les-Apt. Aix-en-Provence** is 55 km (34 mi) away, and **Avignon** is 53 km (33 mi). Both journeys take just over an hour.

Bus

There are regular buses to Apt from Aix-en-Provence (3 times a day), Avignon (8 times a day), and L'Isle-sur-la-Sorgue (5 times a day). The buses arrive at the **gare routiere,** avenue de la Liberation. The lines run in both directions. Timetables can be downloaded at www.apt.fr/Les-transports.html.

BONNIEUX

Bonnieux is all about the views and forgetting about the inconvenience of building a village on the side of a cliff face. Approached from the south on the D36, the silhouette of the village dominates the skyline. The cypress trees, church spire, and Luberon hills in the background are the perfect subjects for painters' canvases. Another highlight is the view to the north from the upper level of Bonnieux out toward Mont Ventoux. Navigating around Bonnieux, though, is difficult. There is a cluster of shops and restaurants at the top of the village. The central level hosts the supermarket. Farther down the hillside is the boulangerie and more shops and restaurants. Steps and narrow roads link the levels, so a certain degree of physical fitness is a prerequisite for fully exploring the village.

Sights
Pont Julien

route de Pont Julien; free

Compared to other Roman sites in Provence, the Pont Julien can seem insignificant. Its three arches span a miserly 50 or so meters (164 ft) of the Cavalon river and its presence is heralded only by a small parking area. Yet, if you visit early in the morning, you will often find yourself alone and able to contemplate in solitude an evocative monument of Roman Provence. The bridge was built in 3 BCE on the Via Domitia, a key road that connected Roman territories in France with Italy. In the shadow of the Luberon, without a single visible building, it's momentarily easy to step

back in time and imagine horses and carts rolling over the arches. Over the years the horses morphed into cars and, quite incredibly, the bridge still carried traffic up until 2005.

L'Église Louise Bourgeois

rue Aristride Briand; tel. 04 90 75 80 54; www. egliselouisebourgeois.com; July 15-Aug. 14 and Sept. 1-11 daily 10am-1pm and 3pm-6pm; free

Louise Bourgeois was a surrealist artist working during the 20th century. Her output was frequently provocative and linked with the feminist movement. In 1998, toward the end of her career, she installed various works in this deserted chapel in the hills above Bonnieux. Among them is a cage-like confessional, an iron cross that grows hands, and a mother suckling her child in an airtight container. The power of the art to shock is augmented by the religious setting, and viewing the works can be disconcerting.

Sports and Recreation
Foret de Cedres

chemin de la Foret; daily 9am-5pm; free

The cedar forest that covers some 250 hectares (617 acres) at the summit of the Petit Luberon (727 m/2,385 ft) has become a year-round mecca for cyclists and hikers and a popular refuge from the summer heat. The first cedar seeds (taken from the Atlas Mountains in Algeria) were sown in 1861 by botanists. The trees reached maturity in 1920 and have been reproducing ever since. Entering the forest yields an immediate feeling of separation from the traditional Provençal landscape. Pine and rock, and sunshine and shade, give way to towering cedar trees and heavy shadow. Paths wind deep into the forest, occasionally breaking out to offer panoramas of the Provençal landscape below. The sharp-eyed claim to be able to see all the way to the Mediterranean. Popular walks include a 2-hour descent to the village of Lauris in the southern Luberon. For the less active, the forest is a pleasant place to picnic, and the drive up offers one of the best views of the hilltop

Wildlife of Luberon Natural Park

Hikers (particularly quiet ones) might come across a wide variety of wildlife when exploring Luberon Natural Park. Some encounters are thrilling; others may require caution.

MAMMALS

Egyptian vulture

- Most well known and ever more frequently seen in and around villages is the **wild boar.** They tend to be nocturnal, so sightings are most usual at dusk and dawn. In the **Foret de Cedres,** boars snuffle underneath the trees. Boars travel in family troops and can be dangerous to humans, particularly if there are young. It's best to steer clear if at all possible.

- Less threatening is an encounter with the largest species of **beaver** in Europe, weighing 20-40 kg (44-88 lb). You'll know these beavers are present if you see claw markings on trees and branches sharpened to a point like pencils.

- Rarest of all mammals in the Luberon is the Mistouflon goat, with only one or two reported sightings a year. The goat has a blue-tinged coat and large, golden curled horns. It also has an extra pair of hind legs. So extraordinary is this animal that it was the subject of a children's book in the 1970s and now has a gift shop named after it in the village of Lourmarin (Le Mistouflon, 9 rue Henri Savornin; tel. 04 90 68 34 01; www.lemistouflon.com/l-histoire).

BIRDLIFE

- Birdlife is abundant. The stars of the show are the four pairs of **Egyptian vultures** known to nest in the steep, rocky cliffs of the Luberon. In flight the vultures are recognizable by the contrast between their pure white plumage and black wingtips. From September to February the vultures migrate to the Sahara.

- Other exciting sightings could include the **bonelli eagle** and the **eagle owl.** Pheasants and guinea fowl roam wild, and as dusk falls the songs of nightingale pairs fill the treetops. Around the **Fort de Buoux,** birds of prey soar above the intimidating cliff face.

INSECTS AND REPTILES

- The diverse flora provides a refuge for more than 265 species of insect, as well as the **ocelle lizard,** the largest lizard in Europe.

- Turning over too many rocks is not advisable, because the Luberon has two species of **scorpion.** The sting of the dark brown, almost black one (*Euscorpius flavicaudis*) is relatively harmless, but if you get stung by the lighter, yellow-orange scorpion (*Buthus occitanus*), it's best to hurry to the hospital.

village of Bonnieux. Information panels along the main route indicate trails that you can follow and the way to the main viewpoints.

Sun-e-bike

1 avenue Clovis Hugues; tel. 04 90 74 09 96; www.sun-e-bike.com; daily 9am-5:30pm; from €28/day

Bonnieux is an ideal place to locate an electric bike business. In all directions, there are some pretty serious gradients. Only experienced cyclists dream of taking these hills on, and even the riders of the Tour de France looked spent in 2017 when they swept out of the Combe de Lourmarin. But strap a battery to a bike and suddenly the endeavor seems a lot less daunting. The quiet country lanes of the Luberon are perfect for a relaxing morning in the saddle.

Shopping
L'Atelier Duo

11 rue Voltaire; tel. 04 90 74 17 13; Apr.-Dec. daily 10am-7pm

This charming boutique sells a clothing collection designed by its owner, Éléonore Magnani, as well as a collection of jewelry by Gérard Seror. Cotton skirts, wool sweaters, and silver bracelets are the must-have pieces.

Boutique Poetic

2 rue de la Republique; tel. 06 40 92 87 78; Apr.-Oct. daily 10am-1pm and 3pm-7pm

Colorful ceramics and sculptures draw customers back again and again to the shop of Thimothée Humbert. It's rather cramped inside, so it's best to leave pets and children on the street.

Pottery Market

Le Village; Easter weekend

Every Easter the winding, narrow streets of Bonnieux are filled with local artisans selling pottery in a rainbow of colors. Some designs are innovative, some traditional, and some wildly eccentric. You can buy everything from salad bowls to serving plates, vases to great garden pots; each piece is unique and created by hand rather than in a factory.

Food
La Terrazza

Cours Elzear Pin; tel. 04 90 75 99 77; www.laterrazza-bonnieux.com; Wed.-Mon. 11:30am-11pm; €15-22

Indisputably the café with the best view in Bonnieux is La Terrazza. On one side of the small road is the kitchen, and on the other there's a long, narrow parquet terrace that commands a view over the whole valley. An Italian friend swears the pizzas are the best in the Luberon.

Café Capelongue

chemin des Cabanes; www.beaumier.com; daily noon-2:30pm and 7pm-9:30pm; €16-48; reserve online

Set in an old sheepfold on the grounds of a luxury hotel, this café serves up chic takes on brasserie classics. Don't expect the bill to be cheap—a lobster sandwich costs €48—but the location is worth the extra euros.

L'Arome

2 rue Lucien Blanc; tel. 04 90 75 88 62; www.laromerestaurant.com; Fri.-Tues. noon-3pm and 7pm-11pm; €56

Even without an outside space, talented chef Jean Michel Page keeps drawing the crowds with an eclectic menu marrying fresh ingredients from the Provençal markets with exotic spices from around the world. The menu is slightly tilted toward fish dishes, the chef's specialty.

Accommodations
Le Clos de Buis

rue Victor Hugo; tel. 04 90 75 88 48; www.leclosdubuis.fr; €175 d

The light, spacious rooms here are attractively decorated, and many have views out across the valley. The garden is mature and well-kept with a small swimming pool surrounded by shady trees and bushes. A summer kitchen where guests can prepare picnics is a welcome touch.

Domaine de Capelongue

chemin des Cabanes; tel. 04 90 75 89 78; www.beaumier.com; €450 d

The Capelongue is an outpost of the Beaumier luxury hotel group. It offers pared-back luxury and a pool with a picture-postcard view of Bonnieux. There's a gastronomic restaurant with a similar view and a more casual café-restaurant.

Getting There and Around

Once you're in Bonnieux, be aware that the village can be challenging to negotiate. Steps carved into the stone between the houses link the various levels of the village. It's best to wear sturdy walking shoes.

Car

From **Avignon,** take the D900 and then the D36. The journey of 54 km (34 mi) takes around 50 minutes. From **Apt,** take the D3. The journey of 15 km (9 mi) takes 20 minutes. From **Lourmarin,** take the D943 and then the D36. The journey of 13 km (8 mi) takes 20 minutes.

Bus

The **915** bus runs five times a day from Avignon to Apt, stopping at the Pont Julien bridge at the foot of the village. The journey takes 1 hour. It's a long uphill walk of at least an hour to the village from the stop. **Line 909** runs three times a day between Apt and Aix-en-Provence, taking 2 hours and stopping in the center of Bonnieux. Full timetables can be found at https://zou.maregionsud.fr.

Zou line 18, between Apt and Cavaillon, runs three times a day and stops at Bonnieux. It costs €2.60 and takes 20 minutes. Consult www.apt.fr/Les-transports.html or https://zou.maregionsud.fr for the latest timetable.

LOURMARIN

It's hard not to fall in love with Lourmarin. The village seems to have everything: It boasts a rich literary history, an imposing 15th-century château, and streets lined with cafés, restaurants, shops, and art galleries. The ultimate test is to mention the village to Parisians. The French urban elite tend not to get overexcited about much, but take a token Parisian, drop Lourmarin into the conversation, and a smile will inevitably come over their face. "J'adore Lourmarin," they will confess.

Sights
Grave of Albert Camus
Off the D27, Cimetière

In works such as *The Stranger* and *The Fall,* Camus tackled fundamental questions about the meaning, or rather the lack of meaning, in life, arguing that man should live by embracing the absurd rather than searching for explanations through faith.

Camus died in a car accident at age 46 in 1960 and was buried in the cemetery in Lourmarin, where he lived. His grave is a simple stone marked with his name. To find it, head south out of the center of the village on rue Henri de Savornin until you reach the D27, turn right, and continue for 500 m (547 yd). Cross the road away from the village and you will see a sign for the cemetery. Go to the back of the first section of the cemetery and walk left along the wall, turn left at the end of the path, walk past five graves, and you will come upon the small stone right next to the path.

Château de Lourmarin
24 avenue Laurent Vibert; tel. 04 90 68 15 23; www.chateau-de-lourmarin.com; May-Sept. daily 10:30am-6:45pm, Apr. and Oct. daily 10:30am-12:45pm and 2:30pm-5:45pm, Nov.-Mar. daily 10:30am-12:45pm and 2.30pm 5:15pm; €7.50, ages 6-16 €3, under 6 free

Unbelievably, this 15th-century castle was nearly knocked down at the beginning of the 20th century. It had descended into ruins and was saved by Robert Laurent Vibert, a cosmetics businessman. Visiting is a rewarding experience. The entrance terraces afford the best available view over the Luberon and the village of Lourmarin. Inside, visitors explore the 15th-century château and the 16th-century château neuf. There's a fine-art collection, an impressive double-spiral staircase, a collection of 16th-19th-century furniture, and the largest fireplace you are ever likely to see. The château is a center for classical music concerts and variety of festivals.

Sports and Recreation
Station Bee's
Station Bee, D27 opposite the tennis club; tel. 06 50 69 49 53; www.stationsbees.com; €25/half day

A great way to see the southern Luberon is to hire an e-bike. Take the D56 toward Vaugines, stop beside the etang in Cucuron for lunch, and return via the D27 to Lourmarin, taking a break at Domaine de Cavale for a wine-tasting.

Entertainment and Events
Yeah Festival
Château de Lourmarin, 24 avenue Laurent Vibert; https://festivalyeah.fr; first weekend in June; day pass €35, weekend €95

One of the best small dance music festivals in Europe takes place the first weekend of June in Lourmarin. The main stage is high up on the terrace of the Château de Lourmarin, where 500 or so ticket holders party the nights away to a collection of DJs headed by local resident Laurent Garnier. The real appeal of the festival, though, is the way it takes the party out into the streets of the village. Throughout the weekend, DJs play sets outside the village

cafés, at the tennis club, and even within the walls of the primary school.

Classical Music
Château de Lourmarin, 24 avenue Laurent Vibert; tel. 04 90 68 15 23; www.chateau-de-lourmarin.com; July-Aug.; adults €28, under 25 €12

Throughout the summer the château hosts classical music recitals. In addition to the pleasure of hearing renowned musicians perform, guests enjoy the ability to picnic on the terrace of the château before performances.

Art Galleries
Gérard Isirdi
4 rue Henri de Savornin; tel. 04 90 08 50 96; www.isirdi.com; daily 10:30am-12:30pm and 3pm-6pm

Enter anyone's second home in Provence and there's a good chance that you will see a framed picture of a man sitting in a café with a glass of pastis reading *La Provence* newspaper. This iconic image is part of painter Gérard Isirdi's back catalog, and prints, posters, and paintings of this and other humorously Provençal images are on sale in his gallery.

Dan Adel
Place Barthélémy; tel. 06 84 59 82 80; https://daniel-adel.format.com; daily 10am-7pm

Château de Lourmarin

Illustrator, photographer, portrait artist, and painter, the multitalented American Dan Adel is a central figure of the Lourmarin art scene. His work has graced the cover of magazines such as *The New Yorker, Vanity Fair,* and *Rolling Stone.* Dan's latest series of paintings employs a revolutionary technique known as dimensional abstraction. The resulting canvases are mesmerizing.

Markets
★ Lourmarin Market
boulevard du Rayol; May-Sept. Fri. 8am-1pm
From May to the end of September, Friday morning is party time in Lourmarin. Buskers fill the streets, the café terraces overflow, and the main boulevard becomes busier than a toy shop on Christmas Eve as people try to squeeze through the narrow spaces between the traders' stalls. There's plenty of fresh fruit and vegetables and takeaway food in the form of spit-roasted chickens, couscous, and even Chinese noodles. The other stands offer an eclectic mix of goods from sweet-smelling lavender sachets to creams guaranteed to remove the scratches from a car's paintwork. Shopping amid the crowds can be difficult, and the best advice for people intending to visit in high season is to come either early (before 10am) or late (after 12:30pm).

Sunday Artisan Market
Place Henri Barthélémy; May-Sept. Sun. 8am-6pm
The premise is simple: In order to have a stand at this market you have to make what you sell. The resulting market hosts an interesting group of artisans, including painters, sculptors, jewelers, and even furniture makers. Items for sale include storage baskets made in the traditional fashion from dried twigs, and iron fish sculptures perfect for decorating a bathroom or hanging beside a swimming pool.

Food
Tuesday Farmers Market
Fruitière Numerique, avenue du 8 Mai; mid-Mar.-Dec. Tues. 5:30pm-9pm

In the space of a few years, the farmers market that runs every Tuesday evening from mid-March until December has become a village institution. Residents and tourists mingle, and many buy cheese and bread and sit down for an impromptu supper with wine purchased from the adjacent bar run by the village wine cooperative. The atmosphere is convivial, and around 7pm a local chef gives a cookery demonstration.

Café L'Ormeau
2 Place L'Ormeau; tel. 04 90 68 02 11; www. cafedelormeau.fr; daily 9am-10pm; €20-35
The food at Café L'Ormeau is more bistronomique than café and has a heavy Italian influence. In the summer months there's often live music. The café's village-center location makes a place to soak up the unique atmosphere of Lourmarin.

Da Ora
1 chemin de Ferrailles; tel. 04 90 68 89 98; www.daora. fr; daily 11:30am-1pm; €26-38
This upmarket Italian restaurant is set amid spacious grounds on the outskirts of the village. The menu is small and based on classic Italian dishes such as tagliata of beef.

Le Puy Verre
4 Place Jean Moulin, Puyvert, tel. 01 88 55 52 47; https://lepuyverre.com; Fri.-Tues. noon-2pm and 7:30pm-10pm; from €40
Just minutes up the road from Lourmarin in a neighboring hamlet, Le Puy Verre offers an imaginative "surprise" menu that changes daily. For food lovers it is not to be missed.

Accommodations
Ajoucadou
9 rue Juiverie; tel. 06 98 83 93 16; www.ajoucadou. com; €90
The Ajoucadou apartments are a real labor of love. Pascal Aussud, the owner, renovated this old town house by hand, creating three bright, light, and spacious apartments. Two have three bedrooms, and one is a studio.

The Best Provençal Markets

lavender stand at the Lourmarin market

For food, the best Provençal market in the region is undoubtedly **Coustellet's Sunday morning farmers market.** Coustellet sits just beneath the famous village of Gordes. It's nowhere near as pretty as its famous neighbor, but its sprawling market attracts gourmands from miles around. There are hundreds of market traders offering fresh seasonal produce from their own fields. An entire morning can easily be spent perusing, tasting, and buying the finest products for a delicious meal.

For Provençal products such as soaps, tablecloths, and dresses, it's hard to beat **Lourmarin's Friday market.** Arrive by 9am to beat the crowds and enjoy the picturesque avenue of traders who set up beneath the centuries-old plane trees. By midday in the summer the market gets overcrowded and it's best to retreat to the center of the village, where there is usually a jazz band playing.

Forlcaquier's Monday market is probably the ideal Provençal market in that it offers a winning combination of the fresh produce of Coustellet and the more eclectic offerings of Lourmarin. It does not get as busy as Lourmarin market, and there's a wonderful old town to explore after a little shopping. Try buying a picnic in the market and climbing to the top of the Citadelle to eat and enjoy the view.

Markets open around 8:30am and shut around 12:30pm.

LUBERON MARKET DAYS

- Monday: Cadenet, Cavaillon, Goult, Lauris

- Tuesday: Lacoste, Gordes, Cucuron, La Tour d'Aigues

- Wednesday: Gargas, Merindol

- Thursday: Goult, Menerbes

- Friday: Bonnieux, Lourmarin, Lagnes

- Saturday: Apt, Pertuis

- Sunday: Ansouis, Coustellet

There is a minimum three-night stay during peak season.

Côté Lourmarin

Impasse du Pont de Temple; tel. 06 09 16 91 80; www. cotelourmarin.fr; €195 d

This is an upmarket bed-and-breakfast on the picturesque main street. The soft furnishings and bed linens are as fine as you will find in any five-star hotel. There's no need for a restaurant, because there's a good brasserie and three cafés just 15 m (49 ft) away across the road. For a romantic stay in the Luberon this is a top choice.

Bastide de Lourmarin

route de Cucuron; tel. 04 90 07 00 70; www. hotelbastide.com; €200 d

The decor here is slightly edgy and challenging, and the restaurant is only open in the summer months. However, the big plus for this hotel is that it has a small pool and is a 1-minute walk from the village. There's a spa offering a range of well-priced beauty treatments. All rooms have Wi-Fi, air-conditioning, and en suite bathrooms.

Le Moulin de Lourmarin

rue du Temple; tel. 04 90 68 06 69; www.beaumier. com; €300 d

The village-center location of the Moulin is its greatest draw. On the doorstep are cafés, restaurants, art galleries, and shops. The decor is inspired by rural Provence, with a lot of wicker-backed chairs and clay tiles.

Les Hautes Prairies

route de Vaugines, Lourmarin; tel. 04 90 68 02 89; www.campasun-lourmarin.eu; €30

Just outside Lourmarin, this campsite has a large pool and a separate waterslide area. There's also a play area for children and a snack bar. As well as pitches for tents, it has cabins for rent. The campground is a 5-minute walk from the village.

Getting There and Around

The village is largely pedestrianized. There is plenty of parking near the center.

Car

Take the D27 from **Lauris** (8 min; 5 km/3 mi) or **Cucuron** (11 min; 8 km/5 mi); the D56 from **Vaugines** (8 min; 5 km/3 mi); or the D943 from **Cadenet** (8 min; 4 km/2.5 mi) or **Bonnieux** (22 min; 13 km/8 mi).

Bus

There is a twice-daily bus (**line 909;** www. apt.fr/Les-transports.html and https://zou. maregionsud.fr) that runs between Apt and Aix-en-Provence, stopping at Lourmarin. There is also a bus (**line 919;** https://zou. maregionsud.fr) to and from the neighboring villages of Vaugines and Cucuron, which runs seven times a day.

CUCURON

There's much more to Cucuron than its enchanting plane tree-ringed water basin (etang). Visitors can easily lose half a day meandering the charming, narrow streets. Although not on the same scale as Lourmarin, there are still plenty of shops and artists' ateliers to dip your head into. It's perfect for those who wish to escape the crowds that can beset more well-known Luberon villages.

Sights
L'Etang

Place de l'Etang, Cucuron

The etang (water basin) in Cucuron is one of the best-kept secret sights in Provence. It's roughly the size of an Olympic swimming pool and surrounded by hefty chunks of old stone that residents use as benches. As on the Cours Mirabeau in Aix, the unique atmosphere is thanks to the centuries-old plane trees that line the etang, which was once a watering point for animals. Over the years the trees have been pruned so that the uppermost branches stretch toward the heavens like the arms of supplicants. The light beneath these trees is a dappled mixture of sunshine and shade, perfect for sitting in a café and enjoying a glass of rosé. Every Tuesday morning one of the most appealing markets in Provence takes place around the etang.

Sports and Recreation
Etang de la Bonde
intersection of D9 and D27, Cucuron

Just outside the village, this large man-made lake surrounded by views of the Luberon mountains is one of the best places to swim in the region. In summer the pebbly beach can get very busy. Out of season, people take their paddleboards and can be seen making serene progress over the still waters. It's also a popular spot with dog walkers. There's a beachside café that serves the local campground, and halfway up the southern side of the lake are two nice spots to stop and eat: La Dolce Villa (for more formal dining) and the Café du Lac (good, simple food).

Circuit l'Ermitage Hike
Distance: *6.7 km (4.1 mi) round-trip*
Time: *2 hours*
Information and Maps: *Cucuron tourist office; www.luberon-sud-tourisme.fr*
Trailhead: *Cucuron tourist office, 11 Cours Pourrières*

This relatively gentle 6.7-km (4.1-mi) walk with a 207-m (679-ft) incline climbs slowly into the Luberon hills. The landscape is dominated by pine trees, rock, and the smell of wild herbs. Halfway around the loop, the 13th-century church of Notre-Dame de Bellevoir, also known as the Ermitage, is the main sight. It sits on a rocky promontory overlooking the valley of the Aigue river, and its chapel includes a small habitable part. A map and information are available from the Cucuron tourist office.

Festivals
La Fete du Mai
rue de l'Eglise; first Sat. after May 21

After the great plague of 1720, the people of Cucuron agreed to cut down a poplar tree as high as the church as an offering to Saint Tulle in return for being saved from the pestilence. Every year since, the people of the village have kept their side of the bargain and hauled a huge felled tree through the streets and up to the church.

the etang in Cucuron

Food and Accommodations
Matcha
18 Montee du Château Vieux; tel. 04 86 78 55 96;
June-Sept. Tues.-Sun. noon-2pm and 7pm-9pm, Oct.-
May Tues.-Wed. noon-2pm, Thurs.-Fri. noon-2pm and
7pm-9pm, Sat. 7pm-9pm; €30-42

This restaurant draws the locals from neighboring villages with its modern French cooking. The menus are good value but dishes can be on the small side. It's also a great place to stay, with rooms above the restaurant and a spa in the basement.

La Petite Maison de Cucuron
Place de l'Etang; tel. 04 90 68 21 99; www.
lapetitemaisondecucuron.com; Wed.-Sun. noon-2pm
and 8pm-10pm; €65

An evergreen stalwart of the village, this one-star Michelin restaurant run by well-loved chef Eric Sapet has a loyal clientele. The menu changes weekly; dishes might include beef cheeks slow-cooked in red wine. There are also very popular cookery courses on offer.

Hotel de l'Etang
Place de l'Etang; tel. 04 90 77 21 25; www.logishotels.
com; €100 d

It is hard to imagine a more quintessentially Provençal spot for a hotel than overlooking the etang in Cucuron. It's so scenic that photos of the hotel grace the local postcards. Rooms are clean but simply furnished, and most have terraces giving onto the etang where guests can often be seen lost in peaceful contemplation.

★ Pavillon de Galon
chemin de Galon; tel. 04 90 77 24 15; https://
pavillondegalon.com; €350 d

This old hunting lodge has been converted into a luxury bed-and-breakfast. There are three immaculately furnished suites, a remarkable formal garden, and an infinity pool. It's an expensive choice, but the property offers seclusion, total privacy, and pampering within walking distance of Cucuron.

Getting There and Around

Cucuron is a small village best visited on foot.

Car
Take the D27 from **Lourmarin** (11 min; 8 km/5 mi) or **Sannes** (6 min; 5 km/3 mi), or the D56 from **Vaugines** (5 min; 2.5 km/1.5 mi) or **Ansouis** (7 min; 5 km/3 mi).

Bus
Bus **line 919** (https://zou.maregionsud.fr) runs to and from the neighboring villages of Vaugines and Lourmarin seven times a day. The journey from Lourmarin takes 25 minutes.

ANSOUIS

Like Cucuron, Ansouis is a little off the beaten tourist track. If it were located anywhere but in the Luberon, the village would be a star attraction. However, with so much competition from neighboring tourist gems, it can be overlooked. This makes it more enjoyable to visit, and it's a pleasant place to use as a base for exploring the region and also south to Aix-en-Provence. The village is built on a relatively steep hill, so reaching the château and church at the summit requires a little effort.

Sights
Château d'Ansouis
rue de Cartel; tel. 04 90 77 23 26; www.chateauansouis.
fr; July-Aug. Sat.-Sun. 3pm and 4:30pm, Apr.-June
and Sept.-Oct. Sat.-Sun 3pm, guided 1.25-hour tour in
French, reserve in advance; €13, under 15 €10

The owners who conduct the tour (only in French) of Château d'Ansouis share details of their daily lives as custodians of the castle and collectors of antiques. The château began life in the 10th century as a military fortress guarding the Aigue River. The 12th and 13th centuries saw further building work before, in the 17th century, the property was transformed into an elegant family home surrounded by formal gardens. Of particular interest is the richly decorated Provençal interior.

Choose Your Luberon Village

Goult

If you like . . .

ART GALLERIES

Head to **Lourmarin.** From American Daniel Adel on the Place Barthélémy, who produces canvases inspired by the rhythms of classical music, to Provençal painters such as Gérard Isirdi (rue Henri Savornin), whose broad brush-stroke interpretation of the Provençal countryside is enduringly popular, Lourmarin is filled with interesting galleries.

SPECTACULAR VIEWS

For views and spectacular sunsets head to the northern Luberon villages of **Gordes** and **Goult.** The rock outside Gordes on the D15 is the best place to savor the improbability of hill-village engineering, whereas the Jerusalem windmill in Goult offers the best panoramic view of the petit Luberon. As dusk falls and the light fades, the hills of the Luberon take on a purple hue.

E-BIKING

If you like e-biking, then head to **Bonnieux** and rent at Sun-e-bike. The D3 between Bonnieux and neighboring Menerbes is one of the most beautiful stretches of road in the Luberon.

ESCAPING THE CROWDS

To escape from the braying herds of second homeowners who can occasionally overrun the smaller Luberon villages, head to **Apt.** The regional capital is often overlooked but is a good base for a more authentic Provençal stay.

Eglise d'Ansouis

5114F Grande Rue

Continue to climb past the château another 100 m (109 yd) or so to the highest point of Ansouis: the village church. The records of the cathedral in Aix point to a date of establishment around the 11th century, although it is unlikely that any of the current building predates the 13th century. Relatively small in size, the church is a peaceful place to visit. The courtyard at the foot of the steps affords an expansive view out over the Luberon.

Food and Accommodations
Le Grain de Sel

Grande Rue; tel. 04 90 09 85 90; www.facebook.com/ legraindeselrestaurant; Wed.-Sun. noon-2pm and 7pm-10pm; €38-42

This great restaurant is perched at the top of Ansouis. Dishes are full of Provençal flavor and can be enjoyed on the small terrace overlooking the Luberon. The restaurant is an excellent value given its idyllic location, and it's loved by locals and tourists alike.

La Closerie

boulevard des Platanes; tel. 04 90 09 90 54; www. lacloserieansouis.com; Fri.-Tues. noon-1:30pm and 7:30pm-9pm, Sun. lunch only; lunch €45, dinner €65-85

Located halfway up the hill at the entrance to Ansouis, this restaurant has earned a reputation as one of the best in the southern Luberon. It serves refined, perfectly executed dishes based on the freshest local ingredients. There's a small terrace, and tables outside need to be reserved well in advance.

Un Patio en Luberon

route du Grand Four; tel. 04 90 09 94 25; https:// info611076.wixsite.com/patioenluberon; €200 d

Ansouis is a great place to stay for those wishing to escape the crowds that can overwhelm some Luberon villages. This two-bedroom bed-and-breakfast has bright, well-decorated rooms and a small terrace. Dating back to the 17th century, it's full of characterful stone

arches and old wooden beams. There's even a small fountain in the courtyard.

Getting There and Around

The village is pedestrianized. There is a small **parking lot** halfway up the hill where the village is built, and there's more parking at the foot of the hill.

Car

Take the D56 from **Cucuron** (8 min; 5 km/3 mi) or **Pertuis** (11 min; 8 km/5 mi), or the D135 from **Lourmarin** (14 min; 10 km/6 mi).

Bus

From Pertuis, **line 908** (https://zou. maregionsud.fr) runs nine times a day. The journey takes 25 minutes.

TOP EXPERIENCE

★ LAVENDER FIELDS OF HAUTE PROVENCE

More remote than the rest of the Luberon, the villages and towns of the Haute Provence have a gentler feel, and, in the countryside, it's the lavender that draws the crowds. From late June to the beginning of August the landscape becomes a patchwork of purple. The higher you go, the later the lavender blooms. Then, on August 15, a crowd that would fill a sports stadium converges on the town of Sault for harvesting competitions and the celebration of all things purple.

The roads in this region are narrow, windy, and often hilly. It can take some time to travel from place to place, particularly if you get stuck behind a tractor or camper van. Sault and Forcalquier are the pick of the places to choose as a base for your stay. Forcalquier is a small town with a vibrant artistic community, and Sault is a large village that has grown prosperous thanks to lavender tourism.

Sights
Sault

Although plenty of other villages would like the title, Sault can rightfully claim to be the center of the Provençal lavender industry. It is surrounded by field upon field of the purple bloom. In the village there are, of course, plenty of shops selling lavender-related products, and, on August 15, the village goes lavender crazy when its hosts nearly 30,000 people for the annual lavender festival.

Château Simiane-la-Rotonde

Le Village; tel. 04 92 73 11 34; www.simiane-la-rotonde. fr; Mar.-Apr. 14 Sat.-Sun. 10:30am-1pm and 2pm-5:30pm, Apr. 15-Apr. 28 and Oct. Wed.-Sun. 10:30am-1pm and 2pm-5:30pm; Apr. 29-June daily 10:30am-1pm and 2pm-6pm, July-Aug. daily 10:30am-6:30pm, Sept. daily 10:30am-1pm and 2pm-5:30pm; €5.50, youth €4.50, under 12 free

If there was a prize for having the narrowest streets in Provence, Simiane-la-Rotonde would be one of the top contenders. Fields of lavender lap at the base of this medieval village, which receives its name from the **Château des Simiane-Agoult.** Built around the 11th century on the site of a former Roman settlement, the château features a round keep (rotonde) that can be seen for miles around. Visitors can drive to the top of the village and park next to the château; a better option is to park next to the entrance to the old village and climb through the atmospheric cobbled streets to the château at the summit.

La Citadelle/Notre-Dame de Provence

30 boulevard Raoul Dufy, Forcalquier

The Chapel of Notre-Dame de Provence, built in the 19th century on the ruins of the old citadel, exerts a siren song on visitors. The gold statue of the Virgin Mary glints in the sunshine, challenging all comers to join her near the heavens. The walk up through Forcalquier old town, through charming squares and past restaurant terraces and fountains, becomes progressively more difficult. The last 5 minutes of the 20-minute climb are up an uneven cobbled path. In a way it's like a minipilgrimage; the journey itself is as important as the end destination. Upon arrival there's a feeling of satisfaction and a wonderful

panoramic terrace. It was from here in the 12th century that the Counts of Forcalquier exerted dominion over lands stretching in a 100-km (62-mi) radius around them. The main route to the citadel is up rue Saint Mary from the old town, but it's possible to just discard the map and head upward.

Salagon Gardens

La Prieure, Mane; tel. 04 92 75 70 50; www.musee-de-salagon.com; Feb.-Apr. and Oct.-Dec. 15 daily 10am-6pm, May-Sept. daily 10am-7pm; €8, children €6, under 6 free

Human occupation of the site at Salagon dates back 2,000 years. There are traces of Neolithic huts and a Roman villa, succeeded by a Christian site and finally the church that was built in the 11th century. Construction of additional buildings continued until the 19th century. As well as admiring the architectural heritage and enjoying the museum, the principal reason to visit is the gardens. There are four different themed zones: village garden, medieval garden, contemporary garden, and the garden of the senses.

Plateau de Valensole

Just off the D6 from Manosque in the direction of Valensole are some of the most extensive fields of lavender in Provence. Continue on past Valensole on the D8 for more sightings. Then take the D953 toward Puimoisson, returning on the D56 toward Valensole. This circular driving route offers plenty of opportunities to stop and stroll out into the lavender for photo opportunities. The pattern of purple changes every year because the lavender is a rotation crop. Look out for fields of sunflowers as well.

Recreation
Sault Historic Walk

Tourist office, avenue de la Promenade; tel. 04 90 64 01 21; www.ventouxprovence.fr; Nov.-Mar. Mon.-Fri. 9am-12:30pm and 2pm-5pm, Apr.-June and Aug.-Oct.

1: Simiane-la-Rotonde **2:** lavender fields **3:** Salagon Gardens **4:** Forcalquier

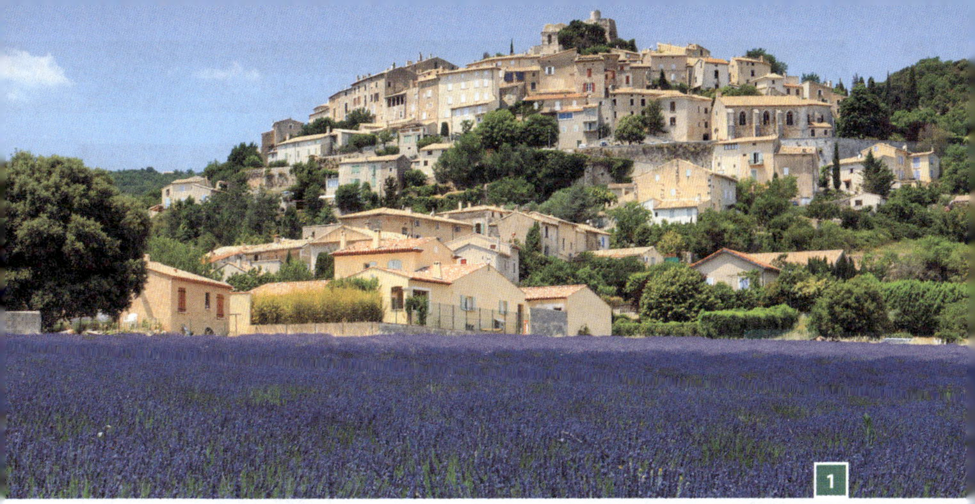
1

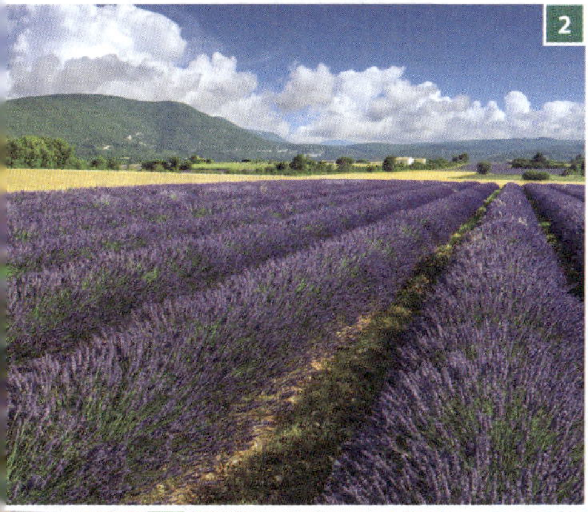
2

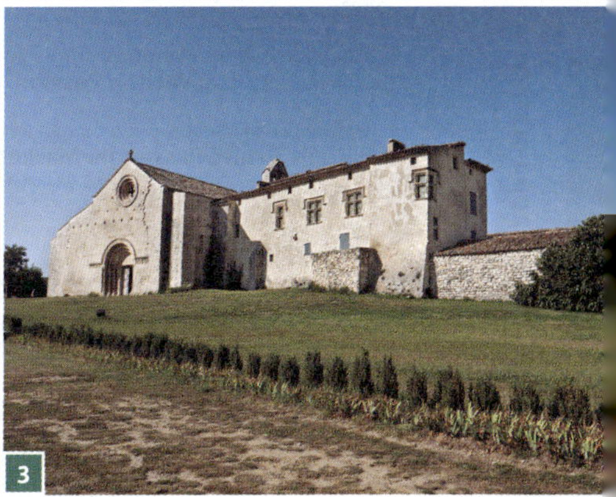
3

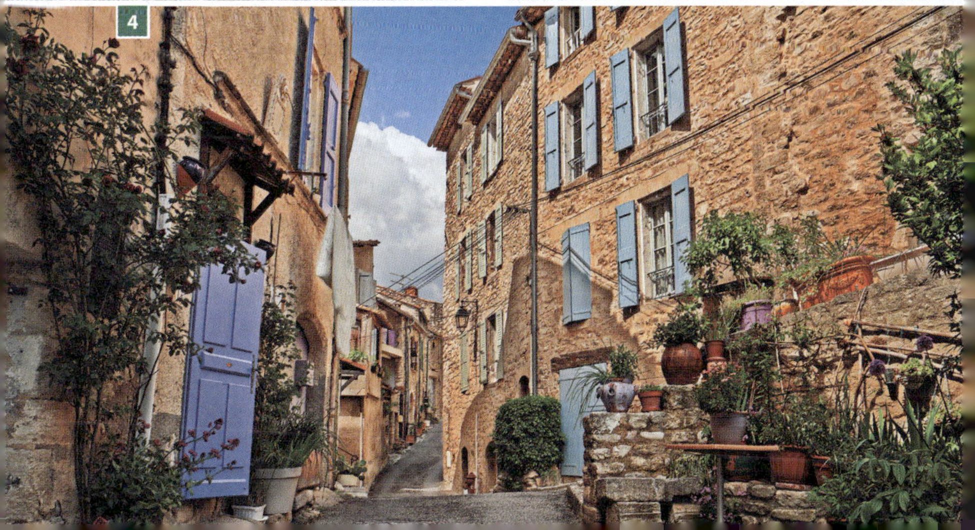
4

Lavender Beyond Haute Provence

Haute Provence is rightly considered the best place to see lavender. Rows of the lavender roll like waves over the horizon, creating a shimmering purple haze. Besides Haute Provence's best spots (page 171), here are a few other ways to experience lavender throughout the region:

- Taste lavender ice cream in almost any ice cream shop. In particular, look out for the sumptuous lavender ice cream produced by artisan maker Ravi in shops across Provence, including those in Lourmarin and L'Isle-sur-la-Sorgue.

- **Bastide,** a shop in Aix-en-Provence, specializes in simple, clean beauty products made from Provençal ingredients, such as lavender honey bodywash (page 133).

- Paraglide over the Verdon's purple lavender fields with the experts at **Verdon Passion** (page 183).

lavender soaps and sachets

- Even in urban Marseille, you can get a whiff of lavender at the **Cruise Passenger Market,** where it's stacked by the bundle along the quayside (page 208).

Mon.-Sat. 9:30am-12:30pm and 2pm-6pm, July-Aug. daily 9:30am-12:30pm and 2pm-6:30pm

Pick up a brochure at the tourist office and follow the arrows through Sault old town that lead past the ruins of the château and the Notre-Dame de la Tour church.

Festivals and Events
Lavender Festival

Sault; tel. 04 90 64 01 21; www.fetedelalavande.fr; Aug. 15; free

It's hard to imagine, but every August 15 the village of Sault welcomes around 30,000 people to its Lavender Festival. It's the size of crowd more commonly associated with sporting events or pop concerts, not Provençal festivals. At the event there is, of course, plenty of lavender and lavender products. What makes the festival unique is its attempt to explore the broader cultural associations with lavender. Musicians play old folk songs, authors sign books, and lavender-harvesting races are held using modern and ancient tools.

Shopping
Forcalquier Market

Place Bourguet

Every Monday morning, Forcalquier hosts one of the largest and most famous markets in Provence. Place de Bourguet and the parking area off rue des Écoles are adorned with the multicolored parasols of the traders. The smell of spit-roasted chicken, bundles of freshly harvested lavender, earthy mounds of dried mushrooms, and steaming plates of couscous and paella are more than enough to justify the town's reputation as a center of the flavors and the senses.

The Boyer Nougat Factory

Place de l'Europe, Sault; tel. 04 90 64 00 23; www.nougat-boyer.fr; daily 8am-7pm; free

The Boyer family have been making nougat in Sault since 1887. Their shop is the oldest of its kind in the Vaucluse, and their nougat some of the tastiest around. Ingredients include lavender honey from the slopes of Mont Ventoux and almonds from trees across the region.

Food and Accommodations
Le Provençal

rue Porte des Aires; tel. 04 90 64 09 09; www. restaurant-le-provencal-sault.fr; Wed.-Sun. noon-2pm and 7pm-9pm, Mon. lunch only; €22-25

Le Provençal is a popular stop in the center of Sault. It offers full flavored, hearty food that is perhaps a little lacking in finesse. The menus in particular are an excellent value at €22 for three courses. This being Sault, expect lavender to crop up somewhere on the menu.

Hôtel Le Pré Saint Michel

425 Montee de la Mont Imbert; tel. 04 92 72 14 27; www.presaintmichel.com; €95 d

This simple, small hotel is a good base for exploring the town of Manosque and visiting the Valensole lavender fields. There's a heated pool and a good restaurant, and guests are only a few minutes' drive from the center of Manosque.

Hostelerie Val de Sault

Ancien chemin d'Aurel; tel. 04 90 64 01 41; www. valdesault.com; €155 d

High in the hills next to Sault, overlooking the lavender fields, is this regional stalwart.

It's known for its friendly staff and restaurant offering good local produce. Rooms have been nicely updated over the years, and there's a pool, tennis court, and—somewhat incongruously—a resident peacock.

Getting There
Car

From Apt to Sault take the D943, covering 30 km (19 mi) in 42 minutes.

For Valensole, exit the A51 motorway at the Manosque junction and take the D6, covering 15 km (9 mi) in 20 minutes.

Bus

From Apt to Sault take the **916 bus.** The journey takes just over an hour and costs €2.10. For timetables visit https://zou.maregionsud.fr.

From Manosque to Valensole take the **133 bus.** The journey takes 40 minutes. For timetables, visit www.mobilite.dlva.fr.

Getting Around

If you do not have a car, opt to see the lavender fields in and around Sault. They can be seen from the village and as you arrive by bus. It is possible to walk from the center of Valensole out toward the lavender fields on the D8. The length of the walk will depend on which fields have been planted in the year you visit, but expect to walk at least 2 km (1.2 mi).

The Gorges du Verdon

TOP EXPERIENCE

You've seen the photos, you've planned the trip, you know what to expect, and yet the Gorges du Verdon still delivers a visceral thrill, a stomach-churning, head-spinning surge of excitement that slowly gives way to silent awe and head scratching. The 700-m-deep (2,300-ft-deep), 25-km-long (16-mi-long) gorge cut by the Verdon river in the limestone cliffs seems to defy logic. How could such a small river cut through so much rock? At this point, geologists might like to note that the limestone rock deposits laid down by a prehistoric sea in the Jurassic period were heavily weakened by large-scale tectonic activity during the Cretaceous period. Non-geologists will probably just continue shaking their heads at the improbability of it all.

In any event, nature has created the most wonderful playground for tourists of all persuasions. Motorists, e-bikers, and hikers can soak up the vertiginous views from either the Route des Crêtes on the north side of the gorge or the Corniche Sublime on the south side. Nature lovers taking the same routes will hope to see one of three species of vultures that inhabit the gorge. Adrenaline junkies can bungee jump, climb, canyon, and white-water raft.

Farther downstream man has lent nature a helping hand. The construction of the Sainte-Croix dam in the early 1970s created France's third-biggest lake. Bordered by pretty villages with beaches and boat hire facilities, the lake now draws just as many crowds as the gorge. The best place to hire a boat is probably the Pont du Galetas (Galetas Bridge). From here it's possible to head up the gorge, enjoy a dip in the clear waters, and stare up in wonder at the cliffs.

ORIENTATION

The cliffs are at their highest in the **upper gorge,** where **Lac de Sainte-Croix** is the main base for water sports. **Castellane** and **La Palud-sur-Verdon** are the main entry points for the northern side of the upper gorge; **Moustiers-Sainte-Marie** and **Aiguines** are good bases from which to explore Lac de Sainte-Croix and the southern side of the upper gorge. The entire circuit of the upper gorge in a car is 130 km (80 mi), and because of the narrow and windy roads

takes a day. Most visitors choose to take either the **Route des Crêtes** on the north side or the **Corniche Sublime** on the south side, both of which offer spectacular viewpoints. The main access road from Marseille, Aix, and Avignon is the **A51** motorway to Saint-Paul-lez-Durance and then the **D952.** Allow 2 hours to drive from Aix to the Gorges du Verdon, and just under 3 hours to get there from Marseille or Avignon.

THE UPPER GORGES DU VERDON (NORTH)

On the north side of the Gorges du Verdon, the main base and departure point for activities is La Palud-sur-Verdon. It's an attractive village with a mountain feel, and a good place to stop overnight, or just to stretch your legs after a drive around the Route des Crêtes. La Palud offers a wide range of sports and activities to set your adrenaline surging, including canyoneering where you don a wetsuit and throw yourself down narrow cascades between rocks. Aquaphobes can opt for traditional climbing or a roped course across the rocks, featuring zip lines and other heart-thumping challenges. Finally, for a real bird's-eye view there's paragliding. An alternative, larger base for the north side of the upper

Gorges du Verdon and Lac de Sainte-Croix

The Gorges du Verdon

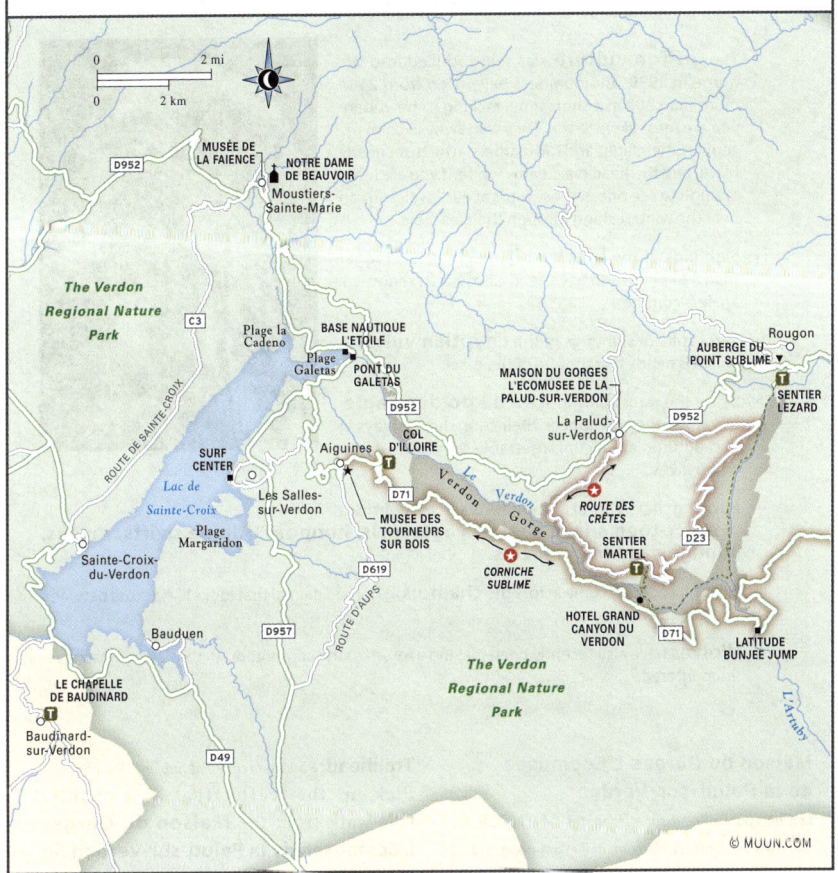

0 2 mi

0 2 km

D952

MUSÉE DE LA FAIENCE

NOTRE DAME DE BEAUVOIR

Moustiers-Sainte-Marie

The Verdon Regional Nature Park

C3

Plage la Cadeno

BASE NAUTIQUE L'ETOILE

Rougon

AUBERGE DU POINT SUBLIME

MAISON DU GORGES L'ECOMUSEE DE LA PALUD-SUR-VERDON

SENTIER LEZARD

Plage Galetas

PONT DU GALETAS

D952

ROUTE DE SAINTE-CROIX

COL D'ILLOIRE

La Palud-sur-Verdon

D952

Aiguines

Le Verdon

Le Verdon

D71

Gorge

ROUTE DES CRÊTES

SURF CENTER

Lac de Sainte-Croix

Les Salles-sur-Verdon

MUSÉE DES TOURNEURS SUR BOIS

SENTIER MARTEL

D23

Plage Margaridon

D619

Sainte-Croix-du-Verdon

CORNICHE SUBLIME

ROUTE D'AUPS

Bauduen

D957

HOTEL GRAND CANYON DU VERDON

D71

LATITUDE BUNJEE JUMP

The Verdon Regional Nature Park

LE CHAPELLE DE BAUDINARD

L'Artuby

Baudinard-sur-Verdon

D49

© MUUN.COM

gorge is the town of Castellane to the east of La Palud on the D952. Here you will find more life in the evenings with a bigger choice of restaurants. Castellane is also the place to stop for tourist information, with two visitor centers offering maps and guidance on everything from activities to accommodations.

Sights

★ Route des Crêtes

Route des Crêtes, off the D952, departing La Palud-sur-Verdon

This 24-km (15-mi) circular route from La Palud runs alongside the gorge for much of its length. At places the drop is a head-spinning 700 m (2,300 ft) to the valley below. There are 14 different points where you can park, admire the views, and perhaps spot vultures and eagles. Thanks to the altitude of the road, it's possible to enjoy some real eye-to-eye encounters. It's important to follow the Route des Crêtes in a clockwise direction, as indicated in the village of La Palud-sur-Verdon; a small part of the road is one-way, and you'll have to turn back if you set off in a counter-clockwise direction.

Wildlife of the Gorges du Verdon

short-toed snake eagle

- The **griffon vulture** was reintroduced into the gorge in 1999, and numbers have risen from 24 to more than 300 in a short time, making it the vulture you are most likely to see. They are tawny in color except for their flight and tail feathers, which are nearly black, and the head is a creamy white. If you are lucky enough to see one of these birds at eye level, you can hear the wind rushing through their feathers.

- Sightings of the **black vulture,** which is bigger than the griffon and has a nearly 3-m (10-ft) wingspan, are less common.

- Rarer still are sightings of the **Egyptian vulture,** which is nearly half the size of the black.

- By far the rarest bird to see is the **golden eagle,** but when they do appear, their courtship displays in mid-autumn are an unforgettable spectacle of giddying dives.

- Bird-watchers can also look forward to sightings of **short-toed snake eagles, eagle owls, hoopoes, Alpine swifts, ravens, choughs,** and **crag martins.**

- Hikers frequently come across the **chamois,** a horned animal that looks like a cross between a goat and an antelope.

- **Wolves** have also recently been making a return to the area, although they are almost never encountered.

Maison du Gorges L'Ecomusee de la Palud-sur-Verdon

Le Château, La Palud-sur-Verdon; tel. 04 92 77 38 02; www.verdontourisme.com; daily 9am-noon and 2pm-6pm; €4

This pleasant small museum concentrates on man's changing relationship with nature in the Gorges du Verdon, starting from prehistoric times. There's plenty of information on the flora and fauna of the region and the unique geology that gave rise to the gorge.

Hiking

Sentier Lezard

Distance: *5 km (3 mi) round-trip*
Time: *2 hours*
Information and Maps: *Maison du Gorges L'Ecomusee de la Palud-sur-Verdon (Le Château, La Palud-sur-Verdon; tel. 04 92 77 38 02)*

Trailhead: *Parking area, Point Sublime, Rougon*

Pick up the leaflet that accompanies this walk from the **Maison du Gorges L'Ecomusee de la Palud-sur-Verdon** for a detailed description of the flora and fauna you will encounter on this 5-km (3-mi) circular walk. The gravel trail is signposted from the parking lot, and unusually for the Gorges du Verdon, the walk is not too hilly. The path passes between the cliffs from sun to shade and offers spectacular views. Allow 2 hours to complete the circuit. When you return, Rougon is a pleasant village that has a couple of cafés where you can relax and enjoy the view.

Sentier Martel

Distance: *14 km (9 mi) one-way*
Time: *7 hours*

Information and Maps: *www.verdontourisme.com/sentier-blanc-martel*

Trailhead: *D23 Chalet de la Maline, La Palud-sur-Verdon*

For more experienced hikers, the Sentier Martel is a 14-km (9-mi) trail that leads down into the gorge. There's a celebrated descent of 200 m (660 ft) or so down a rickety old staircase, as well as, at one point, a wobbly bridge over the river. Both keep minds focused. The route is rocky and difficult at times and passes through the odd small cave, so keep a flashlight handy. Picnic spots are scarce, and the best is a large beach you will encounter beside the river. It is recommended to park your car in La Palud-sur-Verdon. A bus then takes you to the Chalet de la Maline start point and also picks you up at the Point Sublime end point 6 hours and 45 minutes later. Reserve the bus at https://navette.parcduverdon.fr.

Cycling
Verdon E Bike
1 route de Moustiers, La Palud-sur-Verdon; tel. 06 88 10 91 73; http://verdonebike.pagesperso-orange.fr; May-Sept. daily 9am-5pm; €28, reserve in advance

The Gorges du Verdon is just too hilly for all but superhuman cyclists. However, hiring an e-bike is an excellent way to enjoy the 24-km (15-mi) Route des Crêtes.

Adventure Sports
Maison des Guides du Verdon
rue Grande, La Palud-sur-Verdon; tel. 04 92 77 30 50; www.escalade-verdon.fr; book in advance

This outfit offers canyoneering, climbing, roped climbing courses, and hiking. The business has been running since 1984 and caters to all levels from beginners to experts. Excursions can last anything from a couple of hours to the entire day. Prices depend on the activity but are generally about €45 per person per half day.

Aventures et Nature
La Palud-sur-Verdon; tel. 06 82 18 45 71; www.aventures-et-nature.com; daily 9am-7:45pm; from €45/half day

In addition to canyoneering, climbing, and roped climbing courses, Aventures et Nature also offers aqua rando: hikes with lots of dips in the water along the way.

Des Guides Pour l'Aventure
meeting point at Auberges des Crêtes, La Palud-sur-Verdon; tel. 06 85 94 46 61; www.guidesaventure.com; May-Sept. daily 9am-6pm; from €45/half day

Check out Des Guides Pour l'Aventure for white-water rafting. They also offer canyoneering, climbing, roped climbing courses, and hiking.

Roc n Vol
Les Alaves, La Palud-sur-Verdon; tel. 06 89 30 75 74; www.rocnvol.com; May-Sept. daily 9am-6pm; from €45/half day

Roc n Vol offers paragliding in addition to canyoneering, climbing, roped climbing courses, and hiking.

Food
Le Mur des Abeilles
Rougon; tel. 04 92 83 76 33; May-Sept. daily noon-10pm; €8-15

Inventive sweet and savory crepes are served on a spectacular terrace overlooking the Gorges du Verdon. Try goat's cheese and honey or eggplant. There are only six or seven tables outside, so it is sensible to reserve.

Auberge du Point Sublime
D952, Rougon; tel. 04 92 83 69 15; www.auberge-pointsublime.com; May-Sept. daily noon-2pm and 7pm-10pm; €22-36

Simple, locally sourced food is served on a large terrace overlooking the Gorges du Verdon. The clientele is a mixture of cyclists and hikers taking a well-earned break. Hearty portions include grilled rosemary-crusted lamb cutlets. Reserve in advance. There are also comforable rooms for an overnight stay.

Accommodations
Le Perroquet Vert
rue Grande, La Palud-sur-Verdon; tel. 04 92 77 33 39; www.leperroquetvert.com; €65 d

Book early to get a room in this popular four-bedroom guesthouse in the center of the Gorges du Verdon gateway village of La Palud. Rooms are simple with no air-conditioning.

Nouvel Hotel du Commerce

Place Marcel Sauvaire, Castellane; tel. 04 92 83 61 00; www.hotel-du-commerce-verdon.com; €100 d

This is a comfortable 30-bedroom hotel with an outdoor swimming pool and a welcoming, relaxed atmosphere. The former coaching inn is located in the heart of the picturesque town of Castellane, and there are plenty of local restaurants to choose from in the evening. It's ideal for those who want to visit the Gorges du Verdon but also like the busier scene offered by a town in the evenings.

★ Hotel and Spa des Gorges du Verdon

route de la Maline, La Palud-sur-Verdon; tel. 04 92 77 38 26; www.hotel-des-gorges-du-verdon.fr; €260 d

This beautiful luxury retreat sits at an altitude of just under 1,000 m (3,280 ft). There are 30 rooms. Families will like the suite, which sleeps six. The hotel is committed to the environment and is one of two French pilot hotels for the European Union's Nearly Zero Energy Hotels initiative. There's on-site parking, a spa, and a heated outdoor pool.

Information and Services

There are two visitor centers in Castellane: the **Maison Nature & Patrimoines** (Relais du Parc; Place Marcel Sauvaire, Castellane; tel. 04 92 83 19 23; www.maison-nature-patrimoines. com; July-Aug. daily 10am-1pm and 3pm-6:30pm, May-June, Sept.-Oct. Wed., Sat.-Sun. 10am-1pm and 3pm-6:30pm) and the **Verdon Tourist Office** (rue National, Castellane; tel. 04 84 32 04 04; www.verdontourisme.com; May-June and Sept. Mon.-Fri. 9am-noon and 2pm-6pm, Sun. 10am-1pm, July-Aug. daily 9am-6:30pm). The Verdon Tourist Office offers a broad range of services, with

information on everything including hotels, 14 restaurants, activities, and hikes. The Maison Nature has a narrower focus on providing information about accessing the Gorges and its flora and fauna.

Getting There and Around
Car

From **Marseille,** take the A51 to Saint-Paul-lez-Durance and then the D952, covering 144 km (89 mi) in 2 hours 20 minutes.

From **Aix,** take the A51 to Saint-Paul-lez-Durance and then the D952, covering 108 km (67 mi) in 1 hour 50 minutes. From **Avignon,** take the A51 to Saint-Paul-lez-Durance and then the D952, covering 183 km (114 mi) in 2 hours 34 minutes.

From **Nice** and **Cannes,** take the A8 and then the D54 from Le Muy and the D955, covering 147 km (91 mi) in 2 hours 20 minutes.

Bus

There are no direct buses from either Marseille or Aix-en-Provence. Instead take **line 67** (https://zou.maregionsud. fr) to Riez and then **line 450** to La Palud and Castellane. Timetables can be found at www.verdontourisme.com/infos-pratiques/comment-venir-dans-le-verdon.

THE UPPER GORGES DU VERDON (SOUTH)

On the southern side of the Gorges du Verdon, perched above Lac de Sainte-Croix, the village of Aiguines is the main base for exploring the southern gorge. It offers easy access to the beaches of the lake, the Corniche Sublime, and some of the best hiking and climbing in the region, including the challenging Sentier Imbert, which takes walkers into the depths of the gorge. Like most of the villages in the region, Aiguines is vibrant in the summer months but quickly becomes sleepy out of season.

1: view on the Route des Crêtes **2:** view of the Gorges du Verdon near Aiguines **3:** Artuby Bridge **4:** Moustiers-Sainte-Marie

Sights

Musée des Tourneurs sur Bois

Place de la Resistance, Aiguines; tel. 04 94 70 99 17; https://museedestourneurssurbois.com; July-Aug. daily 10:30am-6:30pm, May-June Mon.-Sat. 10am-6:30pm, Sept.-Oct. Mon.-Fri. 10am-6:30pm; €3.50, under 12 free

This is a small but interesting museum that charts the changing craft of wood turning in the village since the 16th century. Beginning with the use of old-fashioned tools to create wooden pieces for games, the museum shows how wood turning has evolved to the methods that artisans now use to create sculptures and designer furniture.

★ Corniche Sublime

D71, Aiguines to Comps-sur-Artuby

Constructed in 1947 to open up the Gorges du Verdon to motorists, the Corniche Sublime is the glamorous name for the stretch of the D71 between Aiguines and Comps-sur-Artuby. Along the way there are plenty of places to stop and admire the imposing topography of the gorge. As with the Route des Crêtes, there's the chance of spotting vultures and eagles. The most picturesque spot is approximately 10 km (6 mi) from Aiguines. Here the road becomes increasingly sinous and the drops ever more vertiginously. Among the best viewpoints to stop at are the Belvedere du Plan, the Belvedere Aiguines, and the Balcon de la Mescla.

Hiking

Col d'Illoire

Distance: *5 km (3 mi) round-trip*
Time: *2 hours*
Information and Maps: *www.alltrails.com/fr/randonnee/france/var/aiguines-col-d-illoire*
Trailhead: *Aiguines*

This 5-km (3-mi) round-trip hike takes about 2 hours to complete. The departure is from the center of the village, and the trail crosses the Corniche Sublime to the spectacular Col d'Illoire viewpoint at 967 m (3,172 ft). The terrain is quite steep and is best for experienced hikers.

Climbing

The southern side of the upper gorge offers a variety of climbs for people with the gear and know-how to make their way up on their own. A list of available climbs, including difficulty levels, can be found on the **Tourism Office website** (www.verdontourisme.com/activites/escalade).

La Corditelle

chemin de la Pinede, Aiguines; tel. 06 10 49 51 92; www.lacorditelle.com; from €25

La Corditelle offers a range of climbing activities for all levels and all ages (starting as young as 5). There are roped courses across the cliffs, and climbing initiation classes are offered.

Adventure Sports

Latitude Bungee Jump

D71, Artuby Bridge; tel. 04 91 09 04 10; www.latitude-challenge.fr; weekends 9am-noon, mandatory online reservation 24 hours in advance; €140/jump

Among the highest bungee jumps in Europe at 182 m (597 ft), the Artuby Bridge is not for the fainthearted. It's on the south side of the gorge and is best accessed from either Castellane or Aiguines.

Food and Accommodations

Hotel Vieux Château

Place de la Fontaine, Aiguines; tel. 04 94 70 22 95; https://hotelchateauverdon.fr; €104 d

A good base for exploring the region, the Hotel Vieux Château has 10 simply furnished rooms. The terrace of the hotel spills out onto the pretty Place de la Fontaine. The hotel's restaurant has a good reputation for regional food.

Hotel Grand Canyon du Verdon

Falaise des Cavaliers, Aiguines; tel. 04 94 76 91 31; www.hotel-canyon-verdon.com; €175 d

Seemingly clinging to the edge of the cliff face, this hotel is not for anyone who suffers from vertigo. Terraces from some of the rooms look directly out over the canyon. All are simply decorated. The restaurant offers an affordable three-course menu for €30, and there are panoramic views from the dining room.

Information and Services

The **Aiguines Tourism Office** is located at 1 Place de la Marie, Aiguines (tel. 04 94 70 21 64; www.lacs-gorges-verdon.fr; Apr.-Sept. Mon.-Fri. 9am-midday and 2pm-6pm, July-Aug. Mon.-Sat. 9am-midday and 2pm-6pm, Sun. 9am-12:30pm, Oct.-Mar. Mon.-Fri. 9am-12:30pm). It offers information about restaurants, accommodations, and all the activities available in the Gorges.

Getting There and Around
Car

Aiguines is accessed by the D71 road, which runs from Lac Sainte-Croix through Aiguines to Comps-sur-Artuby. Moustiers-Sainte-Marie is 15 km (9 mi) away.

MOUSTIERS-SAINTE-MARIE

The most prosperous village in the area is Moustiers-Sainte-Marie. Built on either side of a mountain stream, it has charm to spare with its arching bridges and winding, boutique-lined streets. It's certainly the place to go for a gastronomic meal in the region, with an Alain Ducasse hotel and restaurant on the outskirts and plentiful options for foodies in the village. Maps and friendly advice are available from two helpful tourist offices.

Sights
Musée de la Faience

rue du Seigneur de la Clue; tel. 04 92 74 61 64; https://musee-moustiers.fr; Apr.-end of Oct. Wed.-Mon. 10am-12:30pm and 2pm-5pm, Nov. Sat.-Sun. 10am-12:30pm and 2pm-5pm; €7, under 16 free

Thanks to the quality of the local clay and the abundance of water, Moustiers-Sainte-Marie has a long tradition of making fine glazed ceramic pottery (faience). The museum includes pottery dating back to the 17th century, and as you proceed through the different rooms the decorative fashion of the plates changes dramatically with the decades.

Notre-Dame de Beauvoir

7136 Place Pomey

A set of 262 steps leads out of the eastern side of the village up to the chapel of Notre-Dame de Beauvoir. The church itself is dark and bare, but the climb is worth it for the view back down over Moustiers. Farther on up the trail, at some 300 m (1,000 ft) above the village, a star is suspended between the two cliff faces that frame the village. Now in its 17th incarnation, the star is said to have been initially hung by the knight Blacas to give thanks for his safe return from the Crusades in the 13th century.

Sports and Recreation
Des Guides Pour L'Aventure

Magnans, avenue de Lérins; tel. 06 85 94 46 61; www.guidesaventure.com; May-Sept. daily 9am-6pm; from €40

The company offers 3- to 6-hour walks out into some of the lesser-known trails of the gorge. They also offer canyoneering and rafting.

Verdon Passion

rue de Courtil; tel. 06 08 63 97 16; www.verdon-passion.com; in season daily 9am-7pm; from €40

Paragliding is the specialty of Verdon Passion, with flights for all levels departing from nearby Hamel de Vincel. The instructors will take you soaring over Lac de Sainte-Croix. Canyoneering, white-water rafting, and white-water kayaking are also offered.

Equiverdon

Hameau de Vincel; tel. 06 10 91 31 72; https://equiverdon.fr; in season daily 9am-7pm; 2-hour trek €50

Equiverdon offers horseback riding for all levels in the countryside surrounding Moustiers-Sainte-Marie. Pony tours are also available for small children.

Shopping
L'Atelier Bondil

Place de l'Eglise; tel. 06 37 59 90 81; www.faiencebondil.fr; daily 10am-7pm

Supplied exclusively by the local Bondil Faience atelier, this shop offers both traditional ceramics with decorative blue motifs and more contemporary colorful pieces. They are right in the center of Moustiers-Sainte-Marie.

Food and Accommodations
Restaurant les Tables du Cloitre
Place du Presbytere; tel. 04 92 74 64 31; Thurs.-Sat. noon-2pm and 7pm-8:30pm, Sun.-Wed. noon-2pm; €15-25

As the name suggests, this is an atmospheric place for a meal set beneath old stone arches. The food is simple but excellent value and booking is advisable.

La Ferme de Sainte Cecile
route des Gorges du Verdon; tel. 06 31 61 72 40; Thurs.-Sat. noon-1:30pm and 7pm-8:30pm, Sun. and Wed. noon-1:30pm; lunch €30, dinner €39

Enjoy top-quality regional produce in this spacious restaurant with a large terrace overlooking verdant gardens. Dishes are beautifully presented and full of flavor. Try the fish of the day dressed with a distinctive algae cream.

Hotel Le Colombier
Quartier Saint Michel; tel. 04 92 74 66 02; www.le-colombier.com; €150 d

Located on the outskirts of Moustiers, this simple hotel has large grounds, a tennis court, and a small swimming pool. The rooms can be on the small side, but it's a good base from which to explore the Gorges du Verdon and Moustiers-Sainte-Marie.

★ La Bastide de Moustiers
chemin de Quinson; tel. 04 92 70 47 47; www.bastide-moustiers.com; restaurant Thurs.-Mon. 12:30pm-1:45pm and 7:30pm-8:45pm, Wed. 7:30pm-8:45pm; restaurant menu from €65, rooms from €360 d

One of the best-known addresses in Provence, La Bastide de Moustiers was bought in 1994 by chef and hotelier Alain Ducasse. The bastide was transformed into a 13-bedroom luxury retreat with one of the finest restaurants in the South of France. Each morning, the gardener picks the freshest vegetables from the hotel potager and leaves them at the kitchen door. The chef then creates the menu.

Information and Services
Moustiers-Sainte-Marie boasts two excellent visitor centers: the **Parc Naturel Régional du Verdon office** (Maison du Parc, Domaine de Valx; tel. 04 92 74 88 00; www.parcduverdon.fr; Mon.-Fri. 9am-12:30pm and 2pm-5:30pm) and the main village **tourist office** (Maison Lucie, Place de l'Eglise; tel. 04 92 74 67 84; www.moustiers.fr; Mon.-Fri. 9:30am-7pm, Sat.-Sun. 9:30am-12:30pm and 2pm-7pm). The tourist office has the broader range of services, offering advice on restaurants and accommodations as well as how to best access the nature park.

Getting There and Around
Car
From **Aix, Avignon,** and **Marseille,** the main motorway is the A51. Exit at Manosque and then follow the signs to Valensole, and subsequently Moustiers-Sainte-Marie. From Marseille, the journey is 120 km (75 mi) and takes just under 2 hours; from Aix the journey is 90 km (56 mi) and takes 90 minutes. From Avignon the journey is 150 km (93 mi) and takes 2 hours 20 minutes.

Bus
Public transportation in the region is very limited. Zou **line 67** (https://zou.maregionsud.fr) runs from Marseille and Aix to Riez, from where **line 450** runs to Moustiers-Sainte-Marie. The route takes 3 hours from Aix and 3 hours 40 minutes from Marseille, with buses running several times a day and costing around €20.

On Foot or by Bike
A **cycle and walking route** links Moustiers-Sainte-Marie with Lac de Sainte-Croix. The route is 5 km (3 mi) long and takes either 40 minutes or 90 minutes, depending on whether you are riding or walking.

LAC DE SAINTE-CROIX

Lac de Sainte-Croix is France's third-biggest lake. It is 10 km (6 mi) long and 2 km (1.2 mi) wide and was created in 1973 by the construction of the Sainte-Croix Dam. This glittering expanse of turquoise water has white sand and stone beaches that call to mind the Caribbean, but the three villages that give onto the lake—Les Salles-sur-Verdon, Bauduen, and Sainte-Croix-du-Verdon—are typically French, with plentiful outdoor cafés. The lake draws water sports enthusiasts from across France, and on the beaches you'll find multiple providers renting out boats and other recreation equipment. Petrol (gasoline) motors are forbidden, however, so all watercraft must be powered by by wind, oars, or electricity. To drive around the lake in a car takes about 1 hour. The best place to rent paddleboats and kayaks and head up into the gorge is at the easternmost point of the lake at Pont du Galetas.

Beaches

Plage Galetas

Pont du Galetas, Moustiers-Sainte-Marie

Immediately adjacent to the Pont du Galetas, near Moustiers-Sainte-Marie on the southern side of the lake, this beach is very busy in the summer months. It marks the point where the gorge opens up into the Lac de Sainte-Croix. The small beach is a pleasant place to relax after a paddleboat ride up the gorge.

Plage La Cadeno

route du Lac, Saint-Saturnin, Moustiers-Sainte-Marie

Just off the D957 on the northern side of the lake on route du Lac in Saint-Saturnin, there is a beach next to the Base Nautique La Cadeno. The pebble beach is close to Moustiers and is convenient for a quick swim.

Plage Margaridon

route Margaridon, Les Salles-sur-Verdon

Numerous little beaches line the shore off route Margaridon near the village of Les Salles-sur-Verdon. Fringed by oak trees, with sand and pebbles glowing in the sun, each beach is fit for a postcard. However, Plage Margaridon is the pick of the bunch. Boat hire facilities are available, and there is a large parking area.

Sainte-Croix-du-Verdon

Coming around the corner as you descend into Sainte-Croix-du-Verdon, this beach appears shimmering in the hot sunshine, curving sinuously away to one side of the town. Unsurprisingly, the pebble-and-sand beach

paddleboarding in the Gorges du Verdon

is very popular in the summer. There's a large parking lot, boat hire, and a lifeguard on hand during the main tourist season.

Kayaking and Paddling
Base Nautique l'Etoile

Pont du Galetas, Moustiers-Sainte-Marie; tel. 07 68 94 17 87; https://base-nautique-etoile.fr; Apr.-Sept. daily 10am-6pm; from €10/hour for a kayak

This is the best spot for helping you explore the gorge by boat. On offer are canoes, kayaks, paddleboats, and electric boats. It is not possible to reserve in advance. Gazing up at the cliffs can be mesmerizing and leads to the odd gentle collision!

Base Nautique La Cadeno

route du Lac, Saint-Saturnin, Moustiers-Sainte-Marie; tel. 07 82 10 67 03; www.moustiers.fr; Apr.-Sept. daily 10am-6pm; paddleboats from €14/hour

Located on a pretty sandy beach, the Base Nautique rents out sailboats, paddleboards, and windsurfing equipment, as well as the usual paddleboats, canoes, and electric boats.

Surf Center

Plage Margaridon, Les Salles-sur-Verdon; tel. 07 62 61 06 24; www.facebook.com/ surfcenterlessallessurverdon; from €8/day

Kayaks, paddleboats, and small sailboats can be rented on this picturesque beach. A paddleboat with a slide keeps the kids amused for hours.

Alize Electric Location

Bauduen; tel. 04 92 75 44 69; www.location-bateau-verdon.fr; Apr.-end of Sept. daily; from €35/hour

This provider specializes in electric boats that seat up to eight and have batteries that last up to 8 hours. Alize allows reservations to be made in advance, and in high season they are essential.

Lac Loc

La Plage, Sainte-Croix-du-Verdon; www.lacloc-saintecroix-verdon.com; Apr.-end of Sept. daily 10am-6pm; paddleboats from €14/half hour

Kayaks, canoes, electric boats, and paddleboats are all available. There's also a small snack bar for a refreshing drink after all the paddling. No reservations.

Hiking
Sentier des Muletiers

Distance: *5 km (3 mi) round-trip*
Time: *2 hours*
Information and Maps: *www.cheminsdesparcs. fr/api/fr/treks/54240/le-sentier-des-muletiers.pdf*
Trailhead: *La Plage, Sainte-Croix-du-Verdon*

This 5-km (3-mi) walk, marked with yellow trail markers, starts from the Sainte-Croix-du-Verdon beach. Facing the lake, turn left and follow the dirt track. The circular walk takes 2 hours, during which the gradient rises by 200 m (660 ft). At times the path is uneven and you'll need good walking shoes. The track passes through lavender fields (which will be in bloom between the end of June and early August) and offers good views back over the lake. Shade is not plentiful, so wear a hat and take plenty of water.

Le Chapelle de Baudinard

Distance: *3 km (1.8 mi) round-trip*
Time: *30 minutes*
Information and Maps: *Search Baudinard to Chapelle Baudinard on Google Maps*
Trailhead: *Baudinard sur Verdon*

Families will enjoy this 3-km (1.8-mi), 30-minute walk along a dirt track with a climb of 100 m (330 ft) to a viewpoint over the lake. Park at the entry to the village of Baudinard and take the track that passes the tennis courts. After 300 m (980 ft), take the left-hand track, which climbs up to the chapel and a viewpoint with an orientation table. Afterward, there is a small café (L'Auberge, 37 Grande Rue) in picturesque Baudinard where you can pause and enjoy a drink and light meal.

Moustiers-Sainte-Marie to Lac de Sainte-Croix

Distance: *5 km (3 mi) one-way*
Time: *1.25 hours*

Information and Maps: *www.moustiers.fr/sites/ moustiers.fr/files/blog/2295/lake_of_sainte-croix_by_ quinsons_path_-_hiking_map.pdf*

Trailhead: *chemin de la Maladrerie*

A 5-km (3-mi) walk from Moustiers takes you down to the lake in approximately 1 hour and 15 minutes. Take chemin de la Maladrerie, cross the road, and follow the path to Quinson. At the end of the path, cross the stream and follow the signs to the Lac. The trail comes out at Pont du Galetas.

Food and Accommodations
Café du Midi

rue du Cours, Bauduen; tel. 04 94 70 08 94; daily 9am-10pm; €12.50-22

Bauduen is a great spot for lunch when touring the lake. This is a simple, kid-friendly restaurant with good food and a terrace overlooking the lake. Try king prawns fried in pastis or one of the copious salads.

★ L'Actuel

Le Cours, Sainte-Croix-du-Verdon; tel. 06 68 59 77 49; daily noon-1:30pm and 7pm-8:30pm; €25-35

In the heart of Sainte-Croix-du-Verdon, L'Actuel welcomes a crowd of in-the-know tourists. The food is excellent. Even so, the views (particularly from the small terrace) frequently distract from the plate in front of you. Try beef filet topped with foie gras and a truffle gravy.

Les Cavalets

D71 route du Barrage, Bauduen; tel. 04 98 10 62 40; www.lescavalets.com; €115 d

The rooms, all with en suite bathrooms, need a little updating, but the setting is idyllic with a garden and a small swimming pool overlooking the lake. The restaurant has a large, panoramic terrace and offers good-value menus for around €25 as well as à la carte mains (€17-28).

Getting There and Around
Car

From **Marseille,** take the A51 to Manosque and then the D6 and D8, following directions for Moustiers-Sainte-Marie and then Lac Sainte-Croix, covering 150 km (92 mi) in 2 hours 18 minutes.

From **Aix,** take the A51 to Manosque and then the D6 and D8, following directions for Moustiers-Sainte-Marie and then Lac Sainte-Croix, covering 110 km (68 mi) in 1 hour 50 minutes.

From **Avignon,** take the A7 and then A51 to Manosque and then the D6 and D8, following directions for Moustiers-Sainte-Marie and then Lac Sainte-Croix, covering 182 km (113 mi) in 2 hours 40 minutes.

From **Nice and Cannes,** take the A8 and then the D1555 from Le Muy to Draguignan and then the D955, covering 136 km (85 mi) in 2 hours 20 minutes.

Bus

Public transportation is very limited. Sainte-Croix-du-Verdon can be reached by taking the bus to Riez on **line 67** from Marseille or Aix (https://zou.maregionsud.fr) and then a taxi.

Marseille and the Var Coast

Marseille: You either love it or you hate it. The city, which began life as a small Greek trading settlement in 600 BCE, is today a big, brash, graffiti-splattered metropolis and port. For country bumpkins, the shock of arrival can be an unexpected slap to the face. When visitors recover, they discover a city rich in culture. Hidden amid the cars, the trams, and the crowds are neighborhoods so picturesque it's difficult to believe they belong to the same city.

Marseille offers visitors an eclectic mix of major sights, vibrant neighborhoods, and picturesque seaside spots. The MuCEM building, the centerpiece of the Euroméditerranée project, links past and future with its modernist bridge crossings to Fort Saint-Jean and the old port. The Cathédrale de la Major and Notre-Dame de la Garde are

Highlights

Look for ★ to find recommended sights, activities, dining, and lodging.

© MOON.COM

★ **Vieux Port:** Under the gaze of Notre-Dame de la Garde, this is the heart of the city, where man, metropolis, and the Mediterranean meet, and where any tour of the city should begin (page 198).

★ **Notre-Dame de la Garde:** High on a hill, Notre-Dame de la Garde is more than just a church; the Marseillais believe a whispered prayer to the huge statue of the Virgin Mary crowning the dome will save them. Wherever you are in Marseille, you can sense the statue glinting in the sun (page 199).

★ **MuCEM:** Architect Rudi Riciotti's MuCEM hosts a permanent exhibition about Mediterranean culture, but the star is the building itself. It's a work of futuristic fantasy that enlivens anybody's day (page 200).

★ **Vallon des Auffes (Endoume):** The port of this city is nestled in a rocky coastal inlet. Fishers mend nets and small waves ripple into the

harbor. The feel is very much Greek island rather than big city. Yet the old port is only a 5-minute drive away (page 204).

★ **Cassis:** Sitting at the foot of the highest cliff in Europe, surrounded by vineyards producing the best white wine in Provence, within walking distance of the first of the calanques, and with arguably the most beautiful harbor in the South of France, Cassis is simply a must-visit (page 215).

★ **Les Calanques:** This area offers the opportunity to travel back in time and experience France's Mediterranean coast, free from the curse of mass tourism. Largely inaccessible to cars, this dreamy stretch of pine-fringed seaside is reached on foot or by boat (page 220).

★ **Plage Anse de Renécros:** This beach is an almost perfect half-moon of sand located on the opposite side of the headland to Bandol's port and main shopping drag (page 227).

Marseille and the Var Coast

avant-garde ecclesiastical constructions that bring a smile to the face.

Then there are simpler joys, like walking around the old port and visiting its fish market, or discovering neighborhoods like Le Panier and the Cours Julien, which are incubators for youthful artisan talent. It's also possible to slip away from the bustle of the city into seaside inlets, such as Vallon des Auffes, that call to mind the fishing villages of small Mediterranean islands.

The discoveries continue to the east of

Marseille, where there are two of the most scenic stretches of coast in the South of France. Les Calanques, on the route to Cassis, is a much-feted national park. Mostly accessible on foot or by boat, the coastline transports visitors back to the Mediterranean of the past. Farther east still is the Var coast. Resorts such as Saint-Cyr Les Lecques, Bandol, and Sanary-sur-Mer are mostly ignored by foreign tourists in favor of their more famous cousins on the Riviera, yet they are a delight.

Previous: boats lining the harbor in Marseille's old port; the MuCEM building; the sparkling waters of Les Calanques.

© MOON.COM

PLANNING YOUR TIME

Before you arrive, prioritize what you want to see and plan how to get from one sight to another. Buy the **City Pass** from the tourist office (adults €26 for 24 hours, €35 for 48 hours, and €42 for 72 hours; www.marseille-tourisme.com). Not only do you get free access to all the city's museums and some sights, but also it serves as a transit pass on the public transport network.

If spending some time in a big city appeals to you, then allocate at least a couple of nights to Marseille. However, if you are attracted by some of the world-class sights but don't want the associated hassles of urban life, then it's possible to dip into Marseille for a day, say, before or after your flight from **Marseille Provence Airport.** You can easily get a feel of the place by visiting the **old port** and **MuCEM,** and enjoying a lunch by the seaside. Try to allow a day to visit **Cassis** and **Les Calanques** as well, and, if possible, a third day to enjoy the **Var coast,** which can be visited by train.

Itinerary Ideas

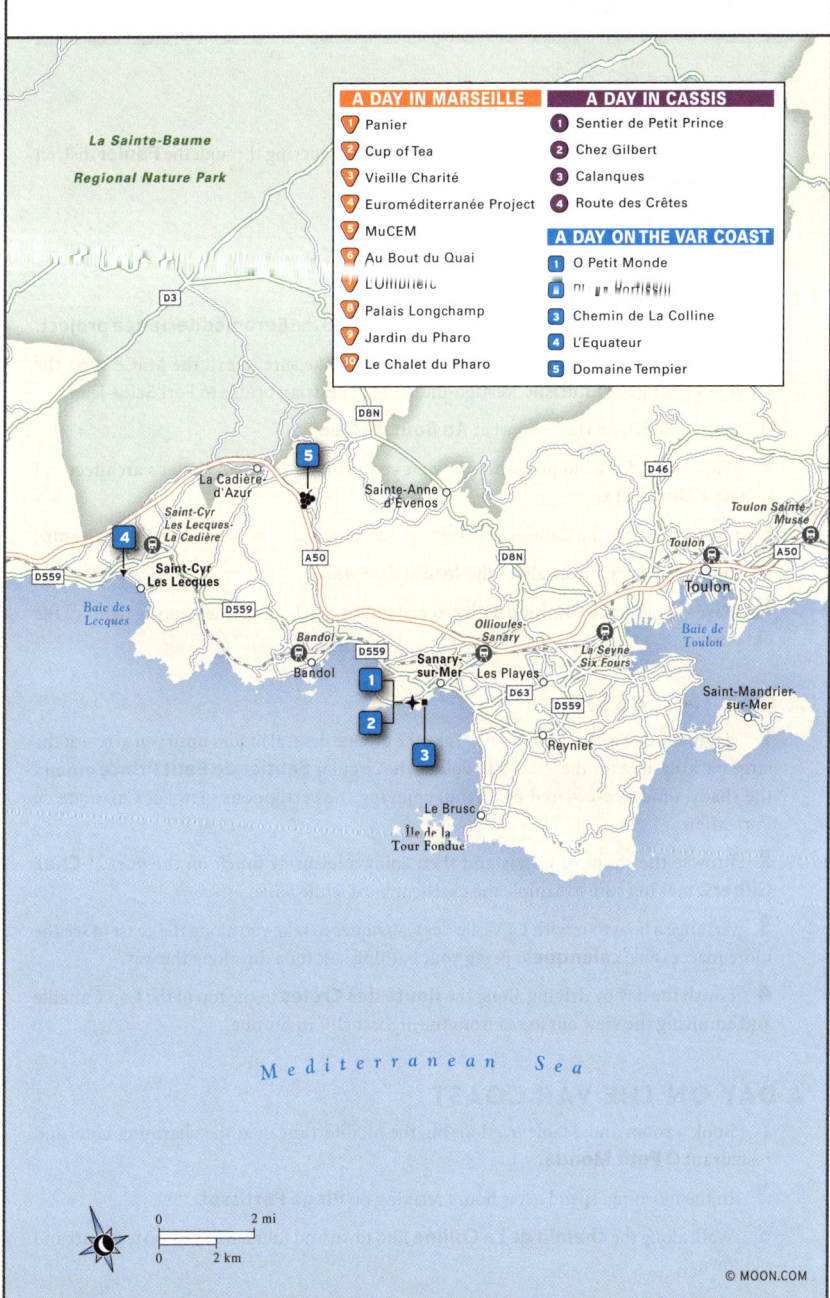

A DAY IN MARSEILLE
1. Panier
2. Cup of Tea
3. Vieille Charité
4. Euroméditerranée Project
5. MuCEM
6. Au Bout du Quai
7. L'Olibrius
8. Palais Longchamp
9. Jardin du Pharo
10. Le Chalet du Pharo

A DAY IN CASSIS
1. Sentier de Petit Prince
2. Chez Gilbert
3. Calanques
4. Route des Crêtes

A DAY ON THE VAR COAST
1. O Petit Monde
2. Pl. ¤ Hortişıll
3. Chemin de La Colline
4. L'Equateur
5. Domaine Tempier

MARSEILLE AND THE VAR COAST

© MOON.COM

Itinerary Ideas

A DAY IN MARSEILLE

1 Begin the day with a riot of wonderful color by wandering through the **Panier** district and admiring the street art.

2 Stop at **Cup of Tea** for a light breakfast.

3 Continue the climb up through the neighborhood to enjoy the architectural splendor of **Vieille Charité.**

4 Make your way down and out of the Panier toward the **Euroméditerranée project.**

5 Tour the architecturally impressive **MuCEM.** Make sure to exit the MuCEM on the 4th floor, taking the futuristic, vertigo-inducing pedestrian bridge to Fort Saint-Jean.

6 Stop for lunch on the old port at **Au Bout du Quai.**

7 Walk around the old port and see what you think of Norman Foster's architectural fantasy **L'Ombrière.**

8 Catch a tram up the Canebière to see the cascading waters of the **Palais Longchamp.**

9 Finish the day with a walk in the **Jardin du Pharo.**

10 Have supper in **Le Chalet du Pharo** restaurant overlooking the glinting lights of the old port.

A DAY IN CASSIS

1 Begin the day with a gentle walk. Take the avenue des Calanques until you arrive at the large parking area for the Presqu'il. Follow the circular **Sentier de Petit Prince** around the shady, pine tree-covered headland, enjoying the vertiginous views of Calanque de Port-Miou.

2 Browse the shops of Cassis and then enjoy a leisurely lunch on the port at **Chez Gilbert,** making sure to sample the excellent local white wine.

3 Arrange a boat tour with La Visite des Calanques to take you along the coast to see the more inaccessible **calanques.** Bring your bathing suit for a dip along the way.

4 Finish the day by driving along the **Route des Crêtes** to the top of the Cap Canaille and admiring the view out to sea from the highest cliff in Europe.

A DAY ON THE VAR COAST

1 Book a room and a table overlooking the Mediterranean at the charming hotel and restaurant **O Petit Monde.**

2 In the morning, spend a few hours relaxing on **Plage Portissol.**

3 Walk along the **chemin de La Colline** into town and tour Sanary-sur-Mer's port and old town.

4 Drive along the D559 for 20 minutes, hugging the coast until you arrive at Saint-Cyr-sur-Mer for lunch at the **L'Equateur** bar.

5 After a lunch spent soaking up the view, indulge the palate with a spot of wine-tasting a few minutes away at **Domaine Tempier** just outside Bandol.

Marseille

ORIENTATION

Marseille is a big city, divided into **16 arrondissements.** It can easily take an hour or more to walk between different neighborhoods. To simplify matters, this guidebook divides them into five areas:

Vieux Port

Head to the old port and the adjacent hill to the south, upon which sits the church of **Notre-Dame de la Garde,** to get an instant feel for the atmosphere of the city. It's where old meets new, as fishers unload their catch in centuries-old fashion in the shadow of the daring modernist **L'Ombrière,** designed by Norman Foster. Climb away from the port toward Notre-Dame de la Garde, and the streets of the city immediately clasp you in their multicultural embrace, with ethic supermarkets and Provençal brocante shops sharing the sidewalks.

Le Panier

The oldest residential district of Marseille is adjacent to the old port facing Notre-Dame de la Garde. Le Panier is a warren of narrow cobbled streets that climb up toward the impressive **Vieille Charité** building. Head here for boho boutiques and colorful street art.

Euroméditerranée Project

The quays adjacent to the **Fort Saint-Jean** at the entrance to the old port were redeveloped as part of the Marseille 2013 Capital of Culture celebrations. This neighborhood is dominated by the **MuCEM** building. The grand public spaces are the perfect antidote to the tight streets of the rest of the city.

Vieux Port

Marseille

To
Centre Commercial
les Puces

LE R2
BRITISH
CONSULATE
LES DOCKS &
TERRASSES DU PORT
LE FRAC

BEST WESTERN
LA JOLIETTE

BOULEVARD DE PARIS

AV. CAMILLE PELLETAN

AVENUE GÉNÉRAL
LECLERC

AV. CAMILLE PELLETAN

BOULEVARD DES DAMES

EUROMÉDITERRANÉE
PROJECT

LE PANIER

RUE DE LA RÉPUBLIQUE

MUCEM

VIEUX PORT

JARDIN &
PALAIS DU
PHARO

LE CHALET
DU PHARO

TUNNEL DU VIEUX PORT

COURS PIERRE PUGET

Plage des
Catalans

LES BORDS
DE MER

CORNICHE DU PRÉSIDENT JOHN FITZGERALD KENNEDY

See "Vieux Port" Map

BOULEVARD TELLENE

BOULEVARD NOTRE DAME

CHEZ FONFON

NOTRE-DAME
DE LA GARDE

RUE FORT DU SANCTUAIRE

RUE D'ENDOUME

© MOON.COM

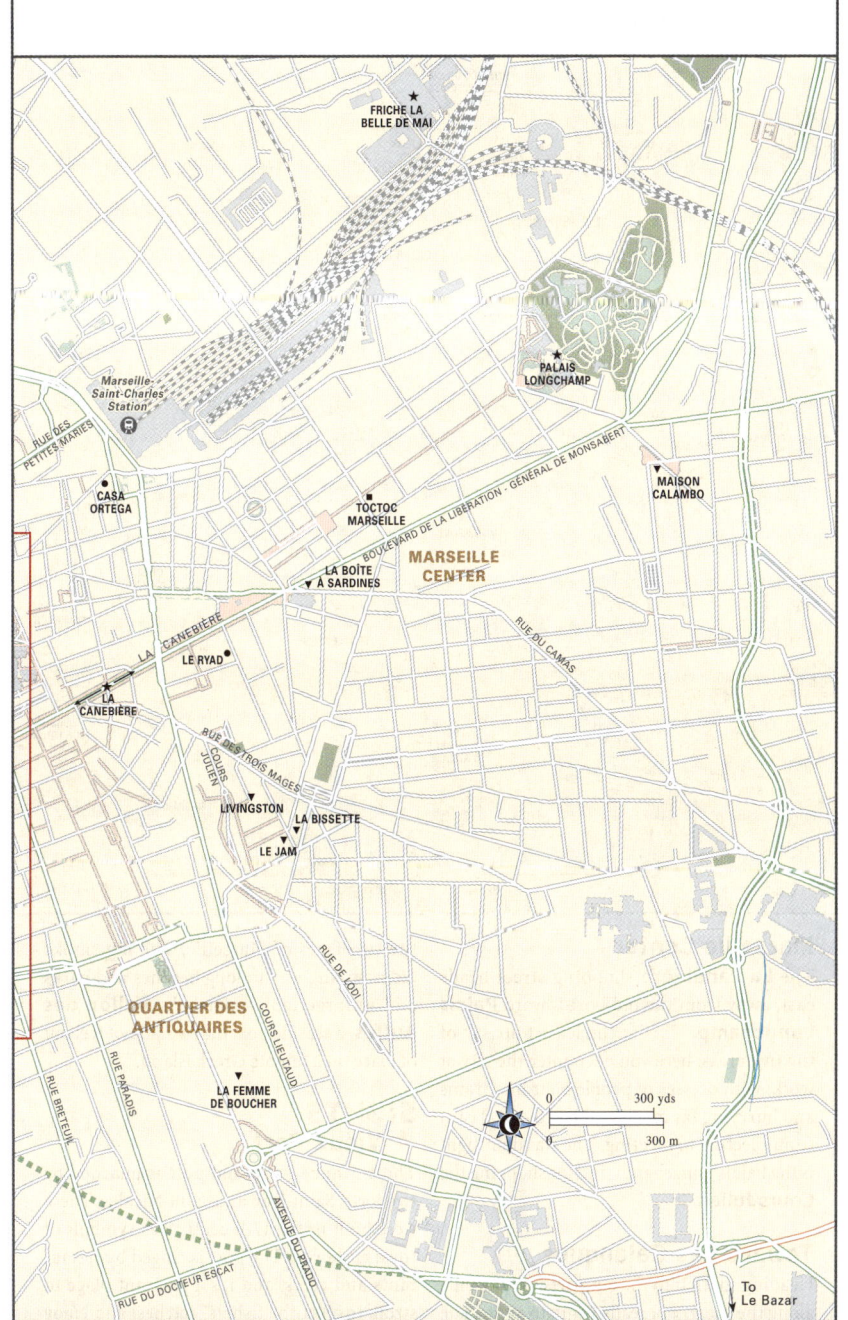

FRICHE LA
BELLE DE MAI ★

Marseille-
Saint-Charles
Station

RUE DES
PETITES MARIES

CASA
ORTEGA

TOCTOC
MARSEILLE

BOULEVARD DE LA LIBÉRATION - GÉNÉRAL DE MONSABERT

PALAIS
LONGCHAMP ★

MAISON
CALAMBO

LA BOÎTE
À SARDINES

**MARSEILLE
CENTER**

LA CANEBIÈRE

LE RYAD

RUE DU CAMAS

LA
CANEBIÈRE

RUE DES TROIS MAGES

COURS
JULIEN

LIVINGSTON

LA BISSETTE

LE JAM

RUE DE LODI

**QUARTIER DES
ANTIQUAIRES**

COURS LIEUTAUD

RUE PARADIS

RUE BRETEUIL

LA FEMME
DE BOUCHER

0   300 yds

0   300 m

RUE DU DOCTEUR ESCAT

AVENUE DU PRADO

To
Le Bazar

Vieux Port

HOSPICE DE LA VIEILLE CHARITÉ
RUE DU PETIT PUITS
RUE DE L'ÉVÊCHÉ
PLACE DE LA MAJOR
RUE COLBERT
CATHÉDRALE DE LA MAJOR
MAISON DE LA BOULE
AU VIEUX PANIER
RUE MERY
QUAI DE LA TOURETTE
LE PANIER
RUE DE LA RÉPUBLIQUE
RUE HENRI BARBUSSE
EURO MÉDITERRANÉE PROJECT
INTERCONTINENTAL HOTEL DIEU
BAZAR DE CÉSAR
GRAND RUE
COSQUER MÉDITERRANÉE
MUSÉE REGARDS DE PROVENCE
LES NAVETTES DES ACCOULES
CUP OF TEA
RUE DE LA LOGE
RUE SAINT-LAURENT
MARSEILLE TOURIST OFFICE
MUCEM
AVENUE DE SAINT-JEAN
RUE DE LA LOGE
LA RESIDENCE DU VIEUX PORT
FORT SAINT-JEAN
AU BOUT DU QUAI
QUAI DU PORT
QUAI DES BELGES
TUNNEL DU VIEUX PORT
VIEUX PORT
L'OMBRIÈRE
VIEUX PORT MEDICAL CENTER
LA POSTE
CHEZ LOURY
SON DES GUITARES
QUAI DE RIVE NEUVE
TROLLEYBUS
CAFÉ SIMON
RUE SAINTE
RUE BRETEUIL
BOULEVARD CHARLES LIVON
SAINTE-CATHERINE
RUE FORT NOTRE DAME
RUE GRIGNAN
MAISON MONTGRAND
RUE NEUVE
RUE DU PETIT CHANTIER
RUE DE LA CROIX
RUE DES TYRANS
RUE RIGORD
LES CHAMBRES D'ABBAYE
RUE ROBERT
FORT SAINT-NICOLAS
BLVD DE LA CORDERIE
GASPARD
COURS PIERRE PUGET
TUNNEL DU PRADO CARENAGE
To NOTRE-DAME DE LA GARDE and VALLON DES AUFFES (ENDOUME)
0   200 yds
0   200 m
© MOON.COM

Marseille Center

The **La Canebière** shopping street heads east, away from the old port toward **Palais Longchamp.** This is the least touristy of the five areas; here you encounter the city at work, with presses of people boarding trams and hurrying (as much as the Provençal can) from meeting to meeting. The standout sight is the Palais Longchamp, and for shopping the **Cours Julien.**

Toward Les Calanques

Heading south into the outskirts of Marseille along the coast, you really begin to appreciate

the two faces of Marseille. The big smoke fades, giving way to long beaches and parks and unforgettable inlets like **Vallon des Auffes,** a seaside port that would not feel out of place on a remote Greek island.

SIGHTS
★ Vieux Port

The entrance to the old port is guarded by two forts, Saint-Jean and Saint-Nicolas. These were built in the 17th century as symbols of royal authority. The port is ringed by restaurants and cafés, and it's a pleasant place to stroll, inspect the fishers' catches, and enjoy the atmosphere of the city. Norman Foster's L'Ombrière has added a touch of glinting modernism to quai de Rive Neuve on the old port and is a symbol of the city's forward-facing nature. The parasol-cum-pavilion, whose name literally means "the shade giver," was installed as part of the 2013 City of Culture urban-renewal program. It's an arresting installation, 48 m (157 ft) long and 22 m (72 ft) wide, composed of mirrored sheets that have the effect of inverting the surrounding world. Just opposite is the daily morning fish market on the quai du Port. Fish simply does not get any fresher, going straight from the boat to the shopping bag.

Château d'If and Iles de Frioul

Depart from 1 quai de la Fraternité, Vieux Port; tel. 04 96 11 03 50; www.lebateau-frioul-if.fr; June-Aug. daily every 30 min 6:30am-8:30pm, out of season hourly 6:30am-8:30pm; €11-23

The most famous of the four Iles de Frioul, thanks to Alexandre Dumas's story of *The Count of Monte Cristo,* is the Isle of If. Dumas's fictional hero Edmond Dantes was imprisoned for 14 years in the Château d'If before escaping. The château was built by Francis I in 1524, enabling the monarch to guard the new Royal Fleet at anchor outside

Marseille and to keep an eye on the city. It quickly became a prison. For literature enthusiasts, the tour of the château contains plenty of allusions to Dumas's book. The cells with graffiti scribbled on the walls leave no doubt as to the desolation felt by prisoners. A more uplifting experience is the view over the water back toward Marseille. Note that the château opens at 10:30am and shuts at 6pm (5:15pm Oct.-Mar.) and walking shoes are advisable. The ferry also stops at Port Frioul, which gives access to the other islands of the archipelago.

★ Notre-Dame de la Garde

rue Fort du Sanctuaire; tel. 04 91 13 40 80; www. notredamedelagarde.com; church daily 7am-6pm, museum Tues.-Sat. 10am-5pm

Sitting proudly atop the highest point in Marseille, Notre-Dame de la Garde is the most recognizable symbol of the city. The 11.2-m (37-ft) golden statue of the Virgin Mary and child that caps the dome of the church is a visible reminder to the Marseillais that someone is always watching over them. Built in 1852, the church incorporates pink and red marble columns, mosaic domes, and gold cornices inlaid with glinting stones. Everywhere there are small ex-voto messages pinned to the walls to give thanks to La Bonne Mere (Good Mother),

MARSEILLE AND THE VAR COAST · MARSEILLE

Notre-Dame de la Garde

as the church is affectionately called by the locals. A small museum contains some of the most colorful ex-votos, which range in style from simple plaques to inscribed model boats.

Le Panier
Hospice de la Vieille Charité

2 rue de la Charité; tel. 04 91 14 58 80; https://musees. marseille.fr; Tues.-Sun. 9am-6pm; free

It's something of a climb up to La Vieille Charité hospice, but the building and associated museums merit the effort. The hospice was built between 1671 and 1749 to treat the poor. Incredibly, after World War II this architecturally harmonious and arresting building was scheduled for demolition before being saved at the last minute. It now houses two excellent museums: one of Mediterranean archaeology that contains the largest Egyptology collection in France outside Paris, and the other of African, Oceanic, and American arts.

Euroméditerranée Project

TOP EXPERIENCE

★ MuCEM

7 promenade Robert Laffont; www.mucem.org; July-Aug. Wed.-Mon. 10am-8pm, Sept.-Oct. and May-June Wed.-Mon. 10am-7pm, Nov.-Apr. Wed.-Mon. 10am-6pm; €11, City Pass

The MuCEM is as much an architectural statement as a museum. It's a futuristic glass box, clad in a web-like concrete skin that the architect (Rudi Riciotti) claims was inspired by the ocean floor. Minimalist concrete bridges connect the box to Fort Saint-Jean and on from there to the Vieux Port. It's undeniably mesmerizing to look at and walk around. It's possible to visit without paying the entrance fee to the exhibitions, which, almost inevitably, fail to live up to the spectacle of the building that houses them. There are two permanent exhibitions: one focused on Mediterranean agriculture and food through the ages, and the other on Mediterranean cities in the 16th and

17th centuries. In addition there is always a nonpermanent exhibition with a broader non-Mediterranean concept.

Fort Saint-Jean

Vieux Port or MuCEM; July-Aug. Wed.-Mon. 10am-8pm, Sept.-Oct. and May-June Wed.-Mon. 10am-7pm, Nov.-Apr. Wed.-Mon. 10am-6pm; fort free, exhibitions with MuCEM pass (€11) or City Pass

The entrance to this fort, built in 1660, is across modernist pedestrian bridges, either from the MuCEM building or from the old port. Once inside visitors are free to wander around the battlements, duck through tunnels in the rock, and climb up to guard posts. The Mediterranean Garden of Migration, which is filled with herbs from Jewish, Islamic, and Christian medicinal traditions and planted with olive trees, is a pleasant place to picnic. The entrance to MuCEM includes a themed five-stage walk through the history of the fort beginning in the Galerie des Officiers. Temporary exhibitions are also staged in the George Henri Rivière building.

Cosquer Méditerranée

promenade Robert Laffont; www.grotte-cosquer.com; daily 9am-9pm

Constructed as part of the 2013 European City of Culture celebrations, the Villa Méditerranée is a striking building that has only just found its purpose, with the opening in 2022 of a replica of **La Grotte Cosquer,** a subterranean cave located in the nearby Calanque de Morgiou. In this cave, the earliest evidence of Stone Age humans in Provence was found, along with paintings dating back to 27,000 BCE. However, only experienced divers can reach the cave, hence the replica, which is surprisingly successful at re-creating the diving experience. Visitors descend below sea level in an elevator, and then tour the re-created cave on armchairs fixed on rails. It sounds weird but somehow the experience works. Back above the surface the visit is

---

1: Le Panier **2:** Fort Saint-Jean in the Vieux Port **3:** the MuCEM building

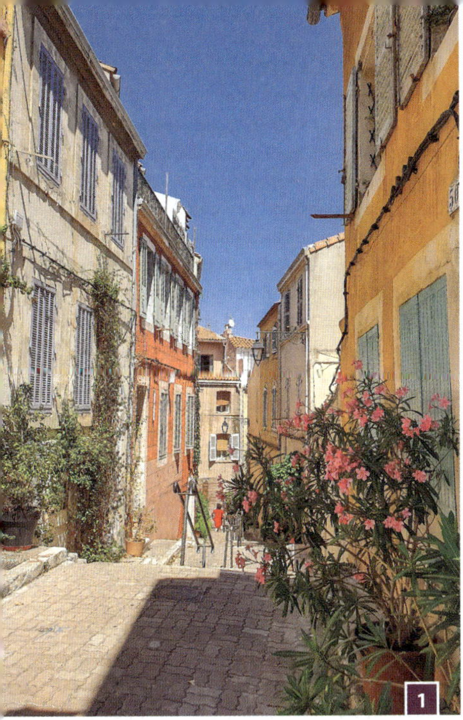
1

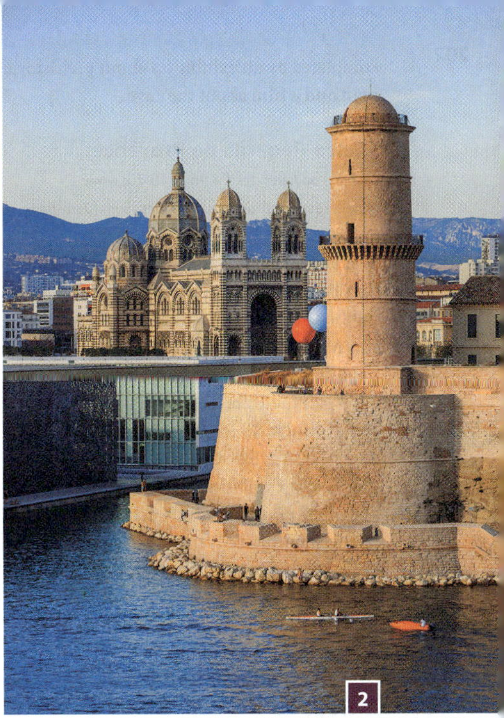
2

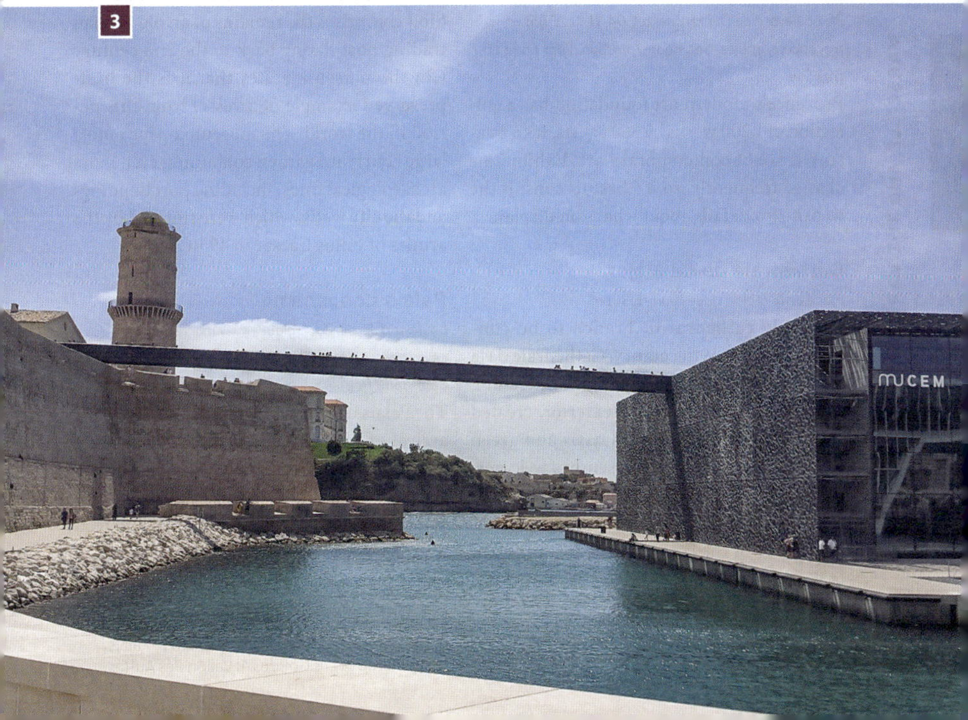
3
MUCEM

completed by an exhibition about prehistoric man and a film about the cave.

Musée Regards de Provence

avenue Vaudoyer; tel. 04 96 17 40 40; www. museeregardsdeprovence.com; Tues.-Sun. 10am-6pm; €7.50, City Pass

The aim of the Regards de Provence foundation is to assemble and make available to the public the artistic heritage of Marseille and the Mediterranean. Works in the permanent collection are displayed on two floors and span the 18th century to the present day. Artists include Pierre Ambrogiani, Louis Audibert, Auguste Chabaud, and Raoul Dufy. Temporary exhibitions change frequently and are often photographic. Facing the sea, opposite the MuCEM, the old sanitary station building that houses the collection is flooded by the Mediterranean light that stars in so many of the artworks. Viewing the collection is an uplifting experience.

Le Frac

20 boulevard de Dunkerque; tel. 04 91 91 27 55; www. fracpaca.org; Tues.-Sat. noon-7pm, Sun. 2pm-6pm; €5, free Sun.

Provence's modern art foundation has a collection of 1,200 works by 560 artists. It's a treasure trove of contemporary art. Exhibitions change frequently, so it's best to consult the website for details about what's on display.

Cathédrale de La Major

Place de la Major; daily 10am-6:30pm

The only cathedral in France to be constructed in the 19th century, Cathédrale de la Major is a wonder of byzantine architecture. The playful stripes on the exterior, created by alternating stone from Cassis and green Florentine marble, convey a sense of welcome. Inside there are yet more stripes, alternating a warmer colored marble with stone. The cathedral is one of the largest in the world, 140 m (459 ft) long with 60-m-high (197-ft) towers, and seating 3,000 people.

Marseille Center

Friche La Belle de Mai

41 rue Jobin; tel. 04 95 04 95 95; www.lafriche.org; exhibitions Wed.-Fri. 2pm-7pm, Sat.-Sun. 1pm-7pm, La Friche public spaces always open; €5 exhibitions

La Friche is constantly evolving and changing its identity. For the moment it's a cultural center where artists, dancers, and performers create; it's also a neighborhood center where children come to play and go to the cinema; it's a social housing project; it's an exhibition and event center; it's a place to hang out, drink a coffee, meet friends, and eat; above all it's a place of tremendous positive energy of which Marseille is rightly proud.

Musée Histoire de Marseille

2 rue Henri Barbusse; tel. 04 91 55 36 00; https:// musees.marseille.fr; Tues.-Sun. 9am-6pm; free

The museum delves into Marseille's rich past, telling the story of France's oldest city from its prehistoric origins to contemporary urban developments, with a focus on the personal stories of its inhabitants. The marquee exhibit comprises the remains of an old Roman trading boat dating back to the 2nd century CE; the museum states that it is the best-preserved example of a vessel from this period in the world. The museum opens onto a large courtyard garden containing even more archaeological finds, including port buildings and the city walls, which were stormed by the armies of Julius Caesar in 49 BCE.

Palais Longchamp

boulevard Jardin Zoologique; www.marseille.fr/ environnement/parcs-et-jardins/parc-longchamp; daily 8am-8pm; free

The Palais was opened in 1869 to celebrate the arrival of the Durance canal, which brought running water to the city. Water cascades from a decorative fountain at the summit of the immense monument known as the Château d'Eau, or water castle, into a middle basin, and then an artificial lake at the base. Climbing

---

1: the Cosquer Méditerranée museum **2:** Palais Longchamp **3:** Les Goudes **4:** Vallon des Auffes

to the top of the ornamental staircase takes you to a pleasant parkland square. Descending another set of steps takes you into a formal garden that used to be a zoo. To either side of the Château d'Eau two wings of the Palais Longchamp house the Musée des Beaux Arts and the Muséum d'Histoire Naturelle.

Toward Les Calanques
Mac

69 avenue d'Haifa; tel. 04 91 25 01 07; https://musees. marseille.fr; Tues.-Sun. 9am-6pm; permanent exhibitions free

Modern art needs a minimalist setting, and the pure white walls of Mac, Marseille's show-piece for contemporary art, certainly deliver. Works by local southern French sculptors and artists, such as César Baldaccini and Benjamin Vautier (known as Ben), mix with better-known names such as Warhol.

★ Vallon des Auffes (Endoume)

Corniche Kennedy, Endoume

This little port is surrounded by the pastel-colored houses of fishers. Small boats are pulled up at anchor, and workers sit mending nets while the Mediterranean gently laps around their feet. This face of Marseille couldn't be more different from the graffiti-marked streets of some areas. There are a couple of good restaurants where you can kick back, relax, and wonder whether the other-worldly big city will still be there when you climb back up the steps. In addition to Vallon des Auffes, Endoume itself has plenty more to offer. The neighborhood is a mixture of old fishing huts and mansion houses. There are three rocky inlets/calanques in Malmousque (Calanque du Cuivre, Anse de Maldormé, Anse de la Fausse Monnaie), and there's a coastal peninsula just past Vallon des Auffes where you can swim off the rocks.

Unité d'Habitation Cité Radieuse

280 boulevard Michelet; tel. 08 26 50 05 00; www. marseille-citeradieuse.org; daily 9am-6pm

After World War II, France faced a desperate shortage of affordable housing. The state asked renowned French-Swiss architect Le Corbusier to come up with a new way of building to solve the problem. Notorious for a 1925 plan to demolish much of central Paris and replace it with 60 skyscrapers surrounded by landing areas for planes, Corbusier had also developed less contentious proposals for the construction of uniform apartment buildings using modern materials. Marseille gave him his first public commission, and his response was a brutalist concrete block divided into modular housing. The exterior was so shocking that the Marseillais quickly christened the building the Maison de Fada, which roughly translates as the "nuthouse." Limited access to the building is free when you sign the visitor register at the entrance, or you can reserve a €15 tour at www.marseille-tourisme.com/en/experience/cite-radieuse-le-corbusier-marseille-en-3573874.

Les Goudes

chemin des Goudes

Les Goudes sits on the boundary between Marseille and Les Calanques. Technically still part of the city—it belongs to the 8th arrondissement—the small fishing port set back amid the rocks has a much more relaxed vibe. The Marseillais like to joke that the port exists at the end of the world, perhaps revealing a fear of what lies beyond their urban environment. Depending on whether you share the Marseillais attitude, Les Goudes is either as far as you are willing to venture or a staging point for a hike out into the wilds of Les Calanques. Whatever conclusion you come to, it's a nice place for a long lunch.

TOURS
Toctoc Marseille

28 boulevard Longchamp; tel. 06 26 89 24 78; www. visiteinsolitemarseille.com; from €55

Marseille is a large, busy city, and it is often overwhelming. For those short on time, the easiest way to explore all the best parts of the city is with a guide. Laurianne Collange is a charming local who tailors her tours to her clients and knows Marseille as well as anyone.

SPORTS AND RECREATION

Parks

Jardin and Palais du Pharo

58 boulevard Charles Livon; www.marseille.fr/environnement/parcs-et-jardins/jardin-du-pharo; Mon.-Fri. 7am-9pm

Perhaps the best way to appreciate the geography of the old port is to climb the adjacent hill to the Jardin du Pharo. High on the headland, this magnificent garden houses the Palais du Pharo, which was intended as a residence for Napoleon III. Construction of the Palais started in 1852, but unfortunately for Napoleon he fell from power before he could take up residence. The building now has a more prosaic use as a conference center. On the lawn in front of the Palais is a moving monument to sailors lost at sea in World War I.

Parc Borely

avenue du Parc Borely; tel. 04 91 76 59 38; www.marseille.fr/environnement/parcs-et-jardins/parc-borely; daily 6am-9pm

This park consists of a beautiful series of formal gardens: Chinese, rose, botanical, and English. There's a pleasant lake that brings much-needed freshness, and for those who are not yet sated by Marseille culture, the 18th-century Château Borely houses a museum of decorative arts. The park is adjacent to the Plage de la Pointe Rouge.

Beaches

The Marseillais are a sun-loving lot. From early spring to late autumn, they hit the beaches. Expect skimpy swimwear, bling sunglasses, and crowds of youngsters playing beach volleyball with rap music booming in the background. Head away from the public beaches and take to the rocks of Endoume if you are looking for a more peaceful experience.

Plage des Catalans

Opposite 6 rue de Catalans

Almost in the city center, in the Pharo neighborhood to the south of the old port, this small sandy beach fills up quickly in the summer. Even in midwinter, you'll find some brave locals soaking up the sun in a sheltered spot before an icy dip. The beach is served by bus line 82 or 83 from the Vieux Port.

Plages du Prado

Intersection of promenade Georges Pompidou and avenue du Prado

This series of artificial sandy beaches, located to the south of the city center in the 8th arrondissement, was created in the 1970s to offer the Marseillais better access to the sea. They are now very popular and back onto a large park (Park Balneaire du Prado). There are more than 1,000 parking spaces, plus the beaches are served by bus line 83 from the Vieux Port.

Plage de La Pointe Rouge

Opposite 39 avenue de Montredon

Farther south along the coast from the Prado beaches, La Pointe Rouge is another busy beach. It's bordered by cafés and restaurants and is seemingly always lively. There's a yachting and kitesurfing school. The beach is served by bus line 583 from the Vieux Port.

Water Sports

Yachting Club de la Pointe Rouge

Port de la Pointe Rouge; tel. 04 91 73 06 75; www.ycpr.net; Tues., Wed., and Fri. 9am-12:30pm and 2pm-6:30pm, Thurs. 8am-noon, Sat. 9am-12:30pm and 2pm-6pm

This welcoming club offers a range of sporting activities, including sailing, diving, paddleboarding, and windsurfing. There's an on-site restaurant and courses for kids in the summer, including a five-day morning sailing school for €200.

Marseille Kite Klub

Entre no 3, Digue Ouest, Port de la Pointe Rouge; tel. 06 18 73 29 93; www.marseillekiteclub.com; daily with reservation; €130/half day

On the other side of the port from the yachting

club, this kitesurfing school offers courses for all levels.

Cycling
Fada E Bike Tours
Meeting point 40 rue Plan Fourmiguier; tel. 07 82 00 73 47; https://gestion.fada.bike; Mar.-Oct. year-round for groups; from €49 per person

E-bikes are a great way to take in a lot of Marseille in just one day. This company offers various guide-led tours, including all the main sights of the city, as well as longer trips out to Les Calanques and Cassis.

Pétanque (Boule)
Mondial La Marseillaise
Throughout Marseille; www. mondiallamarseillaiseapetanque.com; second week in July

The Mondial Marseillaise is the pétanque equivalent of the soccer World Cup. Teams travel from all over the world to take part. There's one important difference, though: Anyone can enter. Sign up on the website and you'll be allocated a pitch number somewhere in Marseille (there are more than 1,000 pitches). Then turn up at the appointed time, and you're off. The competition is a knockout format. After four days of intense competition, the surviving players meet in the televised final, which takes place in Marseille's old port.

Maison de La Boule
4 Place des 13 Cantons; tel. 04 88 44 39 44; www. laboulebleue.fr; Apr.-Aug. daily 10am-7pm, Sept. Wed.-Sun. 10am-7pm, Oct.-Mar. Wed.-Sun. 10am-6pm

Part shop, part museum, part sports club, the Maison de La Boule is the place to go if you are interested in pétanque. There's an indoor pitch if you can't wait to test your new purchases.

FESTIVALS AND EVENTS
Jazz des Cinq Continents
Various locations, Marseille; www.marseillejazz.com; mid-July; from €20

For just over a week in July, the city celebrates jazz music with large open-air concerts in spectacular venues such as outside the Palais Longchamp, or at the entrance to MuCEM. The festival builds on a long tradition of jazz in Marseille. At the end of World War II, *Jazz Hot* magazine christened the city the "new capital of French jazz."

Festival de Marseille
Various locations, Marseille; www.festivaldemarseille. com; mid-June-early July; from €15

Like the city, this festival is open to the world, celebrating all cultures and all forms of expression. Dance, music, opera, mime—the shows make the most of Marseille's dramatic cityscape as a backdrop to performances.

SHOPPING
Shopping Districts and Streets
Le Panier
www.lepanierdemarseille.com/shopping

Le Panier is Marseille's oldest quarter. It's an atmospheric area to shop with its mix of cobbled narrow streets and walls covered in street art. The shops are small and tend to be owner-run. They offer individual creations, including works of art, ceramics, clothes, santons (nativity figures), and soaps.

La Canebière
Marseille's high street is a broad, tree-lined avenue running away from the coast. It's a little run-down, and the shops are now a mixture of chain stores and ethnic grocery stores. Its purpose is more to serve the practical needs of the population, rather than tourists.

Cours Julien
Once an edgy neighborhood, the Cours Julien is now one of the trendiest in Marseille. Walls are, of course, covered in graffiti, but that's Marseille. As in Le Panier, there are lots of small boutiques offering artisan clothes, jewelry, and handbags. If you are looking for an original piece, it's the perfect place to shop.

A History of Pétanque and How to Play

HISTORY

Back at the turn of the 20th century, in the seaside town of La Ciotat, which neighbors Marseille, poor Jules Lenoir was so afflicted by rheumatism that he could no longer play his favorite game: jeu provençal. The three leaps required before throwing his boule toward the target were too much for him. A local café owner took pity and suggested a version of the game where players plant their feet before throwing. Because leaping around in the hot sun is rather tiring, the new game, which was named pétanque (meaning "stuck feet" in Provençal), quickly took off.

pétanque

RULES

The rules are simple: A formal match has four players, two on each side. Play begins with the drawing of a circle in the gravel from which the target ball (cochonette) is thrown. The teams then try to get their boule the closest to the cochonette. A team continues to throw until they manage to get a boule closer than the opposing team. Play then switches to the opposing team, until they get closer. An end is reached when all players have thrown all their boules (customarily three each). One point is awarded for each boule that is closer to the cochonette than any of the boules from the opposing side. The circle is redrawn in the gravel and the cochonette thrown once again. A match is won when one side gets to 13 points.

Beware, if you lose 13-0 at pétanque (also often simply called "boule") you are required to "Kiss the Fanny." This tradition is believed to have originated in the same café in La Ciotat where the game was invented. Fanny was a notably fetching barmaid with a penchant for wearing sexy stockings. As recompense to whitewashed losers, she lifted her skirt and allowed them to kiss her bottom. There is usually a statue of Fanny in a boule club, with an amply pert posterior to kiss.

WHERE TO PLAY

There are plenty of spots across Marseille to play boule. The rule of thumb is that if you find a suitable area of gravel where there aren't too many pedestrians, then you're fine to start playing. One such scenic spot is immediately to the right of the **Cathédrale la Major** as you stand facing it.

Les Docks and Terrasses du Port

quai du Lazaret; www.lesterrassesduport.com
These two new shopping centers near the Euroméditerranée project are filled with all the high-end brands. There's nothing original here, but it's all very practical and air-conditioned.

Antiques and Brocante
Quartier des Antiquaires

rue Edmond Rostand; www.antiquairesmarseille.com; hours of shops vary, most closed Sun.

Since the 1950s this quarter of Marseille has been known for its antiques. The eclectic range of shops offers everything from guitars to paintings, oriental antiques, and Provençal furniture. In the South of France, this quarter of Marseille is second only to L'Isle-sur-la-Sorgue for the range of items available.

Centre Commercial Les Puces

130 chemin de la Madrague de la Ville; www. centrecommerciallespuces.com; Tues.-Sun. 8:30am-7:30pm

If you love shopping and don't mind crowds, this 20-year-old flea market is a must-visit. Part indoor, part outdoor, the center is filled with objects and furniture from house clearance sales. It's the place to pick up a real bargain.

Crafts
Cruise Passenger Market

Vieux Port; May-Oct. 3pm-midnight

Although it's not the most auspicious name for a market, the Cruise Passenger Market is quite enjoyable. Stroll along the quayside in the warm summer air, browsing artisan-made local products. Expect plenty of bundles of lavender, jewelry, olive wood boards, salad bowls, and so forth.

Marseille Specialties
Bazar de César

4 Montée des Accoules; tel. 06 19 70 95 76; www. lebazardecesar.com; Mon.-Sat. 9:30am-7pm

Soapmaking began in Marseille in 1307, and the city is still famous for it today. By law, Marseille soap must contain 72 percent oil, making it exceptionally pure and kind to sensitive skin.

Les Navettes des Accoules

68 rue Caisserie; tel. 04 91 90 99 42; www.les-navettes-des-accoules.com; Mon.-Sat. 9:30am-7pm, Sun. 10am-6pm

You'll find navettes across Provence. They are biscuits prepared with a dash of fleur d'oranger and other spices. The city that gave birth to them was Marseille in 1781. Their boat-like shape was supposedly designed to recall the landing of Mary Magdalene in the Camargue.

FOOD

Marseille is the most exciting place to eat out in Provence. There is a huge variety of restaurants to choose from and a constantly evolving scene with young chefs opening innovative concept restaurants. Seafood is, of course, big in Marseille; look out for towering plates of shellfish (fruits de mer) on ice, any dish with sardines in it (the fish is the unofficial symbol of the city), and, of course, the king of fish soups: bouillabaisse.

Vieux Port
Café Simon

28 cours Honoré d'Estienne d'Orves; tel. 04 91 33 05 14; daily 8am-11pm; €14-22

A little back from the old port, Café Simon is another spot for fish and shellfish. The prices are competitive, and the service is welcoming. It's probably the best restaurant on the attractive Place Estienne d'Orves.

Le Chalet du Pharo

58 boulevard Charles Livon; tel. 04 91 52 80 11; www. le-chalet-du-pharo.com; daily noon-2:30pm and 7pm-10:30pm; €25-38

Located on the hillside park above the old port, this restaurant affords beautiful views across the harbor. It's particularly atmospheric at night when the city lights wink back. Cod filet with a crab bisque is excellent.

Chez Loury

3 rue Fortia; tel. 04 91 33 09 73; www.loury.com; Mon.-Sat. noon-2:30pm and 7:30pm-10:30pm; €19-48

A traditional-style restaurant with 35 covers and white tablecloths, Chez Loury is known for its excellent bouillabaisse, served either in the traditional way over three courses, or for those with less time, as a simple one-course soup.

Le Panier
Cup of Tea

1 rue Caisserie; tel. 04 91 90 84 02; Sat.-Tues. 10am-6pm; €4

A popular stop on the way up the hill to the Vieille Charité, this combination café and bookshop has a shady terrace and an interior overflowing with books to buy.

Bouillabaisse

bouillabaisse, a signature Marseille dish

The Marseillais are known for exaggerating to the very limit of credibility, and so it's no surprise that it was this city that transformed a humble fish soup into a dish famed around the world. The story begins with the local fishermen perfecting a dish from the unsellable leftovers of their catch. Bony, practically fleshless rockfish were added to boiling water, and from the resulting stock is made the most intense fish soup imaginable.

Local restaurateurs got hold of the recipe and popped in some more delectable morsels of fish at the last moment. Croutons were another addition, toasted and then spread with a powerful garlic, pepper, and saffron sauce. What had once been a dish for the poor was rechristened as the reason God invented fish.

Eating bouillabaisse is expensive. It usually costs more than €50 per person, but the experience is dramatic and tasty. Waiters parade the raw fish on a silver platter for customers to check the freshness. Each ladle is carefully measured into the bowl with solemn reverence. Delicate, perfumed filets are then doused with even more soup.

WHERE TO FIND BOUILLABAISSE

- **Chez Loury** in the old port (page 208).
- **Chez Fonfon** near the Vallon des Auffes (page 210).

Euroméditerranée Project
Au Bout du Quai

1 avenue Saint-Jean; tel. 04 91 99 53 36; Thurs.-Sat. and Mon.-Tues. noon-2:30pm and 8pm-10:30pm; €25-36

Another exception to the "don't eat on the old port" rule, this restaurant, right at the tip of the port next to Fort Saint-Jean, serves good fresh fish. The service is welcoming and the crowd is a mixture of locals and tourists. Reservations are usually not necessary.

Marseille Center
★ Maison Calambo

2 avenue Maréchal Foch; tel. 04 91 34 56 85; www. maison-calambo.fr; Tues.-Sat. 9am-7:30pm; €8-20

Outside this family-run seafood institution, icy trays of oysters, mussels, and prawns sit ready for takeaway orders. Inside, the vibe is relaxed as locals and tourists enjoy seafood platters, licking their fingers as they go. Reserve to be sure of a table.

La Boîte à Sardines

2 rue de la Libération; tel. 04 91 50 95 95; www. laboiteasardine.com; Tues.-Sat. noon-3pm; €20-35

There's no doubt about the specialty of the house: sardines. If you're not in the mood, then there's fresh fish of the day and shellfish on ice. The nets and starfish hanging from the ceiling play up the nautical vibe of this Le Panier institution.

La Femme de Boucher

10 rue du Village; tel. 04 91 48 79 65; Mon. and Thurs.-Fri. noon-2pm and 8pm-10pm, Tues.-Wed. noon-2pm; €28-35

As the name suggests, this is one for meat lovers. Laeitita Visse is the eponymous chef, and her restaurant has quickly become a destination for Marseille foodies looking for traditional French cooking in a relaxed, unfussy setting.

Livingston

5 rue Crudere; tel. 04 96 10 00 00; www. livingstonmarseille.com; daily 7pm-10:30pm; €39

This funky new venue has a talented young chef and a menu dedicated to locally sourced seasonal ingredients. It's definitely one for natural wine-loving oenophiles: The wine list is filled with plenty of trendy orange wines (white wines vinified like red wines).

Toward Les Calanques

★ Chez Fonfon

140 rue du Vallon des Auffes; tel. 04 91 52 14 38; www. chez-fonfon.com; daily noon-2pm and 7pm-10pm; seafood €12 per 100 g, bouillabaisse €62 per person

Since 1952 Chez Fonfon has been offering what is widely considered one of the best bouillabaisses in Marseille. Against the idyllic backdrop of Vallon des Auffes, the king of fish soups is served to diners by a coterie of waiters with the pomp and ceremony usually reserved for visiting dignitaries. Reservations are necessary.

L'Écaillerie

26 rue Endoume; tel. 06 16 06 64 73; www. lecailleriemarseille.fr; Mon.-Sat. 6pm-11pm, Sun. 11am-11pm; €8-55

Having lunch on the square while enjoying a glass of biodynamic white and a plateau of shellfish on ice: L'Écaillerie offers a little slice of Med heaven. The menu majors on shellfish, but there are also interesting options like taco ceviche.

BARS AND NIGHTLIFE

Whether it is dancing the night away in a rooftop bar, swaying by a pool, shivering by an ice bar, or sipping cocktails in a chic club, the Marseillais love a good night out. Big-name DJs get the crowds going, and girls and boys drip with gold chains and rings. Wearing sunglasses late into the night is encouraged (perhaps to protect sensitive eyes from the stray rays of a strobe light).

Beach Bars and Clubs
Sport Beach Café

138 avenue Pierre Mendès; tel. 04 28 31 24 56; www. sportbeach.fr; Mon. noon-3pm, Tues.-Thurs. 7pm-11pm, Fri.-Sat. 7pm-2am; drinks from €6

This beach club is a popular spot for an evening drink during the week, but it really gets going on Friday and Saturday nights, with a glamorous crowd drinking and dancing around the pool.

Le R2

Les Terrasses du Port, 9 quai du Lazaret; tel. 04 91 91 79 39; https://lerooftopdesterrasses.com; nightly 7pm-2am; admission from €15

At Le R2 you can dance the night away or simply chill out on a sofa overlooking the sea. The R2 is a rooftop bar and nightclub that draws in the crowds during the summer with different nightly themes.

Bars

Gaspard

*7 boulevard Notre-Dame; tel. 06 88 23 86 66; www.
facebook.com/bargaspard; Tues.-Sat. 6pm-1am;
cocktails €10-12*

One of Marseille's best-known cocktail bars
is Gaspard. It has a lively ambience, with
the bartenders indulging in plenty of showy
throwing of cocktail shakers. Taking a stool
at the wood-paneled bar is a great start to an
evening out on the town.

La Bissette

*17 rue André Poggioli; tel. 06 75 00 39 03; Tues.-Sat.
6pm-2am; drinks from €5*

This funky cocktail bar is enjoying its moment in the sun thanks to its crowd-pleasing
mix of tapas, cocktails, and DJ sets.

Nightclubs

Trolleybus

*24 quai de Rive Neuve; tel. 04 91 54 30 45; http://
letrolley.com; Tues. and Thurs.-Sat. midnight-6am; €10
entry*

Located in an old arsenal, this nightclub consists of three separate rooms: dance club,
whiskey bar, and private club. The old brick
ceilings and long, narrow rooms give it an intimate feel.

Le Bazar

*90 boulevard Rabatau; tel. 06 58 52 15 15; www.
bazarmarseille.com; Thurs.-Sat. midnight-6pm; €25
entry*

It's all going on at Le Bazar. In the summer
there's an open-air club, and in the winter
an ice bar. Dress smart because the bouncers
working the door are known to be very picky.

Live Music

Son des Guitares

*16 rue Corneille; tel. 07 82 48 31 41; Thurs.-Sat.
11:30pm-5am; drinks from €10*

Be warned, you might need a good word from
a contact to get past the doorman at Son des
Guitares. Try asking the hotel concierge or a

tour guide. Once you're in, the atmosphere is
unique, with passionate Corsican musicians
playing live.

Le Jam

*42 rue des Trois Rois; tel. 06 09 53 40 41; http://lejam.
unblog.fr; doors 7:30pm, show 8:30pm; €13*

This intimate bar attracts top jazz artists, and
the tables and chairs are pushed right up close
to the performers. It's best to reserve by text
message in advance. See the website for show
details.

ACCOMMODATIONS
Vieux Port

Maison Montgrand

*35 rue Montgrand; tel. 04 91 00 35 20; www.maison-
montgrand.com; €100 d*

A good choice for families, the Montgrand has
a wide selection of rooms, including two that
sleep four and one that sleeps six. It's a few
streets back from the old port, with a modern
boutique hotel vibe. There's a funky bar and
even a concept store.

★ Les Chambres d'Abbaye

*8 rue Petit Chantier; tel. 06 09 75 87 12; www.
leschambresdelabbaye.fr; €135 d*

If you are looking for something a little bit
different, consider this three-bedroom bed-
and-breakfast just a minute's walk from the
old port. My favorite room is the Chambre
Ninon, which opens onto the small courtyard garden.

La Residence du Vieux Port

*18 quai du Port; tel. 04 91 91 91 22; www.hotel-
residence-marseille.com; €240 d*

This property offers color-filled rooms with
terraces overlooking the old port and Notre-
Dame de la Garde. The suites are particularly
large (up to 60 sq m/645 sq ft) and have stunning picture windows. There's an excellent
on-site restaurant to enjoy when your feet just
can't carry you any farther. Private parking is
available on request.

Le Panier
Au Vieux Panier

13 rue du Panier; tel. 04 91 91 23 72; www.auvieuxpanier. com; €100 d

This is a good place to stay to experience the unique atmosphere of the Panier quarter of Marseille. Rooms come in three sizes: small, medium, and large. The rooms are redecorated every year by an upcoming artist. There are fans but no air-conditioning.

InterContinental Hotel Dieu

1 Place Daviel; tel. 04 13 42 42 42; www. intercontinental.com/marseille; €330 d

Facing Notre-Dame de la Garde, the palace-like building dominates one side of the Vieux Port at the entrance to Le Panier district. Even if you don't stay, drinks on the terrace are a must. Private parking is available.

Euroméditerranée Project
Best Western La Joliette

49 avenue Robert Schuman; tel. 04 96 11 49 49; https:// hotel-joliette.com; €160 d

This location is perfect for those wishing to dip into Marseille for a visit to the Euroméditerranée project. It's on the main street that runs parallel to the quayside. Rooms are decorated in a contemporary style and private parking is available on request.

Marseille Center
Casa Ortega

46 rue des Petites Maries; tel. 06 80 62 53 21; www. casa-ortega.com; €100 d

Caroline Contoz is a welcoming host. Full of local knowledge, she's always available to help you get the very best out of your stay in Marseille. The property is decorated with an eclectic mix of furniture from brocantes.

★ Le Ryad

16 rue Sénac de Meilhan; tel. 04 91 47 74 54; www. leryad.fr; €120 d

Le Ryad is an oasis of calm just off Marseille's bustling main shopping street, La Canebière. Rooms have delightful arched recesses, and the courtyard garden is the perfect place to relax to the sound of water bubbling away in the fountain. Note there is no air-conditioning.

Toward Les Calanques
Nhow Marseille

200 Corniche President JF Kennedy; tel. 04 91 16 19 00; www.nh-hotels.com; €250 d

A picture of minimalist modernist design right on the seafront, the Nhow is a great choice for those wishing to enjoy a little sightseeing before retreating to a large pool with a sumptuous view of the Med.

★ Les Bords de Mer

52 Corniche President JF Kennedy; tel. 04 13 94 34 00; www.lesbordsdemer.com; €250 d

Five minutes from both the first calanque and the old port, Les Bords de Mer has an unequaled location right on the Plage des Catalans. A restorative morning dip in the sea before the crowds arrive is available just outside the door.

INFORMATION AND SERVICES
Visitor Information
Marseille Tourist Office

11 La Canebière; tel. 08 26 50 05 00; www.marseille-tourisme.com

This is the best place for maps, tickets (including the City Pass), and general advice. The **City Pass** can be purchased for 24, 48, or 72 hours (€26, €35, and €42, respectively). It entitles holders to free use of the city's transport system (buses, trams, ferries, metro), free entry into the city museums, reduced entry prices into events and exhibitions, and many other benefits. Full details can be found at www.marseille-tourisme. com.

Post Office

The **La Poste** nearest the old port is located at 50 rue de Rome. It's open Monday and Wednesday-Friday 8:30am-6:30pm, Tuesday 8:30am-noon and 2pm-6pm, and Saturday 8:30am-12:30pm. There are numerous other post offices throughout the city.

Foreign Consulates

The **British Consulate** is located at the docks of Marseille-La Joliette (10 Place de la Joliette Atrium 10, 31st Floor; tel. 04 91 15 72 10). The **U.S. Consulate** is located at 12 boulevard Paul Peytral (tel. 01 43 12 48 85). Consulates for other countries are located in Paris only.

Health and Safety
Police

Police headquarters are located on Marseille's busy central shopping street at 66 La Canebière; tel. 04 88 77 58 00.

Hospitals and Pharmacies

To see a doctor, go to **Vieux Port Medical Center** (48 rue de la Republique; tel. 04 91 90 65 64; Mon.-Sat. 9am-6pm, Sun. 9am-noon). In a medical emergency, go to **Hôpital de la Timone** (264 rue Saint-Pierre; tel. 04 13 42 97 00; daily 24 hr).

Pharmacies are plentiful in Marseille. They all display a green cross symbol, which usually juts out into the street. On the old port, the pharmacy is located at 4 quai du Port. It is open daily 9am-8pm.

Communications
Wi-Fi

Large public spaces such as the Gare de Marseille-Saint-Charles and Marseille airport offer free Wi-Fi. The city is committed to offering free Wi-Fi on public transport, but it is not universally available yet. When taking public transport, check your phone for available networks. Restaurants, bars, and cafés in Marseille may offer their customers free Wi-Fi. Ask before you order.

Newspapers

The regional paper *La Provence* publishes a Marseille edition, concentrating on news from the city and across Provence.

GETTING THERE
Air
Marseille Provence Airport

Marseille Provence Airport is located 20 minutes out of town. It serves more than 100 destinations with scheduled direct flights, and 40 European destinations with flights from its low-cost terminal. **Air Transat** and **Air Canada** fly from Montreal to Marseille airport, but most North Americans will have to fly to a European hub and connect with a flight to Marseille. Transavia offers a limited service from Dubai to Marseille. There are 20 daily connections between Paris and Marseille. There are also plentiful connections between Marseille and the UK as follows: **EasyJet** flies from London Gatwick and Bristol to Marseille; **Ryan Air** flies from London Stansted, East Midlands, Manchester, Edinburgh, Glasgow, Shannon, and Dublin to Marseille; and **British Airways** flies from London Heathrow to Marseille.

A **shuttle bus** runs from the airport to the city center every 15 minutes between 4:30am and 12:10am, costing €8 for a one-way trip and and €12 for a round-trip. A **taxi** from the airport to the center of Marseille costs €65.

Car

From **Marseille Provence Airport,** take the A55, covering 24 km (15 mi) in 20 minutes.

From **Aix,** take the A51 and A7, covering 32 km (20 mi) in 35 minutes.

From **Nice,** take the A8 for 200 km (124 mi) in 2.25 hours. From **Cannes,** take the A8 for 180 km (112 mi) in 2 hours 5 minutes.

From **Barcelona,** take the AP-7 and A9, covering 500 km (311 mi) in 4.75 hours.

Train
Marseille-Saint-Charles Station

Marseille-Saint-Charles Station is just over 3 hours from Paris by TGV (bullet train). There is service nearly every hour, and most TGVs also stop at both Avignon and Aix-en-Provence. The price of TGV tickets varies widely depending on the day, time of travel, and how far in advance you book. A TGV ticket to Paris can be purchased for as little as €25 or can cost over €200. Ouigo tickets are the cheapest; these are basic, low-cost TGV trains without buffet cars.

Bus

The Marseille **bus station** is located at rue Jacques Bory, next to the Marseille-Saint-Charles train station.

From Aix-en-Provence, bus **line 50** (www.lametropolemobilite.fr) runs every 5 minutes between the Aix and Marseille bus stations. The journey time is 30 minutes and the cost €6.

BlaBlaCar (www.blablacar.fr) and **Flixbus** (www.flixbus.fr) operate regular services between Avignon and Marseille; the trip takes 1.25 hours and prices vary from €8 to €40.

Flixbus runs an hourly service from Nice airport to Marseille-Saint-Charles train station. Prices vary from €20 to €50.

From near Cannes, **BlaBlaCar** runs a service four times a day from Route du Grasse Antibes; it takes 2.5 hours and costs around €20.

From Barcelona, **BlaBlaCar** offers service from various providers, with 11 buses a day departing from different locations in Barcelona. Prices are around €50.

GETTING AROUND

Marseille is famous for its heavy traffic and for the eccentric—or dangerous, depending on your point of view—driving of its inhabitants. Avoid driving in Marseille if possible. It's also better to avoid the bus network, because you will inevitably become snarled in traffic. Check distances before walking between the sights listed in this book. Marseille is a big city, and walking from place to place takes a long time. Fortunately, there are excellent metro, tram, and ferry services. All information about public transport can be found at www.rtm.fr.

Transit Passes

By purchasing a **City Pass** (www.rtm.fr) to the city's museums and sights, tourists visiting Marseille can use public transport for free. Visitors without a City Pass should purchase a **Carte 10 voyages** (solo) for €15 from ticket offices in tram and metro stations. This carte allows multiple journeys across the network (except ferries) within a 1-hour time period.

Visitors planning a long stay should consider applying for a **Carte de Transport** by bringing a piece of identification and a passport photo to any of the network's ticket offices. The carte is validated every journey and can be recharged in metro and tram stations.

Metro

The Marseille Metro (www.rtm.fr/sites/default/files/planaxeslourds.pdf) has two lines. **M1** almost—but not quite—completes a circle of central Marseille. **M2** runs north to south roughly parallel with the coast.

Tram

Marseille has three tram lines. They are modern, efficient, and do not get held up by the terrible traffic. **T1** runs diagonally across the center of the city from east to west. **T2** and **T3** depart just to the north of the Euroméditerranée project and run through the city center. T2 turns east toward the Palais Longchamp, whereas T3 continues to the south, ending just to the east of Notre-Dame de la Garde.

E-bike/Scooter

The international e-bike/scooter company **Lime** operates in Marseille. Reserve an e-bike or scooter online using the Lime app. Prices start at €1 to unlock and then cost €0.15 per minute thereafter.

Taxi/Uber

Les Taxis Marseillais can be booked by calling (tel. 04 91 92 92 92) or by downloading their app from your app store. Rides are metered. **Uber** also operates within Marseille.

Boat Shuttle
La Navette

Ferries run in both directions from May to the end of September between:

The **Vieux Port** and **Pointe Rouge,** every hour 8am-7pm.

The **Vieux Port** and **L'Estaque,** every hour 8:30am-7:30pm.

Pointe Rouge and **Les Goudes,** every hour 8:50am-7:30pm.

Tickets are purchased on board and cost €5. Check for timetable updates at www.rtm.fr/lanavette.

The Ferry Boat

By far the best way to get across the old port is by taking the legendary Ferry Boat. The ferry made its first crossing in 1880, traveling from the Place aux Huiles (quai de Rive Neuve) on the south side to the Mairie on the north side. Today the ferry runs every 20 minutes in both directions from 7:30am to 8:30pm. The journey costs 50 cents (free with City Pass) and takes 10 minutes (www.rtm.fr/ferry-boat).

Les Calanques National Park

Christianity tells us that one is supposed to ascend into heaven and descend into hell. God must have forgotten this small nuance when he created Les Calanques. Access by road is restricted, and most visitors hike down mountain paths to visit the string of rocky inlets that punctuate the limestone cliffs between Marseille and Cassis. The deserted beaches and turquoise waters that await resemble many people's vision of paradise. Cassis, the seaside town that sits at the easterly limit of Les Calanques, is one of the most attractive resorts in the South of France.

ORIENTATION

Cassis is located 25 km (16 mi) to the east of Marseille. The coast between the seaside resort and the city constitutes **Les Calanques National Park.** The inland limit of the national park is delineated by the **D559** road, which runs between Cassis and Marseille. The largest calanques of **Morgiou** and **Sormiou** are accessed off this road, which runs into the suburbs of Marseille. The other calanques are accessed either by boat or on foot by taking the **GR98-51** walking trail that runs between the Calanque

of **Callelongue** (which is a few kilometers to the east of the **Port des Goudes** in Marseille) to Calanque de **Port-Miou,** 1 km (0.6 mi) to the west of Cassis.

★ CASSIS

The pastel-colored houses that surround the half-moon harbor of Cassis are reminiscent of Saint-Tropez. From here, boat trips depart regularly into Les Calanques. The more active can follow in kayaks. However, many travelers find Cassis, with its beaches and narrow cobbled shopping streets, captivating enough, particularly since the two nearest calanques—Port-Miou and Port Pin—are easily accessible on foot from a parking area adjoining the town.

Sights
Cap Canaille

Route des Crêtes

A good pit stop on the way into or out of Cassis is Cap Canaille, France's highest sea cliff. Follow the Route des Crêtes (D141) from Cassis to the parking area at the top of the cliff. From here, there is a marvelous sweeping view of the Mediterranean, the cliffs of Les Calanques, and the Bay of Cassis.

Les Calanques

To Central Marseille

D559

LES BAUMETTES

CHEMIN DE MORGIOU

AVENUE DE LUMINY

LUMINY

CHEMIN DE SORMIOU

Parc National

des Calanques

Mont Puget ▲

BAR NAUTIC ▼

LE CHÂTEAU ▲

★ CALANQUE SORMIOU

★ CALANQUE MORGIOU

Cap Sugiton

Pointe du Vaisseau

Cap Redon

Pointe du Figuier

Pointe de la Voile

✪ **LES CALANQUES**

Mediterranean

0   0.5 mi

0   0.5 km

Wineries
Domaine Paternel

1 route Pierre Imbert; tel. 04 42 01 77 03; www. domainedupaternel.com; daily 9:30am-noon and 2:30pm-6pm; tasting free, full vineyard visit €20

Provence is not well known for its white wines. The one exception is Cassis. The steep hillsides around the port produce the region's best white. Thanks to its citrus fruit notes, Cassis is considered one of the best wines to accompany seafood in France, and it can be found on wine lists in Paris's top restaurants.

Domaine Paternel is one of the most respected producers, and its vineyard and tasting room make an easy stop on the way into or out of town.

Beaches
Plage de La Grande Mer

Esplanade Charles de Gaulle

Cassis's main beach is a mixture of stone and sand. As the name suggests, it's open to the sea. As a result, the waves can be powerful and children need to be watched

© MOON.COM

carefully. The beach is to your left when standing in the port and looking out to sea.

Plage du Bestouan

avenue de l'Amiral Gauteaume

Just around the headland, on the opposite side of the port from the Plage de La Grande Mer, is the small cove Plage du Bestouan. It's picturesque and sandy, with lagoon-like water. As you would expect, it gets very busy in the summer.

Hiking

Sentier de Petit Prince

Distance: *2 km (1.2 mi) round-trip*

Time: *1 hour*

Information and Maps: *www.alltrails.com/fr/ randonnee/france/bouches-du-rhone/sentier-du-petit-prince*

Trailhead: *Port-Miou parking lot*

If you are in Cassis and don't have the time for a full excursion into Les Calanques, opt for the Sentier de Petit Prince, a marked trail around the headland that runs

Cassis

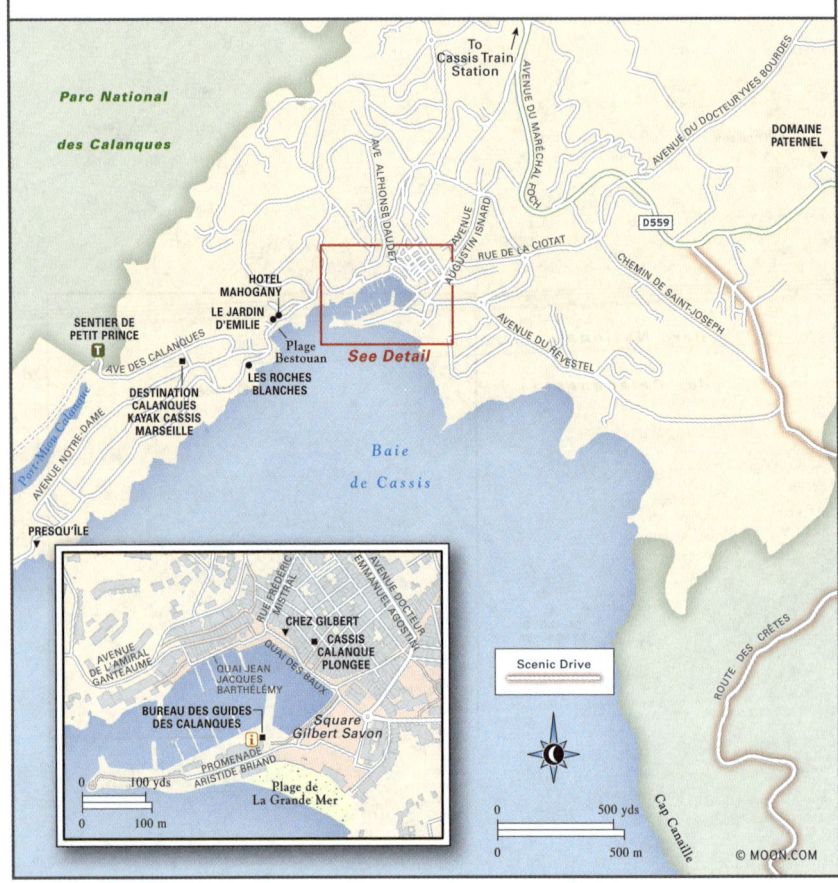

parallel to the Port-Miou calanque for half its length. This 2-km (1.2-mi) walk takes about 60 minutes to complete. From the center of Cassis, walk or drive along avenue des Calanques until you come to the Port-Miou parking. From here, you can pick up the sentier.

Bureau des Guides des Calanques

rue Séverin Icard; tel. 06 61 50 38 48; www.guides-calanques.com; €25 per person/half day

These experienced local guides help you get the best out of Les Calanques. They are experts in the local flora, fauna, and wildlife,

and they know the hidden routes off the main hiking trails. Excursions can be for a couple of hours or a whole day. They also offer climbing classes for beginners.

Sailing and Boat Tours
La Visite des Calanques

quai Saint-Pierre; www.lavisitedescalanques.com; see website for times; from €19

A flotilla of medium-size boats (capacity around 50) run frequently into the calanques. Visits can include three, five, eight, or nine calanques, and can include swimming stops. Tickets are purchased at the kiosk on the portside. You should arrive to purchase your ticket around 20 minutes before the scheduled departure. Timetables listing all departures are on the website.

Water Sports
Cassis Calanque Plongee

3 rue Michel Arnaud; tel. 06 71 52 60 20; www. cassis-calanques-plongee.com; daily 8:30am-noon and 2:30pm-6:30pm; from €45

Clear water and plentiful diving sites with a great variety of marine life, coral, and underwater fauna make this area a diving mecca. There are also plentiful caves and the odd submerged cannon.

Destination Calanques
Kayak Cassis Marseille

avenue des Calanques; tel. 06 15 63 86; https://cassis-kayak-marseille.fr; daily 9am-6pm; €45/half day

Hugging the coast in a kayak is a wonderful way to see the calanques, offering the freedom to stop where and when you want. This company offers full- or half-day guided rentals from Cassis. Reserve a week in advance for the busy summer months.

Food
Chez Gilbert

19 quai des Baux; tel. 04 42 01 71 36; www.chezgilbert. net; Thurs.-Tues. noon-2:30pm and 7pm-10pm; €30-50

Dining at Chez Gilbert is a theatrical experience. The locals come here to be seen. Fish is ordered by the 100 grams, so waiters dutifully proffer the catch of the day on a silver tray before cooking. Shellfish arrives on icy platters, and portside promenaders stop and stare. Meanwhile, diners keep their shades on and look, seemingly oblivious to the attention, out to sea.

★ Presqu'île

Quartier Port-Miou; tel. 04 42 01 15 40; www. restaurant-la-presquile.fr; Tues.-Sun. noon-2pm and 7:30pm-10pm; weekday lunch €49, weekend and dinner €72

This is another restaurant where it's hard to know what's more enjoyable, the view or the food. Scallops with chestnuts and a foie gras emulsion melt in the mouth, but equally alluring is the pine tree-shaded terrace with views of the sparkling sea. There are few better settings for a meal.

Accommodations
Le Jardin d'Emilie

23 avenue de l'Amiral Ganteaume; tel. 04 42 01 80.55; www.lejardindemile.fr; €165 d

This charming small hotel is nestled under pine trees, a short walk from the Plage Bestouan. The room furnishings are just a little old-fashioned, but given the competitive prices and sea views, it's hard to quibble.

Hotel Mahogany

Plage du Bestouan, 19 avenue de l'Amiral Ganteaume; tel. 04 42 01 34 82; www.hotelmahogany.com; €240 d

On the same headland as Les Roches Blanches, Hotel Mahogany overlooks the Bestouan beach. It's an idyllic spot, looking straight out to sea but only a few minutes' walk from the port. There are only six rooms so book ahead. One of the suites even has its own private pool. All have Wi-Fi, air-conditioning, and en suite bathrooms. Private parking is available and must be reserved.

★ Les Roches Blanches

9 avenue des Calanques; tel. 04 42 01 09 30; https:// roches-blanches-cassis.com, €500 d

A short walk from the center of town, out on the headland leading to the calanques, Les Roches Blanches is the standard-bearer for Cassis hotels. Recently redecorated with an eye toward the hotel's 1930s art deco origins, Les Roches Blanches is a dreamy romantic getaway with all the services you would imagine from a luxury small hotel.

Information and Services

The Cassis **tourist office** (tel. 08 92 39 01 03; www.ot-cassis.com) is located at Place Baragnon. It offers information on where to stay and what to do in Cassis, as well as maps and advice on how best to access Les Calanques.

Getting There and Around

The center of Cassis is small and easily walkable.

Car

From **Aix-en-Provence,** take the A52. The journey is 50 km (31 mi) and takes 55 minutes. From **Marseille,** take the A7 and A51, covering 30 km (19 mi) in 40 minutes. Parking is difficult in the peak season in Cassis. The port car parks quickly fill up and the best option is often the park-and-ride car park at avenue des Gourgettes (Parking les Gourgettes).

Train

Cassis **train station** is located at Quartier de La Gare. Trains run almost every hour from **Marseille Saint-Charles Station.** The walk to the harbor takes about 30 minutes. Luckily, it is downhill. One-way tickets are €2.70 and up.

★ LES CALANQUES
Sights
Calanque de Port Pin

Access from Port-Miou, 50 avenue des Calanques, Cassis

Continuing past the ancient quarry at Port-Miou on the GR98-51 for 2 km (1.2 mi) on foot takes you to Port Pin, named after the pine trees that cling to the cliffs on all sides. There's a small sandy beach and plenty of rocks where you can spread out your towel and relax.

Calanque Morgiou

Les Baumettes, Marseille

Morgiou is one of the larger calanques. The attractive pastel-colored houses that nestle behind the port area are testament to its past as a busy fishing village. Outside the summer season, it's possible to drive all the way into the calanque. In summer when the fire risk is high, the path to the calanque departs from two residential districts of Marseille: Luminy and Les Baumettes. The 6.5-km (4-mi) walk into the calanque takes about 90 minutes. Allow a little more time for the return trip.

Calanque Sormiou

Les Baumettes, Marseille

Neighboring Morgiou and joined to it by a coastal path, Sormiou is another of the larger calanques. It has a port, small village, and a restaurant. A small sandy beach gives access to crystal-clear waters that are frequented by scuba divers. Outside the summer season, it's possible to drive all the way into the calanque. In summer, when the fire risk is high, visitors have to park at Les Baumettes. The 4.5-km (2.8-mi) walk takes 1 hour. If you reserve at the Château restaurant, you will be given an access code that allows you to drive into the calanque at all times of year.

Marseilleveyre

Callelongue, Marseille

Gone are the pine trees that characterize the calanques closest to Cassis. Instead, there's bare rock, grass, and the odd fig tree that can cope with the arid climate. Park at the Calanque de Callelongue on the outskirts of the Les Goudes district of Marseille, and then walk 4 km (2.5 mi) (50 min on the GR98-51) or catch a boat. When you arrive, there's a small sand-and-stone beach and a restaurant that receives all its deliveries by sea.

Calanque de La Mounine

Callelongue, Marseille

A 2-km (1.2-mi), 30-minute walk along the GR98-51 from Calanque de Callelongue, Mounine is one of the smallest calanques. There's no beach, but you can stand on the rocks and watch the schools of fish in the clear water below.

Hiking
Marseille to Cassis Hike

Distance: *35 km (22 mi) one-way*
Time: *7.5 hours*

1: view of Cassis town, Cap Canaille, and the Mediterranean Sea from the Route des Crêtes **2:** Plage de La Grande Mer, Cassis **3:** Calanque de Port Pin

Information and Maps: *IGN Map of the Calanques (available at tourist offices in Marseille and Cassis)*

Trailhead: *avenue de la Madrague de Montredon, Marseille, or avenue Victor Hugo, Cassis*

If you are fit and determined, it's possible to walk the 35-km (22-mi) GR 98-51 coastal path between Cassis and Marseille in a day. Pick up the IGN Map of the Calanques from either the Marseille or Cassis tourist office, wear sturdy walking shoes and a hat, and take plenty of water with you. The route climbs 400 m (1,300 ft) total and is marked, but it is rocky and sometimes slippery. Along the way there are good lunch spots at Calanques Morgiou and Sormiou, and there are plenty of places for a swim. At the Marseille end of the walk there is no shade, but nearer Cassis the rocks are covered with pines. Depart from either Cassis or Les Goudes in Marseille.

Cycling
E-bike Tours

Meeting point A Criée/Café des Arts, 30 quai de Rive Neuve, Marseille; tel. 07 82 00 73 47; https://fada.bike; Mar.-Oct., year-round for groups; €45/half-day, €99/full day

This day trip from Marseille takes in some of the city's main sights, such as Notre-Dame de la Garde, and then heads along the coast to Les Goudes and the Calanque de Callelongue. Some tours ride as far as Cassis. The trips give you a taste of Les Calanques but do not really take you out into the wilds. For this, you'll need to put on your walking boots.

Sailing and Boat Tours
Bleu Evasion

Port de la Pointe Rouge, Marseille; tel. 04 91 06 18 87; www.bleuevasion.com; see website for times or to make a private booking; from €89

Bleu Evasion operates small motorboats and a catamaran, both departing from Marseille (Port de la Pointe Rouge). They can be booked privately or you can join a scheduled trip to visit the calanques closest to Marseille. There are stops for swimming.

Food
Bar Nautic

Calanque Morgiou, Marseille; tel. 04 91 40 06 37; Tues.-Sun. 10am-10pm; €15-25

Bar Nautic boasts a large terrace overlooking the sea. It provides fresh grilled seafood and salads to weary hikers. The experience is as much about the location as the food, which although good is not that sophisticated. Cash only.

Chez Les Belges

Marseilleveyre, Marseille; www.facebook.com/chezlebelge13; July-Aug. Fri.-Mon. noon-2:30pm and 7pm-10pm, Apr.-June and Sept.-Oct. Fri.-Mon. noon-2:30pm, opening times may vary, check Facebook page; €15-25

What could be more Provençal than a restaurant that does not have a phone and whose opening hours have been known to be so irregular that the owners put up information at the hike departure point in Calanque de Callelongue? The food is simple (the restaurant gets its supplies by boat), with plenty of fresh fish, as well as a hearty Provençal stew served with a baked potato. Cash only.

★ Le Château

route du Feu de la Calanque de Sormiou, Marseille; tel. 04 91 25 08 69; https://lechateausormiou.fr; Mon.-Sat. noon-2:30pm and 7:30pm-9:30pm, Sun. noon-2:30pm; €25-35

Probably the best-known and most serious restaurant in Les Calanques, the Château has an unparalleled position, looking straight out over the Calanque Sormiou to the open sea. In season, when the road to the port is closed, you can phone ahead to the restaurant and be given a password to allow you through.

Accommodations

The calanques are a national park with restricted access at certain times of year. It is possible to rent holiday apartments in some of the larger calanques but there are no hotels. The closest places to stay are listed in the Cassis and Marseille accommodations sections.

Information and Services

The best places for information on the ca-lanques are the Cassis and Marseille tourist offices.

The **Cassis tourist office** (Place Baragnon; tel. 08 92 39 01 03; www.ot-cassis.com) offers maps and advice on how best to access Les Calanques. The **Marseille Tourist Office** (11 La Canebière; tel. 08 26 50 05 00; www.marseille-tourisme.com) has maps, tickets (including the City Pass), and general advice.

Getting There and Around

The main access points to Les Calanques are from **Port-Miou** in Cassis and **Calanque de Callelongue** in Marseille. From these points visitors must continue on foot or bike. Port-Miou can be reached by following **avenues des Calanques** (either on foot or in a car) out of Cassis. The road leads to the Port-Miou parking. Calanque de Callelongue can be reached by following **boulevard Alexandre Delabre** (either on foot or in a car) out of the **Les Goudes** neighborhood of Marseille.

Car

It's possible to reach the larger calanques of **Sormiou** (chemin de Sormiou) and **Morgiou** (route du Feu de la Calanque de Morgiou), but note that access is closed in the summer due to the fire risk. To find both these roads, turn off the D559 at the Mazargues roundabout; there is an obelisk in the middle. Follow signs to Sormiou and Morgiou.

Boat

Boats to the calanques depart frequently from Cassis and from Port de La Pointe Rouge in Marseille. Try the following operators: **Visite des Calanques** (quai Saint-Pierre, Cassis; www.lavisitedescalanques.com; see website for departure times; from €19 for a visit to three calanques) and **Bleu Evasion** (Port de la Pointe Rouge, Marseille; tel. 04 91 06 18 87; www.bleuevasion.com; see website for times or to make private booking; from €89/half day in the calanques).

The Var Coast

The resorts of Saint-Cyr Les Lecques, Bandol, and Sanary-sur-Mer are part of a succession of seaside resorts that stretch east along the Var coast toward Toulon. For sun worshippers, all three have beach clubs where you can rent a sun lounger, sip on a drink of choice, and soak up the rays while the Mediterranean laps at your feet. Sporty types can indulge in a full range of water sports, including sailing, diving, and windsurfing. Coastal paths connect the resorts, and the railway network is good. Outside of the worst heat of the peak summer season, it's a joy to take a clifftop hike followed by lunch and then the train back home.

ORIENTATION

Saint-Cyr Les Lecques is located 40 km (25 mi) east of **Marseille.** Heading eastward from Saint-Cyr, visitors will encounter **Bandol** and then **Sanary-sur-Mer.** All resorts are accessible off the **A50** autoroute, which runs parallel to the coast.

SAINT-CYR LES LECQUES

Saint-Cyr and Les Lecques are actually two different places that are commonly referred to as one. Saint-Cyr is an inland town, a couple of kilometers from the sea. Its palm tree-fringed suburbs stretch all the way to the wide, sandy bay of Les Lecques. Here, there's a port with water sports activities, a long promenade for a seaside stroll, and a child-friendly beach.

Beaches
Plage des Lecques
boulevard de la Plage

The Var Coast

Stretching for several kilometers, this sandy, shallow bay is ideal for children. There's a pedestrian promenade where you can walk and gaze out at the sparkling sea. Out of season, the promenade is heavily used for Rollerblading and jogging. There are a couple of private beach clubs with sun loungers for hire. Try the Plage de Sophie (www.laplagedesophie.com).

Water Sports
Lec Surf Club

promenade Rose; tel. 06 64 61 58 57; www.lecsurfclub.com; year-round activities, open daily during school

holidays 9:30am-5pm, outside school holidays Sat.-Sun. 9:30am-5pm; board rental from €15

The largest paddleboard and surfing club in Provence takes advantage of the ideal conditions offered by the bay of Les Lecques. During school holidays, there are weeklong courses available for children of all ages.

Wake Sensation

Vieux Port des Lecques; tel. 06 63 46 10 71; http://wakesensation.fr; May-Oct.; from €30 for 10 min

Wake Sensation runs courses for all skill levels, and in the summer offers weeklong programs for children. The main activities are wakeboarding and waterskiing. There's also the option of sitting on an inflatable ring and bouncing around on the surf behind a speedboat.

Hiking
Sentier du Vignes
Distance: *8 km (5 mi) round-trip*
Time: *2 hours*
Information and Maps: *www.saintcyrsurmer. com/incontournables/un-patrimoine-naturel/les-sentiers-de-randonnee/le-sentier-du-littoral*
Trailhead: *La Madrague Port*

The Sentier du Vignes takes you through the Bandol vineyards from La Madrague Port in Les Lecques to the Port d'Alon Calanque. The 4-km (2.5-mi) walk takes just over an hour on a marked path. As always, wear a hat and take plenty of water. Consider also booking a table for lunch at Chez Tonton Ju, situated on the beach of the Port d'Alon Calanque. A restorative lunch with a few glasses of wine make the return leg much easier. Alternatively, from the Port d'Alon Calanque you can continue on the Sentier du Littoral to Bandol, and take the bus back to Saint-Cyr.

Food and Accommodations
Chez Tonton Ju
Saint-Cyr Les Lecques; tel. 04 94 26 20 08; Apr.-Oct. daily noon-8pm; €10-18

This relaxed beach bar comes with a view over the sparkling waters of the Calanque d'Alon. Salads and plates of prawns and oysters dominate the menu. It's a place to kick your shoes off and relax with your feet in the sand.

L'Equateur
9 avenue du Port; tel. 04 94 26 20 02; daily 8:30am-7:30pm; €15-30

Right in the center of Les Lecques' bay, on the promenade behind the beach, this bar-restaurant is filled with a mix of old sea dogs sipping pastis at the bar and tourists enjoying the food. The service is friendly, the vibe relaxed, and the portions copious. Moules frites (mussels with fries) overlooking the Mediterranean has rarely tasted better.

La Plage de Sophie
Plage des Lecques; tel. 06 82 57 89 57; www. laplagedesophie.com; Apr.-Oct. daily 9am-8pm; €17-35

This friendly beach bar has sun loungers to rent and a decent restaurant serving salads, seafood, and burgers. In summer a reservation is imperative.

★ Grand Hotel
24 avenue du Port; tel. 04 94 26 23 01; https://grand-hotel-les-lecques.com; €260 d

This 60-bedroom hotel with large gardens is just a minute's walk from the seafront. The decor is a nice mix of period glamour and modern style. It has two restaurants: one poolside and the other on a terrace overlooking the sea.

Getting There and Around
Once you have arrived at Saint-Cyr, it is best explored **on foot.**

Car
Saint-Cyr Les Lecques is off the A50 coastal autoroute that runs between Toulon and Aubagne. From **Aix-en-Provence,** take the A52 and then the A50, covering 62 km (39 mi) in 50 minutes.

Bus
Zou Bus line 880 (https://zou.maregionsud. fr) connects the coastal villages and towns of Bandol, Saint-Cyr, Le Castellet, La Cadière d'Azur, and La Ciotat. It runs three times a day. In Saint-Cyr, it stops at the train station and also the beach. A single ticket is €3. Traveling the whole length of the line takes 1 hour. It is no more than 20 minutes between adjacent resorts.

Train
Trains run hourly from Saint-Cyr to **Marseille.** Note that the **train station** (avenue de La Gare) is inland, approximately 2.5 km (1.6 mi) from the beach. The journey from Marseille costs €8 and takes 30 minutes. Consult the timetable at www.sncf-connect.com.

1

2

3

BANDOL

Bandol is a busy, bustling resort filled with shops and restaurants. It's large enough to have its own casino and has several beaches. Undoubtedly, the best of them is Plage de Renécros. But it's wine that has made Bandol famous. The rosés are among the best in France.

Sights
Les Îles Paul Ricard

Departs from Le Brusc; www.lesilespaulricard.com; boats daily every 15 min; €18.50, ages 3-12 €10.50

Until recently the Isle of Bendor was the more visited of the two private islands purchased by Paul Ricard in the 1950s. Bendor, however, is closed for renovation until 2026, leaving Embiez to shine. The island is pedestrian and can be visited on foot, by e-bike (hire on arrival), or on a small tourist train. There are picturesque beaches and numerous restaurants.

Beaches
★ Plage Anse de Renécros

Corniche Bonaparte

One of the most attractive beaches on the Var coast, Plage Anse de Renécros is an almost perfect half-moon of sand located on the opposite side of the headland from Bandol's port and main shopping drag. A couple of hotels and beach clubs line one side of the bay, and on the far side dreamy villas peek from beneath pine trees. The sea is sheltered and shallow for a good 15 m (49 ft).

Plage Capelan

Presqu'île de Capelan

Around the headland from Plage Anse de Renécros, this small beach is the place to go to escape the crowds. There are two small parking areas, and the beach is accessed by a set of steps.

Water Sports
Bandol Plongee

Silhouette quai Consigne°7, Allee Vivien; tel. 06 07 45 27 81; www.bandol-plongee.com; initiation dive €70

This established scuba diving school offers dives along the coast for all levels.

Hiking
Sentier du Littoral

Distance: *12 km (7.5 mi) round-trip*
Time: *4 hours*
Information and Maps: *www.bandoltourisme. fr/mon-sejour/explorer/promenades-et-randonnees/ le-sentier-du-littoral*
Trailhead: *avenue Albert 1er, Anse de Renécros*

This coastal path between Bandol and Saint-Cyr Les Lecques is closed at one section due to instability. Thankfully, it's still possible to walk along much of the coast, to Port d'Alon, which is a good spot for lunch and a swim. From Port d'Alon either take the same path back to Bandol or cut inland and take the 4-km (2.5-mi) Sentier du Vignes to Saint-Cyr.

Wineries
Domaine Tempier

1082 chemin des Fanges; tel. 04 94 98 70 21; http:// domainetempier.com; Mon.-Fri. 9am-noon and 2pm-6pm; from €20

The reference domaine for Bandol wine, Domaine Tempier still operates out of a small tasting room at the back of the family home. For serious tasters, it's all about the reds; an aged tempier has an aroma of fine cigar, and swirls of rich, long-lasting flavor hit the back of the throat.

Food
La Chipote

12 Corniche Bonaparte; tel. 04 94 29 41 62; www. restaurant-lachipote.com; daily noon-9:30pm; €14-35

The Var coast is full of restaurants on idyllic beaches; however, La Chipote, with its sweeping view over the Renécros beach, is up there with the best. The food is a cut above the average with inventive salads and fresh seafood dishes. The garlic and parsley squid could not be fresher.

1: the port at Saint-Cyr Les Lecques **2:** grapes in Bandol **3:** Plage Portissol, Sanary-sur-Mer

Le Mediterranee

41/67 avenue Georges V; tel. 04 94 07 59 19; www.
restaurant-le-mediterranee.com; daily 10:30am–
midnight; €15-40

Located just outside Bandol on the Plage
Capelan, this restaurant is set in an idyllic location. The dress code is smart, so don't show
up straight from the beach. The food is as
pretty as the view.

L'Espérance

rue du Dr-Louis-Marçon; tel. 04 94 05 85 29; www.
lesperance-bandol.com; Tues. 7:15pm-9:15pm, Wed.-Sun.
noon-1:15pm and 7:15pm-9:15pm; from €48

This is an address for foodies for whom sublime food is more important than a sublime
view. Kick off the flip-flops and dress up to
enjoy dishes like roasted half pigeon with
cêpes and a port sauce.

Accommodations
Hotel du Golf

10 Corniche Bonaparte; tel. 04 94 29 45 83; www.
golfhotel.fr; €175 d

A 24-bedroom family hotel on Bandol's best
beach, Anse de Renécros, the Hotel du Golf
is a good budget option for a few days on the
Mediterranean. Rooms are simply furnished
but have air-conditioning, Wi-Fi, and en suite
bathrooms. Opt for a room with a terrace overlooking the sea if at all possible. Private parking adjacent to the hotel is included in the price.

★ Hotel Ile Rousse

25 boulevard Louis Lumière; tel. 04 94 29 33 00; www.
thalazur.fr/bandol/hotel; €400 d

A budget buster with even the simplest rooms
coming in at nearly €300, the Hotel Ile Rousse
has a spectacular infinity pool overlooking the
Anse de Renécros Plage. The rooms are modern with terraces overlooking the sea. Many
clients come for the spa treatments and wander around cosseted in fluffy white bathrobes.

Getting There and Around

Upon arrival the best way to get around is on
foot.

Car

Bandol is located off the A50 coastal autoroute, which runs between Toulon and
Aubagne. From **Aix-en-Provence,** take the
A52 and then the A50, covering 75 km (47 mi)
in 58 minutes.

Train

Trains run hourly from **Marseille** (45 min)
to Bandol (avenue de La Gare). Note the
train station (www.sncf-connect.com) is
inland, approximately 2 km (1.2 mi) from the
beach. Single tickets cost around €10.

Bus

Zou Bus line 880 (https://zou.maregionsud.
fr) connects the coastal villages and towns of
Bandol, Saint-Cyr, Le Castellet, La Cadière
d'Azur, and La Ciotat. It runs three times a
day. In Bandol, it stops at the train station and
also the beach. A single ticket is €3. Traveling
the whole length of the line takes 1 hour. It is
no more than 20 minutes between adjacent
resorts.

SANARY-SUR-MER

You can drive through Sanary-sur-Mer
and think there's not much to it. The port
area is wide and attractive, but the real
charm of the town lies hidden. The narrow streets and squares of the old town are
tucked away, and the resort's best beaches
are not immediately accessible from the
center of town. Consequently, Sanary re-
veals its beauty slowly and only to determined visitors. The cliff-side path from
port side to Plage Portissol is a delight.

Sights
La Chapelle Notre-Dame de Pitié

chemin de La Colline; daily 9am-6pm

Take the steps on the western side of the port and climb onto a pedestrian path that leads around the headland. High on the hill you encounter the Chapel Notre-Dame de Pitié. The building started life in the 18th century as a lookout point during storms and for spotting enemy boats. Over the years, it was used to house plague victims and the wounded from the Franco-Prussian War. Juxtaposed with the simplicity of the chapel, the beauty of the location conjures a sense of divinity.

Beaches
Plage Portissol

1-131 boulevard Frédéric Mistral

This popular half-moon bay is linked to the Port of Sanary by the **chemin de La Colline.** There are a couple of beach clubs and a hotel-restaurant. The beach itself is stony, but the waters are clear and it's a lovely place to swim.

Water Sports
Sanary Location

quai Levant; tel. 06 29 47 41 02; www.sanary-location-bateau.fr; daily 9am-6pm; from €140/half day

Sanary Location is a great boat-rental service that offers speedboats, including less powerful boats that do not require a boat pilot license. Clients have the opportunity to explore up and down the coast. Water sports equipment is also available to rent, including wakeboards, water skis, and inflatables to tow. For those wishing to stick to dry land, there are also e-bikes to rent.

Food and Accommodations
★ O Petit Monde

Plage Portissol; tel. 04 94 34 39 59; www.opetitmonde.com; restaurant €24-39; rooms €250 d

O Petit Monde on the Plage Portissol in Sanary-sur-Mer is a restaurant with rooms. The restaurant is located on the ground floor, with tables close enough to the sea that you can feel the spray. The five rooms, all with en suite bathrooms, are on the upstairs terrace. They are spacious and each has outside space with sun loungers and tables. Guests wake up to an empty beach, breakfasting with the priceless sensation of having the Mediterranean to themselves. Bookings for the summer season need to be made several months in advance.

Getting There and Around

Upon arrival, Sanary is best explored on foot.

Car

Sanary is off the A50 coastal autoroute, which runs between Toulon and Aubagne. From **Aix-en-Provence,** take the A52 and then the A50, covering 77 km (48 mi) in 1 hour.

Train

Trains run hourly from **Sanary Ollioules** (around 3 km/1.8 mi from the center of Sanary) to **Marseille** (45 min). Timetable details can be found at www.sncf connect.com. The cost of a single journey from Marseille is €12.

Nice and Antibes

Vibrant, cosmopolitan Nice has more than 2,000 cultural, sporting, and festive events throughout the year. Helped by the plans of an ambitious mayor, the self-styled capital of the Côte d'Azur has become a modern, elegant, eco-friendly city. It has electric bikes for hire, hosts a large student population, and gives off a cool vibe—tattoo parlors, vegan restaurants, and escape rooms as well as a marina of superyachts and some excellent restaurants. In 2021, Nice's city center became a UNESCO World Heritage Site.

Connecting Nice to Menton, the last town in France before the Italian border, are the three Corniches—distinct, scenic roads that run parallel with the coastline. They link beautiful coastal towns like Villefranche-sur-Mer (directly east of Nice), Beaulieu-sur-Mer,

Highlights

Look for ★ to find recommended sights, activities, dining, and lodging.

★ **Promenade des Anglais:** Nice's iconic seafront boulevard, loved by joggers and cyclists, is famous for its belle epoque villas, hotels, and sculptures (page 240).

★ **Vieux Nice:** A tangle of narrow lanes and sunny squares, the heart of old Nice has a seductive range of art galleries, cafés, boutiques, ice cream parlors, and hidden chapels (page 240).

★ **Musée National Marc Chagall:** Opened by the Russian-born artist in 1972, this light, airy museum houses a collection of his extraordinary Biblical Message canvases (page 248).

★ **Villa Ephrussi de Rothschild:** This sumptuous, rose-pink villa is set among landscaped gardens and fountains in Saint-Jean-Cap-Ferrat (page 272).

★ **Haut-de-Cagnes:** This medieval village, set on a steep hill a few kilometers from the sea, has art galleries, great restaurants, and fantastic views from its château rooftop (page 284).

★ **Juan-les-Pins Beach Bars:** Kick back and enjoy a tropical drink at this line of private clubs on the French Riviera's best strip of sand (page 307).

Nice and Antibes

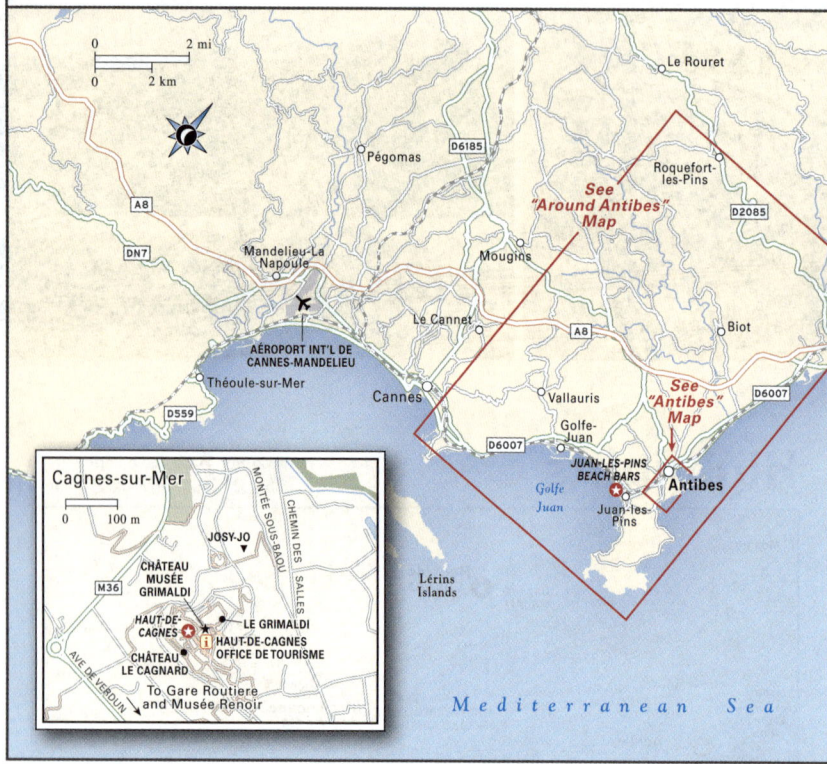

See "Around Antibes" Map

See "Antibes" Map

Le Rouret

Roquefort-les-Pins

D6185

D2085

Pégomas

A8

DN7

Mandelieu-La Napoule

Mougins

A8

Biot

Le Cannet

AÉROPORT INT'L DE CANNES-MANDELIEU

Théoule-sur-Mer

Cannes

Vallauris

D6007

D559

Golfe-Juan

D6007

JUAN-LES-PINS BEACH BARS

Antibes

Golfe Juan

Juan-les-Pins

Lérins Islands

Cagnes-sur-Mer

0   100 m

MONTÉE SOUS-BAOU

CHEMIN DES SALLES

JOSY-JO

M36

CHÂTEAU MUSÉE GRIMALDI

HAUT-DE-CAGNES

LE GRIMALDI

HAUT-DE-CAGNES OFFICE DE TOURISME

CHÂTEAU LE CAGNARD

AVE DE VERDUN

To Gare Routière and Musée Renoir

Mediterranean   Sea

Cap-d'Ail, Èze, and La Turbie. Roman emperors, composers, philosophers, and playwrights all ended up on this stretch of the Riviera's rugged coast.

The coast between Nice and Cannes is dominated by Antibes, which began as a Greek trading post in the 5th century BCE and is now the sailing hub of the Riviera. For a week in June, the westernmost edge of the Baie des Anges is filled with yachts and sailing vessels attending the Voiles d'Antibes boat show. To the south of Antibes is the Cap d'Antibes, a 6-km-long (3.7-mi-long) idyllic peninsula, home to some of the most exclusive villas, breathtaking views, and fantastic seaside walks on the Riviera.

Juan-les-Pins, just west of the Cap d'Antibes, is the party capital of the Côte d'Azur, a mass of beach bars, boutiques, and nightclubs. It's also a great place for water-skiing, fishing, or just relaxing on the sand, and it's famous for its jazz festival every July. Farther west is Juan-les-Pins's calmer neighbor, Golfe-Juan, where Napoleon landed after returning from exile in 1815 and which still holds a festival to mark the occasion. Biot has the world's only museum dedicated to the artist Fernand Léger and also specializes in glassblowing, with its own peculiar, bubble-filled form of glass. The town of Vallauris, farther west, is known for its ceramics, and inspired Pablo Picasso to take up that art form when he visited in the 1940s.

In between Antibes and Nice, Cagnes-sur-Mer was the home of artist Pierre-Auguste Renoir, and contemporary painters still flock

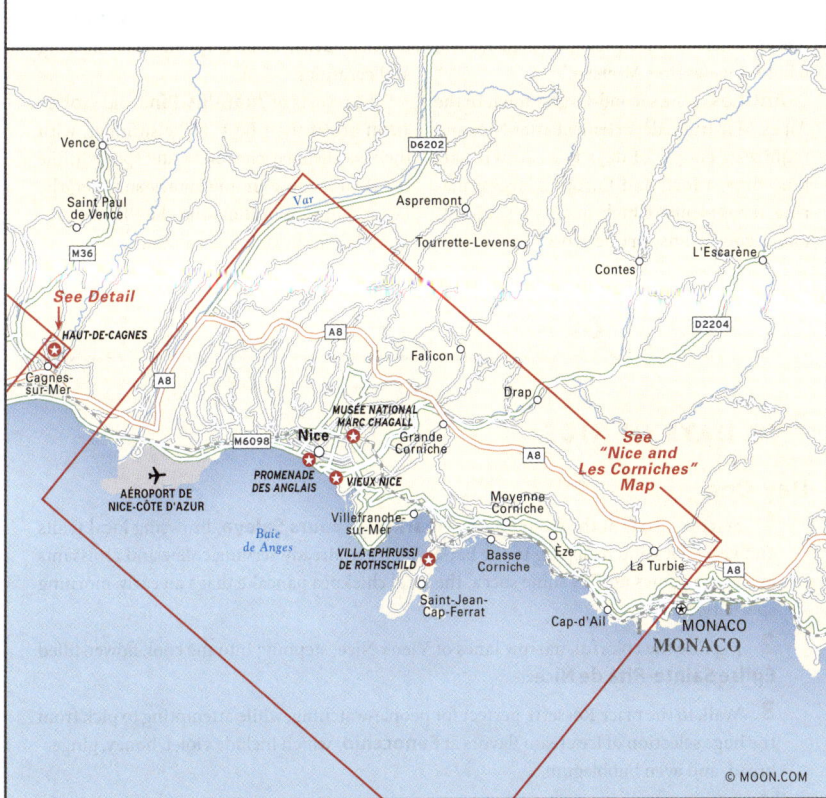

there to sketch the thousand-year-old olive trees in his garden. Cagnes has a string of seafood restaurants above its pebbly beaches and a horse race track that holds meets on warm summer evenings.

PLANNING YOUR TIME

Nice is a great base from which to visit the Riviera. Even for visitors who would prefer to spend their time on the beaches of Cannes, visiting hilltop villages, or watching the Monaco Grand Prix, Nice is worth at least a few days for its art museums, old town, and excellent restaurants. Until the 1930s, the city was predominantly a winter resort, but today

it is popular **year-round,** great for water sports and sunbathing, with a **jazz festival** in July and the two-week **carnival** at the end of February. You can still eat outside in December, the sky is nearly always blue, and there are ski slopes less than 2 hours away.

Besides the A8 autoroute, the three main roads heading eastward from Nice are known as the Corniches. The **Basse Corniche** is closest to the sea, and it is worth stopping off at **Villefranche-sur-Mer, Beaulieu-sur-Mer,** and **Cap-d'Ail** to visit the beaches, museums and restaurants. The **Moyenne Corniche** is a more vertiginous drive and can include an afternoon stopover in the village of

Èze. Driving the **Haute Corniche** is the most spectacular way to see the coastline, with **La Turbie** offering some great places to eat and fabulous views over Monaco.

Antibes is the second-largest town in the Alpes-Maritimes département after Nice and requires a couple of days to visit, with art museums, a fort, and Europe's largest marina. It has plenty of high-quality restaurants and a good transport network. For a more peaceful place to stay, try the perched medieval village of **Haut-de-Cagnes** or the town of **Biot,** known for its glassblowing and ceramics.

The resorts of **Juan-les-Pins** and **Golfe-Juan** are at their best in the summer, with their excellent beaches and water sports, while the Renoir museum in Cagnes-sur-Mer, the Léger museum in Biot, and the Picasso museum in Antibes are open all year.

Itinerary Ideas

TWO DAYS IN NICE

Day One

1 Start your day at the fresh produce **market** on **cours Saleya,** browsing local fruits and vegetables and dropping in the bars, which are already serving coffee and croissants as the sun comes up. Try some socca, the local chickpea pancake that's an early-morning specialty.

2 Explore the colorful, narrow lanes of Vieux Nice, stepping into the cool, flower-filled **Église Sainte-Rita de Nice.**

3 Walk to the place Rossetti, perfect for people-watching, while attempting to pick from the huge selection of ice cream flavors at **Fenocchio,** which include violet, honey, gingerbread, and even bubblegum.

4 Head through the old Jewish quarter of **rue Benoît Bunico,** peeking in some of the galleries on the rue Droite and the fashion boutiques on rue Centrale.

5 Walk up through the marble Escalier de la Porte Fausse and cross over the tram tracks into the **promenade du Paillon,** Nice's urban park, strolling past an esplanade of dancing fountains and places to sit among the trees and sculptures.

6 Heading northeast, the park leads toward **MAMAC,** the contemporary art museum. Don't forget to visit the roof garden for the amazing views.

7 Walk back through the old town and take the steps up to the ruins and the park atop **Colline du Château** (Castle Hill), where there are fantastic views of the Baie des Anges.

8 Stroll along rue Ségurane in the Quartier des Antiquaires (antiques district) and cross the **place Garibaldi** toward rue Bonaparte, where there are restaurants catering to all tastes and mesmerizing trompe l'oeil.

Day Two

1 Board bus number 5 behind Galeries Lafayette department store, which will take you

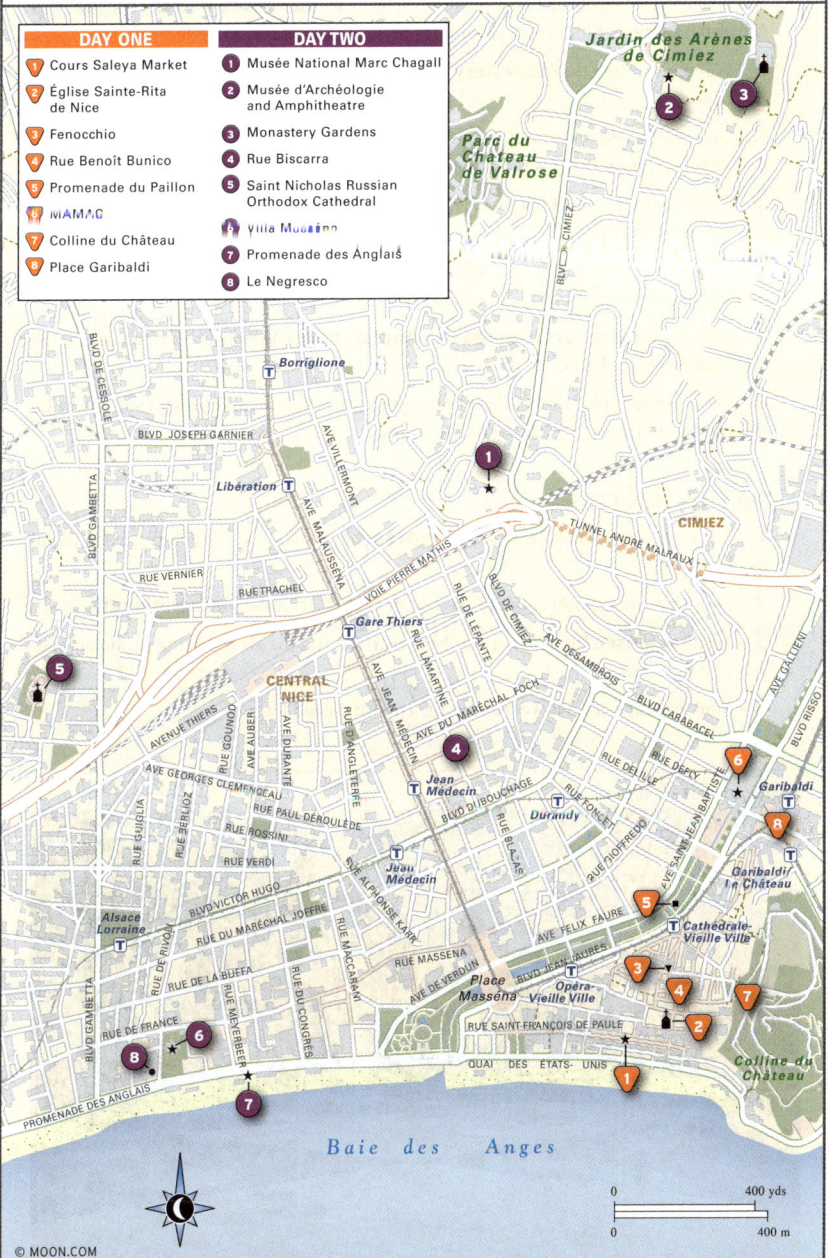

Two Days in Nice

DAY ONE
1. Cours Saleya Market
2. Église Sainte-Rita de Nice
3. Fenocchio
4. Rue Benoît Bunico
5. Promenade du Paillon
6. MAMAC
7. Colline du Château
8. Place Garibaldi

DAY TWO
1. Musée National Marc Chagall
2. Musée d'Archéologie and Amphitheatre
3. Monastery Gardens
4. Rue Biscarra
5. Saint Nicholas Russian Orthodox Cathedral
6. Villa Masséna
7. Promenade des Anglais
8. Le Negresco

Jardin des Arènes de Cimiez

Parc du Château de Valrose

CIMIEZ

TUNNEL ANDRÉ MALRAUX

BLVD DE CESSOLE

BLVD JOSEPH GARNIER

Borriglione

AVE VILLERMONT

Libération

AVE MALAUSSÉNA

BLVD GAMBETTA

RUE VERNIER

RUE TRACHEL

VOIE PIERRE MATHIS

Gare Thiers

RUE DE LÉPANTE

BLVD DE CIMIEZ

AVE DÉSAMBROIS

CENTRAL NICE

AVENUE THIERS

AVE JEAN MÉDECIN

RUE LAMARTINE

AVE DU MARÉCHAL FOCH

BLVD CARABACEL

RUE GOUNOD

RUE AUBER

AVE GEORGES CLEMENCEAU

AVE DURANTE

RUE D'ANGLETERRE

Jean Médecin

RUE DELILLE

RUE DÉFLY

Garibaldi

AVE GALLIENI

BLVD RISSO

RUE GUIGLIA

RUE BERLIOZ

RUE PAUL DÉROULÈDE

RUE ROSSINI

BLVD DUBOUCHAGE

Durandy

RUE FONCET

AVE JOFFREDO

RUE SAINT JEAN BAPTISTE

RUE VERDI

Jean Médecin

RUE BLAVET

Garibaldi le Château

BLVD VICTOR HUGO

Alsace Lorraine

RUE DE RIVOLI

RUE DU MARÉCHAL JOFFRE

AVE FÉLIX FAURE

Cathédrale-Vieille Ville

RUE DE FRANCE

RUE DE LA BUFFA

RUE MACCARANI

RUE MASSÉNA

Opéra Vieille Ville

Colline du Château

BLVD GAMBETTA

RUE MEYERBEER

RUE DU CONGRÈS

AVE DE VERDUN

Place Masséna

BLVD JEAN JAURÈS

RUE SAINT FRANÇOIS DE PAULE

PROMENADE DES ANGLAIS

QUAI DES ÉTATS-UNIS

Baie des Anges

0   400 yds
0   400 m

© MOON.COM

up to Cimiez. The first stop ascending the hill is the **Musée National Marc Chagall,** which also has a café and shady gardens to sit in.

2 From there, walk (or hop on the same bus) to the **Musée d'Archéologie and Amphitheatre,** behind which is Cimiez's park and the Musée Matisse.

3 Wander through the **Monastery Gardens** for a view of the Paillon valley on the far east of Nice.

4 Grab the bus (or take the 20-minute downhill walk) back to the center, where you can have lunch at one of the covered bistros along **rue Biscarra.**

5 Continue to the **Saint Nicholas Russian Orthodox Cathedral,** followed by a stroll through the Quartier des Musiciens, with its art deco and belle epoque villas.

6 If you want to see inside one, **Villa Masséna** has a permanent exhibition on the history of Nice and sumptuously decorated salons on the ground floor.

7 Join the sea at the **promenade des Anglais** to sit on the pebbles or on one of the city's famous blue chairs.

8 Have dinner at a fish restaurant along the cours Saleya, and finish your day at **Le Negresco,** which has a great bar and live music.

A DAY IN ANTIBES

1 Start the day at the local **Marché Provençal** for a look at the colorful selection of local produce.

2 Now that you have an appetite, have a late breakfast on the **place Nationale.**

3 Visit the **Musée Picasso** on Antibes's seafront and the glassblowing studio on boulevard d'Aguillon.

4 Stop for lunch at **Côté Terroir** on the edge of the old town.

promenade du Paillon, Nice

A Day in Antibes

A DAY IN ANTIBES
1. Marché Provençal
2. Place Nationale
3. Musée Picasso
4. Côte Terroir
5. Chemin Tire-Poil
6. La Passagère
7. Absinthe Bar

© MOON.COM

5 Fill the afternoon with a bracing walk along the **chemin Tire-Poil** at the end of the Cap d'Antibes.

6 Have supper in the art deco dining room of the Belles Rives Hotel's **La Passagère** restaurant in Juan-les-Pins.

7 Finish off with something stronger at **Absinthe Bar** back in Antibes's old town.

Nice and Les Corniches

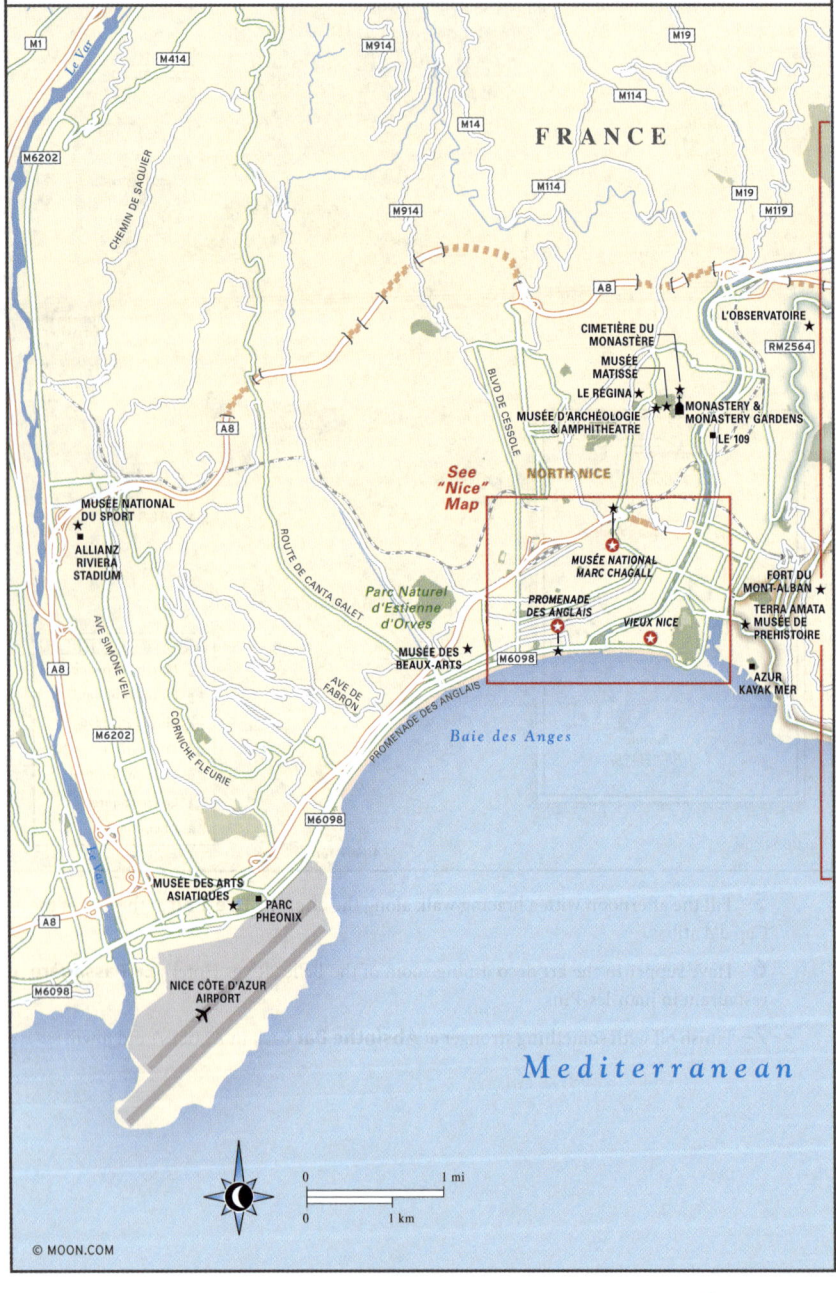

M1

Le Var

M414

M6202

CHEMIN DE SAQUIER

M914

M14

M114

M19

FRANCE

M114

M19

M914

A8

L'OBSERVATOIRE ★

RM 2564

BLVD DE CESSOLE

CIMETIÈRE DU
MONASTÈRE

MUSÉE
MATISSE

LE RÉGINA ★

MUSÉE D'ARCHÉOLOGIE
& AMPHITHEATRE

MONASTERY &
MONASTERY GARDENS

■ LE 109

NORTH NICE

See
"Nice"
Map

MUSÉE NATIONAL
DU SPORT

ALLIANZ
RIVIERA
STADIUM

ROUTE DE CANTA GALET

Parc Naturel
d'Estienne
d'Orves

MUSÉE NATIONAL
MARC CHAGALL

PROMENADE
DES ANGLAIS

VIEUX NICE

FORT DU
MONT-ALBAN

TERRA AMATA
MUSÉE DE
PREHISTOIRE

MUSÉE DES
BEAUX-ARTS

M6098

AVE SIMONE VEIL

AVE DE FABRON

A8

M6202

CORNICHE FLEURIE

PROMENADE DES ANGLAIS

AZUR
KAYAK MER

Baie des Anges

M6098

M6098

A8

MUSÉE DES ARTS
ASIATIQUES

PARC
PHEONIX

NICE CÔTE D'AZUR
AIRPORT

Mediterranean

0   1 mi

0   1 km

© MOON.COM

Nice

ORIENTATION

The bulk of the tourist sites in Nice are squeezed into a relatively small triangle, bordered by the seafront **promenade des Anglais,** a 6-km (3.7-mi) boulevard that comes to an end at the windy outcrop of **Rauba Capeu** (local dialect for "hat stealer"), the **Port,** and the newly developed **promenade du Paillon,** a 12-hectare (30-acre) swath of urban parkland that begins at **place Masséna.** The narrow lanes of **Vieux Nice,** the old town, sit snugly within that triangle.

Moving outward from Vieux Nice, Nice's main shopping street, **avenue Jean-Médecin,** heads directly northward from place Masséna. It's overlooked by the fashionable hillside suburb of **Cimiez,** first settled by the Romans and now famous for its museums. Halfway up avenue Jean-Médecin heading west is **La Gare,** the main railway station. South from there, toward the sea, is the **Quartier des Musiciens,** full of belle epoque and art deco villas, which merges into what real estate agents call the **Carré d'Or** (Golden Square), full of elegant apartment blocks, shady squares, and smart tourist hotels.

In **Central Nice,** streets widen and buildings become more grand, housing enormous landmarks like the Villa Masséna and the Russian Orthodox Cathedral. **West Nice** is home to many of Nice's grandest museums, such as the Musée des Beaux-Arts.

The views from the Colline du Château offer a spectacular orientation to the immediate surroundings of the city, from the **Baie des Anges** (Bay of Angels), which stretches all the way to the Cap d'Antibes, to the long valley created by **Le Paillon** river and, due east, **Mont Boron,** a hill separating Nice from the beach resort and deepwater port of Villefranche-sur-Mer.

SIGHTS

★ Promenade des Anglais

The promenade des Anglais is the wide, sweeping boulevard that flanks the Mediterranean along Nice's seafront. It was built by the large English colony in the 1820s who wanted a paved route to stroll from the old town to their church. The coastal path, which now carries their name, has taken on a mythical quality. It's the first place visitors see when they arrive in Nice, and it's lined with the city's most celebrated buildings, a historic collection of styles and eras, from the 17th-century fishermen's cottages to the belle epoque **Le Negresco,** the art deco facade of the **Palais de la Méditerranée, seaside sculptures,** and even its own Statue of Liberty, a tiny copy of the original.

The palm tree-lined promenade is always full of dog-walkers, joggers, Rollerbladers, visitors, and anyone out for an early-evening stroll. Despite six lanes of traffic and a two-way cycle path, it manages to safeguard its legendary qualities: a place for twilight cocktails, Rollerblading, busking, and people-watching.

★ Vieux Nice

Hemmed in by the arches of the **rue des Ponchettes** and the steep slopes of the Colline du Château, Vieux Nice is the picturesque hub of the city, where bars, art galleries, boutiques, and ice cream parlors coexist inside ancient doorways and vaulted rooms. The area has been gradually gentrified, but most of the facades are still crumbling, and the charm of the place lies in its shabby, meandering, and ancient feel. The narrow lanes appear to be held together by lines of washing and iron balconies. Steep staircases rise up the hill to artists' studios and vintage shops. It's hard to

1: promenade des Anglais **2:** place Rossetti and Cathédrale Sainte-Réparate **3:** cours Saleya **4:** Nice port

Chapelle de la Miséricorde

Galerie des Ponchettes

1

2

3

4

Nice

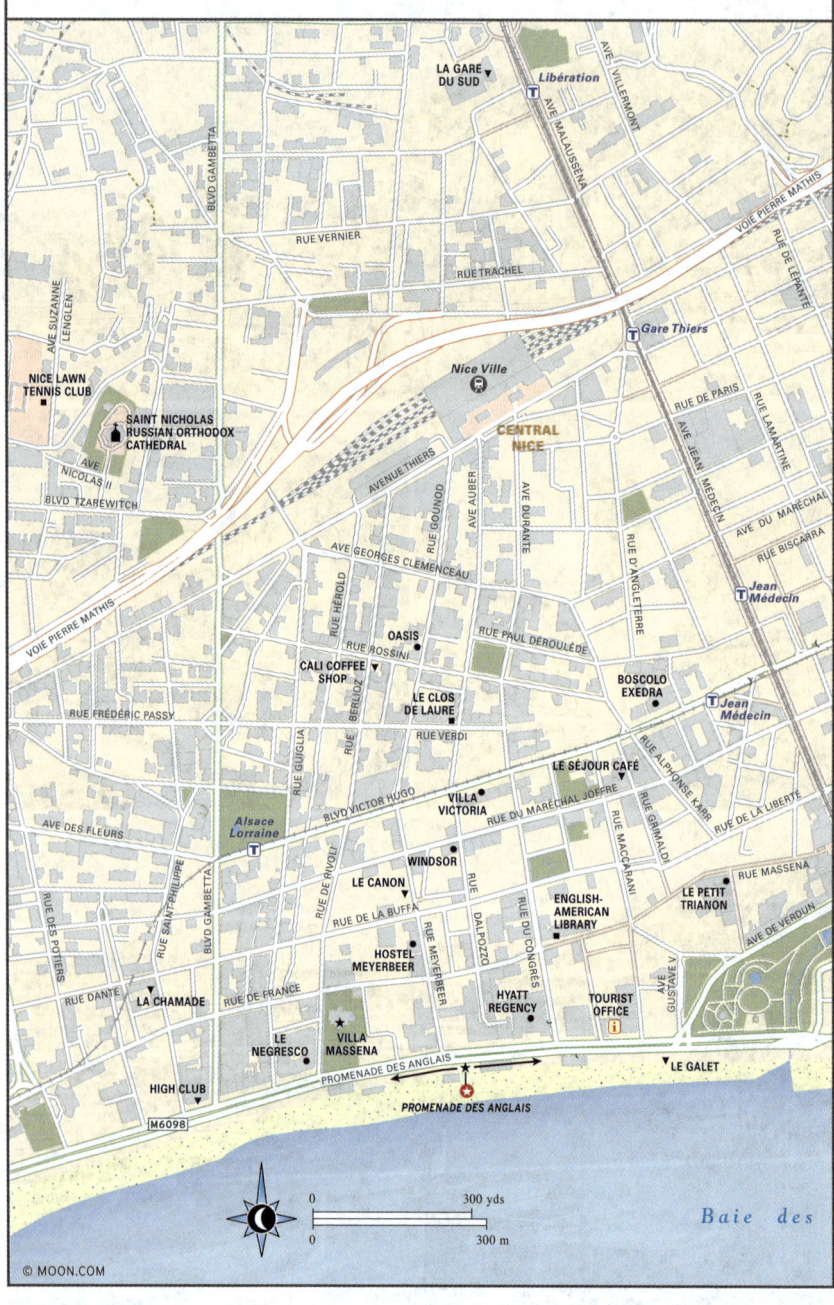

LA GARE DU SUD ▾

Libération Ⓣ

AVE VILLERMONT

BLVD GAMBETTA

RUE VERNIER

VOIE PIERRE MATHIS

RUE DE L'EPANTE

AVE MALAUSSENA

RUE TRACHEL

Ⓣ *Gare Thiers*

NICE LAWN
TENNIS CLUB

Nice Ville ◉

RUE DE PARIS

RUE LAMARTINE

SAINT NICHOLAS
RUSSIAN ORTHODOX
CATHEDRAL

**CENTRAL
NICE**

AVE
NICOLAS II

AVENUE THIERS

AVE JEAN MÉDECIN

AVE DU MARÉCHAL

BLVD TZAREWITCH

RUE GOUNOD

AVE AUBER

AVE DURANTE

RUE BISCARRA

AVE GEORGES CLEMENCEAU

RUE D'ANGLETERRE

Ⓣ *Jean
Médecin*

VOIE PIERRE MATHIS

RUE HÉROLD

OASIS ■

RUE PAUL DÉROULÉDE

RUE FRÉDÉRIC PASSY

RUE ROSSINI

**CALI COFFEE
SHOP** ▾

**BOSCOLO
EXEDRA** ■

RUE BERLIOZ

**LE CLOS
DE LAURE** ■

Ⓣ *Jean
Médecin*

RUE VERDI

RUE GUIGLIA

RUE ALPHONSE KARR

LE SÉJOUR CAFÉ ■

RUE
MASSA

**VILLA
VICTORIA** ●

RUE DU MARÉCHAL JOFFRE

RUE GRIMALDI

RUE DE LA LIBERTÉ

AVE DES FLEURS

BLVD VICTOR HUGO

*Alsace
Lorraine* Ⓣ

RUE MASSENA

WINDSOR ●

RUE DE RIVOLI

**LE PETIT
TRIANON** ■

RUE DES POTIERS

BLVD GAMBETTA

RUE SAINT-PHILIPPE

LE CANON ■

RUE DE LA BUFFA

**ENGLISH-
AMERICAN
LIBRARY** ■

AVE DE VERDUN

RUE MEYERBEER

RUE DALPOZZO

RUE DU CONGRÈS

**HOSTEL
MEYERBEER** ■

RUE DANTE

▾ **LA CHAMADE**

RUE DE FRANCE

**HYATT
REGENCY** ■

**TOURIST
OFFICE** ⓘ

AVE
GUSTAVE V

**LE
NEGRESCO** ■

**VILLA
MASSENA** ★

HIGH CLUB ■

PROMENADE DES ANGLAIS

▾ **LE GALET**

M6098

PROMENADE DES ANGLAIS ◎

0   300 yds

0   300 m

B a i e   d e s

© MOON.COM

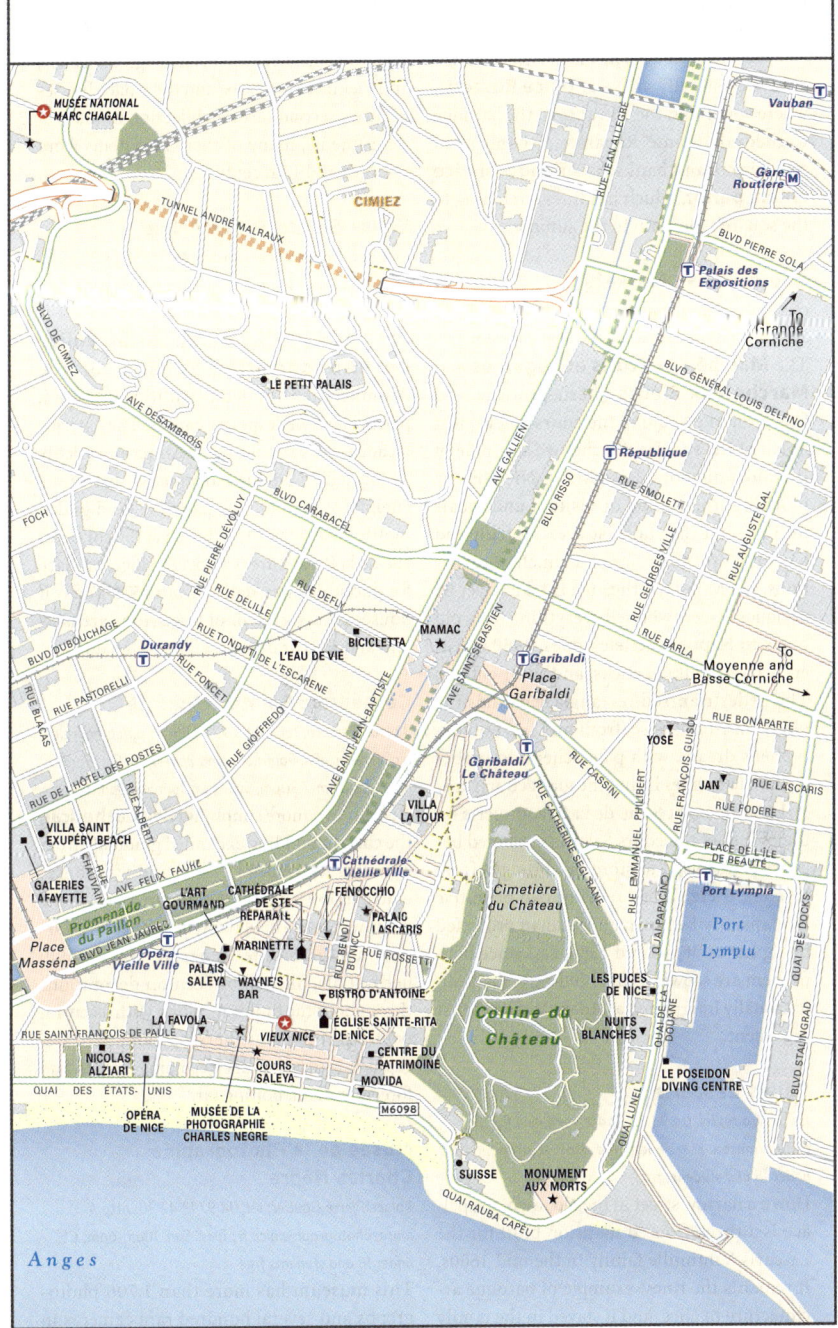

MUSÉE NATIONAL MARC CHAGALL

CIMIEZ

Vauban

Gare Routière

TUNNEL ANDRÉ MALRAUX

BLVD PIERRE SOLA

RUE JEAN ALLEGRE

Palais des Expositions

To Grande Corniche

LE PETIT PALAIS

AVE DE CIMIEZ

AVE DESAMBROIS

BLVD GÉNÉRAL LOUIS DELFINO

République

FOCH

BLVD CARABACEL

RUE SMOLETT

AVE GALLIENI

BLVD RISSO

RUE PIERRE DEVOLUY

RUE DELILLE

RUE DEFLY

RUE GEORGES VILLE

RUE AUGUSTE GAL

BLVD DUBOUCHAGE

Durandy

RUE TONDUTI DE L'ESCARENE

L'EAU DE VIE

BICICLETTA

MAMAC

RUE BARLA

To Moyenne and Basse Corniche

RUE PASTORELLI

RUE FONCET

AVE SAINT-SÉBASTIEN

Garibaldi

Place Garibaldi

RUE BONAPARTE

RUE BLACAS

RUE GIOFFREDO

RUE DE L'HÔTEL DES POSTES

RUE ALBERTI

AVE SAINT-JEAN-BAPTISTE

Garibaldi/ Le Château

RUE CASSINI

YOSE

RUE FRANÇOIS GUISOL

RUE LASCARIS

JAN

RUE FODERE

VILLA SAINT EXUPÉRY BEACH

RUE CHAUVAIN

VILLA LA TOUR

RUE EMMANUEL PHILIBERT

RUE CATHERINE SÉGURANE

PLACE DE L'ÎLE DE BEAUTÉ

GALERIES LAFAYETTE

AVE FELIX FAURE

Cathédrale-Vieille Ville

Cimetière du Château

Port Lympia

L'ART GOURMAND

CATHÉDRALE DE STE-RÉPARATE

FENOCCHIO

PALAIS LASCARIS

Port Lympia

Promenade du Paillon

BLVD JEAN JAURÈS

Place Masséna

Opéra-Vieille Ville

MARINETTE

RUE BENOIT BUNICO

RUE ROSSETTI

QUAI DE LA DOUANE

QUAI DES DOCKS

PALAIS SALEYA

WAYNE'S BAR

BISTRO D'ANTOINE

Colline du Château

LES PUCES DE NICE

QUAI DE LA PROVIDENCE

LA FAVOLA

ÉGLISE SAINTE-RITA DE NICE

NUITS BLANCHES

BLVD STALINGRAD

RUE SAINT-FRANÇOIS DE PAULE

VIEUX NICE

NICOLAS ALZIARI

COURS SALEYA

CENTRE DU PATRIMOINE

LE POSEIDON DIVING CENTRE

QUAI DES ÉTATS-UNIS

MOVIDA

M6098

QUAI LUNEL

OPÉRA DE NICE

MUSÉE DE LA PHOTOGRAPHIE CHARLES NEGRE

SUISSE

MONUMENT AUX MORTS

QUAI RAUBA CAPÉU

Anges

imagine there is any space for churches, but there are, in fact, over a dozen.

Wherever you walk in Vieux Nice, you will always eventually arrive at **place Rossetti,** where you'll find the entrance to the baroque Cathédrale Sainte-Réparate, a handful of restaurants and banks, and **Fenocchio ice cream parlor,** which has lines stretching to the square's fountain in the summer.

Cours Saleya Markets

cours Saleya; flower, fruit, and vegetable market Tues.-Sun. 6am-1:30pm, antiques market Mon. 7am-6pm

The **Marché aux Fruits et Légumes** and **Marché aux Fleurs** (fresh produce and flower markets) along the cours Saleya, just north of the rue des Ponchettes, are some of the most memorable sights in Nice. Every morning, thousands of visitors and locals mill around the stalls, squeezing fruit and haggling over the price of a bunch of carnations. On Mondays, the area is taken over by antiques dealers who sell everything from old maps to Napoleonic telescopes, 1950s posters, carved walking sticks, and silver cutlery. The pedestrianized rectangle separates the old town from the seafront and is a photographer's dream, with pale ocher and umber Italianate facades lining the giant courtyard.

Looking up at 8 rue de la Poissonnerie on the corner of the cours Saleya (so named because it was used to store salt brought from Hyères), there's a bas-relief of two primitive humans, a naked man and woman protected by fig leaves. It is dated 1584, and since both of them are scowling and holding clubs aloft, it's usually interpreted as Adam and Eve's first argument.

Palais Lascaris

15 rue Droite; tel. 04 93 62 72 40; www.nice.fr/fr/culture/musees-et-galeries; Wed.-Mon. 10am-6pm; €10, under 18 and students free

Down a narrow street in the old town, the palace is a three-floored mansion built for the Lascaris-Vintimille family in the mid-1600s. It remains the finest example of baroque architecture in Nice, and it stayed in the family

of aristocrats until the French Revolution. Its ornate vaulted staircase is adorned with frescoes of family crests, mythological scenes, and Flemish tapestries, and the palace houses France's second-largest collection of musical instruments, many of them rare items from the 16th and 17th centuries.

Église Sainte-Rita de Nice

1 rue de la Poissonnerie; www.sainte-rita.net; Mon.-Sat. 7am-noon and 2:30pm-6pm, Sun. 8am-noon and 3pm-6pm

This church, also known as the **Église de l'Annonciation** (Church of the Annunciation), is dedicated to Sainte-Rita, patron saint of desperate causes. The exterior facade is very simple, but the interior is richly decorated in Niçois baroque, with trompe l'oeil false marble, fine paneling, and gilded motifs. Just outside the church at the corner of rue de la Préfecture is a covered loge, or place for public debates, dating from 1574. Today it houses an assortment of large stone artifacts and architectural relics, all behind bars.

Cathédrale Sainte-Réparate

3 place Rossetti; tel. 04 93 92 01 35; http://cathedrale-nice.fr; Mon.-Fri. 9am-noon and 2pm-6pm, Sat. 9am-noon and 2pm-7:30pm, Sun. 9am-1pm and 3pm-6pm

Also known more simply as **Nice Cathedral,** the current building expanded gradually from a small 13th-century chapel and was consecrated in 1699. It is the largest sanctuary in the old town. It has held the relics of the 15-year-old martyr Sainte-Réparate since 1690; her initials, *SR,* can be seen on decorations throughout the building, along with a central painting of Réparate herself, about to be decapitated—having survived being burned and forced to drink boiling tar.

Musée de la Photographie Charles Nègre

1 place Pierre Gautier; tel. 04 97 13 42 20; http://museephotographie.nice.fr; Tues.-Sun. 10am-6pm; €5, under 18 and students free

This museum has more than 1,700 photographs and several hundred rare cameras in its private collection. It holds 2-3 exhibitions each year, plus a special photography exhibition by local children. The building used to be a factory for making electricity bobbins and still has a tremendous industrial feel, with high ceilings and steel gangway stairs. There's a 1-hour guided tour every Friday at 2pm (in French; €6.20) about the current exhibition.

The Port

To the east of the quai Rauba Capeu beneath the Colline du Château is Nice's port where, since the mid-1970s, giant ferries have departed for Corsica from the deep Bassin du Commerce. Cutting inland in an almost perfect rectangle are the Bassin des Amiraux and Bassin Lympia, full of pleasure boats and old-fashioned pointu fishing craft. It's a bustling, noisy part of the city, with the trendy rue Bonaparte behind its northern edge, the former customs houses and flea market (les Puces de Nice) on the western quai de la Douane, and on its eastern edge the quai des Docks, lined with restaurants, chandlers, and the Café du Cycliste, a meeting point for long-distance riders in spandex.

If you don't want to walk around the harbor, or you simply want a more genteel way to chug among the boats and superyachts, Nice Ville runs a free passenger ferry service (quai Entrecasteaux; mid-May-Sept. daily 10am-7pm) across the port, Lou Passagin, in an old-fashioned fishing boat called a pointu (both ends are pointed).

Terra Amata Musée de Préhistoire

25 boulevard Carnot; tel. 04 93 55 59 93; www.nice.fr/fr/culture/musees-et-galeries; Nov.-Apr. Wed.-Mon. 10am-5pm, May-Oct. Wed.-Mon. 10am-6pm; €5, under 18 and students free

Set at the base of a 1970s-looking apartment block, Terra Amata ("loved land") is regarded as prehistory's first-ever "home": a fixed settlement where hunters brought back animals to cook. Some 28,000 prehistoric objects were unearthed at the 400,000-year-old site during excavations in the 1960s. Museum highlights include the imprint of a right foot in limestone, flint choppers, hand axes, and other stone tools, along with slightly dusty dioramas and bones from bears, elephants, and aurochs (an extinct wild ox).

Colline du Château
Le Château

Colline du Château, access via the elevator or stairs at the end of the quai des États-Unis, by foot on rue Rossetti, rue Ségurane, or the Montée Menica-Rondelly; www.explorenicecotedazur.com; Oct.-Mar. daily 8.30am-6pm, Apr.-Sept. daily 8.30am-8pm; free

Having withstood numerous assaults beginning in the 11th century, Nice's hilltop citadel was finally taken by the French in 1705 and destroyed by Louis XIV's soldiers. Visitors can wander around the outside of the ruins, but little remains of the original castle. Today, the surrounding grassy plateau is usually full of strollers and skateboarders and is a great place to see the sweep of the Baie des Anges, planes taking off from the airport, and the hills beyond Cannes. On the other side of the hill, you can look down at the port and along the coast to the lighthouse on Saint-Jean-Cap-Ferrat.

There's a nice café at the very top of the hill, and another one beside the picnic zone and children's play area. The artificial waterfalls, Cascade Dijon, are best viewed from the promenade des Anglais, superbly illuminated after dark. In 1861, a Scottish winter visitor, Sir Thomas Coventry-More, paid to have a cannon set up on Castle Hill and fire a salute at noon every day to remind his wife to return home to prepare his lunch. It goes off to this day (not the same cannon, but an explosive device) at exactly noon, except on April Fools' Day, when it is fired an hour earlier. Behind the château ruins, heading down toward the old town on a walkway covered in colorful mosaics of Greek heroes, is the city's monumental **cemetery,** divided into three sections for Jews, Catholics, and Protestants.

Monument aux Morts

quai Rauba Capeu, place Guynemer

Built into the 90-m (295-ft) outcrop of gray limestone rock beneath the château is Nice's monument to the fallen. It is dedicated to the 2,000 inhabitants of Nice killed in World War I. The first stone was laid in 1924, and it took almost four years to build. At 32 m (105 ft) high, it's one of the largest war memorials in France and a masterpiece of art deco architecture. The five stone steps represent the five years of the war, and the bas-reliefs on either side of the eagle are symbolic of War and Peace, Freedom, Power, Sacred Flame, and Victory on one side, and Work, Home Life, and Fertility on the other side.

L'Observatoire

96 boulevard de l'Observatoire; tel. 04 92 00 30 11; www.oca.eu; tours Wed. and Sat. 10am-11:30 and 2pm-4pm; €12, children €6, under 6 free

Built in 1881 by Charles Garnier (of Paris Opera fame) and with a dome designed by Gustave Eiffel, Nice's observatory is one of the city's most celebrated (but least visited) landmarks. High up on Mont Gros, it has fantastic views of Nice and featured prominently in Woody Allen's 2014 film *Magic in the Moonlight*, which was set on the French Riviera in the 1920s. A guided tour lasts 1.5 hours in the morning and 2 hours in the afternoon and includes a walk through the woods around the observatory and into the Universarium. Public tours are only in French and must be booked via the website (www.observatorium.oca.eu/visites.php). Private tours are available in English via visitenice@oca.eu. Bus number 84 stops at l'Observatoire on the way to Beaulieu from Riquier station.

Fort du Mont Alban

chemin du Fort du Mont Alban; tel. 04 92 00 41 90; www.explorenicecotedazur.com; tours mid-July-mid-Sept. Wed., Fri., and Sun. 10am, 11am, and noon; €6.10, under 18 free

This solid-looking white-stone military fort was built between 1557 and 1560 and is one of the best preserved of its kind in France. Consisting of a central square and defensive towers at each of the four corners, the fortress was part of a line of fortifications that included Villefranche's citadel and Nice's castle.

It was opened to the public for the first time in 2010, and the surrounding park is a popular destination for picnickers and joggers. Note that visits to the fort can only be done as part of a guided tour through the Centre du Patrimoine (14 rue Jules Gilly, Nice) or via the website (https://centredupatrimoinevdn.tickeasy.com).

Place Garibaldi

One of Nice's most historic districts was given a €2.3 million facelift in 2012, providing the streets and squares between the port and the promenade des Arts with a new lease on life. Cafés, bars, shops, and even a movie theater bustle amid the trompe l'oeil murals on all four sides of the square. To the north is the **Crypte de Nice** (http://centredupatrimoinevdn.tickeasy.com/Offres.aspx; tours last 55 min), the underground remains of a medieval village discovered by chance when foundations were dug for the new tram line. On the opposite side, **rue Cassini** leads down to the port and through the **Quartier des Antiquaires** (antiques district), while **rue Bonaparte** has become a hub of Nice nightlife.

MAMAC

place Yves Klein; tel. 04 97 13 42 01; www.mamac-nice.org; Nov.-Apr. Tues.-Sun. 11am-6pm, May-Oct. Tues.-Sun. 10am-6pm; €10, under 18 and students free

Four floors of art, photos, and sculptures from the 1950s to today, MAMAC is the most enjoyable museum in Nice, located in a huge, airy building that forms an elevated glass-framed circle with four giant blocks around it. Its rooftop walkway has fantastic views of the hills around Nice, the old town, Cimiez, and Mont Boron. The permanent collection of around 1,300 works covers the main art movements of the last 70 years, particularly new realism and pop art with substantial collections by Niki de Saint Phalle, Yves Klein, and Ben, who died in June 2024.

1: cemetery on Colline du Château
2: L'Observatoire **3:** Musée National Marc Chagall
4: Cimetière du Monastère

Architecture in Nice

Few cities in the world can boast the range and quality of architecture of Nice. Having attracted the rich and famous since the early 1800s, there has always been plenty of money and desire around to pay for celebrated architects, extravagant structures, and the use of unusual materials.

BELLE EPOQUE

Belle epoque is the dominant architectural style in Nice. The sculpted, highly decorative style included iron balconies, tiled roofs, and servants' quarters on the top floor behind tiny square windows. It was developed in the 1860s and lasted until the First World War.

Notable examples of the more opulent belle epoque architecture are:

- The **Palais Meyerbeer** at 45 boulevard Victor Hugo.

- The **Boscolo Exedra Hotel** at 12 boulevard Victor Hugo.

- The iced-wedding-cake style of **Le Negresco Hotel** at 37 promenade des Anglais.

- **Le Régina** at 71 boulevard de Cimiez.

ART NOUVEAU

At the turn of the 20th century, art nouveau began to take hold. This was a more curvilinear, nature-inspired style that employed decorations in the shape of a whiplash and was centered on teasing, asymmetrical curves and the female form.

There are a few wonderful examples around Nice:

- The wave- and peacock-inspired facade at **15 rue Gounod.**

- The **Villa Rosalia** at 8 rue Berlioz.

- The rust-colored facade of **Palais Pauline** at 2 rue de Lépante.

- **Villa Collin-Huovila** at 139 promenade des Anglais, a pink-domed folly of sinewy curves and sensual ornamentation, made more striking by the provocative, white, minimalist, art deco Villa Monada that was built right next door at number 137.

Cimiez

Scottish writer Tobias Smollett, who came to Nice in 1764, was enthralled by Cimiez. "Nothing could be more agreeable and salubrious," he wrote of the hill, which at that time was still dominated by the Roman amphitheater and ruins. While the seafront and rejuvenated old town have attracted the jet set down from the hillside, the fin-de-siècle villas and wide, airy boulevards make Cimiez a peaceful, elegant haven, with fantastic views of the city.

★ Musée National Marc Chagall

36 avenue Dr Ménard; tel. 04 93 53 87 20; www.musee-chagall.fr; May-Oct. Wed.-Mon. 10am-6pm, Nov.-Apr. Wed.-Mon. 10am-5pm; €10, under 26 (EU members) free, free first Sun. of the month

Nice's most-visited museum is halfway up the steep boulevard toward Cimiez and houses the largest public collection of Marc Chagall's works. The Russian-born artist's Biblical Message series of 17 works is on display in a purpose-built white gallery opened by Chagall himself in 1973. Besides

ART DECO

By 1914, the craze for art nouveau had been crushed by the advent of the First World War and the arrival of a new style, art deco, which was to take Nice by storm.

Rues Rossini, Verdi, and Déroulède, and boulevard Gambetta, are flanked by art deco apartment blocks with futurist, zigzag forms and cruise-liner design, which fit in well with Nice's seaside credentials.

mosaic details on a house in rue Verdi

- The majestic mint-green **Le Gloria mansions** at 123-125 rue de France and the brick **post office** (21 avenue Thiers) are both superb examples of art deco design.

- The Cleopatra motif at the **Bel Azur building** at 49 promenade des Anglais was inspired by the discovery of Tutankhamen's tomb in 1922.

- Nearby at number 13 is the white granite **Palais de la Méditerranée,** the most sumptuous and well-preserved art deco facade in Europe.

TROMPE L'OEIL

Long before it became part of France in 1860, Nice had a tradition of trompe l'oeil, visual deceptions painted on facades, in Vieux Nice, Carabacel, and Cimiez. This technique is used to excellent effect in transforming plain walls into colorful street scenes or false shutters and window frames.

These buildings might not be quite what they seem:

- the **Musée Matisse**

- the buildings around **place Garibaldi**

- It's almost impossible to tell which windows are not painted on at one side of the **Esplanade Georges Pompidou,** and the artist himself is up a ladder above the Di Più restaurant at 85, quai des États-Unis.

---

the Biblical Message series, which is based on the Old Testament, there are over 400 of his works: oil paintings, gouaches, drawings, sketches, and pastels. A 50-minute film shown in the museum auditorium details Chagall's life. There's also a gift shop, a garden with a pool reflecting the artist's mosaics, and a café.

Musée d'Archéologie and Amphitheatre

160 avenue des Arènes de Cimiez; tel. 04 93 81 59 57; www.musee-archeologie-nice.org; May-Oct. Wed.-Mon. *10am-6pm, Nov.-Apr. Wed.-Mon. 10am-5pm; €5, under 18 and students free*

Augustus proclaimed Cemenelum (what the Romans named the hill of Cimiez) the capital of the Roman province of Alpes Maritimae in 14 BCE. Visitors can wander around the paved Roman streets, a Paleo-Christian complex dating from 5 CE, and look over the 3rd-century baths, which demonstrate a clear class structure of who was able to bathe where. The museum is full of Greek and Roman artifacts from the archaeological digs that took place at the site between 1950 and 1969, plus bronzes

Outdoor Art in Nice

Among the city's hundreds of outdoor sculptures are 223 works set out along the tram tracks, which Nice City Hall proudly calls its open-air museum. Statues by more established contemporary artists, all of whom have a strong connection with Nice, are featured along the **promenade des Anglais** and **promenade du Paillon.** The following are some of the more dramatic and eye-catching sculptures in the city.

- **Arc 115° 5:** In the Jardin Albert 1er is this giant curve surging out of the park, carved by Nice-born sculptor Bernar Venet. The angle of the arch corresponds to the sweep of the Baie des Anges.

- **La Chaise de SAB:** One of Nice's iconic blue chairs by artist Sabine Geraudie is on a plinth opposite the Jardin Albert 1er on the promenade des Anglais.

- **Lignes Obliques:** Just behind *La Chaise* is another sculpture by Venet, nine 30-m-high (98-ft-high) girders that join at the top and represent the nine valleys of Nice. It was installed to celebrate the 150th anniversary of the annexation of Nice to France.

Le Pouce, a giant thumb by César, outside Nice's town hall

- **Conversation à Nice:** Seven steel pillars topped by a kneeling or sitting human figure form this piece, created by Spanish artist Jaume Plensa. The illuminated statues, one for each continent, slowly change color at night around the place Masséna.

- **Le Pouce:** a giant thumb by César, one of the French Nouveaux Réalistes, is outside Nice's town hall.

and milestones from all over the region. Most impressive, however, is the amphitheater (arènes), which was used as a cohort training arena and eventually enlarged to provide entertainment for the locals, holding around 5,000 people.

Musée Matisse

164 avenue des Arènes de Cimiez; tel. 04 93 53 81 08 08; www.musee-matisse-nice.org; Nov.-Apr. Wed.-Mon. 10am-5pm, May-Oct. Wed.-Mon. 10am-6pm; €12, under 18 and students free

Overlooking the ruins of Cimiez's hilltop Roman settlement is the Matisse museum, a grand 17th-century Genoese-style villa devoted to the works of the French artist, who lived in Nice from 1917 until his death in 1954. He's buried in the Cimetière du Monastère nearby. Matisse remained

in Nice during the war, but was bedridden for several months after a serious illness. During this time he developed a technique using scissors to cut out shapes from large sheets of gouache painted by his assistants. Thirty-eight of these impressive cut-outs are exhibited in the museum, alongside 31 paintings, 454 drawings and engravings, and 57 sculptures. Matisse was a great collector of art and surrounded himself with items that inspired him in his studio: furniture, textiles, and artist's tools, all of which are also in the museum and offer a unique, personal portrait of the artist.

Cimiez Monastery and Monastery Gardens

place Jean-Paul II; tel. 04 93 81 08 08; Thurs.-Tues. 9am-6pm

Franciscan monks took over this monastery from their Benedictine brothers in 1546 and have been living there ever since. The decorated cloisters are open to the public and are host to the occasional yard sale. The adjoining **Musée Franciscain** (place du Pape Jean-Paul II; tel. 04 93 81 00 04; Mon.-Sat. 10am-noon and 3pm-5:30pm; free) details the life of Franciscan friars in Nice from the 13th-18th centuries. There are reconstructed cells, a frescoed chapel and a mural depicting Saint Francis of Assisi.

The gated **Monastery Gardens** (Apr.-Sept. daily 8:30am-8pm, Oct.-Mar. daily 8:30am-6pm) were once the monks' orchards and vegetable gardens. They can be accessed across the herringbone car park and form a huge panoramic terrace, beautifully laid out in symmetrical rows, with water fountains and plenty of places to sit among the roses and lemon trees.

Central Nice
Saint Nicholas Russian Orthodox Cathedral

avenue Nicolas II; tel. 09 81 09 53 45; www.sobor.fr; Mon.-Fri. 9am-1pm and 2pm-6pm, Sat.-Sun. 9am-6pm; free

Inaugurated in December 1912, the Russian cathedral is one of Nice's most iconic buildings and one of the largest Russian Orthodox religious buildings in the world. Its six gilded onion domes, topped with gold crosses, can be seen poking out from a residential area a few minutes' walk from the railway station. Set in a gated park with a café on the grounds and rabbits running around freely, the cathedral served the large Russian community who had settled in Nice by the end of the 19th century, as well as visitors from the Russian Imperial Court. Emerald-colored mosaics appear on the ocher brick facade beside gilded ceramics. Inside, the single chamber feels surprisingly intimate, with delicately carved wooden ornaments, flower frescoes, and religious paintings taken from the Church of Saint Basil in Moscow.

Villa Masséna

65 rue de France, access also through the gardens at 35 promenade des Anglais; tel. 04 93 91 19 10; www.nice.fr/fr/culture/musees-et-galeries; Nov.-Apr. Wed.-Mon. 11am-6pm, May-Oct. Wed.-Mon. 10am-6pm; €10, under 18 and students free

The jewel of Nice's belle epoque architecture, this enormous seafront villa, built between 1898 and 1901, houses a huge collection of paintings, furniture, costumes, and carnival posters that offer a glimpse into the highest echelons of aristocratic life in Nice in the 19th century. Among the artifacts on display are Napoleon's death mask, a mother-of-pearl tiara worn by Empress Josephine, and maps and landscapes of the old city. The building was given to Nice in 1919 by the great-grandson of the original owner, Niçois André Masséna, and the museum opened in 1921. In the southeast corner of the gardens, facing the sea on a stepped plinth, is the monument to the victims of the July 14, 2016, Nice terror attack.

West Nice

West Nice is dominated by the airport and the Var river, whose mouth opens into the sea at the western boundary of the city. Opposite the rows of private jets is the Parc Phoenix, a 7-hectare (17-acre) park, zoo, and botanical gardens that also houses the Musée des Arts Asiatiques. Since hills to the north and east of Nice curtail its expansion, much of the new development, including the construction of a Paris-La Défense-style business and administration complex, is along the boulevard du Mercantour on the eastern bank of the Var river; the development includes a giant IKEA store and the national sports museum.

Musée des Arts Asiatiques

405 promenade des Anglais; tel. 04 89 04 55 20; https://maa.departement06.fr; July-Aug. Wed.-Mon. 10am-6pm, Sept.-June Mon.-Wed. 10am-5pm; free

At the entrance to the Parc Phoenix, this white marble museum appears to be floating on an artificial lake. The building, designed by Japanese architect Kenzo Tange, is based on the two geometrical shapes fundamental to the Japanese tradition: the square, symbolizing the earth, and the circle, symbolizing the sky. Inside the museum, four giant cubes represent the "mother" civilizations of China and India and their transmission toward Japan and Southeast Asia.

Musée des Beaux-Arts

33 avenue des Baumettes; tel. 04 92 15 28 28; www. musee-beaux-arts-nice.org; Nov.-Apr. Tues.-Sun. 11am-6pm, May-Oct. Tues.-Sun. 10am-6pm; €10, under 18 and students free

High ceilings and light, airy spaces make this a wonderfully serene gallery. A sweeping marble staircase leads to rooms dedicated to Asian arts, French 19th-century paintings, and Jules Chéret, who created posters in a jaunty, titillating art nouveau style. The highlight of the collection, at the end of the corridor, is Bronzino's (1503-1572) mesmerizing *Crucifixion,* one of the finest examples of Florentine Renaissance religious art. The mansion that houses the museum was built for the Ukrainian princess Elisabeth Kotchoubey in 1878, but it was sold to American businessman James Thompson in 1883 and bought by the city of Nice in 1925.

Musée National du Sport

Stade Allianz Riviera, 6 allée Camille Muffat; tel. 04 89 22 44 00; www.museedusport.fr; Sept.-May Tues.-Sun. 10am-5pm, June-Aug. Tues.-Sun. 10am-6pm; €8, ages 18-25 €4, under 18 free

France's national sports museum moved from Paris to Nice, reopening in 2014 inside the city's Allianz Riviera stadium. It has a huge collection of sports artifacts and memorabilia dating from the 16th century to the present day, including 18,000 posters, 400,000 documents, 1,000 sports films, bikes, balls, jerseys, trophies, and Olympic medals dating back to 1900. Admission includes access to both permanent and temporary exhibitions, and stadium tours are available with a combined ticket (adults €16, ages 5-18 €10, under 5 free).

BEACHES

Nice has 15 private and 20 public beaches, all of which stretch along the promenade des Anglais from the port to the airport. They can be very crowded in the summer, despite being covered in rounded gray pebbles (galets). Private beaches have restaurants and bars and offer sun loungers, parasols, beach huts, showers, and toilet facilities (€20-40 for a lounger and parasol), generally lowering prices after 3pm. Two of the beaches, Centenaire and Carras, have easy access for disabled people, identified by a Handiplage sign, and there's a beach reserved for dogs (and their owners) on the promenade opposite the Lenval hospital. Lifeguards are on duty April-September.

SPORTS AND RECREATION
Parks and Squares

The center of Nice is dotted with small parks and squares, many of which were created in the 19th century when wealthy residents imported rare species of plants and trees to enhance their private gardens. Both sides of boulevard Victor Hugo, **Square Puccini, Jardin Alziari de Malausséna, Jardin Moreno,** and **place Mozart,** a little way north toward the railway station, are pleasant places to sit—some have fountains and play areas, and all are always full of lapdogs in jackets. The boulevard ends at **Jardin d'Alsace Lorraine,** which has paved circuits and some interesting sculptures and mature trees. The **Jardin Albert 1er** hosts the jazz festival and becomes the promenade du Paillon over the other side of place Masséna.

Promenade du Paillon

plassa Carlou Aubert; www.nice.fr/fr/parcs-et-jardins; Oct.-Mar. daily 7am-9pm, Apr.-Sept. daily 7am-11pm

Known to locals as the Coulée Verte (green passageway), this swath of lawns, monuments, and paved walkways stretches from the Jardin Albert 1er to beyond the MAMAC, and has revitalized a grimy slice of Nice that used to house the city's bus station. The promenade du Paillon starts properly at the edge of the place Masséna, where there is a garden of fountains—jets of water burst up from the floor in unison, surprising teenagers on skateboards and soaking toddlers.

Parc Phoenix

405 promenade des Anglais; tel. 04 92 29 77 00; www. parc-phoenix.org; Apr.-Sept. daily 9:30am-7:30pm, Oct.-Mar. daily 9:30am-6pm; €5.20, ages 12-18 €3, under 12 free

Parc Phoenix consists of 7 hectares (17 acres) of lawns, lakes, tropical greenhouses, and a zoo in an enclosed park opposite the airport. It has 2,500 plant species and 2,000 animals. Called the "green diamond," the park's centerpiece is one of the largest greenhouses in Europe, a 25-m-high (82-ft-high) pyramid replicating six different tropical and subtropical habitats for plants and animals. Free-flying exotic birds tweet at the caged parrots, mynah birds, and buzzards; flamingos flock on the lake; and emus, kookaburras, lemurs, wallabies, otters, and monkeys hide from the sun in giant cages. It's a pleasant place to spend the afternoon and ideal to while away the hours if waiting for a flight.

Hiking
Mont Boron
Distance: *4.2 km (2.6 mi) round-trip*
Time: *2 hours*
Information and Maps: *Nice tourist office*
Trailhead: *La Réserve restaurant, 60 boulevard Franck Pilatte*

This easy hike around Mont Boron can be completed in a couple of hours. The route starts beside La Réserve restaurant on boulevard Franck Pilatte, on the east side of Nice's port where the ferries dock (gare maritime). It begins by ascending a steep staircase set into the hillside along the edge of the Cap de Nice peninsula and up to the top of Mont Boron past belle epoque villas, and ends at the fort on top of the hill, which has fantastic views of Nice (to the west) and Villefranche-sur-Mer (eastward). The descent is partly along the Corniche Bellevue and includes some precipitous stairs, which

lead down to the Basse Corniche and back to La Réserve.

Cycling

Nice is a popular destination for cyclists of all levels, and bicycling is a smart, pleasant way to travel around the city. For people who enjoy a gentle ride, there's a wide, flat cycle path alongside the seafront promenade.

Privately run rental agencies **Lime** (www. li.me) and **Pony** (https://getapony.com) have a combined 2,000 bikes available in Nice, 90 percent of which are electric; many of the larger hotels also hire out bikes.

Serious amateurs and pro cyclists tend to head east up the Haute and Moyenne Corniches toward Menton and Italy.

Bicicletta Shop

9 and 9bis rue Defly; tel. 09 80 39 33 27; www. bicicletta-shop.com; Tues.-Sat. 9am-1pm and 2pm-7pm

This is a retro-style bike shop just north of MAMAC that rents city bikes and restores vintage cycles. A six-speed city bike with basket, lock, and helmet costs €18.50 per day; tandems are €33.50. Electric bike rental, also with helmet, lock, and basket, is €36.50.

Café du Cycliste

16 quai des Docks; tel. 09 67 02 04 17; www. cafeducycliste.com; Mon.-Sat. 8:30am-6pm, Sun. 8am-1:30pm

For serious cyclists, the café rents out carbon-fiber and steel-framed bikes for €50-95 per day. For those keen on accompanying a group on an organized ride, check out the company's website under "rides." These rides normally happen over the weekend, with an 8am start from the café, and head up to the Col d'Eze, Col de la Madone, or Col de Vence, or to Aspremont and Levens.

Water Sports

There are three separate locations along the coast to take part in Nice's many water sports activities: the promenade des Anglais, the port, and beside the Plage des Bains Militaires. Parascending, waterskiing, wakeboarding, stand-up paddleboarding, kayaking, Jet Skiing, and Flyboarding are on the promenade des Anglais beaches. Scuba diving expeditions leave from the port, and kayaking, stand-up paddleboarding, and snorkeling can be done near La Réserve on boulevard Franck Pilatte.

Le Poséidon Diving Centre

quai Lunel, port de Nice; tel. 04 92 00 43 86; https:// poseidon-nice.com; May-Oct.

Le Poséidon runs dives at more than 10 locations, mostly off Villefranche, accessible to all levels. They offer scuba diving courses (first lesson €75, five dives €325), as well as snorkeling for anyone over the age of 10 and PADI courses. Advance booking is required.

Azur Kayak Mer

50 boulevard Franck Pilatte; tel. 06 50 25 18 41; www. azurkayakmer.com; open year-round

Azur Kayak Mer offers sea kayaking and stand-up paddleboarding for half-day or full-day excursions. They also offer team-building courses and bachelor- or bachelorette-party specials. Two-hour kayak rental is €35; a kayak day trip runs €55. Stand-up paddleboarding for a half day is €40 and €70 for a full day. Participants need to be able to swim 25 m (82 ft).

Tennis

Tennis is an extremely popular sport in the South of France; it's part of the Riviera's lifestyle of glamorous pastimes. The city's clubs are designed mainly for local members, but visitors can book courts 24 hours in advance at the Nice Lawn Tennis Club.

Nice Lawn Tennis Club

5 avenue Suzanne Lenglen; tel. 04 92 15 58 00; http:// niceltc.com; daily 9am-9pm; from €25-35/hour for a hard or clay court

Visitors can book courts after becoming temporary members for €30 per hour (weekdays) or €35 per hour (weekends). The club has 12 flood-lit clay courts, 1 exhibition court, and 5 hard courts. Tennis pros are available to give lessons, and the club also has a shop and restaurant. Two outdoor padel courts are available for €24-40 per hour depending on day and time.

Soccer

Allianz Riviera—Stade de Nice

boulevard des Jardiniers; www.allianz-riviera.fr

The Allianz Riviera stadium is home to the city's soccer team, OGC Nice, and was opened in September 2013. It has a seating capacity of just over 36,000 and was one of the venues for the UEFA Euro 2016 football championships, the FIFA Women's World Cup in 2019, the Men's Rugby World Cup in 2023, and the Paris Olympics in 2024. It hosts occasional home rugby matches for RC Toulon, some of the French national football team's friendly matches, and preseason friendlies between European teams. It's also a major concert venue. For access via public transport, take tram Ligne 3 toward Saint-Isidore and exit at stop Stade, from where it's a 5-minute walk.

ENTERTAINMENT AND EVENTS
The Arts
Le 109

89 route de Turin; tel. 04 97 12 71 11; http://le109.nice.fr; Tues.-Sat. 1pm-7pm

Nice's municipal abattoir was saved from demolition and is now the pumping, supercool pôle (hub) of the city's contemporary culture movements. Part of the enormous structure is a venue for parties, performances, and concerts; the rest is given over to workshop space and individual artists' studios. On weekends, DJs play hip-hop, techno, and rock music late into the night, and in the daytime it's more serene, with an architecture library, reading rooms, and a space for alternative art. It's a 15-minute drive from the city center.

Opéra de Nice

4-6 rue Saint-François de Paule; tel. 04 92 17 40 79; www.opera-nice.org; ticket office open Tues.-Sat. 11am-5pm

After the original building was destroyed by

a mid-performance fire in 1881, Nice's opera house reopened in 1885 with Verdi's *Aïda*. Designed by local architect François Aune, it's a majestic belle epoque building with flourishes of the baroque, furnished with plush red velvet seats, slightly cramped boxes, and a gilded interior. It hosts operas, classical concerts, and ballets, with plenty of gala evenings and special concerts for children during the summer. Seat prices range €10–86, although to encourage young people to attend, tickets for all performances (except opera) are only €5 for students under 26 in restricted-view seats.

Festivals and Events
Nice Jazz Festival
Théâtre de Verdure, place Masséna; tel. 04 97 13 55 33; www.nicejazzfestival.fr; July; €40, ages 16-24 and over 65 €32, ages 10-15 €20, 4-day pass €150

One of the world's oldest and most eclectic jazz festivals, also including world music and a touch of soul, the Nice Jazz Festival runs for four days in mid-July. Two giant stages between the ocean and place Masséna host three sets each night with usually a couple of free concerts thrown in as well. Miles Davis, Ella Fitzgerald, Herbie Hancock, and Dizzy Gillespie have all appeared and, more recently, Sir Tom Jones and Japanese pianist Hiromi.

Nice Carnaval
www.nicecarnaval.com; Feb.-Mar.; €28, ages 6-10 €10, under 6 free

Nice's Carnaval is one of the biggest, most spectacular celebrations in Europe, with two weeks of firework displays, floral parades, sequined costumes, flower battles, and giant papier-mâché heads. The carnival tradition in Nice dates back to the 13th century. Today, it still passes through the main **place Masséna,** but the parade is surrounded by solid black, 3-m-high (10-ft-high) barriers. It's a security measure, but it also prevents anyone who hasn't paid from catching a glimpse of the floats. Visitors can see some of the floats and giant heads of the king of the carnival in the daytime, when they are left in the Place Masséna.

The weeks of disruption due to traffic, the giant scaffold for the seating, the barriers, the stern security agents, and the influx of tourists have dampened local spirits. Nevertheless, it's a fun, spectacular event, attracting over a million people to the city and generating almost €2 million in ticket sales and 1,800 jobs. Parallel events include a **Carnaval Swim,** a **Carnaval Fun Run,** the **Queernaval,** a giant **Zumba party,** and the **Battle of Flowers.** The papier-mâché Carnaval King is burned on a huge bonfire in place Masséna on the final night of the event.

SHOPPING
Shopping Districts
Avenue Jean-Médecin
Nice's main shopping street is the avenue Jean-Médecin, which has the tram line passing down the middle and runs from the **Galeries Lafayette** department store in place Masséna to the **place Charles de Gaulle,** where it turns into the avenue Malausséna. Jean-Médecin is lined with all the globally recognized international chain stores plus some one-off boutiques and designer shops. Halfway up is the **Nicetoile** shopping center, and a few hundred meters farther north, the spectacular **Iconic** shopping center, which opened in 2024. The roads leading off Jean-Médecin are also full of shops, cafés, beauty salons, and hairdressers. In place Charles de Gaulle and on both sides of the tram lines there's an excellent fresh produce market every morning except Monday, and the former Gare du Sud railway station has been developed into a food hall.

The roads north of the **promenade du Paillon, rue Defly, rue Delille,** and **rue Tonduti de l'Escarène** have more interesting shops if you're looking for secondhand books, bicycles, original art and crafts, and vinyl records.

Vieux Nice
Shops in the old town tend to be privately run boutiques catering to tourists looking for bundles of lavender, locally manufactured soap,

olive oil, and trendy clothes. **Rue Droite** has some vintage-wear stores among the art galleries, as do **rue Centrale** and **rue Benoît Bunico.** Around **place Rossetti** are plenty of ice cream parlors and local biscuit makers, as well as fashion boutiques, and **rue de la Boucherie** is great for leather goods and swimwear. On the first and third Saturdays of each month there is a secondhand book fair in the square outside the **Palais de Justice.**

Rue de France

The rue de France is pedestrianized from the intersection with **rue de Congrès** until it reaches the **place Masséna,** and it provides a posh shopping experience. The network of roads around **rue Alphonse Karr, rue Longchamp, rue Paradis, avenue de Verdun,** and down to **avenue de Suède** have designer boutiques including Chanel, Dior, and Louis Vuitton, as well as Bensimon, specialist jewelers, and Laguiole knives. It's worth a stroll to **rue Massenet,** where Albert Goldberg, who created Façonnable, has his flagship store Albert Arts, the empire of British chic.

Department Stores
Galeries Lafayette

6 avenue Jean-Médecin; tel. 04 92 17 36 36; www.galerieslafayette.com/magasin-nice; Sept.-June Mon.-Sat. 9:30am-8pm and Sun. 11am-8pm, July-Aug. Mon.-Sat. 10am-9pm and Sun. 11am-8pm

The Parisian-style department store occupies the northeast corner of place Masséna. The five floors of fashion franchises include men's, women's, and children's clothing. Perfume, cosmetics, and luxury-brand handbags are on offer on the ground floor, while the top floor stocks luggage and a large variety of kitchen supplies alongside a café-restaurant called La Table.

Art
Charvin

39 rue Gioffredo; tel. 04 93 92 92 82; https://charvin-arts.com; Mon.-Sat. 9am-7pm

The original Charvin art shop has been on the Riviera since 1830 and supplied Cézanne, Bonnard, and Ambrogiani with their paints. Today, run by Bruno Charvin, it's a beautiful shop of paint boxes, portable easels, pastels, chalk, smocks, watercolor sets, handmade brushes, and wall-to-wall oils, gouaches, and acrylics.

Antiques and Brocante
Les Puces de Nice

rue Robilant; Tues.-Sat. 10am-6pm

Les Puces de Nice (flea market) is a collection of antique and brocante shops overlooking Nice's marina. It backs onto rue Ségurane, the center of Nice's antiques district. Most other shops in the district have giant chandeliers and oil paintings in the window and require appointments to browse, but Les Puces, right on the waterfront, is more fun and is full of authentic bric-a-brac that fits easily into hand luggage.

Gourmet and Food
La Gare du Sud

35 avenue Malausséna; Tues., Wed., and Sun. 11am-11pm, Thurs., Fri., and Sat. 11am-midnight, until 11pm mid-Oct.-mid-May

The second incarnation of Nice's gourmet food hall opened in July 2023 in the former railway station of the chemin de Fer de Provence. The giant hall retains its industrial appeal, with an iron frame and glass roof high above diners, who can choose from a large range of Mediterranean cuisine, which can be ordered via giant tactile pads.

L'Art Gourmand

21 rue du Marché; tel. 04 93 62 51 79; www.lart-gourmand.com; daily 10am-7pm

This is a sweet shop extraordinaire, with handmade chocolates in the window and long wooden counters of nougat, biscuits, and jelly-fruit fingers. Everywhere are piles

1: Nice Carnaval **2:** secondhand book fair in Vieux Nice **3:** Charvin art shop **4:** exterior of La Gare du Sud food hall

of biscuits, huge bowls of crystallized fruit, and Provence-inspired ice cream. It's self-service, so visitors grab a bag, fill it up with jelly fruits and confectionary, and head to the counter for weighing. Up a flight of stairs is a tiny tearoom whose walls are hand-painted in old-fashioned chocolate designs.

Nicolas Alziari

14 rue Saint-François de Paule; tel. 04 93 85 76 92; www.alziari.com.fr; Mon.-Sat. 9am-7pm, Sun. 10am-7pm

Opposite the town hall is Nice's best-known local supplier of olive oil. Established in 1868, the shop also sells olive tapenade, honey, Provençal sweets, and olive oil soaps, all in tourist-friendly gift boxes and pretty canisters. There's an olive bar at the back of the shop and the company's olive mill is open for visits at 318 boulevard de la Madeleine (Mon.-Fri. 8am-noon and 2pm-6pm).

Le Clos de Laure

10 rue Verdi; tel. 04 93 91 36 53; www.leclosdelaure.com; Mon.-Sat. 10am-7pm

Stop here for a trove of the most delicious jams, tapenades, honeys, pastes, and regional delicacies that go perfectly with yogurt, cereals, roast meats, or as an aperitif on the nearby beach. It's a friendly family business and all products come from the southeast of France. Bestsellers include rose-petal jam, lavender honey, and pear and chocolate spread.

FOOD
Promenade des Anglais
Le Galet

3 promenade des Anglais; tel. 04 93 88 17 23; www.galet-plage.fr; Sun.-Thurs. 9:30am-10pm, Fri.-Sat. 9:30am-11pm; €15-25

Le Galet is the first foray onto Nice's beachfront by a group that runs five of the city's most popular Mediterranean restaurants. Le patron, Philippe Cannatella, claims Le Galet (the pebble) is "not at all show-offy" as some of the other beach restaurants along the promenade are, but it still feels pretty glamorous, with its own magazine to read while you're waiting for your salade Niçoise and glass of rosé. Cocktails are €13, or €10 for nonalcoholic versions.

Vieux Nice
Marinette

16 rue Colonna d'Istria; tel. 04 93 88 29 52; www.restaurantmarinette.fr; Wed.-Sun. 8:30am-11am breakfast, noon-4pm lunch, 4pm-6pm teatime; €9-19

Set on a quiet street behind the cathedral in the old town, the two-floored Marinette serves the best chocolate cake in Nice. It also does a great breakfast, with pancakes, juice, granola, yogurt, and fresh coffee from a roasting house nearby. They serve salads and burgers for lunch, but most people go for the homemade cakes, which are exceptional. The three-arched building was originally a dormitory for the local priests.

★ Bistro d'Antoine

27 rue de la Préfecture; tel. 04 93 85 29 57; Tues.-Sat. noon-1:45pm and 7pm-9:45pm; €13-21

Bistro d'Antoine rightly deserves its glowing reputation for serving fantastic food at very reasonable prices. The menus are simple, pared down to fine ingredients and precious flavors; dishes include green lentils with Perugina sausages (€15) and a magret de canard (€16). Owner Armand Crespo also runs a wine bar around the corner, La Cave du Cours, with cave paintings on the walls and racks of wine displayed around the cellar.

La Favola

13 cours Saleya; tel. 04 93 04 45 23; www.nice-restaurant.fr; daily noon-2:30pm and 7pm-10:30pm; €12-26

The first restaurant you come to on the cours Saleya is probably the best. Huge portions of pasta, pizzas that hang over the side of the plate, fried fish, il Gran Fritto Misto di Pesce, and desserts big enough for four people make up the menu. It's a noisy, lively Italian joint with leather banquettes and a large dining room upstairs. Pizzas run €13.50-17.50, with the truffle pizza costing €21.50. Reservations are not accepted.

The Port
★ Yose

20 rue Bonaparte; tel. 09 83 46 43 21; www.
yoserestaurant.fr; Tues.-Sat. noon-2:30pm and 7pm-
10:30pm; €24-27

Yose offers a taste of Peru in the center of
Nice. The restaurant, which was opened
by two French friends, Benjamin and
Thomas—who trained together in Nice and
worked in a Yosemite restaurant together
(hence the Yose)—specializes in ceviche
starters and Peruvian recipes made with
Mediterranean products. Mains include
grilled octopus with pachamanca sauce
and Peruvian seafood paella. It's an incredibly popular restaurant in the lively
rue Bonaparte, so advance booking is
recommended.

Jan

12 rue Lascaris; tel. 04 97 19 32 23; www.restaurantjan.
com; Tues.-Sat. 6pm-10pm; €165-195 without wine

Head chef Jan Hendrik van de Westhuizen
combines local Mediterranean flavors with
strong hints of his native South Africa to produce genuinely beautiful food. The six-course
JAN tasting menu is at Michelin-star prices
(his first star was gained in 2016), but it is fine
dining at an exquisite level.

Central Nice
La Chamade

17 rue Saint-Philippe; tel. 06 69 52 44 04; https://
pizzerialachamade.fr; Mon.-Fri. noon-2:45pm and
6:30pm-11pm, Sat.-Sun. 6:30pm-11pm; €9-17

Always full of Italians and with silver trophies
on the shelves (for winning pizza competitions), this simply decorated Neapolitan restaurant serves some of the best pizzas in town.
Slices are by the meter—75 cm (30 in) of pizza
Margherita costs €25, and 75 cm of the chef's
special with more difficult toppings €27, but
it's enough to feed four.

Cali Coffee Shop

49 rue Rossini; tel. 07 61 60 40 00; Mon.-Sat. 7am-
5pm; €11-24

Cali serves a great selection of lunchtime poké
bowls and homemade bagels as well as an all-
day breakfast, fresh lemonade, smoothies,
and excellent coffees. There are also vegetarian and vegan options, all within 5 minutes of
the railway station.

Le Séjour Café

11 rue Grimaldi; tel. 04 97 20 55 35; www.sejourcafe.
com; Tues.-Sat. noon-2pm and 7:30pm-10pm; €20-31

The decor here is a combination of immaculate French style and family ambience, with

La Chamade pizza

shelves of photos, model boats, books, and trinkets to look at while you are eating. Main dishes include a saddle of rabbit, scallop risotto, and roast beef with gnocchi. The restaurant has proven so successful that the owners opened its sister, Mon Petit Café, next door, which has more limited choices but is focused on local dishes.

★ Le Canon

23 rue Meyerbeer; tel. 04 93 79 09 24; www.lecanon. fr; Mon.-Tues. and Thurs.-Fri. noon-2pm and 7:30pm-10:30pm, Wed. 7:30pm-10:30pm; €28-38

If you only have time for one meal in Nice, let it be at Le Canon. They feature a dazzling array of dishes, full of intriguing combinations of locally caught fish, pork from Grasse, and beef from Ponclet, with sauces and spices gathered from all over France. A chalkboard of delights is brought over for guests' perusal, but it usually takes a while to decide. With a window full of wine bottles, diners can feel hidden away from nosy passersby on the street.

L'Eau de Vie

11 rue Delille; tel. 09 88 54 23 29; www.restaurant-eaudevie.fr; Mon.-Fri. noon-1:30pm and 7pm-9:30pm, Sat. 7pm-10pm; €23-42

L'Eau de Vie offers a fantastic mix of inspired Mediterranean flavors and Japanese cooking. Tempura sage leaves is an amuse-bouche, followed by French classics with a Japanese twist, and green tea cheesecake for dessert. On the 18th of every month they serve a selection of Japanese tapas. The interior is light and airy, and they have a large outdoor terrace in the summer on a pedestrianized road.

BARS AND NIGHTLIFE
Bars
Movida

41-43 quai des États-Unis; tel. 04 93 80 48 04; www. movida.today; daily 10am-2am

Overlooking the promenade, Movida is a buzzing tapas bar and bodega with plush interiors and a seafront veranda. It's perfect for watching the sports cars overheat in the semipermanent traffic jam beneath, or the giant ferries departing for Corsica in the distance. High stools on the terrace upstairs have the best views of Castel beach, Rollerbladers, and teenagers diving off the nearby rocks.

Wayne's Bar

15 rue de la Préfecture; tel. 04 93 13 46 99; www. waynesbar-restaurant.com; daily 10am-midnight, happy hour 4pm-8pm

By day, Wayne's is a friendly British pub with TVs, framed music posters, and rock memorabilia covering the walls. After dark the giant screen room at the back transforms into a live music venue, with tabletop dancing and crowds of revelers spilling out onto the street. The bar food, mainly burgers, salads, and chips with cheese, is very popular.

Nightclubs
Nuits Blanches

20 quai Lunel; tel. 06 13 68 64 37; https:// nuitsblanchesclub.com; Thurs.-Sat. midnight-5am; free entrance

A discreet entrance on the west side of the port opens in to one of Nice's premier nightclubs for the over-25 set. Cocktails cost €13, while a magnum of Dom Perignon champagne will set you back €1,000. A "nuit blanche" means to stay up all night, and that is the idea at their themed disco and '80s nights where the resident DJs play until dawn.

High Club

45 promenade des Anglais; tel. 07 81 88 42 04; www. highclub.fr; Thurs.-Sat. midnight-5:30am; entrance €10

The nightclub to end all nightclubs, High has three clubs at the same location, each one offering different styles of music, atmosphere, and clientele depending on your mood and persuasions. High is a hyped-up, expensive, blue-neon-lit discotheque with private booths, and VIP areas that peak around 2am. Studio 47 is for the over-30s, with padded walls, cushioned barstools, round tables, and an '80s vibe. Sk'High is the gay- and lesbian-friendly club. A valet is there to park your car; there's a smoking room, a "fooding" area, and a boutique for souvenir T-shirts.

ACCOMMODATIONS

In Nice, the cheaper hotels tend to be located near the railway station, but there are two great hostels for budget stays in the center (Hostel Meyerbeer and Villa Saint Exupéry Beach). Since Nice has such a huge range of hotels, competition is greater and prices are lower than in some of the nearby resorts.

Vieux Nice
Villa La Tour

4 rue de la Tour; tel. 04 93 80 08 15; www.villa-la-tour. com; €130 d

Under the tallest clock tower in Nice, Villa La Tour is perfectly located for exploring the old town, strolling the promenade du Paillon, and visiting the MAMAC art museum. It has 17 rooms with period furniture and modern art. There's also a restaurant-bar called Le VLT, which has its own terrace and is open to non-hotel guests, and is where Mauritian owner Madame Billiard serves her own spicy rum.

Palais Saleya

21 rue du Marché; tel. 04 92 00 09 09; www. palaissaleya.com; €230 d

The Palais Saleya is not cheap, but it's worth it for the location and the spacious, modern, well-equipped family apartments on the edge of the Vieux Nice. All 26 rooms are in a renovated 18th-century mansion overlooking the place du Palais de Justice, including the two-bathroom, 80-sq-m (861-sq-ft) Prestige Mezzanine Suite, which sleeps six and has a washing machine.

Cimiez
Le Petit Palais

17 avenue Emile Bieckert; tel. 04 93 62 19 11; www. petitpalaisnice.com; €200 d

A haven of serenity overlooking the city from Cimiez hill, this boutique hotel is perfect for those looking to escape the noise and crowds of the center. The belle epoque villa was the former home of the French actor and

playwright Sacha Guitry, but has been a hotel since 1947. Today it has 25 rooms, a small heated pool, a relaxation corner, and a garden terrace.

Central Nice
Hostel Meyerbeer

15 rue Meyerbeer; tel. 04 93 88 95 65; www. hostelmeyerbeer.com; €87 d, single bed in mixed dorm from €14

Meyerbeer has legendary status among backpackers and young travelers for its friendly ambience and great location. It has bright, cheery, private bedrooms; four-, six-, and eight-bed dorms from €14 a night; and free Wi-Fi and computers to use. A big breakfast is served in the lobby (€7). Like all good hostels, the staff are upbeat and helpful—they'll store your luggage if needed, and they offer free city tours and even organize a pub crawl around Nice's best watering holes.

★ Villa Saint Exupéry Beach

6 rue Sacha Guitry; tel. 04 93 16 13 45; www. villahostels.com; €120 d, single bed in dorm from €15

Opening the front door, you immediately enter a world of smiles, helpful staff (it's not always like this on the Riviera), organized activities, and global travelers. This is a fun place to stay, and it has a terrific café-bar where guests can buy food, prepare their own, or just chat about surfing and bus timetables on the Côte d'Azur. Private rooms are available, but most of the 212 beds are in dorms. There's no age limit (they've had a 90-year-old staying), and perks include free-to-use beach mats, computers, table tennis, a gym, a sauna, laundry service, and an eternal party vibe.

Oasis

23 rue Gounod; tel. 04 93 88 12 29; www.hotelniceoasis. com; €125 d

Vladimir Oulianov, better known as Lenin, stayed at this former Russian pension in 1911. His countryman, the playwright Anton Chekhov, spent his winters there at the turn of the 20th century and wrote *Three Sisters* in the gardens where guests now eat their breakfast. Set back from the street, the peach-colored hotel has 44 rooms, some in an annex across the courtyard. Halfway between the railway station and the beach, the hotel also has its own car park (rare in Nice).

Le Petit Trianon

11 rue Paradis; tel. 04 93 87 50 46; www.lepetittrianon. fr; €130 d

Taking its name from Marie Antoinette's tiny château on the grounds of the Palace of Versailles, this mother-and-son-run hotel is equally charming, occupying five spacious rooms on the 2nd floor of a building on the poshest street in Nice. Guests can have breakfast brought to their rooms or go to one of the many cafés in the pedestrianized district just downstairs.

★ Windsor

11 rue Dalpozzo; tel. 04 93 88 59 35; www. hotelwindsornice.com; €165 d

In a large stone-fronted villa three blocks from the sea, the Windsor is the self-styled "art" hotel of the city. Rooms have been painted by local and international artists, including Glen Baxter and Peter Fend. It hosts art exhibitions in the foyer and has the best hotel elevator in Nice—it plays the sound of a rocket-launch

countdown—as well as a small pool, a tropical garden, and a great bar. It also runs an annual video art festival (www.ovni-festival.fr).

★ Villa Victoria

33 boulevard Victor Hugo; tel. 04 93 88 39 60; www. villa-victoria.com; €200 d

The Villa Victoria is a handsome hotel along one of Nice's widest and leafiest boulevards. If you're looking to get away from the crowds, this hotel is a good option, hidden among the plane trees and art deco mansion blocks. The best rooms overlook the orange trees and bougainvillea-covered La Rotonde centerpiece in the huge gardens, where guests can take breakfast in the summer and watch the hotel's two cats chase away the seagulls.

INFORMATION AND SERVICES
Tourist Information
Office de Tourisme et des Congrès

5 promenade des Anglais; tel. 0892 707 407; www. nicetourisme.com; July-Sept. daily 9am-7pm, Oct.-Dec. and May-June Mon.-Sat. 9am-6pm, Jan.-Apr. Mon.-Sat. 9am-5pm

The main tourist office is on the promenade des Anglais and has free maps of the city and details about exhibitions, events, concerts,

gardens at the Villa Victoria

and the Nice carnival. They can also help you reserve accommodations and sell the French Riviera Pass (www.frenchrivierapass.com), which offers deals on museums, activities, and transport on the Côte d'Azur.

Bureau d'Information Touristique Nice-Gare

avenue Thiers; tel. 04 92 14 46 14; www.nicetourisme. com; July-Sept. daily 9am-7pm, Oct.-June Mon.-Sat. 9am-6pm, Sun. 10am-5pm

This tourist office is outside the railway station.

Centre du Patrimoine

14 rue Jules Gilly; tel. 04 92 00 41 90; www.nice.fr/ fr/culture/vos-rendez-vous-patrimoine; Mon.-Thurs. 8:30am-1pm and 2pm-5pm, Fri. 8:30am-3:45pm

Nice's heritage center has leaflets, flyers, and details about current exhibitions in the city and information about its historic buildings. It run tours every weekday on topics including the old town, Cimiez, Colline du Château, Nice in the Jazz Age, and the Crypt de Nice.

English-American Library

12b rue de France; tel. 04 93 87 42 67; www.nice-english-library.org; Wed. 3pm-5pm and Sat. 10am-noon; entrance through church gate at 11 rue de la Buffa

Behind the Anglican Holy Trinity church and accessible through a garage on the pedestrianized rue de France is one of the oldest permanent institutions in Nice, the English-American Library. The *Herald Tribune* is delivered daily, and the *Spectator, Time,* and the *New Yorker* magazines arrive every week. Visitors can take out a weekly subscription or are welcome to read in the library.

Services
Post Office

Nice's main post office is on avenue Thiers (tel. 3631; www.laposte.fr; Mon.-Fri. 8am-6pm, Sat. 8am-12:30pm), 100 m (328 ft) to the west of the main railway station. It's set in a large, art deco brick building on the corner of rue Gounod. The last post is at 5pm on weekdays and 11am on Saturdays. Stamps can also be purchased from tabacs, recognizable by the flashing double tapering-carrot sign.

Banks

Main branches of the major French banks can be found on **boulevard Gambetta** and **avenue Jean-Médecin,** with smaller branches and ATMs all over the city.

Health and Safety

Nice's main police station, the **Commissariat de Police,** is at 1 avenue du Maréchal Foch (tel. 04 92 17 22 22; www.pre-plainte-en-ligne.gouv.fr). Most police officers will speak enough English to be helpful to visitors who are in trouble or require assistance. For all assaults, theft, and other crimes, visitors must file a report at the commissariat (police station); it can take many hours but is necessary for insurance claims. If you are the victim, you can fill out a declaration form online, which saves time queuing, but you still have to sign it at the commissariat.

For **lost and found,** head to the police's dedicated department for objets trouvés at 42 rue Debray (tel. 04 97 13 44 10; https://web. nice.fr/formulaires/objets-trouves; Mon.-Thurs. 8:30am-5pm, Fri. 8:30am-3:45pm).

For an English-speaking doctor, contact **Riviera Medical Services** (tel. 04 93 26 12 70; www.rivieramedical.com).

Clinique Saint-George

2 avenue de Rimiez; standard clinic tel. 04 93 81 71 50, emergency clinic tel. 04 92 26 77 77; www.clinique-saint-george.com

The Clinique Saint-George is a private hospital in Cimiez, but the cost of treatment and drugs is a fraction of what it would be in the United States.

Hôpital Pasteur

30 voie Romaine; tel. 04 92 03 77 77; www.chu-nice.fr/ nos-hopitaux/hopital-pasteur

The centrally located public hospital is Hôpital Pasteur, which deals with all medical emergencies except for those involving children, who must be taken to **Hôpital Lenval,**

Nice's Showpiece Hotels

Le Negresco

As with many of the famous Riviera resorts, some of Nice's hotels are out of the price range of the average traveler, but are glamorous enough to be sightseeing destinations in themselves. A few famous accommodations on the **promenade des Anglais, boulevard Victor Hugo,** and beneath the **Colline de Château** are worth walking by or peeking into, perhaps even splurging at one of their luxurious restaurants before returning to your more reasonable accommodation for the night.

LE NEGRESCO

37 promenade des Anglais; tel. 04 93 16 64 00; www.hotel-negresco-nice.com; €700 d
Le Negresco is probably the most recognizable building in Nice, and it's the belle epoque showpiece of the promenade des Anglais. It's worth dressing smartly just to have a look at the enormous ballroom or have a drink in the walnut-lined bar, which has live music on Thursdays. The menu at the Michelin-starred restaurant, Chantecler, is €190 plus another €100 for wine, but the modern brasserie-diner, La Rotonde, serves a weekday two-course lunch for €39. The Negresco beach club restaurant is open May-September.

HYATT REGENCY PALAIS DE LA MÉDITERRANÉE

13 promenade des Anglais; tel. 04 93 27 12 34; https://nice.regency.hyatt.com; €525 d
The extraordinary facade of the Hyatt Regency Palais de la Méditerranée is probably the most magnificent example of art deco architecture in Europe.

HOTEL SUISSE

15 quai Rauba Capeu; tel. 04 92 17 39 00; www.hotel-nice-suisse.com; €340 d
Irish writer James Joyce began his near-indecipherable work *Finnegans Wake* in Hotel Suisse in October 1922. The entrance has changed, but the ginger and cream facade remains the same.

BOSCOLO EXEDRA

12 boulevard Victor Hugo; tel. 04 97 03 89 89; https://nice.boscolohotels.com; €250 d
The Boscolo Exedra is an Italianate palace in shining white, attracting a clientele with a taste for the high life. The lobby, accessed through swing doors up a cobbled entrance, has a life-size rearing horse in bronze and huge chaise-longues, all in white with extravagant floral decorations, plus a white snooker table in a room at the back.

just behind the promenade des Anglais (57 avenue de la Californie; tel. 04 92 03 03 92; https://lenval.org).

Pharmacie Riviera

66 avenue Jean-Médecin; tel. 04 93 62 54 44; https://pharmacie-riviera.fr; Mon.-Sat. 24/7, Sun. 8am-7pm
Pharmacies in Nice are usually well stocked and provide a helpful service to those seeking medical advice. Most pharmacies will have a notice in their windows detailing which ones are open on Sundays.

GETTING THERE
Air
Nice Côte d'Azur Airport

rue Costes et Bellonte; tel. 08 20 42 33 33; https://en.nice.aeroport.fr
Nice (NCE) is France's third-busiest airport (after Paris Charles de Gaulle and Paris Orly). It is a focus city for **Air France** (24 destinations) and an operating base for **EasyJet** (52 destinations). Destinations include all major European cities as well as flights to Montréal-Trudeau (Air Canada Rouge and Air Transat), Beijing (Air China), Atlanta and New York (Delta Air Lines), and Philadelphia (American Airlines). The flight to Nice Côte d'Azur from **Paris** takes about 1.5 hours.

Getting to Nice from the Airport

The best way to get to Nice from the airport is by **tram,** Ligne 2, which takes passengers from both terminals through the center of town and to the port. A return ticket "Aéroport" costs €10. Lignes d'Azur tickets cost €1.70 (€7 for a day pass) but there's also a refundable €2 fee for the "carte." These tickets cannot be bought at the airport tram stop, but can be purchased at the nearby Grand Arénas tram stop. The service runs June-August 4:15am until 1:14 am and September-May from 4:15am to 12:30am (approximately every 5 min). The journey to the center of town (Jean-Médecin tram station) takes 25 minutes.

Taxis charge €32 to the city center. The journey takes 15 minutes, or 25 minutes during rush hour.

Car

Nice is on the A8 autoroute, which runs from Aix-en-Provence to Menton on the Italian border. There are **péages (tolls)** approximately every 50 km (31 mi), with more around Nice and Monaco. The tolls accept credit and debit cards or coins. The journey from Aix-en-Provence to Nice (174 km/108 mi) costs €20 in tolls, and from Menton to Nice (30 km/19 mi) is €3.

The Nice exits (sorties) are numbers 50-55 (50 for the promenade des Anglais).

The drive along the coast roads from Italy or from Marseille is more scenic but can be busy in the summer. Using the A8 autoroute, it takes approximately 2 hours 20 minutes to get to Nice from **Marseille,** and 30 minutes to get to Nice from **Monaco.** Driving along the coast road, it takes around 30 minutes to **Antibes,** 40 minutes to **Monaco,** and 50 minutes to **Cannes.**

Train

The main train station, **Nice Ville** (avenue Thiers; daily 5am-12:30am), is served by frequent trains from Marseille and Ventimiglia. Local services to Menton stop at **Nice Riquier** station (place Auguste Blanqui; ticket office open Mon. Sat. 7am-6:20pm), which serves the east of Nice, while **Nice Saint Augustin** (avenue Maître Maurice Slama; ticket office open Mon.-Sat. 8am-noon and 1pm-4:30pm, Sun. noon-5:30pm) is the closest rail station to the airport and serves the west of Nice (tram stop Grand Arénas).

Fast, direct trains from **Paris** leave several times a day, with journey times to Nice running at just under 6 hours, and one-way tickets from around €25 if booked well in advance. The train from Marseille to Nice takes about 2.5 hours and tickets start around €22.

It is easy to travel to most towns along the coast by TER trains. For example, trains to **Cannes** leave from Nice every 15-30 minutes, getting you there in 20-40 minutes, with tickets starting around €6. You can get to **Monaco Monte-Carlo Station** in just over 20 minutes, with trains leaving every 15 minutes on average, for just €4.90.

Bus

Flixbus (www.flixbus.fr) and **BlaBlaCar buses** (www.blablacar.fr) are relatively cheap long-distance coach services that connect the major cities in France. Both services leave from Nice's **gare routière** (16 avenue des Diables Bleus) and Nice airport. **Phocéens Cars** (www.phoceens-cars.com) runs buses every day from Nice to Aix-en-Provence-Marseille and Nice to Sisteron-Grenoble that depart from the gare routière. **Eurolines** (www.eurolines.fr) operates an extensive coach network from Nice to all over Europe.

GETTING AROUND

Walking the promenade des Anglais is the best way to get a feeling for Nice, and since the city is relatively flat, it is a very pleasant way to discover your surroundings. The streets of Vieux Nice are narrow and can be steep as they ascend the Colline du Château. The attractions in Cimiez—the Musée Matisse, Musée National Marc Chagall, and Roman ruins—are on a **steep hill,** while the attractions near the airport—Parc Phoenix, the Musée International d'Art Naïf, and Musée des Arts Asiatiques—are a 75-minute walk from the center along the promenade des Anglais.

Car

Driving anywhere on the Riviera can be a frustrating experience, with traffic jams, double-parked vehicles, and traffic lights making for frequent stops. Most of Nice is on a grid system, with junctions at the end of each block to negotiate, as well as a large number of one-way streets. However, most drivers are polite, and while few stop for pedestrians, they do tend to give way for cyclists. Street **parking** is free for the first half hour, and there are numerous car parks all over the city (some are also free for the first 30 min).

Visitors really only need to rent a car when they leave Nice since the walkable streets, efficient bike rentals, and tram network make for an eco-friendly and accessible city. All the major car rental companies have offices at **Nice airport,** and many also have railway station and **downtown** locations where they are joined by local car hire companies.

Tram

Nice has three tram lines (http://tramway. nice.fr). Ligne 1 runs north and east of the city, departing every 4-5 minutes at peak times. The second tram line, Ligne 2, runs from the port (Port Lympia station) to both terminals of the airport and passes through the center of the city, with four stations underground. Ligne 3 links the airport to Saint-Isidore via the Allianz Riviera football stadium (Stade station). Tickets are the same for the tram and the local bus network, Lignes d'Azur: €1.70 for a single, €7 for a day pass, and €20 for a 7-day pass with a €2 refundable charge for the ticket. Tram lines 1 and 2 have connecting stations at Jean-Médecin and Garibaldi.

Bus

Lignes d'Azur (www.lignesdazur.com) is the local bus network, with its main office at 33 boulevard Dubouchage. Tickets are valid for the tram and bus network (see above for charges) and valid for 74 minutes (though are not valid for return journeys within this time frame).

For those wanting to go on a tour just outside Nice, **Le Grand Tour** (www. nicelegrandtour.fr; one-day pass adults €23, children ages 4-11 €8, students and seniors €21) is an open-top bus that travels from Nice to Villefranche-sur-Mer, taking in 12 stops where passengers can hop on and off. The route runs along avenue Max Gallo beside the Jardin Albert 1er and heads east along the coast to the Villefranche Citadelle before returning to Nice, where it passes alongside the port, the Russian cathedral, the railway station, and the promenade des Anglais. Tickets can be purchased online or on the bus. Small dogs are welcome, and there is space for one wheelchair.

Bike

There are scores of bike rental shops all over the city. Prices range €10-30 per day, and some of the larger hotels also have city bikes to use. Cycling is easy in Nice since the promenade des Anglais and most of the center is flat with wide, well-maintained bike lanes. However, heading east toward Villefranche-sur-Mer or north to Cimiez are both steep routes suitable only for experienced cyclists or those on e-bikes. Secure locks are advised.

Nice has two privately run bike hire operators, Lime and Pony. American-owned Lime (www.li.me) has 1,000 bikes available in Nice and surrounding areas, all electric with baskets at the front. French-owned Pony (https://getapony.com) has 800 electric bikes available and 200 pedal bikes, all of which can transport a second person behind the cyclist on a padded seat. Bicycles can be picked up from 250 designated zones around the city and use a GPS locating system via the companies' downloadable apps. Pony bikes cost €1 to release with a QR code and then €0.19 per minute (it €3 for two journeys) Lime bikes are slightly more expensive, costing €1 to release and €0.23 per subsequent minute. Both companies offer 24-hour and month-long rental deals. The maximum speed is 25 kmh (16 mph).

Les Corniches

The Corniches are three roads that run parallel to the coast linking Nice to Menton: one beside the sea (the Basse Corniche, or the Corniche Inférieure), one halfway up the mountains (the Moyenne Corniche), and one perched on the edge of the clifftop (the Grande or Haute Corniche).

Traffic on the **Basse Corniche** can slow to a standstill in the summer, but it's an entertaining journey, snaking through the Riviera's top resorts, past casinos and grand hotels and alongside the coastal railway. The Basse Corniche passes through Villefranche-sur-Mer, Beaulieu-sur-Mer, Èze-sur-Mer, Cap-d'Ail, and Monaco.

The **Moyenne Corniche** was constructed in the 1920s when the area began to take off as a vacation destination. It's wider and popular with motorcyclists; it passes through long tunnels, over viaducts, and through Èze.

The **Grande Corniche** is the highest and most spectacular of the coastal routes. Built by Napoleon along the route of the ancient Roman road, the via Julia Augusta, it is covered in mist early in the morning, and the hairpin bends and thick vegetation make it a popular ride with sports-car enthusiasts and cyclists.

VILLEFRANCHE-SUR-MER
(Basse Corniche)

Villefranche-sur-Mer is a seafront town along the Basse Corniche, with steep staircases, huge fortifications, and, because of its deepwater port, a long history of military attacks and political landings. To prevent pirate attacks, a vast star-shaped citadel was built in 1567, which today contains the town hall as well as gardens and museums. The huge stone fortification takes up a large portion of the Villefranche seafront and splits the town into two: the deepwater quay and the touristy old town, which is lined with restaurants and ends in a long sandy beach.

Villefranche is a favorite with holidaymakers for its sandy beaches, peach and ocher facades, and active fishing industry. The resort hit the news in 1971 when the Rolling Stones took over the Villa Nellcôte and recorded their album *Exile on Main Street* in the basement.

Les Corniches

M2204

M2204

LA PROVENÇALE

A8

ASTRORAMA ★

AVE DES DIABLES BLEUS

FRANCE

Le Paillon

M2204

A8

M46

TOURIST
OFFICE

ARC EN CIEL

LA TAVERNE
D'ANTAN

Èze

PERFUMERIE
FRAGONARD

M2564

Jardin Exotique

LA CHÈVRE D'OR

LES CABANONS
DU CHEMIN
DE NIETZSCHE

BASSE   CORNICHE

*Parc du
Vinaigrier*

RM2564

AVE DE LA CONDAMINE
(GRANDE CORNICHE)

MOYENNE CORNICHE

M6007

ANJUNA BEACH

M33

M34

M6098

Plage Petite
Afrique

M6007

Beaulieu-sur-
Mer

Plage des
Marinières

Villefranche-
sur-Mer

M6098

CHAPELLE
ST-PIERRE

VILLA KÉRYLOS

Plage
de la Baie
des Fourmis

*Rade de
Beaulieu*

Plage de la Darse

*Rade de
Villefranche-
sur-Mer*

VILLA EPHRUSSI
DE ROTHSCHILD

**See
"Saint-Jean-Cap-Ferrat
and Beaulieu-sur-Mer"
Map**

M6098

M25

*Jardin
Botanique
Les Cèdres*

Saint-Jean-
Cap-Ferrat

TOURIST
OFFICE

Paloma
Plage

BLVD DU GÉNÉRAL DE GAULLE

M125

Plage des Fosses

Plage des Fossettes

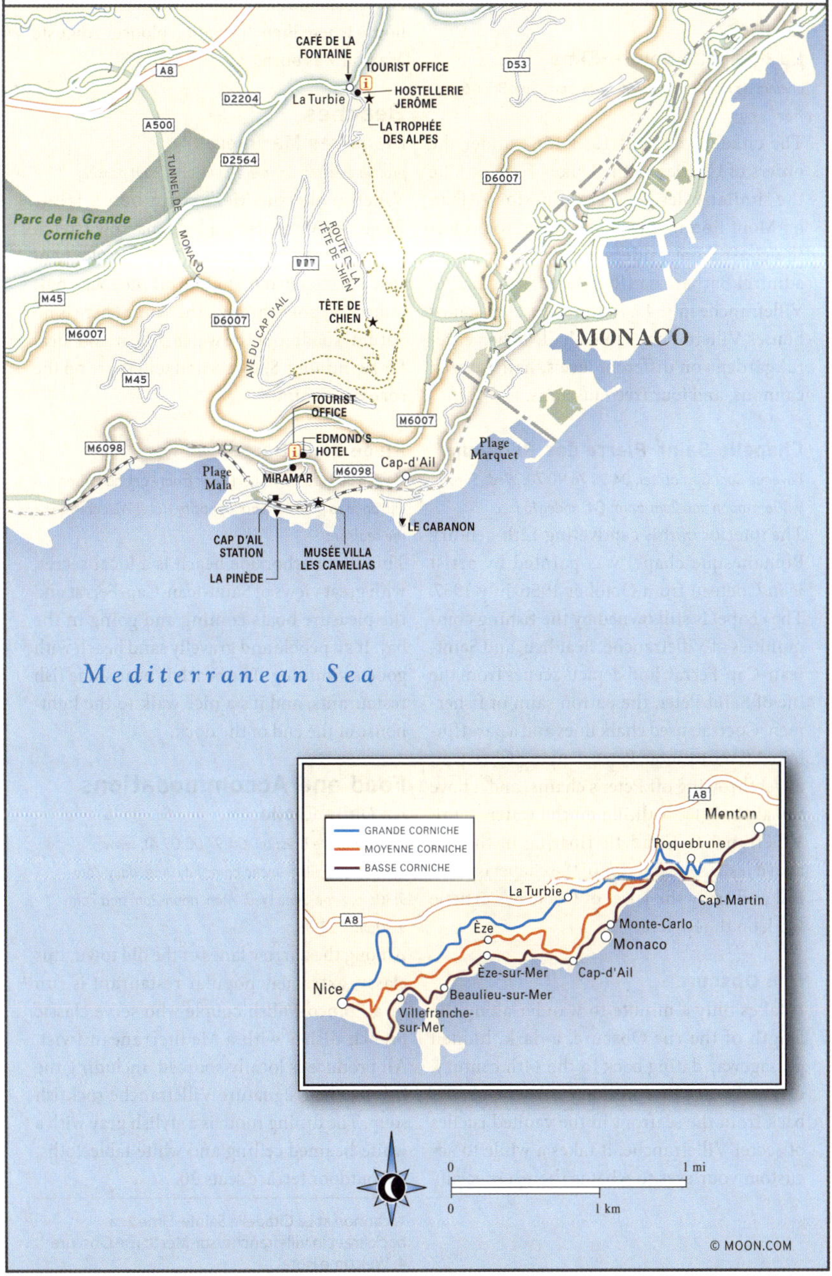

CAFÉ DE LA
FONTAINE

TOURIST OFFICE

D53

A8

D2204
La Turbie

HOSTELLERIE
JERÔME

A500

LA TROPHÉE
DES ALPES

D2564

D6007

TUNNEL DE MONACO

Parc de la Grande
Corniche

ROUTE DE LA TÊTE DE CHIEN

M45

D6007

D97

M6007

AVE DU CAP D'AIL

TÊTE DE
CHIEN

MONACO

M45

TOURIST
OFFICE

M6007

M6098

EDMOND'S
HOTEL

M6098

Cap-d'Ail

Plage
Marquet

Plage
Mala

MIRAMAR

CAP D'AIL
STATION

LE CABANON

LA PINÈDE

MUSÉE VILLA
LES CAMELIAS

Mediterranean Sea

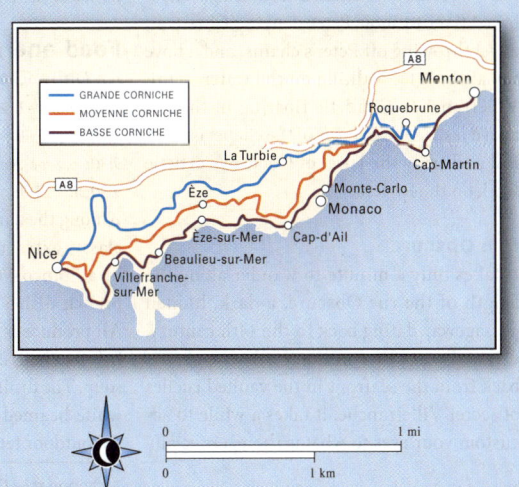

A8

Menton

Roquebrune

GRANDE CORNICHE
MOYENNE CORNICHE
BASSE CORNICHE

La Turbie

Cap-Martin

A8

Monte-Carlo

Èze

Monaco

Nice

Èze-sur-Mer

Cap-d'Ail

Beaulieu-sur-Mer

Villefranche-
sur-Mer

0   1 mi

0   1 km

© MOON.COM

It's not open for visits, but the locals are still recovering from the all-night sessions.

Sights
La Citadelle Sainte-Elme

chemin de Ronde; Mon.-Sat. 8am-6pm, Sun. 3pm-6pm; free

The citadel was built 1554-1567 under the orders of Charles II, the Duke of Savoy. Like the similarly designed Fort du Mont-Alban on Mont Boron, the citadel aimed to protect the coast from raiders following Ottoman admiral Barbarossa (Red Beard)'s sacking of Villefranche in 1543. Restored in 1981, it now houses Villefranche's town hall, some peaceful gardens on different levels, half a dozen cannons, and four free museums.

Chapelle Saint-Pierre des Pêcheurs

1 avenue Sadi Carnot; tel. 04 93 76 90 70; Wed.-Sun. 9:30am-noon and 2pm-6pm; €4, under 15 free

The interior of this captivating 12th-century Romanesque chapel was painted by artist Jean Cocteau from October 1956-July 1957. The chapel is still owned by the fishing communities of Villefranche, Beaulieu, and Saint-Jean-Cap-Ferrat, and depicts scenes from the life of Saint Peter, the patron saint of fishermen. Cocteau used chalk lines and a paraffin-based fixer to show Peter denying Christ, an angel throwing off Peter's chains, and, above the altar, Peter walking on the water—with Villefranche's Citadelle floating in the sky above Jesus. The interior of the chapel is magical and truly the jewel of Cocteau's artistic work on the Riviera.

Rue Obscure

It takes only a minute to wander along the length of the rue Obscure, a dark, hidden passageway dating back to the 14th century, originally used for military maneuvers. Set back from the seafront in the vaulted ruelles of secret Villefranche, it takes a while to accustom your eyes to what is there, especially when the sun is very bright in the summer. Entrance to the top end of the passageway is opposite number 41 rue du Poilu, the tiny house where French aviation pioneer Auguste Maïcon was born.

Beaches
Plage des Marinières

promenade des Marinières; plenty of paid parking

Villefranche-sur-Mer's main beach is the Plage des Marinières, a long, narrow stretch of sand that lines the bay. There is plenty of space to stretch out or jump off the rocky outcrops that poke through the sand. There is no natural shade, but the water is clear and ideal for swimming. Showers and toilets are on the roadway behind.

Plage de la Darse

chemin du Lazaret, a short walk from Parking Wilson; entrance to paid car park opposite Hotel Welcome on the seafront

This tiny harborside beach is a local secret, with great views of Saint-Jean-Cap-Ferrat and the pleasure boats coming and going in the bay. It's a pebble and gravelly sand beach with good swimming. The beach is near some fish restaurants, and it's a nice walk to the lighthouse at the end of the dock.

Food and Accommodations
La Belle Etoile

1 rue Baron de Brès; tel. 04 97 08 09 41; www. labelleetoile-villefranche.com; July-Aug. daily 7pm-9:30pm, Sept.-June Wed.-Mon. noon-2pm and 7pm-9:30pm; €23-33

Among the narrow lanes of the old town, this classy, extremely popular restaurant is run by a Franco-Polish couple who serve classic French dishes with a Mediterranean twist. All produce is locally sourced, including the fish for their signature Villefranche rockfish soup. The dining room is a stylish gray with a white-beamed ceiling and white tablecloths. An outdoor terrace seats 20.

1: cannon at La Citadelle Sainte-Elme **2:** a backstreet in Villefranche-sur-Mer **3:** rue Obscure **4:** Welcome Hotel

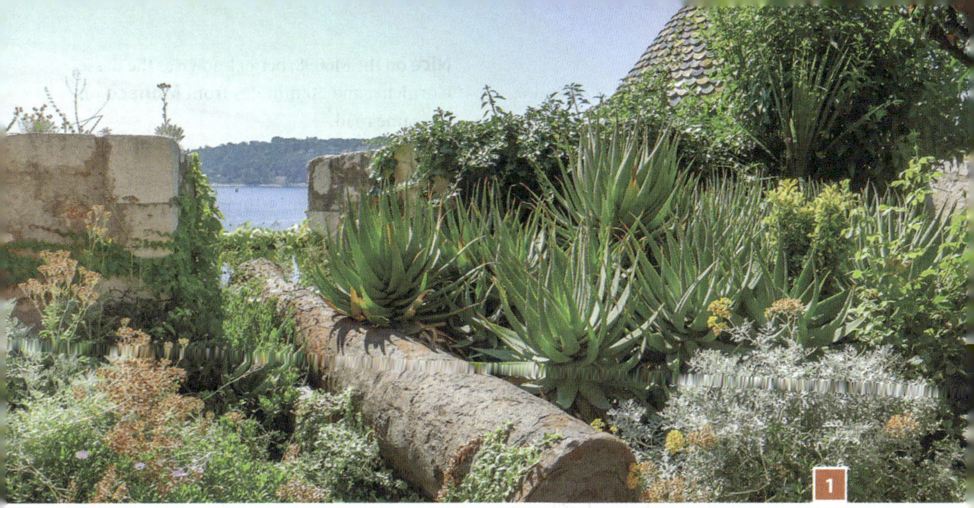
1

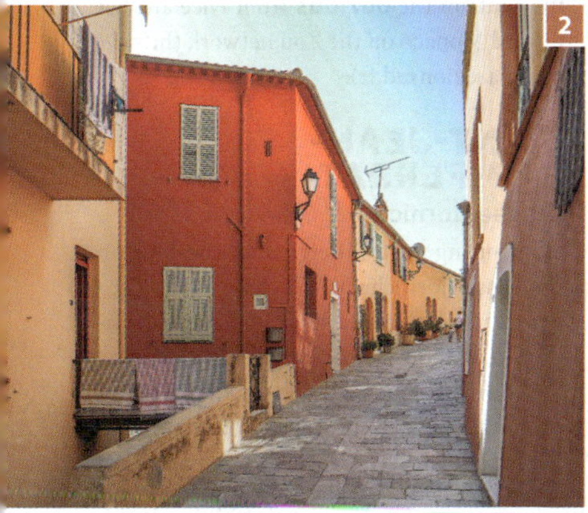
2

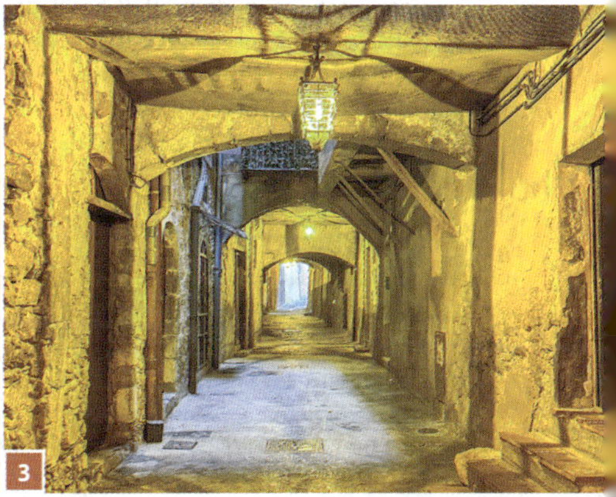
3

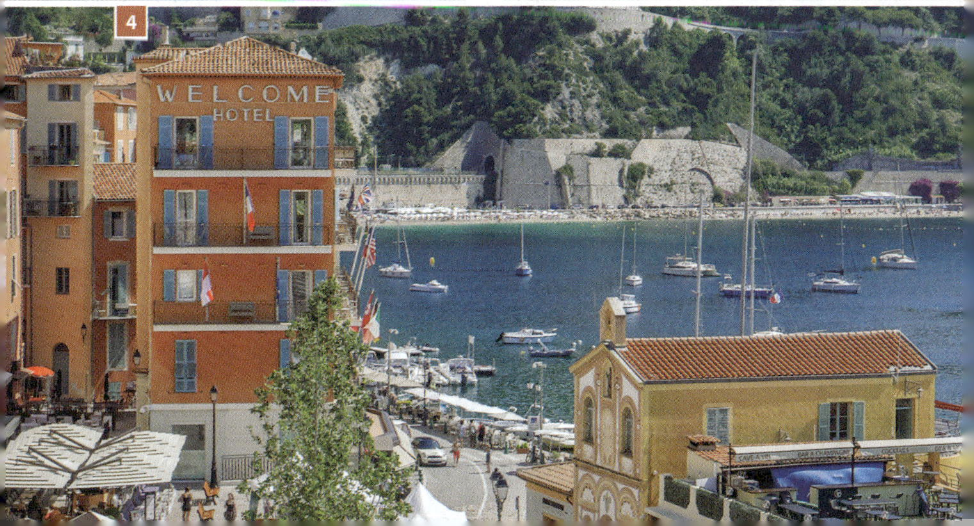

4

La Mère Germaine

*quai de l'Amiral Courbet; tel. 04 93 01 71 39; www.
meregermaine.com; daily noon-2:30pm and 7pm-10pm;
€24-39*

Specializing in seafood, fish, and pasta dishes, this smart waterfront restaurant has been a fixture in Villefranche since 1938. Everything is caught locally, and diners can sit among the fishing nets on the edge of the water during the summer.

Hotel la Regence

*2 avenue du Maréchal Foch; tel. 04 93 01 70 91; www.
laregence-hotel.fr; €100 d*

Opposite the tourist office are the big red awnings of this centrally located hotel with a busy restaurant-bar underneath (Chez Betty, a popular watering hole for cyclists). The road outside is the main coastal route between Nice and Monaco, so it can get busy, especially on Sundays, but the views are great and the service is very friendly.

Welcome Hotel

*3 quai de l'Amiral Courbet; tel. 04 93 76 27 62; www.
welcomehotel.com; €300 d*

The bouillabaisse-colored seafront hotel is perfectly located for a view toward Cap-Ferrat and the railway line to Ventimiglia as it disappears into a tunnel. There's a Jean Cocteau mosaic on the floor of the lobby—he stayed in room 22—and attracts smart-looking tourists who charter boats in the quayside opposite.

Information and Services

Office de Tourisme et de la Culture

*Jardin François Binon; tel. 04 93 01 73 68; www.
villefranche-sur-mer.com; Sept.-June Mon.-Sat. 9am-
noon and 2pm-5pm, July-Aug. daily 9am-6:30pm*

The tourist office is located at the east corner of the Jardin François Binon, a square with monuments and benches that runs alongside the Basse Corniche as it passes through the town. There's a tiny white book-exchange station opposite the office.

Getting There and Around

Villefranche is only a 15-minute drive from Nice on the M6098, better known as the Basse Corniche, and 30 minutes from **Monaco** on the same road.

The **train station** (avenue Georges Clémenceau; ticket office open Mon.-Sat. 8am-1pm and 2pm-4:10pm) lies just above the Marinières beach, and visitors can walk from there directly down to the sand or along the elevated rue du Poilu toward the old town and La Citadelle. The train takes 6 minutes from Nice Ville station (€2.10).

Lignes d'Azur (tel. 0810 06 10 06; www. lignesdazur.com) **bus number 15** departs from the promenade des Arts in Nice and stops in Villefranche at the standard Lignes d'Azur fare of €1.70. Villefranche is also on the route of the **607 bus** from Nice airport toward Monaco on the Zou network (https:// zou.maregionsud.fr).

SAINT-JEAN-CAP-FERRAT
(Basse Corniche)

Continuing past Villefranche-sur-Mer on the Basse Corniche, you'll come to the Cap-Ferrat peninsula and its resort, Saint-Jean-Cap-Ferrat, once a rocky outcrop inhabited by a few fishermen and farmers whose simple homes were congregated around the chapel in Saint-Jean port. In 1876, an artificial lake was created as a water supply, attracting wealthy Niçois to venture and dip their feet in the water. Saint-Jean-Cap-Ferrat is still one of the most desirable locations on the Riviera, and the construction of larger and grander villas continues. Perhaps the most sumptuous of all is the Villa Ephrussi de Rothschild, built in the early 1900s and open for visits today. Full details of all events are on the tourist office website, www. saintjeancapferrat-tourisme.fr.

Sights

★ Villa Ephrussi de Rothschild

*1 avenue Ephrussi de Rothschild; tel. 04 93 01 45 90;
www.villa-ephrussi.com; Sept.-Oct. and Feb.-June daily
10am-6pm, July-Aug. daily 10am-7pm, Nov.-Jan. Mon.-
Fri. 2pm-6pm, Sat.-Sun. 10am-6pm; €16, over 65 €15,*

Saint-Jean-Cap-Ferrat and Beaulieu-sur-Mer

© MOON.COM

ages 7-25 €11, family ticket (2 adults and 2 children) €45, under 7 free

A sumptuous palace in pinks and ivory, this villa was built for Baroness Béatrice de Rothschild—socialite, art collector, and member of the well-known banking family—in the early 1900s on a 7-hectare (17-acre) estate near the entrance to the Cap-Ferrat peninsula. The villa represents the high point of belle epoque on the Riviera and contains her vast collections of porcelain, fine art, and tapestries. Private apartments are open to visitors all year, as are the gardens; each year in early May the villa hosts a rose festival. There's a tea salon and rows of musical fountains, which soar to the sound of classical music three times an hour.

Musée des Coquillages

quai Lindbergh; tel. 04 93 76 17 61; www.musee-coquillages.com; Mon.-Fri. 10am-noon and 2pm-5:45pm, Sat.-Sun. 2pm-5:45pm; free

Over 7,000 shells are on display at Saint-Jean's seashell museum in the old port, the largest collection in the Mediterranean. The tour begins with a 6-minute film and is followed by examining the contents of 33 cases bulging with shells. Microscopes are set up to assist identification, and there are some fabulously large specimens to hold.

Beaches
Paloma Plage

chemin de Saint-Hospice; toilets and showers available Apr.-Sept.

Paloma Plage, like the other beaches on Saint-Jean-Cap-Ferrat, is covered in a light gray and white gravelly grit, but somehow Paloma has that extra bit of cachet. It's a 10-minute walk from the port and faces the cliffs of Beaulieu-sur-Mer and Èze to the east. Thanks to the Paloma Plage restaurant-bar, the beach has attracted the jet set (and boat set) for almost 70 years. The beach was named after Picasso's daughter, Paloma, because the artist and his family and friends used to spend time there.

Plage des Fossettes

avenue Claude Vignon

On the opposite side of Saint-Jean-Cap-Ferrat's extra limb to Paloma Plage, Fossettes is a much smaller beach below the Jardin de la Paix, a public garden where they hold open-air concerts on a stage set among the pines in the summer. It's peaceful even in July and August, and it's a great location for snorkeling because of the surrounding rocks.

Sentier Sous-Marin, located at one end of the Plage des Fossettes, is an underwater pathway that has been designed to allow swimmers to explore the aquatic world near the shore. Swimmers need to bring their own snorkel, mask, and flippers to complete the 200-m (656-ft) route, which goes to a depth of 3 m (9.8 ft). It is officially open from mid-June to mid-September but can be explored any time of the year.

Food and Accommodations
Plage de Passable

chemin de Passable; tel. 04 93 76 06 17; www.plage-de-passable.fr; Easter weekend-Sept. daily 10am-midnight; €18-32

On the west side of the peninsula, facing Villefranche, the Plage de Passable is named after its pine- and palm tree-surrounded beach, the first one to be developed on the peninsula. It has a huge 200-seat outdoor terrace and specializes in big plates of seafood spaghetti and octopus, but also does steak tartare and an excellent lemon tart.

La Frégate

11 avenue Denis Séméria; tel. 04 93 76 04 51; https://hotellafregate.jimdo.com; €100 d

La Frégate is incredibly reasonably priced for a hotel overlooking one of the most swish coastlines in France. Back rooms look out over the garden, where guests can have breakfast, but it's worth paying the small extra charge for the balcony and, of course, the sea view. The decor in the rooms may be a little dated, but there's an unusual painting and sculpture gallery attached to the hotel that adds a bit of artistic verve.

Royal-Riviera

3 avenue Jean Monnet; tel. 04 93 76 31 00; www.royal-riviera.com; €530 d

The Saint-Jean-Cap-Ferrat peninsula entrance is guarded on the sea side by the Royal-Riviera, a majestic building originally named the Panorama Palace when it was built in 1904. Having had several other names and uses (it housed hundreds of war orphans and Serbian refugees during the First World War), it was refitted in 1999. There are two bars, table tennis tables, a wellness center, a fitness room, waterskiing from a private jetty, and a vast heated outdoor pool.

Information and Services

The tourist office is located on quai Virgile Allari (tel. 04 93 76 08 90; Oct.-Apr. Mon.-Sat. 9am-1pm and 2pm-5pm, public holidays 9:30am-1pm, May-Sept. Mon.-Sat. 9:30am-1pm and 2pm-6:30pm, Sun. and public holidays 9:30am-1pm and 2pm-5:30pm).

Getting There and Around

From the **Basse Corniche,** take the avenue Denis Séméria at the roundabout between Villefranche and Beaulieu (just past the service station), where signposts lead you to the harbor **car parks.** Saint-Jean-Cap-Ferrat is a 20-minute drive (10 km/6 mi) from **Nice** on the coast road and a 30-minute drive (16 km/10 mi) from **Monaco.**

The nearest **train station** is Beaulieu-sur-Mer, from which visitors can take the bus number 15 to Saint-Jean-Cap-Ferrat.

Bus number 15 (Lignes d'Azur; tel. 0810 06 10 06; www.lignesdazur.com; €1.70 single) goes from Nice (promenade des Arts) to Port de Saint-Jean every 15 minutes (twice an hour on Sun. and public holidays).

BEAULIEU-SUR-MER
(Basse Corniche)

Beaulieu-sur-Mer is the archetypal Riviera resort, with a sparkling art deco casino, palm trees, tennis courts, superyachts floating in the marina, and plenty of tiny dogs running up and down the promenade. It was popular with the British and Russians in the 1920s, and its belle epoque villas and architecture have been beautifully restored. A stunning example is **La Rotonde,** formerly the dining room of the now-vanished Hotel Bristol; it hosts occasional art exhibitions and can be rented out for weddings and events.

David Niven, the actor, and Gustave Eiffel both had villas in Beaulieu, but neither were as fabulous as the Villa Kérylos, built in the style of a Greek villa at the turn of the 20th century.

Sights
Le Casino

4 avenue Fernand Dunan; tel. 04 92 00 60 00; www.casinodebeaulieu.com; Mon.-Thurs. and Sun. 10am-3am, Fri.-Sat. 10am-4am

Beaulieu's art deco casino is just this side of tasteful, with giant chandeliers, pink lighting, a sweeping central staircase, and showcases full of art and designer sunglasses in the entrance hall. Inside, guests (over 18 years old) can play roulette, blackjack, Punto Blanco, poker, or any of the 75 slot machines. The casino also houses La Belle Époque concert hall, Le Sky Beach club, and Le Baccara restaurant.

Villa Kérylos

impasse Gustave Eiffel; tel. 04 93 01 47 29; www.villakerylos.fr; Apr.-Sept. daily 10am-5pm, May-Aug. daily 10am-6pm; €11.50, under 18 and EU students free

Archaeologist and patron of the arts Théodore Reinach instructed architect Emmanuel Pontremoli to build him a replica of a Greek villa on the promontory to the east of Beaulieu. The result was a stunning tiered construction of sheer white facades and rooftop pergolas—a genuine dream house. Completed in 1908 after six years of construction, Reinach's family (related to Baron Maurice Ephrussi, whose wife Béatrice built the villa on Saint-Jean-Cap-Ferrat) continued to live in the villa until 1967, when it was donated to the Institut de France. Now open to the public, it is listed as a monument historique and shows off Reinach's obsession with the ancient world. The villa is filled with exact copies of ancient Greek furniture, sculpted

columns, sunken baths, and marble tiling. The name *kérylos* means "kingfisher," a good omen in Greek mythology.

Beaches
Plage de la Baie des Fourmis
avenue Fernand Dunan

Swimming is safe and supervised during the summer on the beach directly in front of the casino on Beaulieu's main strip, with views of the eastern coast of Saint-Jean-Cap-Ferrat. The beach has showers, toilets, and easy access from staircases up to the promenade Maurice Rouvier.

Plage Petite Afrique
boulevard Alsace Lorraine

Walking past the Port de Plaisance toward Monaco, you arrive at the Plage Petite Afrique, which has a huge beach volleyball court and a grid of boule courts. There's a snack bar and a posh beach club that beckons sunbathers, and the pleasant, protected bay makes it ideal for families. Toilets and showers are on-site.

Food and Accommodations
Very Valentina
Port de Plaisance; tel. 04 93 80 03 45; Tues.-Sun. 9am-11pm; €16-29

Facing the yachts in the marina, this big Italian-style restaurant is a good place to watch the boats coming and going while eating a heaping plate of pasta. Gnocchi and vegetarian risotto are specialties.

Aristée
48 boulevard Maréchal Leclerc; tel. 04 93 62 20 09; www.restaurantaristee.com; Wed.-Mon. 9am-10:30pm; €22-40

On one of Beaulieu's main thoroughfares, Aristée is open for morning coffee, light lunches, and evening meals such as squid cassolette or seabass with courgettes. With pink and brown floor tiles and bistro furniture, it's

a tastefully decorated place to enjoy a meal in the calm away from the seafront.

Hotel Select
1 rue André Cane; tel. 04 93 01 05 42; www.hotelselect-beaulieu.com; €80 d

Hotel Select is a bright, family-run hotel in a belle epoque building on Beaulieu's market square. The hotel has 19 air-conditioned rooms, including some triples and family suites. It's a reasonably priced and charming option in an expensive location. Breakfast is €10.

La Réserve de Beaulieu
5 boulevard du Maréchal Leclerc; tel. 04 93 01 00 01; www.reservebeaulieu.fr; €900 d

Unflinchingly luxurious, La Réserve is one of the most sumptuous of the palace-hotels on the Riviera. With a huge swimming pool overlooking the sea, a billiard room and a board games lounge, the wood-paneled Gordon Bennett Bar (named after the American press magnate and sports enthusiast who loved the hotel), and several restaurants, it is a prestigious destination even for the well-heeled. Founded in 1880, everyone from Rita Hayworth to Charlie Chaplin, Frank Sinatra, and Sir Thomas Lipton has stayed in the "pink palace."

Information and Services
The **tourist office** is opposite the railway station on place Clémenceau (tel. 04 93 01 02 21; www.otbeaulieusurmer.com; mid-Apr.-June and Sept. Mon.-Sat. 9:30am-1pm and 2pm-5:30pm, July-Aug. Mon.-Sat. 9:30am-1pm and 2pm-6:30pm, Sun. 9:30am-1pm, Oct.-mid-Apr. Mon.-Fri. 9am-12:30pm and 2pm-5pm, Sat. 9:30am-1pm) and is a good source for maps, information, and transport timetables.

Leaving from the Kiss and Drop landing area in the Port de Plaisance, a solar-powered boat service called **SeaZen** (tel. 06 52 73 95 54; www.seazen.fr) can take up to eight people on a 1-hour catamaran tour of the coastline between Nice and Monaco. It can also be rented privately for a 1-hour lesson,

1: Villa Ephrussi de Rothschild, Saint-Jean-Cap-Ferrat **2:** La Rotonde, Beaulieu-sur-Mer **3:** the steep Sentier Frédéric-Nietzsche path in Èze **4:** Villa Kérylos, Beaulieu-sur-Mer

Walking Tour Around Cap-Ferrat

Paloma Plage

Walking around the entire peninsula on the former customs officers' path is a great way to see the beaches, creeks, private villas, and hotels of the cape. The 9-km (6-mi) walk takes 3-4 hours and requires sturdy shoes, since the rocks are rough-edged and there are some steep stairs cut into the crags.

- Begin at the **Hotel Royal-Riviera** on the western edge of Beaulieu and take the **promenade Maurice Rouvier** past place David Niven and toward the port.

- Continue across the quayside and take the **avenue Jean Mermo** toward the Jardin de la Paix, over **Paloma Plage,** and to the headland, where there's a cemetery at **Pointe Saint-Hospice.** On the other side of the chapel is a separate war cemetery for Belgian soldiers who died at the **Villa Les Cèdres,** a palatial home owned by King Leopold II of Belgium that became a military hospital in the First World War.

- Turn right at the headland toward **Pointe du Colombier,** an elbow-like shape jutting off Cap-Ferrat, and then head past **Fossettes** and **Fosse** beaches before heading out to the cape along the **chemin de la Carrière** (near the disused quarry that supplied the stone to build Monaco harbor).

- The path continues past the luxury **Grand-Hotel du Cap-Ferrat** before arriving at the **lighthouse.**

- The coastal path merges with the **Sentier du Littoral,** the path that encircles the main peninsula, past lots of bays, zigzagging around and up and down through rocky staircases along the west side of the cape until the **Plage de Passable,** a good place for a swim and refreshments.

- Rejoin the **chemin de Passable** and head onto **avenue Denis Séméria** to the smaller of the two tourist offices and back to the port.

independent trips, and private tours (details on website).

Getting There and Around

Beaulieu is on the **Basse Corniche,** a 15-minute drive from Nice and 25 minutes from **Monaco** on the coast road.

The Beaulieu-sur-Mer **railway station** (place Georges Clémenceau; www.sncf-connect.com) is on the Nice-to-Ventimiglia line between Villefranche-sur-Mer and Èze. It takes around 10 minutes to reach Nice and the fare is €2.40.

Bus number 607 (https://zou. maregionsud.fr) and **Lignes d'Azur bus 15** (tel. 0810 06 10 06; www.lignesdazur.com; €1.70) pass through Beaulieu-sur-Mer from Nice.

ÈZE

(Moyenne Corniche)

The rocky outcrop between Beaulieu-sur-Mer and Cap-d'Ail has been settled since 220 BCE, and Bronze Age remains can still be seen throughout the modern village of Èze. The lords of Èze built a château in the second half of the 12th century that was destroyed by the French in 1706. The ruins, one of the Riviera's most popular attractions, still perch 500 m (1,640 ft) above sea level, a startling sight. Today, cobbled lanes wind up through the tiny village, with art galleries and gift shops selling handbags, jewelry, leather belts, and fridge magnets in almost every nook and arched doorway.

Sights

Jardin Exotique

rue du Château; tel. 04 93 41 10 30; www. jardinexotique-eze.fr; Jan.-Mar. and Nov.-Dec. daily 9am-4:30pm, Apr.-June and Oct. daily 9am-6:30pm, July-Sept. daily 9am-7:30pm; €7, under 18, students, and groups €4, under 12 free

Much of the village is strewn with bougainvillea, roses, jasmine, prickly cacti, and honeysuckle, but right at the top, among the château ruins, is Èze's exotic garden, a paradise of cacti and spurges. Designed by agronomist

Jean Gastaud in 1949, the garden consists of different vegetation zones: to the south are succulents and xerophytes from desert regions, and to the north, hidden among caves and cascades, are plants from the humid Mediterranean. The subtropical zone has waterfalls, benches, and vaporizers.

Perfumerie Fragonard

158 avenue de Verdun; tel. 04 93 41 05 05; www. fragonard.com/fr/usines/eze; daily 9am-6:45pm; free entrance and tour

Fragonard is one of the "big two" perfumeurs (along with Galimard) from Grasse, the self-styled world capital of perfume, 50 km (31 mi) away in the hills above Cannes, that have outlets in Èze. The Fragonard factory is to the east of the village, a huge site that accommodates coachloads of tourists and has a boutique almost as big as the factory. The free tour (available in English) lasts about half an hour and is an interesting look at how scents are extracted from flowers, many of which come from the local hillsides. Visitors can see the yellow duck soaps and lemon verbena shower gels in the shop being manufactured in the factory, alongside hundreds of perfumes, lip balms, and scented candles.

Hiking

Sentier Frédéric-Nietzsche

Distance: *2 km (1.2 mi) one-way*
Time: *1 hour*
Information and Maps: *Èze tourist office*
Trailhead: *Beside the entrance to La Chèvre d'Or*

The Sentier Frédéric-Nietzsche (Nietzsche's Footpath) begins beside the first-tier entrance to the super-posh La Chèvre d'Or and descends to the Basse Corniche at Èze-bord-de-Mer, the beachfront directly underneath the perched village. When Nietzsche did the walk in the mid-1880s, it would have involved more clambering over rocks and sliding down slopes, but today it has been semi-paved and is an easy (though steep) walk for holiday philosophers pondering whether they have packed a sunhat and have enough water and sun lotion. The 2-km (1.2-mi) path takes about 50 minutes to walk down and just over an hour to walk up, with striking views out to sea. Pine trees clinging to the cliff provide some shade.

Chemin Saint-Michel

Distance: 6 km (3.7 mi) round-trip
Time: 3 hours
Information and Maps: Èze tourist office
Trailhead: Èze-sur-Mer train station

From Èze railway station, take the avenue de Provence and go up the first staircase to join the chemin Saint-Michel through the Lamaro district. Follow the route to the Chapelle Saint-Grat (patron saint of olive protection). The views over Beaulieu-sur-Mer and Saint-Jean-Cap-Ferrat are exceptional. It's a relatively steep route until you reach the plateau Saint-Michel, where the outbound hike ends at the botanical gardens in the small hamlet. Even in the summer, hikers are shaded by pines and oak trees along the way. The route has regular viewing stations over the coast. Return along the same track to Èze-sur-Mer with the spectacularly perched Èze village high above it.

Food and Accommodations
La Taverne d'Antan

6 rue Plane; tel. 04 92 10 79 61; www.lataverne-eze.fr; daily noon-3pm and 7pm-11pm; €12-22

Èze doesn't feel very Italian until you go into the old-style (d'antan) vaulted chamber of La Taverne. There, it's noisy and friendly, and you can have any pizza (€12-18), any kind of homemade pasta (€13-22), or any kind of risotto (including fruits de mer) for €15-22.

Anjuna Beach

28 avenue de la Liberté; tel. 04 93 01 58 21; www. anjunabay.com; end of Apr.-end of Sept. 10am-11pm; €26-65

This tropical, Indonesian-style beach bar has live music on weekends and a relaxed Bali vibe. There's a covered restaurant and bar in turquoise and teak. Their shingle private beach is open from 10am, with umbrellas and sunbeds for rent. There's space for 350 guests on the terrace.

Arc en Ciel

4 avenue du Jardin Exotique; tel. 04 93 41 02 66; www. arcencieleze.fr; €100 d

Located just outside the village, Arc en Ciel (The Rainbow) has six simple rooms in a characterful hotel with a snack bar and a gift shop. It's a great base for walks and visiting the villages along the Corniches.

Les Cabanons du Chemin de Nietzsche

chemin de Nietzsche, La Calanca; tel. 06 14 52 63 61; https://cotedazurfrance.fr; May-Oct.; €1,500 weekly, single nights by arrangement

For an experience far from tourists, traffic, and Riviera cocktail lounges, Penelope Guiauchain and her husband have created two luxury cabins built on stilts. They are a 10-minute walk down Sentier Frédéric-Nietzsche amid the tree canopy beneath Èze. One cabin is 60 sq m (645 sq ft) with a large bedroom, kitchen, and bathroom; the other is half the size. Both have incredible views of the coast, open floor plans, wooden decking, and all the standard modern conveniences.

Information and Services
Office Municipal de Tourisme

place Général de Gaulle; tel. 04 93 41 26 00; www. eze-tourisme.com; Jan., Mar., Nov., and Dec. Mon.-Sat. 9am-4pm, Feb., Apr., May, and Oct. Mon.-Sat. 9am-6pm, June-Sept. daily 9am-7pm

The tourist office is at the bottom of the village on the west side, providing free maps and brochures about Èze's main attractions and events.

Getting There and Around

Èze village is a 25-minute drive (11 km/7 mi) from Nice and 20 minutes (10 km/6 mi) from Monaco, using the **Moyenne Corniche** (M6007). The M45 and M46 link the Grande Corniche to the Moyenne Corniche, while the M45 continues down with its heart-stopping gradient and hairpin bends connecting the Moyenne with the Basse Corniche, which is closest to the sea.

Bus number 82 (Lignes d'Azur; tel. 0810

06 10 06; www.lignesdazur.com) goes from Vauban in Nice to Èze village and the Col d'Èze on the Grande Corniche once an hour on weekdays and Saturdays, with a less frequent service on Sundays. A single fare costs €1.70, and the journey time is 37 minutes. **Line 83** connects Beaulieu-sur-Mer with Èze bord-de-mer, Èze village, and the Col d'Èze. A single fare costs €1.70, and the journey time is 23 minutes. **Zou bus 602** (https://zou.maregionsud.fr) goes from Nice Vauban to Èze village and on to Monaco once every 2 hours.

CAP-D'AIL
(Basse Corniche)

The next stop east on the Basse Corniche, Cap-d'Ail was part of neighboring La Turbie until 1908, by which time Baron de Pauville, financier and founder of the local newspaper, *Le Petit Niçois*, had already built a 150-bedroom belle epoque palace, **Hôtel Eden,** on the hill down to the sea. It quickly became a magnet for the stars, and in the first decade of the 20th century, scores of huge villas sprang up all over the cape. Josephine Baker, Greta Garbo, Colette, Jean Cocteau, Auguste Lumière, and Gertrude Lawrence all partied on the seafront, and a young Princess Elizabeth came to stay in 1936 with her parents (King George VI and his wife, Elizabeth) at Castel Lina.

Today, Cap-d'Ail is still a vibrant seaside resort. The main thoroughfare is the Basse Corniche, which is lined with plenty of cafés, ice cream sellers, and restaurants for lunch. A stroll there can be followed by heading down to one of the beaches or spending an afternoon at the Villa les Camélias or **Villa Le Roc Fleury.** The coastal footpath, **Sentier du Littoral,** runs toward Plage Mala and past a series of extravagant seaside villas.

Sights
Musée Villa les Camélias
17 avenue Raymond Gramaglia; tel. 04 93 98 36 57; www.villalescamelias.com; Apr.-Nov. Tues.-Fri. 9:30am-

12:30pm and 2pm-6pm, Sun. 11am-6pm, Dec.-Mar. Tues.-Fri. 9:30am-noon and 1:30pm-4:30pm, Sun. 10am-4pm; €9, ages 12-18 €5, under 12 free

Among the elegant villas on the winding road from the main route through Cap-d'Ail to the seafront is Les Camélias, now a private local history museum and gallery. The garden has a swimming pool and a gravel terrace with bronze sculptures. Beside the entrance is a pedal-powered pianola capable of playing over 150 different parchment cylinders of jazz and classical music from the 1920s, which visitors are also encouraged to play. Further rooms include period photographs, news clippings, and cabinets dedicated to a history of Cap-d'Ail.

Beaches
Plage Marquet
avenue Marquet, street parking on nearby roads

Principally a beach for families, Plage Marquet has a volleyball court, lots of water sports options, soft gravelly sand, and two beach restaurants.

Plage Mala
allée Mala, street parking on nearby roads (very limited)

Beloved of Italian day-trippers and the Monaco jet set (who always seem keen to get out of Monaco), Plage Mala is one of the best beaches on the Riviera, not for the beach itself (which is gravelly and rocky) but for its deep, clear waters and protected (almost secret) location. Access from the town center is via a steep stone staircase, but the most picturesque route is to head down toward the sea from the railway station and turn westward (in the direction of Nice) and walk along the coastal footpath.

In the summer, canoes and kayaks are available for rent on the public section of the beach. There are grottoes to explore and two private beach clubs, **Eden** (tel. 04 93 78 17 06; https://edenplagemala.com) and **La Réserve** (tel. 04 93 78 21 56; www.lareservedelamala.com/en/homepage), which serve food and rent out sun loungers and parasols.

Hiking
Sentier de Cap-d'Ail

Distance: 5 km (3 mi) one-way
Time: 2.5 hours
Information and Maps: Tourist office in Cap-d'Ail
Trailhead: Plage de la Mala

Popular with joggers and dog-walkers, the Sentier du Littoral (coastal footpath) runs for almost 5 km (3 mi), connecting Plage Mala with Plage Marquet. Access is prohibited when the weather is stormy. The path ends at the eastern end in a little parking lot for Plage Marquet.

Food and Accommodations
Le Cabanon

Pointe des Douaniers; tel. 04 97 14 80 05; http://lecabanoncapdail.fr; Sept.-Oct. Tues.-Sat. 9am-midnight, Sun. 9am-2pm, Mar.-Aug. daily 9am-midnight; €16-32

A laid-back restaurant far away from the concrete and hullabaloo of its surrounding resorts, Le Cabanon is a collection of odd-shaped tables and chairs on the Pointe des Douaniers headland between Monaco and Cap-d'Ail. The "little hut" used to be a fisherman's house, and the owners have tried to re-produce the same heritage of "grilled sardines, a game of boules, and an evening of cards."

La Pinède

10 avenue Raymond Gramaglia; tel. 04 93 78 37 10; www.lapinederestaurant.com; Mar.-Oct. Thurs.-Tues. 10am-midnight; €24-36

La Pinède was also an old fisherman's hut and is famous for its fish dishes, served alongside the coastal footpath and shaded by some twisting pine trees. The restaurant does a bouillabaisse (Marseille's celebrated fish stew) for two, lobster, bream, and shellfish.

Miramar

126 avenue du Trois Septembre; tel. 04 93 78 06 60; www.miramarhotel.fr; €100 d

Bright, crimson-colored Miramar has rooms with terraces overlooking the sea and economy rooms overlooking the village, plus triples and family rooms with bunk beds, all at very reasonable prices. The hotel is across the street from the tourist office.

Edmond's Hotel

87 avenue du Trois Septembre; tel. 04 93 78 01 01; www.capresort.com; €120 d

Managed by the same group that runs La Réserve on Plage Mala, Edmond's is a swish, centrally located hotel that has accommo-dated Winston Churchill, Greta Garbo, and Général de Gaulle. It was once an annex to the Hotel de Paris in Monaco, and some of the furniture comes from the original Monaco hotel.

Information and Services

The **tourist office** is located between the roundabout and car park opposite the Miramar Hotel on the main avenue du Trois Septembre (tel. 04 93 78 02 33; Jan.-May Mon.-Fri. 9:30am-12:30pm and 1:30pm-5:30pm, Sat. 9am-1pm, June and Sept. Mon.-Fri. 9:30am-1pm and 2:30pm-6pm, Sat. 9am-1pm, July-Aug. Mon.-Fri. 9:30am-1pm and 2:30pm-6pm, Sat.-Sun. 9am-1pm).

Getting There and Around

It's an easy drive from **Nice** on the Basse Corniche or Moyenne Corniche (18 km/11 mi); journey time is 30-40 minutes. It's a 10-minute drive (3.5 km/2 mi) to **Monaco.**

Cap-d'Ail **railway station** (avenue de la Gare; www.sncf-connect.com) is on the Marseille-Ventimiglia line. **Monaco** train station is only 3 km (1.8 mi) away (3 min; €2.10), while **Nice** is only 18 minutes (€3.80) away.

Zou Bus number 607 (https://zou.maregionsud.fr; €2.10) runs from Nice Normandie Niemen to Monaco and back every 20 minutes along the coast road. The **602** runs from Nice Vauban to Monaco and back every 2 hours Monday-Saturday. The journey time from Nice is 45 minutes and from Monaco is 15 minutes.

LA TURBIE
(Grande Corniche)

The highest point on the Grande Corniche signaled the border between the Roman Empire and Gaul, and it was here that the Romans built the landmark monument to Emperor Augustus, La Trophée des Alpes. The monument became a fortress for the Counts of Provence in the 11th century and eventually a popular place for the Riviera's winter visitors to wander among the ruins.

Today, the fortified old town of La Turbie, accessed through arched medieval gates, is a network of cobbled lanes and stone houses with street names carved, Roman-style, into the corners. A **market** takes place on Thursday mornings in place Théodore de Bainville, in the center of the town. On the main street—the Grande Corniche—is a huge pale-stone fountain given to the inhabitants by King Charles Felix of Sardinia, which brought water from a source in Peille in the hills above La Turbie over the Roman aqueduct. It's a popular stop for cyclists to fill their water bottles as they head along the Corniche.

Sights
La Trophée des Alpes

avenue Prince Albert ler de Monaco; tel. 04 93 41 20 84; www.trophee-auguste.fr; Sept.-Apr. Tues.-Sun. 10am-1pm and 2pm-5pm, May-Aug. Tues.-Sun. 10am-6pm; €6, under 26 free

Built on the highest point of the via Julia Augusta in 6 BCE, this "trophy" celebrated the victories of Emperor Augustus over the local Alpine tribes. Craftsmen and engineers came from Rome to construct what is, even by Roman standards, a huge monument. It originally had 24 white Doric columns surrounding a tower, topped by a statue of Augustus, the adopted son of Julius Caesar, making the monument over 50 m (164 ft) high. When the Roman Empire fell in 476 CE, the trophy was plundered; much of La Turbie village is made from stones taken from the monument.

Once inside the grounds of the site, visitors can walk around the giant podium and climb up a staircase running to the top of the monument, which has amazing views inland to the Alps and to Sanremo in Italy along the coast. On the village side of the monument is the long inscription dedicated to Augustus, which describes his crushing victories over the 45 tribes "from the Adriatic to the Tyrrhenian." The white limestone blocks were excavated from two local quarries.

Astrorama

route de la Revère; tel. 04 93 85 85 58; www.astrorama.net, shows Apr.-Oct.; 7pm-11pm, Nov.-6pm-10pm; €12, ages 6-25 €9.50, under 6 free

At 650 m (2,132 ft) above sea level on the Grande Corniche between La Turbie and Èze, Astrorama is a planetary observation center open to the public. It offers an introduction to astronomy with the chance to look through telescopes at the night sky. There's a planetarium, study center, terrace covered in huge telescopes, a team of experts providing information, and evening spectaculars. Ticket reservations must be made online; check the website for dates.

Sports and Recreation
Grande Corniche Park

La Grande Corniche, follow signs for the Col d'Èze and then Fort de la Revère, where there is a car park; www.departement06.fr; Apr.-Oct. daily 7:30am-8pm, Nov.-Mar. daily 8am-6pm

Running through this rugged 712-hectare (1,760-acre) park are several walking routes, a 1.4-km (0.9-mi) sports route with outdoor gym equipment, ornithological observation points, a rock face to climb (Gouffre du Simboula), and the Maison de la Nature, a former military building that has been converted into a teaching and information center.

The park houses the **Astrorama** (see above) and the interesting **Fort de la Revère,** a moat-encircled military installation and barracks built in the 1880s as part of the coastal defenses. It was used as a prison by the Vichy regime in World War II, operating under the German powers, to house captured Allied—mainly British—pilots. The building is currently closed to the public.

Tête de Chien

route de la Tête du Chien

The huge promontory of pink-gray rocks above Cap-d'Ail, part of the Grande Corniche Park, is known as the "dog's head," but it could easily be an iguana, monkey, or tortoise. Leaving La Turbie village on the avenue de la Pinède, the route heads along the clifftop to an old military post from where you can see the Golfe de Saint-Tropez, about a 2-hour walk round-trip—and a steep but rewarding trek (past some Second World War German bunkers) up the mountain high above Cap-d'Ail.

Food and Accommodations

Café de la Fontaine

4 avenue du Général de Gaulle; tel. 04 93 28 52 79; www.hostellerie-jerome.com; daily 11am-3:30pm and 6pm-11pm; €18-24

Managed by the same chef, Bruno Cirino, and team that run the Michelin-starred Hostellerie Jerôme, the Café de la Fontaine is right next to the trickling fountain that brought fresh water to the village in 1824. The café-restaurant serves fabulous bistro-style food, which is chalked up on a blackboard and changes every day.

Hostellerie Jerôme

23 rue Compte de Cessole; tel. 04 92 41 51 51; http:// hostellerie-jerome.com; €180 d

Within the walls of the medieval village, this flower-clad, half-hidden hostellerie has five large rooms above the excellent gastronomic restaurant. Breakfast is served on the terrace overlooking the sea. Many people come to La Turbie just for the restaurant, but this is a great little hotel, too.

Information and Services

Point Information Tourisme

2 place Detras; tel. 04 93 41 21 15; www.ville-la-turbie. fr; May-Oct. Tues.-Sat. 9am-noon and 2pm-6pm, Nov.-Apr. Mon.-Fri. 10am-1pm and 2pm-5pm

The tiny tourist office is located between the Cave Turbiasque bar-restaurant and a hairdresser on place Detras. It has brochures, maps, and flyers about events and activities in La Turbie and can also book accommodations.

Getting There and Around

The scenic, exciting drive from **Nice** to La Turbie along the **Grande Corniche** takes around 30 minutes (19 km/12 mi). La Turbie is 20 minutes (8 km/5 mi) from **Monaco.** There is a free **car park** in La Turbie on place Neuve as well as street parking.

Zou bus 603 (tel. 04 13 94 30 50; https:// zou.maregionsud.fr) travels from Nice Vauban to La Turbie (€2.10) and onward from La Turbie to Monaco eight times a day Monday-Friday and six times a day Saturday-Sunday and public holidays.

Cagnes-sur-Mer

Halfway between Nice and Antibes is Cagnes-sur-Mer, a seaside town split into three distinct quarters. Its old quarter, Haut-de-Cagnes, is high on a hill where winding, cobbled streets lead up to the Château Musée Grimaldi, one of the finest examples of a medieval village in the South of France. The new town, Cagnes-ville, has a covered food market and boasts the region's largest horse racing track. Cros-de-Cagnes is a former fishing village with over 3 km (1.8 mi) of pebbly beaches, plenty of seafront restaurants, and the Université Internationale de la Mer overlooking the marina.

SIGHTS
★ Haut-de-Cagnes

Haut-de-Cagnes, the old quarter of

1: Haut-de-Cagnes **2:** square boules **3:** Église Notre-Dame de la Mer **4:** grand workshop of Pierre-Auguste Renoir at Musée Renoir

Cagnes-sur-Mer, was popular with composers, poets, and artists, who gave it a very international complexion. The perched village is still arty today, with local residents painting their letter boxes and installing sculptures made from pebbles and slate on their porches. It's also popular with cyclists and diners. The large château terrace catches the winter sun at midday; benches face the villa-clad hills behind, where there's boules courts and a children's play area. The village also hosts an annual square boules tournament in August in which competitors launch their boules down the steep cobbled slopes beneath the château.

Château Musée Grimaldi

place du Château; tel. 04 92 02 47 35; Oct.-Mar. Wed.-Mon. 10am-noon and 2pm-5pm, Apr.-June and Sept. Wed.-Mon. 10am-noon and 2pm-6pm, July-Aug. Wed.-Mon. 10am-1pm and 2pm-6pm; €4, under 26 free

Up a gentle slope off the main square, this crenellated castle was built around 1300 as a fort for the Grimaldis, but by 1620 it had become more of a countryside palace, with fireplaces, family rooms, and ceiling frescoes. The highlight of the château is the room dedicated to portraits of Suzy Solidor—a singer, muse, cabaret owner, writer, and resident of Cagnes who sported short blonde hair and an androgynous look even in the 1920s. The château's upper level is dedicated to temporary art exhibitions, and the roof has amazing views of the coast and the hills behind Cagnes.

Cros-de-Cagnes
Église Notre-Dame de la Mer

20 avenue Général Leclerc; tel. 04 93 31 01 69; Mon., Wed., Thurs., and Sat. mass 6pm, Fri. mass 9am, Sun. service 11am

Across from Cros-de-Cagnes's gravelly boules court is one of the most unconventional-looking churches in France. The steeple that houses the bells is constructed from metal scaffolding, and the exterior of the church looks like an out-of-town office block built in the mid-1970s. The church, with a pointu fishing boat shipwrecked in the entrance, hosts

gospel concerts and is the starting point for the **Fête de la Mer** in early July.

Cagnes-Ville
Musée Renoir

chemin des Collettes; tel. 04 93 20 61 07; https:// tourisme.cagnes.fr; Oct.-Mar. daily 10am-noon and 2pm-5pm, Apr.-May daily 10am-noon and 2pm-6pm, June-Sept. daily 10am-1pm and 2pm-6pm, garden open year-round daily 10am-6pm; €6, under 26 free, free first Sun. of the month

Artist Pierre-Auguste Renoir bought the magnificent Domaine des Collettes, covered in ancient olive trees, in 1907 and spent the last 12 years of his life there, painting up to the morning he died. Visitors can wander around the grounds and see 14 original paintings as well as 30 sculptures, his state-of-the-art (for 1908) kitchen, a large bathtub, his studio, personal correspondence, wicker wheelchairs, and easels, as well as a film about the artist shown in the garden annex.

SPORTS AND RECREATION
Hippodrome de la Côte d'Azur

2 boulevard JF Kennedy; tel. 04 92 02 44 44; https:// hippodrome-cotedazur.fr; €5, under 18 free, free parking

There can't be many horse race tracks in the world closer to the sea than this 60-hectare (148-acre) track, which hosts flat racing, steeplechases, and the "trot," a popular French form of modern-day chariot racing. The winter schedule has evening and day races, the biggest of which are the **Grand Prix de la Riviera Côte d'Azur** in February and the end-of-season **Grand Criterium de Vitesse de la Côte d'Azur** in mid-March. The summer season, in July and August, has nighttime trot races on Mondays, Wednesdays, and Fridays. The track holds up to 12,000 spectators, and it has three restaurants and several snack bars, lounge terraces, and food stalls. There is no dress code, but most people, especially for the evening meetings, dress smartly (no beachwear).

Musée Escoffier de l'Art Culinaire

the kitchen in Musée Escoffier de l'Art Culinaire

Legendary French chef **Auguste Escoffier,** the godfather of French cuisine, was born in Villeneuve-Loubet, just west of Cagnes-sur-Mer, in 1846. He lived there until he was 13, when he went to work at his uncle's restaurant in Nice, the same year that the city became part of France. He joined up with César Ritz and went on to manage the kitchens of the London Savoy, the Carlton, and the Ritz in London and Paris, and also acted as a guest chef at New York's Pierre hotel.

MUSÉE ESCOFFIER DE L'ART CULINAIRE
3 rue Auguste Escoffier; tel. 04 93 20 80 51; http://fondation-escoffier.org; daily 10am-1pm and 2pm-6pm; adults €6, children under 11 free, free admission first Sun. of the month
The house where he was born is now a museum, the Musée Escoffier de l'Art Culinaire. It's a gem of a museum—not just for cooking fans—and has rooms full of copper pans, jelly molds, and the chef's private artifacts. There's also a cooking library, an 18th-century Provençal kitchen, and one floor given over to more than 2,000 menus from the past, including one from Christmas dinner during the siege of Paris in 1870, when food shortages meant chefs began cooking animals from the zoo, including antelope terrine with truffles and elephant consommé.

Visitors to the museum are served a cup of peach melba (invented by Escoffier to honor the Italian soprano Nellie Melba) during the summer and a coffee in the winter.

FOOD AND ACCOMMODATIONS
Cité Marchande
rue du Marché and rue Giacosa; Tues.-Sun. 7am-1:30pm
The pretty, partially covered market is a great place to buy fresh produce in the center of Cagnes-sur-Mer. It has around 30 stalls full of local fruit and vegetables, organic products, fish, sweets, and cheeses, plus a market bar. While you're there, look up at the art nouveau frieze on the Brasserie des Halles opposite the entrance.

Josy-Jo
2 rue du Planastel, Haut-de-Cagnes; tel. 04 93 20 68 76; https://josy-jo.com; mid-Dec.-mid-Nov. Tues.-Sat. noon-3pm and 7:30pm-10pm; €29
At the northern entrance to the old village, the restaurant was once the studio of artists Modigliani and Soutine. It looks like a smart

farmhouse, and the menu, full of traditional Provençal flavors, specializes in rustic lamb, veal, and beef dishes cooked over a charcoal grill. They also do wild prawns flambéed in pastis (€30) and rockfish soup (€18). Portions are big, and the metal-framed terrace furniture is artistically inspired.

Le Grimaldi

6 place du Château, Haut-de-Cagnes; tel. 04 93 08 67 12; www.hotelgrimaldi.com; €99 d

This property is housed in an impressive 17th-century stone bastide on the château square in the heart of Haut-de-Cagnes. It has four distinctive rooms, all with visible beams and exposed stonework. The Medieval Suite on the top floor has a rooftop terrace. The hotel also has a bistro-restaurant with a large patio on the square that does a decent lunch menu for €16. Breakfast is included.

Château Le Cagnard

54 rue Sous Barri, Haut-de-Cagnes; tel. 04 93 20 73 22; www.lecagnard.com; mid-Feb.-Dec.; €290 d

This luxury four-star hotel is built into the perched village's 13th-century ramparts. The former guard room is now the lobby, and little cubbyholes, curios, old documents, chess sets, and antique furniture can be found throughout the hotel. Half of the 30 rooms have their own private entrances, dotted around the narrow roads that surround the hotel. Le Cagnard also has a classy restaurant with a retractable roof and large outdoor terrace.

INFORMATION AND SERVICES

Cagnes-sur-Mer has three tourist offices:

- **Haut-de-Cagnes Office de Tourisme** (place Docteur Maurel; tel. 04 92 02 85 05; www.cagnes-tourisme.com): This office

has maps for the self-guided tour of artists' homes in the neighborhood.

- **Cros-de-Cagnes Office de Tourisme** (99 promenade de la Plage; tel. 04 93 07 67 08)

- **Cagnes-ville Office de Tourisme** (6 boulevard Maréchal Juin; tel. 04 93 20 61 64)

GETTING THERE AND AROUND
Car

Cagnes is 12 km (7.5 mi) to the west of **Nice** and can be reached either by the M6098 coast road or the A8 (exit number 48, 15 min).

There is plenty of **parking** in the center of town and along the seafront. Parking at the **Planastel car park** (12 rue du Château; tel. 04 89 98 26 76; 2 hours €3.60) in Haut-de-Cagnes is the best option for driving to the village. Cars are parked automatically in one of the 158 spaces via a series of machine transfers—it's definitely worth watching your car disappear into the depths of the hill, which sinks down 14 levels.

Train

Cagnes-sur-Mer station (avenue de la Gare; www.sncf-connect.com) is on the railway line between Marseille and Ventimiglia. Journey time to Nice is 14 minutes (€3.30) and to Cannes 23 minutes (€5.20).

Bus

Zou bus number 650 (tel. 04 13 94 30 50; https://zou.maregionsud.fr) travels from Nice Parc Phoenix to Grasse and stops at Cagnes railway station. A free **shuttle bus** (number 44) does a round-trip to the village of Haut-de-Cagnes seven days a week, departing every 15-20 minutes 7am-12:15am from the bus station in square Bourdet in Cagnes-ville (www.cagnes-tourisme.com).

Antibes

Antibes is one of the liveliest towns on the Riviera, with a handful of interesting museums, a daily Provençal market, and year-long sports and cultural events. Founded in the 5th century as Antipolis, a Greek trading post, it was controlled by the Romans until the fall of the empire, then repeatedly sacked by marauding invaders—Visigoths, Vandals, Saracens, and Barbarians—until it fell under the protection of the Lords of Grasse and eventually the Grimaldi family. French King Henri IV bought it in 1608, constructing the gray-stone Fort Carré as a holdout against Savoy invaders from the east. Today the fort looks out at the superyachts in Port Vauban, Europe's largest pleasure boat marina.

The constant arrival of yacht crews gives Antibes a strong nautical ambience. English is the common language, shops are dedicated to boat supplies, and there's lots of drinking going on. This seafaring feel is set against the city's strong artistic heritage, with artists' studios and workshops all over the city, a long literary legacy, and the fortified château in the medieval quarter now a museum dedicated to Spanish artist Pablo Picasso.

ORIENTATION

Le Vieil Antibes, the picturesque old town, is situated between **Port Vauban** and the **Cap d'Antibes peninsula** to the south. At its northern entrance is a row of vaulted barracks and maritime warehouses along the boulevard de l'Aguillon, **les Casemates,** which have been converted into artists' studios, glassblowing workshops, and sculpture ateliers. On the other side of the boulevard is a line of restaurants and bars.

West of Port Vauban is the modern bus station, the **gare routière,** and the **railway station,** both a 10-minute walk from the old town via the city's main squares, **place du Général de Gaulle** (new town) and the **place Nationale** (old town).

Within the old town, **Le Safranier** district (www.lacommunelibredusafranier. fr) is a self-styled "free community" set up in 1966, with the aim of maintaining local

Le Safranier district

Antibes

Antibes Harbor

Port Vauban

Antibes Train Station

Jardin René Cassin

TOURIST OFFICE

RUE SADI CARNOT

AVE DE LA LIBÉRATION

AVE DU 11 NOVEMBRE

D6098

D6007

D35

D6107

AVE DE VERDUN

BLVD D'AGUILLON

Pré aux Pêcheurs

CASEMATES ART STUDIOS

JACQ KINO
DIDIER SABA

ATELIER DU SAFRANIER

Plage de la Gravette

CHAPELLE SAINT-BERNARDIN

L'ARAZUR

ANTIBES BOOKS

Place Nationale

MARCHÉ PROVENÇAL

RUE SADE

MAISON JAMES CLOSE

ABSINTHE BAR

MUSÉE PICASSO

LE ZINC

VIEIL ANTIBES

GARE ROUTIÈRE

MODERN HOTEL

LA PLACE

CÔTE TERROIR

LE COMPTOIR DE LA TOURRAQUE

LE SAFRANIER

RUE DU GÉNÉRAL VANDENBERG

PROMENADE DE L'AMIRAL DE GRASSE

RUE DE LA TOURRAQUE

MUSÉE ARCHÉOLOGIQUE

Square Albert 1er

Mediterranean Sea

ROYAL ANTIBES

ROYAL SKI NAUTIQUE

To Plage de la Salis and Le Poseidon

RUE DE LA RÉPUBLIQUE

AVE MIRABEAU

AVE TOURNELLI

RUE SADI CARNOT

AVE THIERS

GRAND CAVALIER

PASTEUR

RUE LACAN

RUE VAUBAN

AVE ROBERT

AVE DU SOLEAU

AVE GAMBETTA

BLVD D'ALGER

RUE DIRECTEUR CHAUDON

DUGOMMIER

BLVD GUSTAVE CHANCEL

AVE NIQUET

BLVD DU MARÉCHAL FOCH

AVE GUILLABERT

BLVD ALBERT 1ER

AVE GAZAN

AVE DU 24 AOÛT

AVE DES FRÈRES ROUSTAN

REILLE

AVE MARÉCHAL

AVE GÉNÉRAL MAIZIÈRE

AVENUE LEMERAY

AVE DES DAMES BLANCHES

AVE GASTON BOURGEOIS

AVE SAINT-DONATIEN

BLVD MARÉCHAL LECLERC

BLVD DU PRÉSIDENT WILSON

COURS MASSÉNA

QUAI DES COMMANDANTS D'AFRIQUE DU NORD

QUAI HENRI-RAMBAUD

0   0.1 mi

0   0.1 km

© MOON.COM

traditions in a spirit of freedom, bringing together local residents through parties, outdoor dinners, chestnut fairs, and traditional flower battles. The concept is still going strong, and the whole district, identified by orange and yellow plaques on its street corners, is a maze of pretty houses covered in flowers.

SIGHTS
Port Vauban

tel. 04 93 21 72 17; www.riviera-ports.com

Europe's largest pleasure-boat marina is undergoing a transformation through the end of 2027 to include a new footbridge, a heliport, a port crew center, a pedestrian esplanade, restoration of the ancient Saint-Jaume bastion ramparts, a yacht club restaurant, and an upgrade to the quai des Milliardaires. The harbor has mooring for 1,700 boats and is an entertaining location to watch the comings and goings, the moorings, the carefully maneuvered departures, and the normally secret world of yacht crews and boat owners.

Le Fort Carré

avenue du 11 Novembre; tel. 04 92 90 52 13; www.antibes-juanlespins.com; Tues.-Sun. 10am-5pm; €3, students and over 65 €1.50, under 18 free

On a promontory at the far north of Antibes is the imposing Fort Carré. Built toward the end of the 16th century by King Henry II at a time when Antibes was the easternmost port in France, it was part watchtower and part military base. Louis XIV's chief military engineer, Vauban, further fortified the structure at the end of the 17th century. When Nice joined France in 1860, the fort was demilitarized and eventually became a training center for the army after the Second World War, when a sports complex was built under the fort. Visitors are welcome to walk around the fort and up the stairs to the raised courtyard, which provides great views of the marina and out to sea.

Vieil Antibes
Chapelle Saint-Bernardin

14 rue du Docteur Rostan; Wed. 9am-noon and 1:30pm-5pm, Tues., Thurs., and Sat. 9am-noon and 1pm-5pm, Fri. 9am-10am (mass) and 1pm-5pm

Antibes's neo-Gothic chapel, built in the 16th century for the Pénitants Blancs, is dedicated to Saint Bernardino of Siena, a Franciscan missionary. Badly damaged by fire in 1970 (photos are on display beside the entrance), it reopened to the public in June 2008 after more than 11 years of restoration. The hand-painted ceiling and walls, featuring a deep-blue starry sky and depictions of the four evangelists, the Virgin Mary, and Jesus above the central nave, are in gleaming condition.

Musée Picasso

place Mariejol; tel. 04 92 90 54 26; www.antibes-juanlespins.com; mid-Sept.-mid-June Tues.-Sun. 10am-1pm and 2pm-6pm, mid-June-mid-Sept. Tues.-Sun. 10am-6pm; €8, under 18 free

Built on the remains of a Greek encampment, the château that now houses this museum was occupied by the Grimaldi family until 1385, after which it became the town hall in 1792, a military barracks in 1820, and finally a history museum in 1925. Pablo Picasso visited the museum in 1946 and was offered the chance to use it as a workshop. He turned the 1st floor into a studio and thanked the city for its generosity by donating 23 paintings, 44 sketches, and a further 78 ceramics that he had made at the nearby Madoura pottery in Vallauris. In December 1966, the château was officially inaugurated as the Picasso museum; the collection was enlarged in 1991 by donations from Jacqueline Picasso, the artist's last wife, whom he met at the Madoura pottery.

Musée Archéologique

Bastion Saint-André 1, avenue Général Maizière; tel. 04 92 90 53 31; www.antibesjuanlespins.com; Nov.-Jan. Tues.-Sat. 10am-1pm and 2pm-5pm, Feb.-Oct. Tues.-Sun. 10am-12:30pm and 2pm-6pm; €3, students and over 65 €1.50, under 18 free

Housed in the two vaulted galleries of the Bastion Saint-André, a late 17th-century guards' depot built by Vauban, the archaeology museum contains relics and remains from excavations around Antibes and its surrounding waters. Dozens of shipwrecks from Etruscan, Greek, Phoenician, and Roman times have delivered a wealth of ceramics, jewelry, encrusted amphorae, mosaics, bronze coins, and colored glass. The museum also has a monumental marble fountain, sarcophagi, urns, funereal stelae, and a collection of huge lead anchors.

BEACHES

Antibes has always been more of a pleasure port than a destination for beach holidays, but Plage de la Salis and Plage de la Gravette are both pleasant-enough stretches of sand to spend a few hours sunbathing or watching the yachts.

Plage de la Gravette

promenade Amiral de Grasse; closest parking lots are the Parking du Port Vauban and Esplanade Pré des Pêcheurs

Access to this fine-sand beach with a view of the old town's seafront is through the fifth stone archway on the right via an iron gate that runs behind the quai des Pêcheurs, on the ocean side of the Port Vauban. It's a nice public beach, with a snack bar open from April to the end of September and public showers. There's another tiny beach to the north of the concrete, but it's very rocky.

Plage de la Salis

boulevard James Wylie; parking in the Parking Salis or Parking du Ponteil

Antibes's most popular beach runs along the length of boulevard James Wylie, where the Cap d'Antibes peninsula begins. The narrow beach has thick, golden sand and several snack bars open from April to September. It's a public beach with no sunbed or parasol rental, and it can be very crowded on weekends in July and August, but it's great for games of beach volleyball, and it has a seasonal lifeguard and first-aid hut for severe cases of sunburn, heatstroke, and the occasional jellyfish sting.

SPORTS AND RECREATION
Water Sports
Royal Ski Nautique

Pointe d'Ilet, Le Ponteil; tel. 06 62 03 24; mid-June-early-Sept. daily 9am-6:30pm; €70/30 minutes

This waterskiing center, which is located only 100 m (328 ft) from Plage de la Salis, also offers wakeboarding, wake-surfing, paddleboats, and inflatable banana boats.

Cycling

As one of the Riviera's main transport hubs, Antibes's roads can be very busy, so it's not a relaxing place to cycle. The old town is cobbled and narrow, so the nicest nearby routes are those around the Cap d'Antibes. The coastal road eastward toward Cagnes-sur-Mer is a popular ride but can be congested in the summer.

Côte-Ebike

1200 chemin des Combes; tel. 04 92 90 61 61; www. cote-ebike.com; Tues.-Sat. 9am-noon and 2pm-7pm (until 6:30pm Sat.); €14-39/day

This company builds its own electric bikes and rents them out from a shop in Antibes and one in Villeneuve-Loubet. They also have a large selection of road and mountain bikes, tandems, and electric scooters. Half-day rental for an electric bike is €25 and €112 for four days. Special deals are available for multiple and weekly rentals.

Top Cycle

1465 chemin des Combes; tel. 04 93 74 08 03; www. topcycle.fr; Mon. 3pm-7pm, Tues.-Sat. 10am-7pm; €70/day

Top Cycle has 200 sq m (2,152 sq ft) of floor space to show off its new and used bicycles just north of Antibes. There's a repair workshop that has an excellent reputation, and the owners organize mountain bike rides every Sunday morning (details on the shop's Facebook page).

FESTIVALS AND EVENTS

Voiles d'Antibes

www.voilesdantibes.com; first week of June

This yachting event is a season-opener for the Mediterranean boating crowd, bringing together some of the finest sailing vessels on the seas for five days of match racing and quayside events. In the water, boats compete in different classes: yachts built pre-1950, classic yachts built before 1976, and boats that sailed in the Americas Cup from 1958 to 1987. On dry land, the regattas are free to watch, and there are exhibitions, cocktail parties, and yachting events from 9am to 10pm.

SHOPPING

Maison James Close

30 rue James Close; tel. 07 61 76 46 92; https:// maisonjamesclose.com; Tues.-Sat. 10:30am-1pm and 2pm-7pm, Sun. 10:30am-1pm

On Antibes's coolest shopping street, Maison James Close sells a big range of household artifacts, home decorations, and curios. It's great for original jewelry, tiles, lamps, tea trays, framed prints, and unconventional gifts.

Antibes Books

13 rue Clémenceau; tel. 04 93 61 96 47; www. antibesbooks.com; daily 10am-7pm

Popular with locals who want to practice their English, the town's huge Anglophone community, and passing yacht crews, Antibes Books has a large selection of new editions as well as postcards, recipe books, local history volumes, stationery, jigsaw puzzles, and board games. The shop also holds book signings, often pertaining to histories of the French Riviera and French romances written by local residents and Francophiles.

Casemates Art Studios

boulevard d'Aguillon; hours vary between galleries, most are open Tues.-Sat. 10am-12:30pm and 2:30pm-6:30pm, Sun.-Mon. 2:30pm-6:30pm during high season

The former military barracks and warehouses lining the eastern flank of the boulevard d'Aguillon have been turned into artists' studios, galleries, and workshops. All are open to the public, who are welcome to look around and make a purchase. Standout studios belong to assemblage artist **Jacq Kino** (www.jacq-kino.com) at number 29 and glassblower **Didier Saba** (www.saba.verrerie. com) at number 27; Saba also runs glassblowing courses at his workshop. The **Atelier du Safranier** (http://ateliersafranier.chez.com) has a large printing press once used to print etchings by Dalí and Picasso.

FOOD

French

Côté Terroir

3 place Gare des Autobus; tel. 04 92 90 06 45; Tues.-Sat. noon-2pm and 7pm-9pm; €13-15

Christelle and Thibault Brillon have created a great little restaurant specializing in local produce. Thibault is a former national dessert champion, so it's worth saving some space for the strawberry cake. The restaurant has a decked terrace and a small boutique selling locally produced beer, olives, honey, and tins of sardines.

Le Comptoir de la Tourraque

1 rue de la Tourraque; tel. 04 93 95 24 86; Thurs.-Tues. 7:30pm-10:30pm; €22-33

Just 50 m (165 ft) from the Provençal market on the road toward Cap d'Antibes, Le Comptoir de la Tourraque is a cozy, Bohemian-looking restaurant with a chalkboard built into its outdoor wall. Main courses include shoulder of lamb with mashed potatoes and seared tuna flavored with Asian and Provençal spices. The menu is small, with a classy selection of wines.

L'Arazur

8 rue des Palmiers; tel. 04 93 34 75 60; www.larazur.fr; Oct.-May Thurs.-Sun. noon-2pm and 7pm-10pm, June-Sept. Mon.-Sat. 7pm-10pm; €36-42

L'Arazur is a mecca for fine dining in Antibes. Chef Lucas Marini's menu might include starters like blue lobster with mushrooms, black bread, and parsley or seabass tartare with green mango, basil, and ginger. You can follow a main of pork belly with eggplant purée, roasted lemon sauce, and mixed vegetables with a chocolate and orange blossom sponge cake with calamansi lime sorbet for dessert. L'Arazur also runs a high-end catering outlet, La Maizon, for gourmet lunchtime dishes at 3 avenue Pasteur.

Le Zinc

15 cours Masséna; tel. 04 83 14 69 20; Tues.-Sun. noon-2pm and 7pm-10pm; €17-33

Always full of marketgoers and locals, Le Zinc (named after the old-style zinc-topped café tables) serves French food with a Mediterranean twist. Located on the southern end of the Marché Provençal, creative specialties include sautéed squid and chorizo, octopus salad in red wine, and eggplant caviar.

Snack Bars
Le Poseidon

8 boulevard James Wylie; mid-Mar.-mid-Nov. daily 8:30am-8pm; €5-12

On the promenade above the sandy Plage du Ponteil, Le Poseidon is a beach hut/snack bar serving almost everything, from a salmon and chicken poké bowl to ice cream, amid a friendly ambience and with great views of the sea and Antibes's ramparts. There's a takeaway service, or there is seating at their tables and chairs under pink parasols for around 20 people.

Markets
Marché Provençal

cours Masséna; tel. 04 22 10 60 01; June-Aug. daily 6am-1pm, Sept.-May Tues.-Sun. 6am-1pm

Best approached from Port Vauban to the north, Antibes's Provençal covered market is one of the region's best fresh produce sources, with local fruit and vegetable stalls, cheese, fish, charcuterie, and flowers mixed up with tables of lavender honey, nougat, and barrels of olives. Perfect for picnic supplies, it's also a good location to find unusual produce such as purple carrots and Italian spiced sausages. At lunchtime in the high season, the area is washed down and replaced by modern art, sculpture, and handmade jewelry. The high roof keeps the place nice and cool in the summer.

It is up a few steps behind a statue of Championnet, the French Revolutionary army commander who died of typhus in Antibes in 1800—a good meeting point if you get lost among the fruits and cheeses.

BARS AND NIGHTLIFE
★ Absinthe Bar

25 cours Masséna; tel. 04 93 34 93 00; Thurs.-Sat. 9am-12:30am, Sun. 9am-6pm, Mon.-Wed. 9am-7:15pm

Hidden in a 9th-century cellar beside the Provençal market is an authentic absinthe bar complete with art nouveau posters, original adverts, and cabinets full of glass absinthe dispensers. The brainchild of owner Frédéric Rosenfelder, the bar is accessed down a winding staircase at the back of his shop, which sells liqueurs and local products. Patrons are advised to have a maximum of three glasses of the "green fairy," as it was known during Vincent van Gogh's time. Today's absinthe may not be the rough-edged green poison it was at the turn of the 20th century, but it is still a powerful drink, made all the more pleasurable when poured over a sugar cube in this authentic-feeling bar where period hats line the shelves and live piano music fills the air.

ACCOMMODATIONS
Under €100
Modern Hotel

1 rue Fourmillière; tel. 04 92 90 59 05; www.modernhotel06.com; €62 d

---

1: Le Comptoir de la Tourraque **2:** Absinthe Bar
3: Port Vauban

1

2

3

A cheap option with character in the old town's pedestrianized quarter, the Modern Hotel, opened by the Fechino family over 100 years ago, is kept modern with frequent refurbishments. It's small and friendly, and all 17 rooms are air-conditioned and soundproof. Some rooms also have kitchenettes. Breakfast is €7.

€100-200
La Bastide du Bosquet

14 chemin des Sables; tel. 06 51 76 72 73; www. lebosquet06.com; Mar.-mid-Nov.; €135 d

Far and away the best bed-and-breakfast in Antibes, La Bastide du Bosquet is in a residential area on the way to Juan-les-Pins and the Cap d'Antibes. The stone farmhouse has been in the Aussel family for several generations. Guests can stay in the bedroom where Guy du Maupassant wrote part of the novella *Bel Ami* (ask for the Chambre Jaune), the only room overlooking the sea and gardens. There are four double bedrooms and two annex rooms for families, and parking is available. The beach is a 5-minute walk away; sun loungers and towels are provided.

★ La Place

1 avenue du 24 Août; tel. 04 97 21 03 11; www.la-place-hotel.com; €150 d

Owner Bernadette Walberer prides herself on offering guests a personal touch in this contemporary-looking three-star hotel on the edge of the old town. All 14 rooms are stylish, with all the standard modern conveniences and high-quality fittings. Original art decorates the walls of the bedrooms, the stairwells, and the communal areas, and there's a large breakfast lounge that transforms into a tea salon in the afternoon and buffet in the early evening. Third-floor rooms have views of the Alps in the distance.

Over €200
Royal Antibes

16 boulevard Maréchal Leclerc; tel. 04 83 61 91 94; www.hotelroyal-antibes.com; €250 d

With 64 rooms, suites, and apartments, the four-star Royal is one of the largest hotels in Antibes and serves top-end holidaymakers and business clients. It has a spa, fitness rooms, a sauna and a hammam, two meeting rooms, and private parking, with the Café Royal restaurant on its seafront terrace open for breakfast, lunch, and dinner. The Royal Suites are 100-sq-m (1,075-sq-ft) accommodations with hot tubs and duplex terraces, but there are also classic rooms for guests who don't mind a courtyard view.

INFORMATION AND SERVICES
Tourist Information
Office de Tourisme d'Antibes

60 chemin des Sables; tel. 04 22 10 60 10; www. antibesjuanlespins.com; Feb.-Mar. and Oct.-Dec. Mon.-Sat. 9am-12:30pm and 1:30pm-5pm, Sun. 9am-1pm, closed Sun. Nov. and Jan., Apr.-June and Sept. Mon.-Sat. 9am-12:30pm and 2pm-6pm, Sun. 9am-1pm, July-Aug. daily 9am-1pm

The tourist office has free maps and brochures on current events and activities to do in and around Antibes. They can also help you book accommodations, and they offer reductions if water sports activities are booked though the office.

Medical and Emergency Services
Centre Hospitalier d'Antibes Juan-les-Pins

107 avenue de Nice; tel. 04 97 24 77 77, 04 97 24 77 48 (emergencies), or 04 97 24 77 39 (pediatric emergencies)

Antibes's public hospital is a 5-minute drive from the old town heading toward Nice and has a 24-hour emergency service.

Pharmacies take turns staying open 24/7; most will post details in their windows, or you can check https://pharmaciedegarde.co/pharmacie-ouverte-antibes.html.

GETTING THERE
From Nice Airport

Antibes is 16 km (10 mi), a 25-minute drive, from Nice airport on the A8 autoroute or

D6007 coast road. A taxi from the airport to Antibes costs about €45.

Airport Express Bus 82 from Nice airport Terminal 2 to Vallauris (approx. one bus per hour 8:10am-10pm) also stops at the **pôle d'echanges Antibes** (www.niceairportxpress.com; journey time approximately 45 min; adults €9, children under 12 €4.50).

Car

Antibes is 25 km (16 mi; 35 min) west of **Nice** and 15 km (9 mi; 15 min) east of **Cannes,** and is easily reached by the scenic coast road D6007 or the A8 autoroute (take exit number 44 for Antibes). The autoroute is usually faster, but there is a toll (€1.60) and a lot of heavy-goods vehicles to negotiate on the way.

Train

Antibes station (Gare SNCF, avenue Robert-Soleau; www.sncf-connect.com) is behind Port Vauban. It is on the Marseille-Ventimiglia line. Fast trains to and from Nice take approximately 18 minutes, and slow TER trains, which stop at every station, take around 30 minutes (€5.20). The train from Marseille takes around 2 hours 20 minutes (€30).

Bus

Buses in and around Antibes stop at one of two places: the **gare routière d'Antibes** on place Guynemer (tel. 04 93 34 37 60), which is a central location to Vieil Antibes, or the **pôle d'echanges Antibes,** adjacent to the train station on boulevard Général Vautrin.

Zou bus number 620 (tel. 04 13 94 30 50; https://zou.maregionsud.fr) travels from Nice's Parc Phoenix to Cannes, stopping at Antibes's pôle d'echanges (2 per hour; journey takes approximately 40 min; €2.10).

Taxi

Taxi Antibes (tel. 04 93 67 67 67; www.taxiantibes.com) takes passengers as far as Saint-Tropez or Monaco.

GETTING AROUND

Walking in old Antibes is a pleasure because all the sights are close to each other. There is no need to take buses, trains, or taxis unless you are leaving the city.

Car

It is not recommended to take your car into Antibes's old town, where the streets are narrow and partly pedestrianized. Otherwise, driving in Antibes and along the coast generally is easy, with all destinations well signposted.

There are 10 paying **car parks** in the town (www.antibes-juanlespins.com/proximite/stationnement/parkings-de-la-ville), with a 150-space free car park opposite the Fort Carré on avenue du Onze Novembre. Envibus number 14 takes visitors into the center.

Bus

The local bus network, **Envibus** (tel. 04 89 87 72 00; www.envibus.fr), connects all the villages around Antibes, departing from the **gare routière. Bus 14** runs from the free car park opposite the Fort Carré to central Antibes and takes 10 minutes. Tickets cost €1 per single trip, €8 for a 10-trip pass, and €10 for a 7-day pass.

Around Antibes

D204

D2085

D2085

MUSÉE ESCOFFIER
DE L'ART CULINAIRE

Parc
Saint-James
"Le Sourire"

Villeneuve-
Loubet

Parc
Saint-
Véran

D2

Vallon de Mardaric

To Cros-de-Cagnes
Office de Tourisme

D4

A8

D604

D198

La Brague

OFFICE DE
TOURISME

CAFÉ
BRUN

LA VERRERIE
DE BIOT

Biot

Parc des
Bouillides

MUSÉE D'HISTOIRE ET
DE CÉRAMIQUE BIOTOISE

HOTEL-RESTAURANT
LES ARCADES

MUSÉE
FERNAND LÉGER

Parc de
Vaugrenier

D6098

D504

D504

D535

La Valmasque

D4

D103

A8

Parc de la
Brague

Train
Station

D35

A8

Parc des
Semboules

TOP CYCLE

COTE-EBIKE

D704

D6007

D6098

D435

D35

D35BIS

LE FORT
CARRÉ

FORT CARRÉ
CAR PARK

D135

A8

MUSÉE NATIONAL
PABLO PICASSO/
LA GUERRE ET LA PAIX

LE CAFÉ DE FRANCE

OFFICE DE
TOURISME

Vallauris

LA BIGARADE

Antibes

GARE
ROUTIÈRE

MADOURA

D135

See
"Antibes"
Map

Parc Exflora

Golfe
Juan

Plage
Golfe-Juan

La
Grande
Plage

Juan-Les-
Pins

D803

Train Station

LE BISTROT DU PORT

JUAN-LES-PINS
BEACH BARS

D6107

SUBVISION

TOURIST OFFICE

D2559

SANCTUAIRE
DE LA GAROUPE

0   1 mi

0   1 km

See
"Cap d'Antibes"
Map

Golfe Juan

Jardins
d'Eilenroc

© MOON.COM

Around Antibes

Some of the Riviera's most popular destinations are found close to Antibes. The coastal resorts of Juan-les-Pins and Golfe-Juan offer excellent sandy beaches, water sports and nightlife, while, inland, Biot and Vallauris are known for their glassware, ceramics, and artists' studios.

ORIENTATION

Halfway between Cagnes and Antibes, Biot has been the French capital of glassblowing since the mid-1950s. It still has a handful of glass workshops open for visits as well as boutiques selling the famed bubble-glass in the charming old village and on the main route de la Mer.

The 6-km-long (3.7-mi-long) thumb-shaped Cap d'Antibes is the location for some of the most expensive villas in the world. It's almost impossible to see over their high gateways, but their occupants often head to the seafront in **Juan-les-Pins,** one of the prettiest and liveliest promenades on the Riviera. Farther west is the pleasure port of **Golfe-Juan,** and 1.5 km (1 mi) inland is the town of **Vallauris,** which is part of the same commune and which takes its name from Roman times and the Latin *valles aurea*—golden valley.

BIOT

Biot's main tourist attraction is the Fernand Léger museum, but the old village is one of the most authentic in the region, with some great little restaurants and a **daily market.** For walkers and picnickers, the Parc Departémentale de la Brague runs between Biot and Valbonne, with waterfalls and a Mediterranean forest.

Sights
Musée Fernand Léger
255 chemin du Val de Pôme; tel. 04 93 53 87 20; www.musee-fernandleger.fr; May-Oct. Wed.-Mon.
10am-6pm, Nov.-Apr. Wed.-Mon. 10am-5pm; €7.50, free parking

Attracted by Biot's tradition of ceramics, avant-garde artist Fernand Léger bought a small house and large plot of land in the village in 1955 before dying only two weeks later at the age of 74. A museum was built in his honor on the land, inaugurated in 1960 by his wife, Nadia, in the presence of Picasso, Braque, and Chagall. It's the world's only museum dedicated to Léger. The cinema on the ground floor shows Léger's first film, an automated odyssey of images set to a soundtrack by George Antheil, *Le Ballet Mécanique.* The gardens offer views of the giant mosaics on the museum's exterior walls. Everything about the place is large-scale, with stained glass windows spanning two floors. There's a snack bar in the garden among sculptures based on Léger's work. The original house is now the museum's administrative center.

Musée d'Histoire et de Céramique Biotoise
9 rue Saint-Sébastien; tel. 04 93 65 54 54; www.musee-de-biot.fr; July-Sept. Wed.-Sun. 10am-6pm, Oct.-June Wed.-Sun. 2pm-6pm; €4, students and pensioners €2, under 16 free

Opened in 1981 and run by local volunteers, this ceramics and history museum occupies the former 16th-century Saint-Jacques hospital. It tells the story of 2,000 years of history in Biot through ceramics, photography panels, everyday objects, and a typical 19th-century kitchen. In the 17th and 18th centuries there were 40 potters in the village, most of them making giant urns (jarres) used to store olive oil, dried beans, and flour, and exported out of Antibes, Marseille, and Genoa.

La Verrerie de Biot
chemin de Combes; tel. 04 93 65 03 00; www.verreriebiot.com; Mon.-Sat. 9:30am-6:30pm, Sun.

1

2

10:30am-1:30pm and 2:30pm-6:30pm, workshop closes 6pm; free

With its huge, fuchsia-colored archway and metal posts, it's hard to miss Biot's largest glassblowing center. The verrerie organizes guided tours (35 min; 4pm Mon.-Fri.; €6 adults, €3 ages 7-14), which take visitors through the process of blowing, rolling, shaping, and coloring the particular Biot strain of glassware, which contains bubbles called le verre bullé that were actually created by mistake. More than 700,000 people visit the glassblowing center each year.

Sports and Recreation
Parc de la Brague

www.departement06.fr; Apr.-Oct. daily 7am-8pm, Nov.-Mar. daily 8am-6pm

Encompassing 630 hectares (1,550 acres) across the communes of Biot, Antibes, and Valbonne, the Parc de la Brague is a fantastic location for walks, horseback riding, cycling, and picnics. The Brague river runs through it, providing a refreshing avenue during the hot summer. There's also a 9-km (6-mi) riverside walk to observe the herons, ducks, and waterfowl. The park has an abundance of oak and pine trees with scents of cistus, myrtle, and wild rosemary.

To get here from Antibes, follow signs for Sophia Antipolis, then the route des Crêtes toward Biot. There are 11 parking areas on the roads that traverse the park.

Food and Accommodations
Café Brun

44 impasse Saint-Sébastien; tel. 04 93 65 04 83; www. cafebrun-biot.fr; Mon.-Fri. noon-2pm and 7pm-10pm, Sat. 7pm-10pm; €14.50-19.50

Open since 1987 in a shady square just off the old village's main street, this café has a friendly pub atmosphere, with an open fire in the winter and sunny terrace in the summer. Specialties include chicken brochettes, burgers, steak-and-fries, snails, goat's cheese, five

1: Musée Fernand Léger **2:** glassware in La Verrerie de Biot

different salads, and a mixed grill that few can finish for under €20.

★ Hotel-Restaurant Les Arcades

14-16 place des Arcades; tel. 04 93 65 01 04; www. hotel-restaurant-les-arcades.com; restaurant Tues.-Sun. noon-2pm and 7pm-10pm, hotel mid-Feb.-Dec.; restaurant €35, rooms €100 d

Les Arcades, built in 1480, has 12 rooms, all decorated with a touch of the medieval and Provence. Seven have a bath and five have showers only; all have open beams, four-poster beds, fireplaces, and stone sinks. The restaurant serves excellent Provençal fare. For art lovers, the hotel has its own collection of paintings and sculptures from eminent artists such as Victor Vasarely and Georges Braque, both of whom dined under the restaurant's arches.

Information and Services
Office de Tourisme de Biot

4 chemin Neuf; tel. 04 93 65 78 00; www.biot-tourisme.com; Mon.-Fri. 9:30am-12:30pm and 1:30pm-6pm, Sat. and bank holidays 11am-5pm, Sun. during school holidays 11am-7pm

The tourist office has free maps and brochures on current events and activities to do in and around Biot. They can also help you book accommodations.

Getting There and Around
Car

Biot is 25 km (16 mi; 35 min) to the west of **Nice.** It can be reached either by the coast road (D6007 and then D7) or the A8 (exit number 46).

Train

Biot station (route Nationale 7; www.sncf-connect.com) is on the railway line between Marseille and Ventimiglia. Journey time to Nice is 22 minutes (€4.50) and to Cannes 17 minutes (€3.90).

Bus

Zou bus number 620 (tel. 04 13 94 30 50; https://zou.maregionsud.fr) travels from

Cap d'Antibes

BLVD RAYMOND POINCARÉ

Train Station

Juan-Les-Pins

See Detail

JUAN-LES-PINS BEACH BARS

To La Grande Plage

LA PASSAGÈRE

BELLES RIVES HOTEL

BLVD ÉDOUARD BAUDOIN

Plage le Crouton

LE POSEIDON

LA BASTIDE DU BOSQUET

CHEMIN DES SABLES

LA JABOTTE

BLVD DU CAP

BLVD JAMES WYLLIE

Plage de la Salis

LE BACON

BD DE BACON

Bois de la Garoupe

LE BISTRO DU CURÉ

Jardin Botanique de la Villa Thuret

ROUTE DU PHARE

SANCTUAIRE DE LA GAROUPE

AVE MALESPINE

D2559

Mediterranean Sea

Golfe Juan

BLVD DU MARÉCHAL JUIN

BLVD FRANCIS MEILLAND

BLVD DE LA GAROUPE

WATERSNOW

Plage de la Garoupe

CHEMIN TIRE-POIL

TRAVERSE DES NIELLES

BLVD JFK

VILLA FABULITE

AVENUE MRS L.D. BEAUMONT

Jardins d'Eilenroc

VILLA EILENROC

0   0.5 mi

0   0.5 km

To Train Station

Juan-Les-Pins

PLAGE LA JETÉE

AV GUY DE MAUPASSANT

BV DU PRÉSIDENT WILSON

Square Sidney Bechet

VILLA D'ELSA

OFFICE DE TOURISME

i

JUAN-LES-PINS BEACH BARS

CAPUCCINO

PAM PAM

LE COLOMBIER

BLVD ÉDOUARD BAUDOIN

Jardin de la Pinède

Square Frank Jay Gould

0   200 yds

0   200 m

Nice's Parc Phoenix to Cannes via Biot railway station. **Envibus number 10** (tel. 04 89 87 72 00; www.envibus.fr) runs from Antibes bus station to Valbonne via Biot station and Biot village (30 min journey time from Antibes).

CAP D'ANTIBES

Cap d'Antibes is a great place for an afternoon cycle or hike along the coastal footpath to the sumptuous Villa Eilenroc, donated to the city of Antibes and open to the public. It has tremendous views of the coastline and passing yachts and has several creeks and bays beneath it, including the **Plage Croupatassière,** a nice place for a dip.

Sights
Villa Eilenroc
460 avenue Mrs Beaumont; tel. 04 93 67 74 33; Wed. and Sat. 10am-4pm; €2, under 12 free, free Oct.-Mar.

With 11 hectares (27 acres) of gardens, bowers, fountains, ornamental gates, and ponds, the gardens of the Villa Eilenroc are a charming place to walk and imagine you are back in the Roaring Twenties and about to dine with the Fitzgeralds, or maybe the Duke of Windsor and Wallis Simpson, who stayed at the neighboring Château de la Croë in the 1930s. Designed by Charles Garnier (of the Paris and Monte-Carlo Opéra fame), the Villa Eilenroc was built for the Dutch financier Hugh-Hope Loudon in 1867 (the villa's name is an anagram of his wife's name, Cornélie). The villa has its own headland, olive grove, rose and aromatic gardens, and views of the Baie des Milliardaires. It was used as the location for the party scenes and tennis games in Woody Allen's Riviera-set film *Magic in the Moonlight* (2014).

Jardin Botanique de la Villa Thuret
62 boulevard du Cap (entrance on chemin Gustave Raymond); tel. 04 97 21 25 00; https://jardin-thuret. hub.inrae.fr; summer Mon.-Fri. 8am-6pm, winter Mon.-Fri. 8:30am-5:30pm; free, reservations required for groups

Cap d'Antibes's botanical gardens were established in 1865 by botanist Gustave Thuret,

who left behind a large villa and 3.5 hectares (8.6 acres) of ornamental and tropical grounds. The site is now owned by the French state and managed by the National Institute of Agronomic Research (INRA), which has introduced some fantastically exotic plants never before seen in France.

Sanctuaire de la Garoupe
route du Phare; Tues.-Sun. 11am-6pm, Sun. mass 11:30am

At the highest point on the Cap d'Antibes, the Garoupe Hilltop, is the Sanctuaire de la Garoupe, set on a plateau surrounded by oak and pine forests and comprising a chapel, **café,** gift shop, and viewing station alongside the cape's lighthouse.

The sanctuary is made up of three interconnected chapels. The largest, the **Chapelle Notre-Dame de la Garde,** dates from the end of the Middle Ages and was expanded in 1520. The **Chapelle Notre-Dame de Bon Point** dates from the 13th century and is dedicated to sailors and fishermen. Its golden virgin is carved from fig-tree wood. Both chapels are full of interesting ex-votos, models of boats hanging from the ceiling, cabinets of military uniforms, medals, and parts of anchors, bells, crosses, and barrels. Alongside the semaphore station is the tiny **Chapelle du Calvaire.**

It's a popular destination for walkers and cyclists—it's just 2.8 km (1.7 mi) from Antibes along the boulevard du Cap or 3 km (1.9 mi) from Juan-les-Pins via the chemin des Sables.

Beaches
Plage de la Garoupe
chemin de la Garoupe

This lovely northeast-facing stretch of sheltered sand offers clear water and views of the Lérins islands and the Mercantour in the distance. It is best appreciated out of season, because it can be very crowded in the summer. **Plage Keller** (1035 chemin de la Garoupe; tel. 04 93 61 33 74; www.restaurant-plage-cesar-antibes.fr; Mar.-Oct.), a private beach club in the bay, has a gastronomic restaurant,

Le Cesar (mains €21-48), and sun loungers and parasols to rent. There is limited parking above the beach.

Plage du Croûton
boulevard du Maréchal Juin

A small white-sand beach next to the pleasure port of the same name, Plage du Croûton is worth getting to early to see the day's catch at the local fishmonger. The public beach is toward the north end and is a popular night-time destination for teenagers.

Water Sports
Watersnow
Baie de la Garoupe; tel. 06 60 52 60 28; www. watersnow.com; June-mid-Sept.; lessons €65

In front of the Baie Dorée Hotel beside the Plage de la Garoupe, Watersnow offers paddleboarding, waterskiing, wakeboarding, and wake-surfing lessons for anyone over six years old (four years old for waterskiing). Five mornings of waterskiing lessons cost €400, but you can ride on an inflatable dragged behind a speedboat for just €25.

Hiking
Chemin Tire-Poil
Distance: *5 km (3 mi) round-trip*
Time: *2 hours*
Information and Maps: *Tourist office in Antibes*
Trailhead: *Plage de la Garoupe*

The coastal footpath that runs on the southeast headland of the cape is known as the Tire-Poil (hair puller)—such is the strength of the wind. It's a very rugged walk, with steep staircases and occasionally hairy drops, but it's a good way to discover the southeastern tip of the peninsula and the only way to see inside the grounds of some of the larger villas.

The path begins at the Plage de la Garoupe over the rocks, where spiny spurge plants and bonsai-like sea fennel fill the cracks among the rocks, and Aleppo pine trees

---

1: Villa Eilenroc 2: Baie des Milliardaires 3: Jardin Botanique de la Villa Thuret 4: model ships hanging in the chapels of Sanctuaire de la Garoupe

form an arch for hikers to walk through. The path leaves the coast just behind the Villa Eilenroc, where walkers are allowed to swim in **Croupatassière bay** before traveling inland along avenue Mrs Beaumont, through some Holm oak forests, and back to the Plage de la Garoupe inland via the avenue de la Tour Gandolphe and avenue André Sella.

Food
Le Bistro du Curé
A Bon Port; tel. 04 93 61 33 01; http://lebistrotducure. fr; Oct.-Mar. Thurs.-Sun. 10:30am-5:30pm, Apr.-Sept. Tues.-Sun. 10:30am-5:30pm; €6-18

Part of the Garoupe sanctuary, this café-bistro does a very attractive Sunday brunch of eggs, bacon, sausage, soup, cheeseboard, and risotto, perfect for a stopover on a walk or cycle around the cape. The bistro serves mainly Provençal fare on an outdoor terrace and garden, which is planted with the garoupe bushes that gave the place its name.

Le Bacon
664 boulevard de Bacon; tel. 04 93 61 50 02; www. restaurantdebacon.com; daily noon-2pm and 7pm-10pm; €55-148

A firm favorite among fresh-fish gourmets, Le Bacon has been around since 1948, when the Sordello family set up an awning and two makeshift tables to sell pan bagnat, lemonade, and beer on the coast road. The "buffet" became a small restaurant serving Alphonsine Sordello's excellent fish dishes, after which it expanded and became well known, gaining a Michelin star in 1979. It's still run by the same family and serves a well-heeled Riviera-glam clientele.

Accommodations
La Jabotte
13 avenue Max Maurey; tel. 04 93 61 45 89; www. jabotte.com; €77 d

On a side road behind Salis beach, La Jabotte is a popular, fun, and friendly seaside guest house that's a bargain on the edge of the Cap d'Antibes. Most of the nine rooms have terraces, and there's a family room upstairs that sleeps four and includes a TV. There's also a year-round sauna, free bikes, and a sea kayak for rent. Owners Nathalie and Pierre provide beach towels, a tuktuk service around Antibes for guests, and friendly dogs and cats to stroke in a vibrant, flower-filled guesthouse.

Villa Fabulite

150 traverse des Nielles; tel. 04 93 61 47 45; www. fabulite.fr; Apr.-Oct.; €230 d

This one-story hotel is down a quiet lane toward the southern end of the cape. High-quality, spacious, and modern rooms open out into a Mediterranean garden with citrus and olive trees and a pool. Some of the rooms are especially for families, with connecting doors, and most have private terraces. Poolside lunches are served 11am-4pm.

Getting There and Around
Car

The Cap d'Antibes is a 5-minute drive from central Antibes and it takes another 15 minutes to tour the entire cape before joining the boulevard Maréchal Juin on the western side of Juan-les-Pins.

Cars are allowed to park on the roads where indicated, but the only sizable **parking lots** are behind the Plage de la Garoupe and along avenue André Sella, which juts to the southwest from the beach.

Train

The closest train stations (www.sncf-connect. com) are in **Antibes** (take Envibus 2 from the pôle d'echanges) and **Juan-les-Pins** (take Envibus 15 and change at the Palais des Congrès, stop to catch bus 30 or 31, and change again at Hermitage bus stop for bus 2 to the Cap d'Antibes).

Bus

Envibus 2 (tel. 04 89 87 72 00; www.envibus. fr) is the only line that runs down the Cap d'Antibes. It goes from the pôle d'echanges in Antibes (beside the railway station) to the Eden Roc hotel down the central boulevard Francis Meilland on Cap d'Antibes. The

journey time is 15 minutes. Tickets cost €1 for a single, €8 for a 10-trip pass, and €10 for a 7-day pass.

JUAN-LES-PINS

Juan-les-Pins has a very different vibe from Antibes. It's more about holidays and the beach rather than sailing and food. It has no historic old town and can feel quite empty during the low season when its hundreds of holiday apartments are vacant and some of the shops, hotels, and restaurants close.

The pine trees of the Jardins de la Pinède to the southeast of the town center provide a cooling place to sit in the summer. The roads surrounding the gardens lead to Juan-les-Pins's large hotels, the casino, and, toward Antibes, the Palais des Congrès, which houses the tourist office, conference center, and a few cafés.

Beaches

With its sandy beaches, clear water, and adoring sunbathers, this part of the coast is a magnet for water sports enthusiasts. Nearly every type of water activity is available, from being tugged along while clinging to an inflatable banana to having private waterskiing lessons off the Belles Rives pontoon.

TOP EXPERIENCE

La Grande Plage
boulevard Charles Guillaumont

It's called "the big beach," and it is—one long, uninterrupted stretch of fine sand bordering the boulevard Charles Guillaumont along the length of Juan-les-Pins. This beach offers excellent swimming conditions, and there are food kiosks, showers, and lifeguards all along the strip, but no natural shade.

Festivals and Events
Jazz à Juan

place Pinède Gould; www.jazzajuan.com; July; tickets €25-80

Twinned with New Orleans in Louisiana, Juan-les-Pins sees itself as the little French sister of the capital of jazz. It holds its annual festival, Jazz à Juan, in July, billing itself as more prestigious and authentic than the Jazz Festival in Nice. Over the years it has attracted some of the biggest names in the jazz genre—Dave Brubeck, Ray Charles, Oscar Peterson, Sarah Vaughan, Ella Fitzgerald, Keith Jarrett, Shirley Horn, Stevie Wonder, and Gilberto Gil.

The festival is held on a seafront stage erected in the sandy boules courts of **La Pinède,** a small pinewood grove on the seafront toward the east of the resort. Along the **boulevard Edouard Baudoin,** handprints of many of its stars are embedded into the pavement. Alongside the festival proper, there's **Jazz Off**—smaller-scale gigs, impromptu concerts, and jam sessions taking place all over the town. While there are scores of musical events along the Riviera, the pine trees, cicadas, warm nighttime air, and seafront location make Juan-les-Pins a very special place to listen to jazz.

Tickets can be bought from the tourist offices in Antibes and in Juan-les-Pins and are also on sale outside La Pinède Gould venue on boulevard Baudoin starting 1 hour before showtime.

Shopping
Capuccino

6 boulevard Edouard Baudoin; tel. 09 80 43 16 16, Thurs.-Sat. and Mon.-Tues. 10am-6pm

The archetypal Riviera fashion boutique, Capuccino has been around for over 30 years in a prime location overlooking the sea. Clothes are all handmade in Nice, and each piece is unique. Everything has a slightly '70s, hippie-chic, Brigitte Bardot feel, perfect for the beach bars and nightlife on the Riviera. Prices range from €299 for handmade denim shorts to a few thousand euros for dresses that take a few months to make.

★ Food and Beach Bars
Le Colombier

promenade du Soleil; tel. 04 93 61 24 66; www. hotelhelios.fr; beach service Apr.-Sept. or Oct. daily

9am-6pm, lounge bar and evening restaurant service 6:30pm-11:30pm; €18-42

Le Colombier is one of the best of Juan-les-Pins's beachfront restaurants and serves large portions of pasta, fresh fish, steaks, beef Rossini (€35), their famous Jack Daniel's burger, and a full range of pizzas—the Cardinal is topped with lobster and salmon roe (€42). It feels expensive, but there can't be many better places to have a meal with your feet in the sand and a swim afterward. If Le Colombier is full, simply walk 10 m (33 ft) across the sand to Helios Plage (tel. 04 93 61 85 77), which is owned by the same company but offers a slightly smarter vibe, and also has a jetty with loungers, a shower, and wait service.

Plage La Jetée

22 avenue Guy de Maupassant; tel. 04 93 61 16 74; www.plagelajetee.com; €19-30

This spot is huge and encompasses a separate beach restaurant and promenade brasserie as well as a cocktail bar. It's a great place for a salad—Niçoise or Caesar—with local specialties including bagna cauda (raw vegetables with warm anchovy sauce) and goat's cheese with acacia honey dressing. The Assiette de la Jetée is a combined dish of lobster, salmon, smoked salmon, foie gras, crab meat, and smoked duck breast (€29). They also host beach parties and jazz evenings, and have a car-parking (voiturier) service on the weekends and every day in July and August.

Pam Pam

137 boulevard Président Wilson; tel. 04 93 61 11 05; www.pampam.fr; Apr.-Sept. daily 2:30pm-3am, Oct.-Jan. Fri.-Sat. 2:30pm-1:30am

Author Graham Greene makes a passing reference to this famous cocktail bar in his novel *May We Borrow Your Husband?* It's a Brazilian, tropical rhumerie (rum bar) and cocktail lounge with grass skirts and wooden carved parrots everywhere. The place sparkles with live shows, smiling bar staff, and great late-night sundaes.

Accommodations
Villa d'Elsa
17 avenue Docteur Dautheville; tel. 04 93 61 05 10; http://villadelsa.com; €135 d

From its street entrance, the hotel looks more like an urban apartment block than a villa, but it hides an art deco building with 15 nicely decorated rooms behind a busy avenue of boutiques, bars, and restaurants. All rooms have small kitchens. It's a good option for visitors who like to be in the thick of things.

★ Belles Rives Hotel
33 boulevard Edouard Baudoin; tel. 04 93 61 02 779; www.bellesrives.com; Mar.-Dec.; €325 d

In 1925, in an upstairs room at the Villa Saint-Louis, American writer F. Scott Fitzgerald, accompanied by his wife, Zelda, began writing *Tender Is the Night*. The villa was enlarged in 1929 and became the Belles Rives. There is no finer example of glamorous 1920s decor in the South of France, with its walnut dressers, striped walls, cabinets full of ceramic prizes, patterned carpets, and the coast's best hotel bar. The terrace of the Fitzgerald Bar, a palace of animal prints and art deco furniture, overlooks the private beach beyond the Michelin-starred **La Passagère** restaurant.

Information and Services
Office de Tourisme
Palais des Congrès, 60 chemin des Sables; tel. 04 22 10 60 01; www.antibesjuanlespins.com; July Aug. daily 9am-7pm, Apr.-June and Sept. Mon.-Sat. 9:30am-12:30pm and 2pm-6pm, Sun. 9am-1pm, Jan.-Mar. and Oct.-Dec. Mon.-Sat. 9am-12:30pm and 1:30pm-5pm, Sun. 9am-1pm

The tourist office also manages ticket sales for the town's jazz festival.

Getting There
Car
Juan-les-Pins is a 5-minute drive from the center of Antibes. The best place to park is

---

1: Juan-les-Pins beach bar 2: La Grande Plage
3: Gilberto Gil's handprint along the boulevard Edouard Baudoin for the Jazz à Juan festival
4: Capuccino boutique

---

under the **Palais des Congrès,** which houses the tourist office.

The drive from **Nice** via the A8 (exit 44) takes around 40 minutes (27 km/16 mi) and 45 minutes on the more scenic coast road, the D6007 via Antibes.

Train
Juan-les-Pins station (2 avenue de l'Estérel; www.sncf-connect.com), just a 3-minute walk from the beach, is one stop west of Antibes on the Marseille-Ventimiglia line between Cannes and Nice. Journey time from Nice is 29 minutes (€5.70); from Marseille, 2.5 hours (€37).

Bus
Envibus 1 (tel. 04 89 87 72 00; www.envibus. fr) runs from the pôle d'echanges in Antibes (beside the rail station) to Juan-les-Pins (journey time 8 min). Tickets cost €1 for a single, €8 for a 10-trip pass, and €10 for a 7-day pass.

GOLFE-JUAN
Besides a pleasant promenade (the avenue Frères Roustan), sandy beach, and two marinas full of yachts—the **Port de Golfe-Juan** and **Port Camille Rayon**—there's little to see of note in Golfe-Juan out of season. Napoleon Bonaparte landed there from the island of Elba in 1815, marking the beginning of his march to reconquer Paris after abdicating on April 6, 1814. There are two monuments commemorating his arrival.

Beaches and Water Sports
Plage Golfe-Juan
avenue du Frère Roustan; parking along seafront

A continuous stretch of pale sand runs for several kilometers along the boulevard du Littoral between Juan-les-Pins and Golfe-Juan, ending at the Port Camille Rayon marina. The beach may be only 10 m (32 ft) deep, but it's a great place for swimming and sunbathing, although there's no natural shade.

Subvision
quai Saint-Pierre; tel. 04 93 63 00 04; www. antibesjuanlespins.com; lessons €99, 10 dives €360

Subvision runs introduction-to-diving, experienced-diving, underwater-photo, and advanced night-diving lessons from the old marina in Golfe-Juan. Dives takes place all around the coast, all year long, including dives off the Lérins island of Saint-Honorat.

Food
Le Bistrot du Port
53 avenue des Frères Roustan; tel. 04 93 63 70 64; www. bistrotduport.com; Thurs.-Fri. and Sun.-Mon. noon-2pm and 7pm-10pm, Tues. and Sat. 7pm-10pm; €28-36

This is the seafront's posh choice for grilled lobsters, bouillabaisse, freshly caught fish cooked over a wood-fired grill, and sea anemones with herbs. Head chef Mathieu Allinei is one of France's maîtres restaurateurs and prepares the fish according to the variety and the season, which means either with artichokes, à la Provençale (with tomato and garlic), in white wine, baked in a salt crust, or grilled. It's one of the smarter, more sedate choices among the seafront restaurants.

Information and Services
Tourist Office
Parking du Vieux Port, avenue des Frères Roustan; tel. 04 93 63 73 12; www.vallaurisgolfejuan-tourisme. fr; mid-Sept.-mid-June Mon.-Sat. 9am-noon and 2pm-5pm, mid-June-mid-Sept. daily 9am-12:30pm and 2pm-6:30pm

The tourist office has free maps and brochures on current events and activities to do in and around Golfe-Juan. They can also help you book accommodations.

Getting There and Around
Car
Golfe-Juan is midway between **Cannes** (6 km/3.7 mi, 10 min) and **Antibes** (6 km/3.7 mi, 10 min) and can be reached either by the D6007 coast road or the A8 (exit number 44).

Train
Golfe-Juan-Vallauris station (avenue de la Gare; www.sncf-connect.com) is on the railway line between Marseille and Ventimiglia. Journey time to Nice is 32 minutes (€6.40)

and to Cannes 5 minutes (€2.20). The railway station is connected to Vallauris by Envibus number 20 from place Nabonnand.

Bus
Envibus number 8 (tel. 04 89 87 72 00; www. envibus.fr) goes from Antibes bus station on place Guynemer to Golfe-Juan (journey time 15 min). **Bus 20** goes from Golfe-Juan, place Nabonnand, to **Vallauris,** place Cavasse (journey time 8 min). Tickets cost €1 per single trip, €8 for a 10-trip pass, and €10 for a 7-day pass.

Zou bus number 620 (tel. 04 13 94 30 50; https://zou.maregionsud.fr) travels from Nice Parc Phoenix to Cannes, stopping in Golfe-Juan, place Nabonnand (2-3 per hour). The journey takes just over 1 hour (€2.10). **Airport Express Bus 82** from Nice airport Terminal 2, to Golfe-Juan/Vallauris (approx. two buses per hour 8:10am-8:20pm) stops in Golfe-Juan, place Nabonnand (www.niceairportxpress. com; journey time 1 hour; single €9, children under 12 €4.50).

VALLAURIS

Vallauris's exceptional clay soil led it to become famous for its pottery, even before Pablo Picasso arrived with his arty entourage in 1946. There were 32 separate potteries in the town in 1829, and most of the population worked in the industry. There are still over 50 ceramic workshops, pottery collectives, galleries, and artists' studios in the town, mainly along the **avenue Clémenceau.**

Sights
Musée National Pablo Picasso and La Guerre et la Paix
place de la Libération; tel. 04 93 64 71 83; www. musee-picasso-vallauris.fr; Sept.-June Wed.-Mon. 10am-12:15pm and 2pm-5pm, July-Aug. daily 10am-12:30pm and 2pm-6pm; €6, under 18 free

Vallauris's national museum dedicated to Pablo Picasso is actually just one work: a huge mural, *La Guerre et La Paix,* painted in the chapel of a 16th-century former priory of the Abbey of Lérins, which became the Château de Vallauris.

In 1951, Picasso was celebrating his 70th birthday with Vallauris's pottery guild in the deconsecrated chapel and expressed his desire to paint a peace mural on its walls. Vallauris's communist mayor agreed and Picasso began the mural, naming it after Leo Tolstoy's epic novel *War and Peace*. It comprises two panels of over 100 sq m (1,075 sq ft). As visitors enter the chapel through a sliding door and head into the vaulted room, the right-hand side portrays a pastoral scene, while the left-hand side depicts the horrors of war. Picasso added another panel at the far end of the chapel to link the themes, where four figures holding hands are looking up toward a dove to symbolize the cult of peace between nations.

Madoura
rue Suzanne et Georges Ramié; tel. 04 93 64 41 74; Mon.-Fri. 10am-1pm and 2pm-5pm; free

The pottery where Picasso famously worked is now an exhibition space. Besides the Spanish artist, Marc Chagall, Victor Brauner, and Henri Matisse all visited the atelier and created pieces there. Madoura today has some of Picasso's originals and certified copies as well as a permanent exhibition of founder Suzanne Ramié's ceramics.

Food and Accommodations
Marché Provençal
place Paul Isnard; Tues.-Sun. 8am-1pm

Vallauris's fresh produce market takes place around Picasso's sculpture of the man and sheep and is a lively spectacle with local produce and flowers filling the stalls until lunchtime.

Le Café de France
1 place de la Libération; tel. 06 85 08 83 55; Tues.-Sun. noon-2pm, Thurs.-Sat. 7pm-10pm; €10-25

On the square in front of the Picasso museum and the town's war memorial, Le Café de France's terrace is protected by some giant plane trees. Produce comes from Vallauris's Marché Provençal and specialties include lamb shank with orange and saffron and cod in coconut with lemongrass.

La Bigarade
336 montée des Pertuades; tel. 04 93 63 97 78; www.labigarade.com; €105 d

This eco-friendly guesthouse has a solar-heated outdoor pool and an organic citrus orchard. The rooms have sea, garden, pool, or mountain views, and guests are invited to a table d'hôte evening meal by hosts Gislaine and Tony Damiano. Parking is free.

Information and Services
Office de Tourisme
4 avenue Georges Clémenceau; tel. 04 93 63 18 38; www.vallaurisgolfejuan-tourisme.fr; mid-Sept.-mid-June Mon.-Sat. 9am-noon and 2pm-5pm, mid-June-mid-Sept. daily 9am-12:30pm and 2pm-6:30pm

The tourist office has free maps and brochures on current events and activities to do in and around Vallauris. They can also help you book accommodations.

Getting There and Around
Car
Vallauris is a 5-minute drive from **Golfe-Juan** on the D135 (2.5 km/1.3 mi); the village is a 15-minute drive from **Antibes** (6.2 km/3.7 mi) and 40 minutes from **Nice** on the A8 autoroute (exit 44, signed for Vallauris, 25 km/16 mi).

The most convenient place to park if you're visiting the potteries and museums is in the **place Cavasse** outside the town hall.

Train
The train station at **Golfe Juan** also serves Vallauris. Take Envibus 8 or 18 between the two (see above).

Bus
Envibus (tel. 04 89 87 72 00; www.envibus.fr) numbers 8, 18, and 20 go from place Nabonnand in Golfe-Juan to Vallauris, place Cavasse (journey time 5 min). Bus route 5 takes an inland route from Antibes bus station on place Guynemer to place Cavasse in Vallauris (journey time 20 min). Tickets cost €1 per single trip, €8 for a 10-trip pass, and €10 for a 7-day pass.

Cannes, Saint-Tropez, and the Western Riviera

Cannes is the glamorous centerpiece of the
Côte d'Azur and probably what comes to mind when visitors think
of the French Riviera. Its main seafront drag, La Croisette, is lined
with majestic hotels and designer shops above a long, sandy beach,
crowded to bursting point in the summer, with pleasure-boat marinas at both ends. Farther west, Saint-Tropez is a curious blend of extravagant wealth and everyday living in what is still, essentially, a tiny
fishing village.

Cannes may have built its reputation on its film festival, but it has
become the Mediterranean hub for all media festivals. Anyone not in
a swimsuit or sailing cap will invariably have a plastic identification

Highlights

Look for ★ to find recommended sights, activities, dining, and lodging.

★ **Cannes Film Festival:** The most famous movie festival in the world takes place on Cannes's seafront every May (page 324).

★ **Rue Saint-Sauveur:** This little-known gem on the hillside of Le Cannet is full of restaurants and art galleries (page 332).

★ **Îles de Lérins:** These two delightful islands covered in pines, eucalyptus trees, and hiking paths are a short ferry ride from Cannes (page 336).

★ **Château de la Napoule:** Visit an enchanting, fantastical, neo-medieval castle on the seafront in Mandelieu-la-Napoule (page 339).

★ **Cycling the Corniche d'Or:** Enjoy a ride along the picturesque coast road between Mandelieu and Agay (page 341).

★ **Saint-Tropez:** Join film stars in the cool backstreets of the old town over a glass of rosé (page 365).

★ **Plage de Pampelonne:** This 5-km (3-mi) expanse of white sand is one of the best beaches in France (page 383).

★ **Bormes-les-Mimosas:** This unspoiled medieval village overlooking the coast has wonderful views and excellent restaurants (page 386).

★ **Les Îles d'Or:** These real treasure islands are perfect for picnics, walks, and bike rides (page 391).

Cannes, Saint-Tropez, and the Western Riviera

badge around their neck, heading to a conference at the Palais des Festivals on the seafront.

The coastline between Cannes and Saint-Tropez, with Fréjus at its center, is a fascinating area in terms of geology and wildlife. It's largely free of crowds and lacks the ostentation and pretension of the other resort towns. The beaches are sandy and interspersed with the rough crags of the Massif de l'Estérel mountain range, making for a picturesque section of the coastline that is popular with bird-watchers, divers, and motorists.

A visit to Saint-Tropez's Vieux Port is a highlight of any Riviera holiday. Brightly colored fishing boats knock against multimillion-dollar superyachts in a marina surrounded by lavishly decorated café-bars. There's a daily fish market and net-repairing

service in front of boutiques selling Tahitian pearls, handmade watches, and macramé bikinis, and, everywhere, the faint smell of sardines, paraffin, and rosé wine. The peninsula surrounding Saint-Tropez is covered in vineyards, with châteaus offering tours and tastings.

Hyères is the most westerly and oldest resort on the Riviera, a popular destination before Nice was even part of France. Writer Edith Wharton moved into the local château in the 1920s, and the town became a winter resort with health spas, a casino, and belle epoque villas overlooking the sea. When the fashionable season changed from winter to the summer after the First World War, Hyères fell out of favor, but today it is a delightful, unspoiled town.

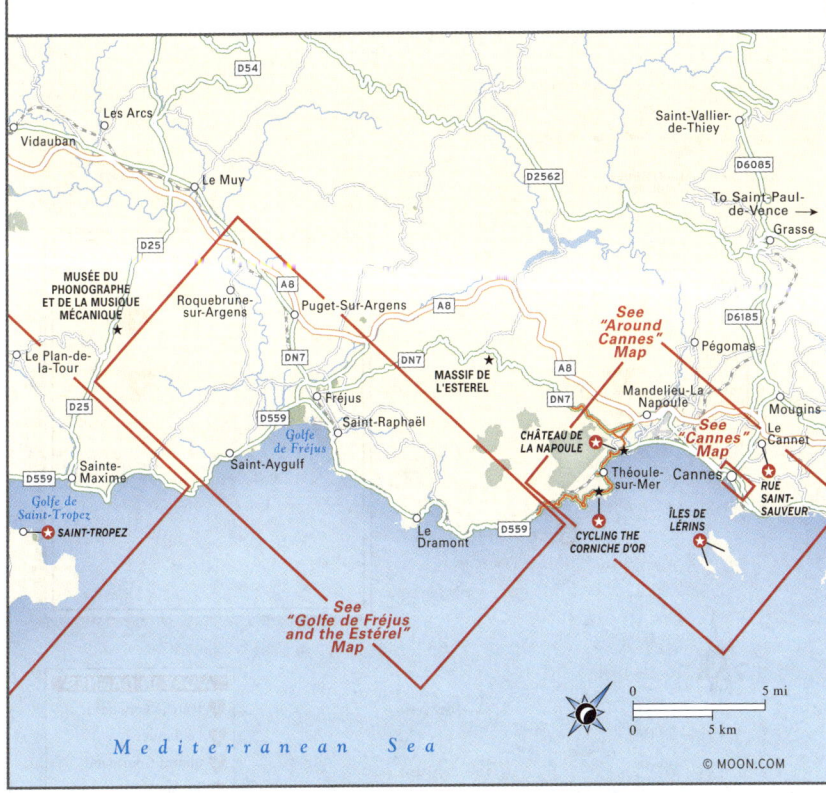

PLANNING YOUR TIME

Though **Cannes** is an expensive place to stay, with limited budget options, the attractiveness of its seafront promenade and the excellent facilities, shops, and transport links make it a good home base. **Two days** is enough to cover most of the sites in Cannes, which should include a trip to one of the **Îles de Lérins** and a visit to the **Pierre Bonnard museum** in **Le Cannet.** Spend a day in **La Mandelieu-la-Napoule** for the château and beaches or drive past **Théoule,** where the red rocks of the **Massif de l'Estérel** provide a stunning backdrop for the coast road.

Plan on spending a full day in **Fréjus,** following the **arts and crafts walk** through the town and visiting some of the **Roman ruins** and **cathedral** complex. Though Fréjus is only 10 minutes' drive from the A8 autoroute, **Saint-Raphaël** is a better place to use as a base, as it's on the Corniche d'Or, the coast road, and the fast-train line from Cannes (23 minutes) and Nice (50 minutes). It's a fun, breezy, more stylish seaside town with excellent **seafood** restaurants around the **Port Santa Lucia.**

A **week** would be ideal to cover this region, including a day in **Saint-Tropez,** a day on the

A Day in Cannes

A DAY IN CANNES

1 Marché Forville
2 Le Suquet
3 Bistrot Gourmand
4 Armani/Caffè
5 Aux Bons Enfants
6 Bâoli Club

© MOON.COM

Plage de Pampelonne, and a visit inland to **Bormes-les-Mimosas.** However, if you only have a few days, make sure you include Saint-Tropez and a trip to **Porquerolles Island** off **Hyères.** Stay in **Sainte-Maxime** where there are regular ferries to all the best destinations, a lively **night market,** and great restaurants.

The region is popular all year round, but quieter and much easier to visit **out of season,** particularly September or October when the sea is still warm.

Itinerary Ideas

A DAY IN CANNES

1 Have breakfast in one of the cafés surrounding the **Marché Forville,** Cannes's fresh produce market.

2 Wander along the steep, narrow streets of **Le Suquet** old town and head to the summit for great views of Cannes.

3 Eat a bouillabaisse for lunch at the **Bistrot Gourmand,** near the market.

4 Spend the afternoon lying on the sands of the Plage de la Croisette or people-watching in the **Armani/Caffè.**

5 Enjoy a Mediterranean-inspired supper at **Aux Bons Enfants** restaurant.

6 Finish the day with a dance at **Bâoli Club** on Cannes's Port Canto seafront.

TWO DAYS AROUND SAINT-TROPEZ

Day One

1 Base yourself in Sainte-Maxime at the comfortable **Hotel Royal Bon Repos.**

2 Take the boat to Saint-Tropez's **Vieux Port** from quai Léon Condroyer in Sainte-Maxime. Ferries leave every 15 minutes in the summer for the 20-minute crossing.

3 Enjoy lunch at the colorful **Zinzin** restaurant behind the port.

4 Walk through the old town and up to **La Citadelle.**

5 Spend the afternoon in the **Musée de l'Annonciade.**

6 Have a drink at **Senequier** on the quayside, watching the superyachts in the marina, before catching the boat back to Sainte-Maxime.

7 Cap off your day visiting Sainte-Maxime's **night markets.**

Day Two

1 Head toward Ramatuelle, a 25-minute drive along the coast, and visit the **Domaine Fondugues-Pradugues** vineyard for a tasting beside the route des Plages.

2 After stocking up on some rosé, drive to **Ramatuelle** and explore the narrow lanes of the medieval village.

3 Stay until dusk on the **Plage de Pampelonne** and walk along the seafront for an aperitif at Indie Beach.

4 Drive back to Sainte-Maxime once the roads are clear for dinner and jazz at **Le Café de France.**

5 Finish the night with a visit to the local **Casino Barrière.**

Two Days Around Saint-Tropez

Sainte-Maxime

See Detail

*Golfe de
Saint-Tropez*

See Detail

*Mediterranean
Sea*

Saint-Tropez

*Parc de
la Citadelle*

© MOON.COM

DAY ONE	DAY TWO
1 Hotel Royal Bon Repos	1 Domaine Fondugues-Pradugues
2 Vieux Port	2 Ramatuelle
3 Zinzin	3 Plage de Pampelonne
4 La Citadelle	4 Le Café de France
5 Musée de l'Annonciade	5 Casino Barrière
6 Senequier	
7 Night Markets	

Cannes

Cannes is a popular stopover for sunbathing tourists looking for a touch of glamour, and it is also Europe's premier festival and conference venue. The best introduction to the place is probably having a drink at one of the cafés along La Croisette seafront promenade. From there, visitors can watch the parade of sports cars and beachgoers, the boats heading to the Îles de Lérins, and the conference attendees walking up the steps of the Palais des Festivals. Expensive designer shops line the seafront and the roads immediately behind, but despite the opulence on display, Cannes still feels very local. The Marché Forville has dozens of stalls selling inexpensive fruit, flowers, and vegetables every morning; there are sandy public beaches on the waterfront (in front of the five-star hotels), and local transport is fast, air-conditioned, and surprisingly cheap.

ORIENTATION

Cannes's old town, **Le Suquet,** is a steep mount of cobbled streets full of restaurants, bars, and boutiques. It rises up to the west of the old port, where the **quai Saint-Pierre** is also lined with fish and pasta restaurants, ending at the **quai Laubeuf,** where boats leave for the **Îles de Lérins.** To the east of the port is the **Palais des Festivals,** centerpiece of Cannes's film festival and business tourism, a huge white steel-and-glass construction that's visible from everywhere in Cannes, and the best starting point for any visit, since it also contains the tourist office.

The **boulevard de la Croisette,** a wide seafront roadway with sandy beaches on one side, designer shops and palatial hotels on the other, starts behind **Square Hahn,** with its old-fashioned carousel. Parallel with the promenade are the **rue Félix Faure, rue d'Antibes,** and, set back a little in front of the **Marché Forville,** the pedestrianized **rue Meynadier,** all great for shopping. Heading

eastward along La Croisette is another marina, **Port Canto,** which ends at **Palm Beach,** a casino and nightclub complex on the Cap de la Croisette.

SIGHTS
Promenade de la Croisette

Originally called the boulevard de l'Impératrice, La Croisette takes its name from the little cross (*crouseto* in Provençal) that marked the place where the pilgrims would gather on the mainland before heading to the monastery on the island of Saint-Honorat, 3 km (1.8 mi) offshore. The cross is in an open monument, facing the Îles de Lérins in the parking lot behind Palm Beach. The promenade today is a place to stroll with your poodle, ponder which beach to land on, and dodge the many joggers and in-line skaters who make the return trip to Palm Beach. The seafront walkway is called the promenade de la Croisette and the adjacent roadway the boulevard de la Croisette, but the two tend to be interchangeable in the minds of locals and tourists alike.

Le Palais des Festivals

1 boulevard de la Croisette; tel. 04 92 99 84 00; https://en.palaisdesfestivals.com

The Palais des Festivals et des Congrès de Cannes is so large that it's best viewed from the ferry to the Îles de Lérins, where the ensemble of gleaming white and glass-fronted structures looks like a just-landed spaceship. Inside is a casino, 18 separate auditoriums, and the tourist office. Each year it holds dozens of events, concerts, shows, and conferences, highlighted by the NRJ Music Awards and the Cannes Film Festival.

During the film festival, stars are dropped off in their limousines at the foot of the red-carpeted staircase and climb up to the main cinema to watch themselves on the big screen. Amid the melee and squeals

Cannes

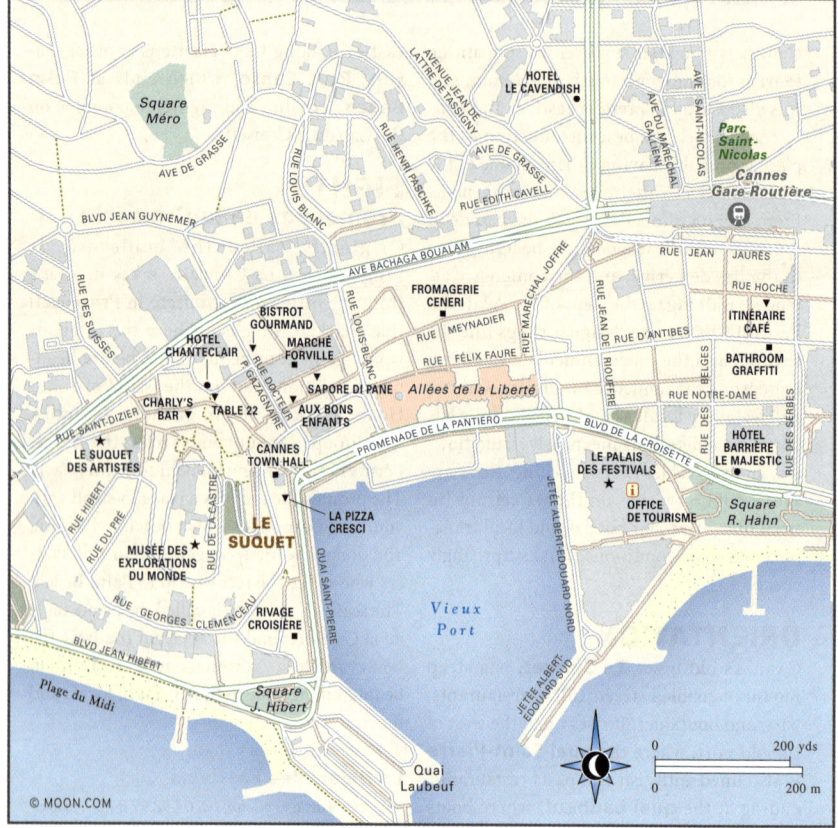

of the festival premieres, the best views of the red carpet are from the terrace of the **Hotel Barrière Le Majestic** opposite or behind the stepladders of the awaiting paparazzi.

La Malmaison

47 boulevard de la Croisette; tel. 04 97 06 45 21; www. cannes.com

La Malmaison is Cannes's center for modern and contemporary art, housed in a detached belle epoque villa surrounded by palm trees on La Croisette. It has a substantial permanent collection and runs 2-3 temporary exhibitions each year featuring 20th- and

21st-century artists, including sculptors and photographers.

At the time of writing the building was under reconstruction but was set to reopen at the start of 2025.

Le Suquet

Le Suquet is Cannes's most touristy district. Cobbled, winding streets, which originally housed fisherfolk, are now full of bars, restaurants, and gift shops. From Marché Forville or the port, it's a 10-minute walk via **rue du Suquet** and **rue Saint-Antoine** to the summit, where there's an 11th-century fortified priory, now the Musée des Explorations

du Monde, with a tower to climb, a chapel, a shady square and a Hollywood-style *Cannes* sign.

Le Suquet des Artistes

7 rue Saint-Dizier; tel. 04 97 06 45 21; www.cannes. com; July-Sept. daily 10am-6pm, Oct.-June Tues.-Sun. 10am-1pm and 2pm-6pm; €4, under 18 free

Le Suquet des Artistes is an arts space with workshops and galleries for local artists to meet and display their works in Cannes's former morgue. It was built in the 16th century and, today, its narrow white corridors bring an enigmatic ambience to the art exhibitions.

Musée des Explorations du Monde

place de la Castre; tel. 04 89 82 26 26; www.cannes. com; Oct.-Mar. Tues.-Sun. 10am-1pm and 2pm-5pm, Apr.-June Tues.-Sun. 10am-1pm and 2pm-6pm, July-Aug. daily 10am-7pm, Sept. Tues.-Sun. 10am-1pm and 2pm-6pm (open until 9pm Wed. June-Sept.); €6.50, under 18 free

The museum, inside the fortified priory at the top of Le Suquet, houses a collection of artifacts from the Himalayas, the Pacific Islands, the Arctic, Mediterranean antiquity, and pre-Columbian ceramics. The original collection was donated to Cannes by Baron Lycklama in 1877 and greatly enlarged from other travelers' private collections. Three rooms are

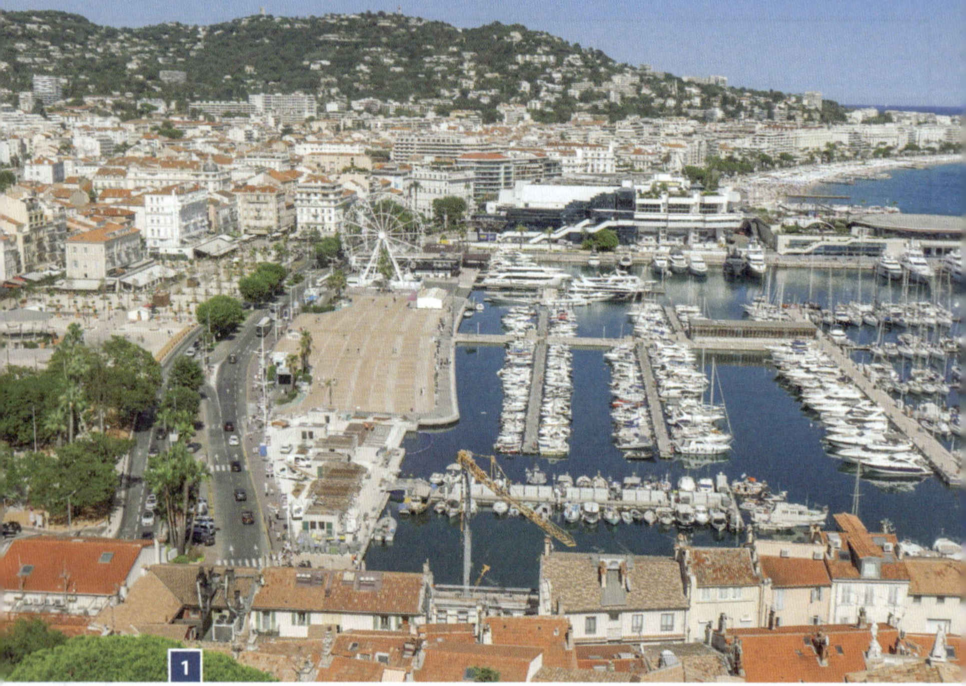

1

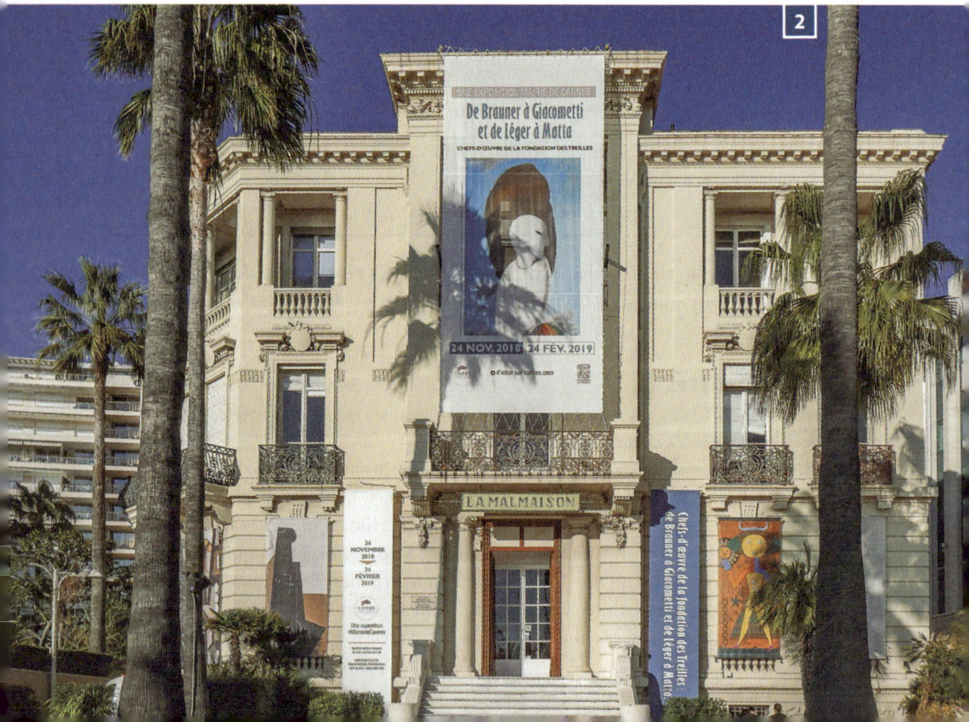
De Brauner à Giacometti
et de Léger à Matta
CHEFS-D'ŒUVRE DE LA FONDATION DES TREILLES
24 NOV. 2018 24 FÉV. 2019
LA MALMAISON
24
NOVEMBRE
2018
24
FÉVRIER
2019
Chefs-d'œuvre de la fondation des Treilles :
de Brauner à Giacometti et de Léger à Matta

2

Best Places to People-Watch

ARMANI/CAFFÈ

42-43 boulevard de la Croisette; tel. 04 93 99 44 05; www.armani.com; daily 9:30am-9pm

The Armani/Caffè, set among the haute couture fashion houses on the north side of La Croisette, is the ultimate place to people-watch. Tables can be hard to come by but it's worth the wait, if only for sipping a cappuccino while gazing at the passing fashionistas and pearlized sports cars.

ITINÉRAIRE CAFÉ

10 rue Hoche; tel. 04 93 99 68 14; www.facebook.com/itinerairecafe; Mon. 8:45am-7:30pm, Tues.-Sat. 8:30am-7pm

This is a proper café for coffee connoisseurs, with baristas preparing hand-roasted varieties and offering a large selection of teas to take away. Outside tables are great for being among the crowds on rue Hoche, which is full of cafés, pastry shops, and fashion boutiques.

Armani/Caffè

SAPORE DI PANE

8 rue du Marché Forville; tel. 04 93 38 66 97; daily 5am-8pm

Sapore di Pane is the perfect place for watching the market-goers and traders in the mornings alongside the Forville market. The Italian-style café also sells bread, cakes, olive focaccia, and pizza slices at reasonable prices. It's a buzzy, lively place where it's easy just to sit there and order breakfast, brunch, and then lunch alongside great coffee and juices.

devoted to 19th-century seascapes and landscapes of Cannes and its surroundings, and the Chapelle Sainte-Anne attached, which was a place of worship for the original monks who occupied the building, has a collection of over 400 musical instruments from around the globe. Walk up the 109 steps of the 11th-century watchtower for excellent views of the Riviera.

Cimetière du Grand-Jas

205 avenue de Grasse; tel. 04 97 06 41 52

One of the most monumental cemeteries in the South of France is the Grand-Jas on the hillside above Cannes. *Grand-Jas* is a Provençal term meaning "the big sheep-pen" or "a place to rest." Building began in 1866.

Former British Lord Chancellor Brougham is buried in the walled English section, as is Sir Thomas Woolfield, who bought up much of the city in the 1840s, encouraging his countrymen to follow him to France's south coast. Woolfield's gardener, John Taylor, who started one of the world's most prestigious real estate agencies, is also buried there, as is Carl Fabergé, creator of the first imperial Easter egg for the Russian court in 1885, and France's inspector-general of historical monuments and author of *Carmen,* Prosper Mérimée. Pablo Picasso's first wife, ballerina Olga Khokhlova, is in the allée coté ouest alongside Cannes's most celebrated artists, singers, mayors, Resistance heroes, explorers, poets, and priests.

1: view of Cannes port from the top of Le Suquet
2: La Malmaison

FESTIVALS AND EVENTS

International Games Festival

Le Palais des Festivals, 1 boulevard de la Croisette; www. festivaldesjeux-cannes.com; Feb.; free

The Games Festival takes place for three days every February in the Palais des Festivals and includes video-gaming as well as more traditional board games, trading cards, construction toys, and war, fantasy, and simulation games. There are hundreds of exhibitors, tournaments, and conferences during the festival, which attracts over 100,000 visitors.

★ Cannes Film Festival

Le Palais des Festivals, 1 boulevard de la Croisette; www. festival-cannes.com; May

For global coverage and media interest, the Cannes Film Festival dwarfs every other event taking place in the South of France. For two weeks in May, Cannes is overtaken by the film competition, with full hotels, full car parks, full restaurants, increased security, and a wild passion for selfies, self-promotion, and wheeling and dealing at the world's largest international film market. If you want to be there during the festival, you need to book a hotel room at least six months in advance, and be prepared for lots of queuing, shoving, overpriced meals, and impatient autograph-hunters. However, the rewards are considerable—photo opportunities of stars on red carpets, blissfully warm evenings, late-night screenings, and chance access to private views and parties in hillside villas. Besides the official selections, there are special screenings, a classics festival, a shorts festival, and **cinéma de la plage,** screenings that are open to the public at 9:30pm on the sand. A huge screen is erected on the beach near the Palais in front of several hundred deck chairs, and tickets are available free of charge in the tourist office on the day of the screening.

Wearing formal evening-wear and hanging around the bottom of the Palais's red carpet can sometimes get you a free pass (from a VIP walking past with a spare ticket). Films in the Semaine Internationale de la Critique (Critics' Week) tend to be more avant-garde, but tickets are easier to come by for nonaccredited members of the public (visit the Espace Miramar at 35 rue Pasteur for details; www. semainedelacritique.com). Tickets are also available for the Quinzaine des Réalisateurs (Directors' Fortnight; www.quinzaine-realisateurs.com). The Théâtre Croisette in the JW Marriott hotel is the fortnight's main screening venue, but films are also shown at several venues around Cannes and open to the general public: the **Cinéma Les Arcades** (77 rue Félix Faure), **Cinéma Olympia** (5 rue de la Pompe), and the **Salle Alexandre 3** (19 boulevard Alexandre 3).

International Fireworks Festival

Le Palais des Festivals, 1 boulevard de la Croisette; www. festival-pyrotechnique-cannes.com; July-Aug.

The International Fireworks Festival, running since 1967, takes place over six nights in July and August and sees countries competing for the title of Best Display. Teams have to present a spectacular firework show for at least 25 minutes with an accompanying musical soundtrack. The displays are launched from five barges a few hundred meters from the shore. The public can watch from the promenade de la Croisette, with many spectators staring skyward from their own boats and superyachts anchored in the bay.

SPORTS AND RECREATION

Beaches

The sandy beaches along **La Croisette** are a mix of private and public. The public beaches are busy by 10am, or if you choose a private beach, a sun lounger and parasol cost €20-40 for the day.

Alternatively, head west from central Cannes down the **boulevard du Midi,** to the long stretch of sand toward Cannes la Bocca. Or, head east toward the Cap de la Croisette and stop at **L'Ecrin Beach** beside Port Canto. For an altogether wilder and often almost deserted swimming experience among the rocks, take the ferry to one of the **Îles de**

Cannes Film Festival

Cannes Film Festival

HISTORY

The idea for an international film festival in France was first broached in 1938 when, at the Venice Film Festival, French diplomat Philippe Erlanger witnessed judges being forced to award the prize to a Nazi propaganda documentary. Determined to inaugurate a free-choice rival to Venice, he received permission from the French Education Ministry to launch a film festival in 1939. One film was shown—*Quasimodo,* by William Dieterle—for which a cardboard replica of Notre-Dame cathedral was built on the beach.

The festival's official inauguration was delayed by the Second World War; it was finally launched in 1946 with films shown in Cannes's municipal casino.

Over the decades, different strands to the festival have been launched—**International Critics' Week,** the **Directors' Fortnight, free late-night screenings** for the public on the beach, a **shorts festival,** and **World Cinema days.**

AWARDS

Altogether around 100 films are shown during the festival fortnight, with several different prizes awarded:

- the **Palme d'Or,** for the winner of the main competition
- the **Grand Prix,** a special jury prize
- *Un Certain Regard,* for more daring works
- **Best actor, actress,** and **director**
- the **Caméra d'Or,** for the best first feature film
- the **Cinef,** for a new movie from a film school
- the **Queer Palm,** for the best LGBTQ+-related film
- the **Palm Dog,** for the best canine performance

Lérins, where the water is bright turquoise and the pine forest touches the beach.

Plage de la Croisette

boulevard de la Croisette; limited parking on La Croisette

The 800-m (2,600-ft) beach beneath La Croisette is one of the most famous strips of sand in the world, having appeared in hundreds of film festival photos. It runs from the Palais des Festivals in the west to Port Canto in the east, with public sections at both ends, and the main expanse in front of all the large hotels given over to private beach clubs. Access to the sand is via steps or a long slope at the far east end of the Plage du Festival. Lifeguards are on duty April-October, and the beach is particularly busy during the film festival (mid-May) and at night in July and August, when Cannes holds its weekly firework displays over the bay.

Plage du Midi

boulevard Jean Hibert; parking along the seafront boulevard

Not usually as crowded as the sand along La Croisette, the Plage du Midi is mainly public and full of families, with just a few private beach clubs cordoning off their chairs and sun loungers. Access is via stone steps that descend regularly onto the sand from the promenade above. The road and cycle lane run the length of the beach, as does the railway between Cannes and Cannes la Bocca.

Zamenhof Plage

boulevard de la Croisette (eastern end); limited parking on the boulevard and at Parc Croisette parking lot

Measuring 300 m (984 ft) long and 30 m (98 ft) wide, Zamenhof Plage is a private beach managed from June to September under a regie municipal system, which means the city rents out beach chairs, parasols, and lockers at a cheaper rate than the private beach clubs. Chairs and parasols can be booked for a morning or afternoon session (8:30am-1pm or 1:30pm-6:30pm; loungers €9 per day/€6.50

half day, parasols €8 per day/€4.50 half day). It's a good option for families: The water is shallow, and there are showers, changing rooms, and free toilets available.

The other municipal "private" beach is **Plage Macé** near the Palais des Festivals.

Water Sports
Rivage Croisière

20 quai Saint-Pierre; tel. 04 92 98 71 31; www.rivage-croisiere.com; May-Sept. Tues.-Sun. 10:15am-5:15pm; full-day cruise €122

Rivage Croisière has a large catamaran cruiser for trips along the coast. A full-day trip (10:15am-5:15pm) costs €122 for adults and €76 for ages 4-10. The price includes sailing in the bay of Cannes, cruising toward the **Îles de Lérins** and **Massif de L'Estérel,** swimming and snorkeling, kayaking, and relaxing on the boat with a buffet lunch. Half days with lunch cost €88 for adults and €65 for children.

Ecomusée Sous-Marin

Île Sainte-Marguerite; www.cannes.com; free access but no boat mooring

Just off the southern side of the île Sainte-Marguerite are six giant underwater sculptures by British artist Jason deCaires Taylor. The "museum" is 84-132 m (275-430 ft) from the coast at a depth of 3-5 m (10-16 ft), and visitors can swim above and among the sculptures if they are equipped with a mask and snorkel. Each 2-m (7-ft) head weighs around 10 tonnes (11 tons) and is intended to act as an artificial reef and refuge for marine life. Ferries leave from Cannes's quai Max Laubeuf to Sainte-Marguerite, from where it's a 20-minute walk to the southern side of the island.

Cycling

The long, steep ascents from the town heading north and the proliferation of traffic do not make Cannes ideal for bikes, but cycling along the coast is fun and popular, and there are intermittent dedicated cycling paths in both directions.

Energy Bike

15 boulevard Général Vautrin; tel. 04 89 82 78 02; https://energy-bike.fr; daily 10am-7pm (until 6pm Sun.); €15/half day, €20/full day, €90/week

Energy Bike rents standard bicycles, electric bikes (€20/€30 for half-day/full-day rental or €175 per week), and electric mountain bikes (€40/€60 for half-day/full-day rental or €350 per week) as well as child seats, baskets, and pavement scooters for zipping around Cannes.

Booking Bikes

19 avenue Maréchal Juin; tel. 04 93 93 30 34; https:// bookingbikescannes.com; daily 9:30am-6:30pm; from €14/day

Booking Bikes has a large range of town bikes, carbon-framed racing bikes (€55/day), and electric bikes (€29/day). They also rent out scooters and motorbikes at competitive rates. Booking and payment is done online.

Golf
Golf Country Club de Cannes-Mougins

1175 avenue du Golf; tel. 04 93 75 79 13; www. golfcannesmougins.com

Super-swish Cannes-Mougins's 18-hole golf course is a private club reserved for members, but the club makes some starting times available for visitors (€160 greens fee) as long as they have the correct shoes, attire, license, and handicap (max 36). Guests can book a round by calling the pro shop.

SHOPPING

Cannes is a magnet for fashion boutiques, shoe shops, jewelry stores, and chocolatiers. While the stores on **La Croisette** are full of designer wear, **rue d'Antibes, rue Meynadier, rue des Serbes, rue Hoche,** and **rue du Commandant André** are best for holiday shopping and high-quality souvenirs.

Marché Forville

6 rue du Marché Forville; www.marcheforville.com; produce market Tues.-Sun. 7:30am-1:30pm, antiques market Mon. 7:30am-5:30pm

Cannes's art deco market was built in the mid-1930s and provides visitors and local restaurateurs with an unparalleled choice of fresh produce. Along with fruits and vegetables, stalls sell olives, fresh meat, charcuterie, cheese, locally caught fish, spices, and cut flowers. The entire area is pedestrianized, and the covered market is surrounded by food shops, small cafés, and bars—it's a lively place to spend the morning and is especially nice in fig season (late autumn).

Bathroom Graffiti

52 rue d'Antibes; tel. 04 93 39 02 32; www. bathroomgraffiti.com; Mon.-Sat. 10am-7pm

The rue d'Antibes is Cannes's prime shopping street, and its coolest shop is Bathroom Graffiti, a spacious interior design store. They sell everything from electronic gadgets, fluorescent power banks, snow globes, board games, and clocks to inflatables, Perspex vases, handbags, beanbags, rugs, lightboxes, T-shirts, and body lotion. The shop has several outlets in Paris, but Cannes is its only store in the South of France.

Fromagerie Ceneri

22 rue Meynadier; tel. 04 93 39 63 68; www. fromagerie-ceneri.com; Oct.-Dec. and Feb.-Apr. Tues.-Sat. 8am-7pm, Sun. 8:30am-12:30pm, May, June, and Sept. Tues.-Sat. 8am-7pm, Sun. 8:30am-12:30pm, Mon. 10am-6pm, July-Aug. Tues. Sat. 8am-7:30pm, Sun. 8am-1pm, Mon. 8am-7pm

Ceneri sells over 300 different cheeses from all over Europe, but its specialty is a brie with truffles that goes for €70 a kilo. The wood-paneled shop has been a fixture in Cannes since 1968, run now by third-generation Hervé Ceneri, who ensures that cheese remains a serious business in the South of France. They have a goat's cheese with basil, paprika-wrapped Boulette, and Camembert with Calvados, and they even sell Irish Cashel Blue.

FOOD

Cannes has restaurants to suit every budget, from fast-food joints and organic ice cream parlors to beachside terraces and Michelin-starred restaurants.

French
Le Jardin

15 avenue Isola Bella; tel. 04 93 38 17 85; www.lejardin-cannes.fr; daily noon-2:30pm and 7pm-11:30pm; €12-26
Run by the same family for over 30 years, in what used to be Cannes's boulodrome, Le Jardin has a particular way of doing things. The dishes are a roll call of French cuisine: herring filets with warm potatoes, magret de canard, leek vinaigrette, and a chocolate mousse or Agen prunes to finish. Dishes are often grilled with panache in front of the diners, and pizzas are cooked in the wood-fired oven. The Darmon family's restaurant is huge, with a large garden terrace that is romantic and always full. A three-course lunch runs €22.

★ Aux Bons Enfants

80 rue Meynadier; www.aux-bons-enfants-cannes.com; Tues.-Sat. noon-2pm and 7pm-10pm; €37
On the pedestrianized rue Meynadier, Aux Bons Enfants offers one of the most authentic and satisfying dining experiences in the city. Many of the rustic recipes have been served since the place opened in 1935. Chef Luc Giorsetti's grandfather, who opened the original restaurant, had secured a job as a garçon de café on a cruise liner in 1912, but went out drinking with his friends to celebrate and woke up too late the next morning to catch his ride. The name of the ship? The *Titanic*! Fresh vegetables come from the nearby Marché Forville, where Luc used to play as a boy. Mains include squid daube, terrine of roast artichoke with cheese and pancetta, black pudding with a mustardy apple sauce, and Andouillette sausages from Troyes. There's no phone, so you can't book ahead, and they do not accept credit cards, but turn up and there's usually room. Just remember to bring cash (there's an ATM around the corner).

La Cave

9 boulevard de la République; tel. 04 93 99 79 87; www.lacavecannes.com; Tues.-Fri. noon-3pm and 7pm-11pm, Sat. and Mon. 7pm-11pm; €38-60
La Cave has been a dining staple in Cannes since 1989. Its brass and pale wood fittings, glass dividers, chalkboard menus, and leather banquettes give it a supremely French bistro feel. Kitchen staff and waiters wear stiff aprons, and the wine list has over 1,000 varieties. Specials include artichoke salad, lamb shank, and haunch of tender veal.

Bobo Bistro

Mediterranean
Bobo Bistro

21 rue du Commandant André; tel. 04 93 99 97 33; https://bobobistro.com; Mon.-Sat. noon-4pm and 7pm-11:30pm, Sun. 7pm-11pm; €18-37

One block from La Croisette, Bobo Bistro is always packed, attracting a mixture of savvy tourists and locals thanks to its great food and friendly ambience. The walls of the dining area are decorated with colored cotton bobbins and mirrors in rope frames, and it's furnished with schoolhouse chairs and tiled tabletops. Ivy overhangs the outdoor terrace and gives it a vintage Paris-meets-the-Mediterranean feel.

★ Bistrot Gourmand

10 rue du Docteur Pierre Gazagnaire; tel. 04 93 68 72 02; www.bistrotgourmandcannes.fr; Tues.-Sat. noon-2pm and 7pm-9:30pm, Sun. noon-2pm; €20-41

Every town in France will have a "bistrot gourmand" somewhere, but few reach the culinary heights of this one in Cannes. Head chef Guillaume Arragon obtains his produce from Marché Forville just around the corner and offers diners local fare, such as a mushroom ravioli with truffle and parmesan cream, sea urchins from Cannes with a chestnut foam, ceviche with coriander and hummus, bouillabaisse fish stew, and an orange and Grand Marnier crème brûlée to finish. A three-course lunch is €33.

La Môme

6 impasse Florian; tel. 04 93 38 60 95; https://lamomecannes.com; daily 7:30pm-12:30am; €35-80

One of the traditional classy choices in Cannes, La Môme is on a pedestrianized street just off La Croisette and specializes in Italian-Mediterranean cuisine, served with a bit of glitz—lots of flambés and tableside carvings. Eat either in the snazzy dining room or on a large outdoor terrace. Restaurant staples include truffle risotto, Wagyu beef, a Corsican surf and turf, duck in maple syrup, and a strawberry and basil pavlova for dessert. It's

flashy and fun, but waiting times can be long, especially during festival times. La Môme also has its own beach club.

Table 22

22 rue Saint-Antoine; tel. 04 93 39 13 10; www.restaurantmantel.com; Tues.-Sat. 7pm-10pm; €90-125

The equivalent of one flight of stairs up Le Suquet hill, Table 22 is a smart, bistronomic restaurant with pastel-toned furniture and soft lighting. The food, meticulously prepared by chef Noël Mantel, is creative and delicately balanced with a touch of Provençal flair, making it the top choice along the rue Saint-Antoine. It's at the pricey end of a traveler's budget, but the service is refined and the restaurant offers vegan, vegetarian, and gluten-free options.

Italian
La Pizza Cresci

3 quai Saint-Pierre; tel. 04 93 39 22 56; www.pizza-cresci-cannes.com; daily noon-midnight; €9-16.50

This huge Italian restaurant overlooking the marina began from humble roots. In 1956, Francis Cresci opened his first restaurant in Nice, with the idea of serving pizzas "in the street," a new concept to France. It was an immediate success, and he opened the Cannes restaurant in 1960, which became part of a small empire of Italian restaurants and hotels on the Riviera. They do 13 different pizzas—served in giant halves and cooked in a show-off oven beside the front entrance. It's an old-school pizzeria, decorated with black-and-white photos and wall murals, making for a lively ambience. They also serve salads, pastas, soups, fish, meat dishes, and Italian desserts. There are rooms upstairs if you find you've had too many margaritas.

BARS AND NIGHTLIFE

Besides the beach bars, which often stay open until dawn, Cannes has a huge range of nightlife, everything from late-opening Irish pubs to neon-lit discotheques full of movie stars.

Bâoli Club

Port Canto, boulevard de la Croisette; tel. 04 93 43 03 43; https://baolicannes.com; Apr.-Oct. restaurant daily 8pm-close, club daily midnight-5am, Nov.-Mar. weekends only

Bâoli is Cannes's most celebrated nightclub, a popular meetup place for the Riviera's pretty young things, but it also attracts its fair share of sheiks and besuited conference-goers. Bâoli also has a restaurant and a rooftop bar, Cloud Nine (open June-Sept.), complete with soft banquettes, a cocktail bar, trays of nibbles, and hanging wicker chairs.

Charly's Bar

5 rue du Suquet; tel. 04 93 68 30 66; daily 5pm-2:30am

Charly opened his bar in a cave in 2006, having previously worked as a car mechanic, toy salesman, singer, and model. There's a wine and champagne bar annex next door—choose depending on what mood you are in. Charly's is lively and noisy, but it's a fun place to have a cocktail and tapas, dance, and listen to live DJs (you can hear the music from the port).

ACCOMMODATIONS

Under €100

Parc Bellevue Camping

67 avenue Maurice Chevalier; tel. 04 93 47 28 97; https://parcbellevue.com; late Mar.-Oct.; mobile homes €270-850/week, sites for camper vans and tents from €33/night

Cannes's only campground is a 5-acre, three-star site within 10 minutes' drive of the center. Bellevue has six types of mobile homes and good-size pitches for camper vans, caravans, and tents in a shady park. It has a large swimming pool, a kids' water feature, boules and volleyball courts, table tennis, and table football. There's a grocery store for provisions, a bar-restaurant that also does takeaway food, free Wi-Fi access, and karaoke and dancing in the evenings.

Hotel Chanteclair

12 rue Forville; tel. 04 93 39 68 88; www.hotelchanteclair.fr; €68 d

Behind the Marché Forville at the foot of Le Suquet, the Chanteclair is a haven of Provençal calm. The green-shuttered building

has 15 rooms—singles, doubles, and triples—with showers and some with private baths. Rooms are on the small side, but the price is unbeatable for central Cannes. It's clean, welcoming, and popular with budget travelers. Breakfast is available in the courtyard, although there are lots of coffee shops around the marina, market, and Le Suquet, all a stone's throw away.

€100-200

Hotel Le Cavendish

11 boulevard Carnot; tel. 04 97 06 26 00; www.cavendish-cannes.com; €175 d

A few blocks from the seafront at the bottom of the boulevard Carnot, four-star Le Cavendish is a listed belle epoque-art nouveau residence and a paean to old-fashioned charm. Owners Christine and Guy Welter have enhanced the place with charming touches such as lavender-scented sheets, real candles on the marble stairwell, parquet flooring, and an antique lift. Guests are invited for complimentary drinks each evening in the lounge bar.

Over €200

★ Hotel Barrière Le Majestic

10 boulevard de la Croisette; tel. 04 92 98 77 00; www.hotelsbarriere.com; €430 d

The closest of the luxurious seafront hotels to the Palais des Festivals is Le Majestic, which has a huge pool on its decked terrace that is covered over during the film festival to guarantee more space at the bar for starlets, directors, and TV journalists. The hotel has 350 rooms and suites, a brasserie, several beach restaurants, and the art deco Bar Fouquet's Cannes, complete with gold-leaf columns, black lacquer tables, and velvet armchairs.

Hotel Martinez

73 boulevard de la Croisette; tel. 04 93 90 12 34; www.hotel-martinez.com; €615 d

Opened in 1929, the Martinez is the art deco gem of La Croisette, a palace dedicated to luxurious stays with a touch of nautical chic. Le Sud is the main dining room, but there's also the two-Michelin-starred La Palme d'Or restaurant on the 1st floor, the Martinez bar (with 22 different types of gin), Le Jardin terrace restaurant, La Plage du Martinez beach bar, and L'Oasis wellness center.

INFORMATION AND SERVICES

Office de Tourisme

Palais des Festivals, 1 boulevard de la Croisette; tel. 04 92 99 84 22; www.cannes-destination.fr; daily 9am-7pm

The tourist office has free maps and brochures on current events and activities to do in and around Cannes. They can also help you book accommodations.

GETTING THERE

Air

Cannes is 27 km (16 mi) from **Nice Côte d'Azur airport** (www.nice.aeroport.fr). **Airport Express Bus 81** from Terminal 2 (every 45 min) stops at **Cannes railway station** (tel. 0800 06 01 06; www.niceairportxpress.com; journey time 50 min; adults €19.40, under 12 €4.50).

All the major **car rental** agencies have offices in front of Terminal 2, including Avis, Budget, Europcar, Hertz, and Sixt.

Cannes-Mandelieu airport (245 avenue Francis Toner; tel. 0820 426 666; www.cannes. aeroport.fr) is used for general and business tourism and is France's second-busiest business travel airport after Paris-Le Bourget. **Palm Bus A** travels from Cannes railway station to Cannes-Mandelieu airport (journey time 20 min, departing approximately every 20 min).

Car rentals are also available at Cannes-Mandelieu airport from **Europcar** (tel. 04 92 19 01 90; www.europcar.com).

Car

Cannes is on coastal road D6007 between Antibes and Mandelieu-la-Napoule. It takes around 40 minutes to drive from the center of **Nice** to Cannes (32 km/20 mi) and 10 minutes from the center of **Antibes** (11 km/7 mi). Take exit 41, Cannes la Bocca, or exit 42, Cannes/Mougins, of the A8 autoroute and drive down the D6285 toward the sea.

Street **parking** is limited, but there are several large car parks. **Parking Palais** (1 boulevard de la Croisette; first 2 hours free on weekdays) is closest to the Palais des Festivals, and **Parking Suquet-Forville** (7 rue Louis Pasteur) is closest to the old town.

Train

Cannes SNCF railway station (Gare de Cannes, rue Jean Jaurès; tel. 36 35; www. sncf-connect.com) is five blocks from the sea. **Ouigo** and **InOui** high-speed TGV trains stop at Cannes on the route between Ventimiglia and Marseille, while local TER trains stop on the Grasse-to-Menton route. Travel time to Nice is 27 (TGV; €9.20) or 38 minutes (TER; €7.90) and to Marseille just over 2 hours (€37) There is a free drop-off and pickup zone (dépose-minute).

Bus

Zou bus line 620 (tel. 04 13 94 30 50; https:// zou.maregionsud.fr) leaves from Nice Parc Phoenix to Cannes railway station every 15 minutes with a less-frequent service on Sundays. Journey time from Nice is around 75 minutes (€2.10 single).

GETTING AROUND

Cannes is easy to walk around and find your whereabouts because there is a grid pattern of roads behind the seafront promenade, with rue Meynadier being mostly pedestrianized. Le Suquet, the old town, is steep but is an easy walk to the top.

Local Bus

Cannes's bus network is called **Palm Bus** (tel. 0825 825 825; www.palmbus.fr), with 26 routes covering the entire Cannes and Pays des Lérins area. **Palm Night** offers five routes on a nighttime service (8pm-2am depending on route). The **bus station** (gare routière) is located at the foot of Le Suquet between the mairie (town hall) and the quai Saint-Pierre. Tickets cost €1.60 for a single journey (€13 for a book of 10). A Pass 1 Jour (one-day pass) costs €4.30, and a Pass 3 Jours (three-day pass) is €8.60.

Around Cannes

ORIENTATION

High on the hillside overlooking the bay of Cannes is **Le Cannet,** a residential quarter with a charming old town full of attractive restaurants and art galleries. It has become an artsy suburb of the seaside resort below, with a local village atmosphere and professionally painted murals on the walls depicting the French artist Pierre Bonnard, who has a museum dedicated to him in Le Cannet.

The **Îles des Lérins** are only a 20-minute ferry ride from Cannes's old port, yet they seem far removed from the glitz and glamour of the city. Scented with eucalyptus and pine trees, they are havens for taking shady walks, sitting down with a picnic, and exploring the rocky coves.

Farther west on the coast road is **Mandelieu-la-Napoule,** with a huge marina, a row of protected sandy beaches, and an interesting château to visit, rebuilt by an imaginative American couple in the 1920s. Past Mandelieu, the color of the coast changes to a magnificent rust red, first visible in **Théoule-sur-Mer,** which has a busy pleasure-boat port, a couple of beaches, and some excellent coastal walks. The coast road at Théoule begins to wind and rise up into the Estérel mountains, providing a picturesque and thrilling drive among the pines, eucalyptus trees, and some huge private villas.

LE CANNET

On first impression, Le Cannet is a hillside residential area with a traffic-clogged boulevard taking visitors down to Cannes, past automobile concessionaires and furniture warehouses. However, its old district is a delight and definitely worth leaving the coast for. Behind the Pierre Bonnard museum, a center dedicated to the French Postimpressionist artist, the gently sloping rue Saint-Sauveur is lined with great restaurants and shops. Some terrific, giant murals

of Bonnard and his dog have been painted by artist Big Ben on the walls behind the museum and outside the Saint-Sauveur car park. Inhabitants of Le Cannet have always felt very different from those of Cannes—the Cannois were fishers and traders while Cannettans were farmers. They persuaded King Louis XVI to sign a decree in 1774 guaranteeing that the two areas be administrated separately.

One of the nicest things to do is wander along the banks of the **Siagne canal** as it passes through Le Cannet. Inaugurated in 1868, it was constructed thanks to Lord Brougham, who saw the need to bring drinking water into Cannes. The canal runs for 43 km (26 mi) to the Siagne river. Follow the Sur les Pas de Bonnard walk (details from the tourist office) to reach the canal.

Sights
★ **Rue Saint-Sauveur**

Rue Saint-Sauveur has, as its name suggests, become something of a savior to Le Cannet's tourist industry. The street is lined with places to eat, jewelry shops, craft stores, ceramic studios, and even a violinmaker. One of the best views on the Riviera is two-thirds up rue Saint-Sauveur as it opens out into the **place Bellevue.** There's a small boules court, benches, and a few dozen tables for the surrounding restaurants. On the western facade is the *Oranger du Patrimoine* (Heritage Orange Tree), a fresco illustrating the 140 founding families of Le Cannet brought in by the Lérins monks to repopulate the town after the plague in the 15th century. It was painted by local fresco artist B Amooghli Saraf in 1990.

Musée Pierre Bonnard

16 boulevard Sadi Carnot; tel. 04 93 94 06 06; www. museebonnard.fr; Sept.-June Tues.-Sun. 10am-6pm, July-Aug. 10am-8pm; €7, students under 26 free, family ticket (2 adults and 2 children) €14

French artist Pierre Bonnard lived in Le Cannet from 1922 to 1947. He rented several villas before buying one in 1926 in the hills above the town, which he named Le Bosquet and where he spent the rest of his life. He produced almost 300 works there, including paintings of his own garden, interiors, nudes, landscapes, and portraits of his beloved dogs and cats. Much of the art is now in the museum, which occupies the Hotel Saint-Vianney, a four-story belle epoque villa. The world's only museum dedicated to Bonnard's works, it has over 50 originals and attempts to purchase another of Bonnard's works every year with the help of the public.

The museum has set up an app called **Sur les Pas de Bonnard** (in the footsteps of Bonnard), which details a cultural trek around Le Cannet. The 2-hour walk begins at the museum and includes the covered Siagne canal, Le Bosquet house, and art panels showing where the artist set up his easel.

Food
Le Coin Gourmet
314 rue Saint-Sauveur; tel. 04 93 45 44 70; https:// lecoingourmet.abc06.fr; Thurs.-Mon. noon-2pm and 7pm-9:30pm, Tues. noon-2pm; €26-33

Occupying a tiny corner building along the rue Saint-Sauveur, this restaurant has outdoor tables on the large terrace of the place Bellevue. The upstairs dining room can seat 25 and serves traditional Mediterranean cuisine based on fresh produce: leek vinaigrette with egg mimosa to start, followed by octopus and chorizo risotto and vanilla crème brûlée to finish.

La Villa Archange and Le Bistro des Anges
rue de l'Ouest; tel. 04 92 18 18 28; www.bruno-oger. com; bistro Mon.-Sat. noon-2pm and 7pm-10pm, Sun. noon-2pm; three-course meal €39-65; gastronomic restaurant Tues.-Thurs. 7:30pm-9pm, Fri.-Sat. noon-2pm and 7:30pm-9pm; €105-350

A little way outside the center of Le Cannet is chef Bruno Oger's bastide, a fine-dining enclave consisting of a chic 190-seat bistro and a two-Michelin-starred gastronomic restaurant. The Villa Archange used to be a manor house, and the dining and reception areas are still in intimate rooms with modern art (by the chef's wife, Hélène) on the walls and refined, bourgeois fittings. Private parking is available for 50 cars.

place Bellevue on rue Saint-Sauveur

Around Cannes

To Mougins and
Golf Country Club de
Cannes-Mougins

D809

D138

D109

D9

A8

D1009

D1109

D9

D109

D92

PARC
BELLEVUE
CAMPING

D6007

OFFICE DE
TOURISME

D6007

*Cannes-
La-Bocca*

OFFICE DE
TOURISME

BOULEVARD
DU MIDI

Cannes
la Bocca

Mandelieu-la-
Napoule

OFFICE DE
TOURISME

CANNES-
MANDELIEU
AIRPORT

A8

CAMPING
LES CIGALES

MANDELIEU2ROUES

D6098

Plage des Dauphins

D2098

Plage de Robinson

D6007

HOTEL LA
CORNICHE D'OR

LA CIGALE
LES BARTAVELLES

*Golfe de la
Napoule*

Mandelieu-la-Napoule

La Napoule

CHÂTEAU DE LA NAPOULE

MANDELIEU WATERSPORTS

Plage du Château
Plage de la Raguette

Plage de la Rague
et des Mineurs

Théoule-sur-Mer

Plage du Suveret

MARCO
POLO

LA CABANE
DU PÊCHEUR

OFFICE DE
TOURISME

Plage de l'Aiguille

PROMENADE PRADAYROL

Pointe de l'Aiguille

Théoule-
sur-Mer

*Parc Naturel
Départemental
de l'Estérel*

CYCLING THE
CORNICHE D'OR

D6098

PALAIS
BULLES

HOTEL
LE PATIO

Miramar

TIARA
YAKTSA

BAT'SKI

Cycle Route
Corniche d'Or

D135

D435

D35BIS

★ RUE SAINT-SAUVEUR
— LE COIN GOURMET
★ MUSÉE PIERRE BONNARD

LA VILLA
ARCHANGE AND
LE BISTRO
DES ANGES
▼

Le
Cannet

Vallauris

D6007

D6098

OFFICE DE
TOURISME

D803

D135

Golfe
Juan

D6007

Cimetière du
Grand-Jas

D6285

Golfe Juan

See
"Cannes"
Map

Cannes

D6007

D6007

Plage
du Midi

Zamenhof
Plage

BÀOLI CLUB

Pointe
Croisette

L'Ecrin Beach

Palm Beach

AVE MARÉCHAL JUIN

LE FORT ROYAL AND
MUSÉE DE LA MER

★ LA GUERITE

Île Sainte-Marguerite

FOURS À BOULETS ★

★ FOURS À
BOULETS

✴ ÎLES DE LÉRINS

LA TONNELLE

Île Saint-
Honorat

⚓ SAINT-HONORAT
MONASTERY

Mediterranean
Sea

0   1 mi

0   1 km

© MOON.COM

Accommodations
La Villa d'Emma

395 Ancien chemin de Mougins, Mougins; tel. 07 82 44 24 32; €215 d

Named after the owners' daughter, this family-friendly bed-and-breakfast is conveniently situated between Le Cannet and Mougins. It has great views of the bay of Cannes and a nice pool surrounded by wooden decking, sun loungers, and a fitness studio. Two upmarket bedrooms have sea views and private terraces, and there's also a modern cottage to rent with a private outside space.

Information and Services
Office de Tourisme

place Bénidorm, 73 avenue du Campon; www.lecannet-tourisme.fr; July-Aug. daily 9am-12:30pm and 1:30pm-6pm, Mar.-June and Sept.-Oct. Mon.-Sat. 9am-12:30pm and 1:30pm-5pm, Nov.-Feb. Mon.-Fri. 9am-12:30pm and 1:30pm-5pm

Le Cannet's tourist office may be the least-accessible tourist office in France, situated in a parking lot between two busy roads. It's a long walk from the historic center, but handy if you are catching the bus to Le Cannet from Nice or Cannes. The tourist office has free maps and brochures on current events and activities to do in and around Le Cannet.

Getting There and Around

Set on the hills above Cannes, much of Le Cannet is steep, but it's easily walkable and provides excellent views of the coast and rooftops of the city below.

Car

For Le Cannet, take exit 42 on the A8 autoroute signed for Cannes/Mougins, then take avenue du Campon toward Cannes before heading north on the boulevard Carnot into Le Cannet center. There is plenty of roadside parking and two central **car parks:** Parking des Orangers (31 rue des Orangers) and Parking Saint-Sauveur (10 Jard de l'Edem). It's a 10-minute drive from the center of Cannes to Le Cannet (4 km/2.5 mi).

Bus

Palm Bus (www.palmbus.fr) routes 1, 4, and 11A go between Le Cannet and the center of Cannes. Fares are fixed at €1.50 (€12.50 for a 10-journey pass), and the ride from Cannes is only 10 minutes.

★ ÎLES DE LÉRINS

Sainte-Marguerite is the largest of the four Îles de Lérins and closest to the mainland. It has a lagoon at its western end and a royal fort and maritime museum where the Man in the Iron Mask was incarcerated for 11 years. It's an easy 8-km (5-mi) walk around the circumference of the island, and there are an additional 20 km (12 mi) of walking trails. There are a few private homes on Sainte-Marguerite, but for the most part visitors can wander almost anywhere on the paths that crisscross the island and swim in the clear surrounding waters.

Some 500 m (0.3 mi) farther south is **Saint-Honorat,** which measures only 1.5 km (1 mi) long and 400 m (1,300 ft) wide. Walking is restricted to the coastal perimeter of this island, which surrounds the vineyards belonging to the monastery. Out of season, you can be sitting in your own rocky inlet under a pine tree watching the cormorants and seagulls. In the summer, however, there are so many private yachts in the waters between the two islands—the Plateau du Milieu—you can almost walk across to Sainte-Marguerite without touching the water.

The two smaller islands, **La Tradelière** and **Saint-Feréol,** are only accessible by small private boat or sea kayak.

Sights
Le Fort Royal and Musée de la Mer

Île Sainte-Marguerite; tel. 04 89 82 26 26; www.cannes-france.com; June-Sept. daily 10am-5:45pm, Oct.-May Tues.-Sun. 10:30am-1:15pm and 2:15pm-4:45pm (until 5:45pm Apr.-May); €6.50, free entry first Sun. of the month

Sainte-Marguerite's royal fort was built in the 1620s by the Duke of Guise as a way of protecting the bay of Cannes from invaders. The island was occupied by the Spanish in 1635, but it was returned to France two years later and Vauban, Louis XIV's chief military engineer, increased its fortifications toward the end of the 17th century. The pentagonal form built on top of a cliff gives it a dramatic appearance from the sea. The fort became a state prison until 1874, when inmate and former army officer Marshall Bazaine managed to escape by climbing down a handmade rope he had hung from one of the fort's gargoyles (a terrace is named after him). Prison cells including one that held the Man in the Iron Mask, are open for visits. The other aspect of the fort is a **museum** dedicated to marine archaeology and sea life; it occupies the older part of the fortifications.

Fours à Boulets

Pointe du Dragon and Pointe du Vengeur, Île Sainte-Marguerite

Of great interest to history buffs are the island's cannonball ovens (fours à boulets), of which there are two at either end of Sainte-Marguerite: one on Pointe du Dragon and the other on Pointe du Vengeur (maps are available at the ferry terminal in Cannes). The ovens were built on the orders of a young Napoleon Bonaparte and were used to protect the bay of Cannes from invading ships. Cold cannonballs were placed at the top of the oven into a channel groove and were heated as they gradually descended into the furnace. After 35 minutes, cherry-red cannonballs would be ready to fire at passing ships, setting alight their wooden decks.

Abbaye de Lérins

Île Saint-Honorat; www.abbayedelerins.com

The first abbey was founded here in 410, and but for a few violent incursions by marauding Saracens in the 700s, it was left in peace until the French Revolution. Soon after, the island was "nationalized" by Napoleon Bonaparte, forcing the Benedictine monks to move to Vallauris while a celebrated French actress known as Mademoiselle de Sainval occupied their monastery. In 1859, the Bishop of Toulon

bought back the island, and a Cistercian brotherhood has been there ever since.

Today there are 22 monks. Visitors can enter the **church,** a starkly beautiful structure, and are welcome to pray with the monks. The order abides by Saint Benedict's dictum of *ora et labora* (pray and work). They pray seven times a day, promoting brotherhood, respect, and tolerance, and they spend much of their days tending 8 hectares (20 acres) of vines, hand-picking the grapes, and aging the celebrated Abbaye de Lérins wines. Bottles are available to buy in the monastery **boutique,** along with soaps, olive oil, Christian-themed key rings, and postcards.

Food
La Tonnelle

Île Saint-Honorat; tel. 04 92 99 54 08; www.tonnelle-abbayedelerins.fr; mid-Feb.-mid-Jan. daily lunch noon and 1:45pm; €24-45

La Tonnelle serves surprisingly good food and is open for lunch every day. Its terrace is protected by large pine trees, with its eponymous barrels (tonnelles) among the tables. Dishes are Mediterranean but there are also burgers and salads. The restaurant is a 5-minute walk from the ferry port and serves wine from the abbey.

La Guerite

Île Sainte-Marguerite; tel. 04 93 43 49 30; www. restaurantlaguerite.com; Apr.-Oct. daily noon-6pm; €38-92

The largest of only a handful of restaurants on Sainte-Marguerite, La Guerite has a 1,000-sq-m (10,000-sq-ft) terrace with views back to the mainland and serves Mediterranean cuisine, prepared Greek-style by chef Yiannis Kioroglou. It's popular with the yachting crowd and is always full (reservations are needed in the midsummer) despite the high prices (part of island life). Vegetarian and gluten-free options available.

Getting There and Around

The crossing from Cannes takes 15 minutes to Sainte-Marguerite and 20 minutes to

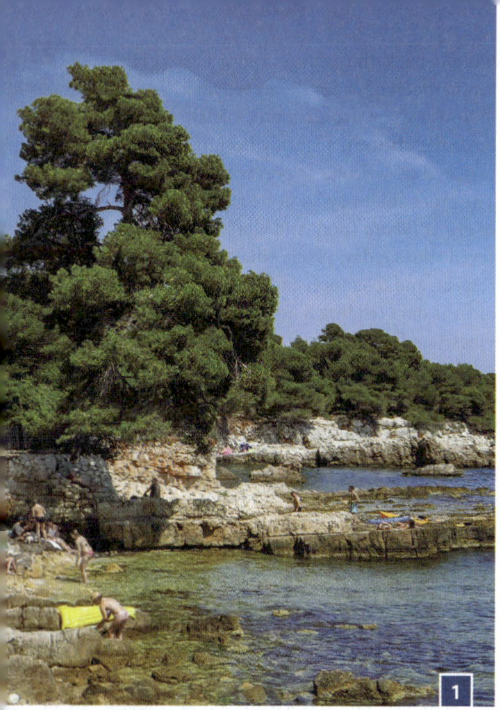
1

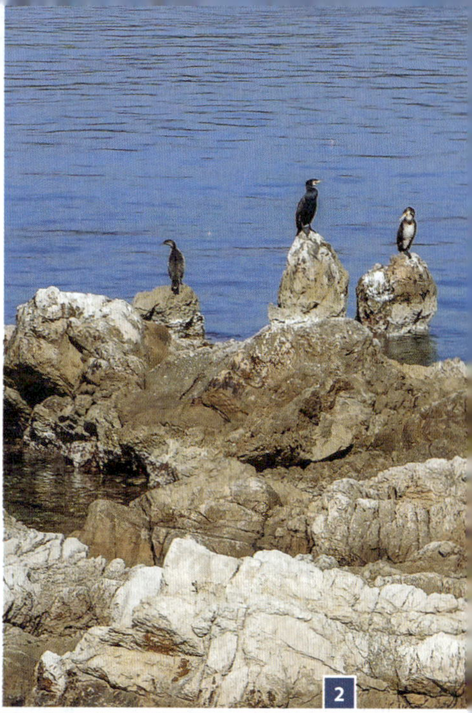
2

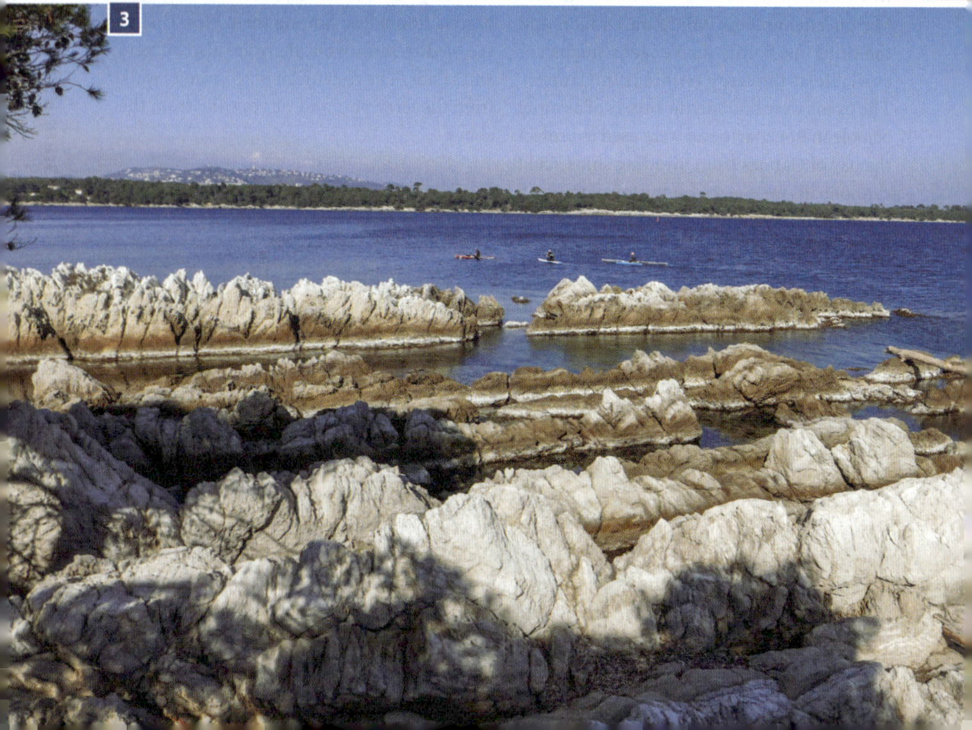
3

Saint-Honorat. Boats depart every day from the far end of the **quai Laubeuf,** 7:30am-5:30pm, returning to Cannes 7:45am-6pm approximately every half hour in high season.

Two ferry companies leave for Sainte-Marguerite: **Riviera Lines** (tel. 04 92 98 71 31; www.riviera-lines.com) and **Trans Côte d'Azur** (tel. 04 92 98 71 30; www.trans-cote-azur.com; return tickets adults €17.50, ages 13-25 €16, ages 5-12 €11, under 5 free).

Planaria runs the service to Saint-Honorat (tel. 04 92 98 71 38; www.cannes-ilesdelerins. com; return tickets adults €16.50, ages 13-18 €15, ages 8-12 €11, ages 4-7 €8, under 4 free).

Maps of the islands can be picked up from the ferry company desks at the departure point. No bikes are allowed on the ferry.

MANDELIEU-LA-NAPOULE

The resort of Mandelieu-la-Napoule is dominated by the ginger-stone, rough-hewn Château de la Napoule on the seafront. The port was founded in the Middle Ages as a fishing harbor called Epulia, and it still has a lot of charm, despite the large marina, which can accommodate over 1,100 boats. The town center is a pleasant enough shopping and eating area; the nicest thing to do is follow the **river Siagne** through Mandelieu to the sea and its protected pretty beaches. The resort is known today mainly for its Mimosa Festival, which happens every year in February, and La Napoule Boat Show, which takes place in mid-April.

Sights
★ Château de la Napoule
boulevard Henry Clews; tel. 04 93 49 95 05; www. chateau-lanapoule.com; Apr.-Sept. daily 10am-6pm, group tours at other times by arrangement; €7, students €4, under 7 free

Set on the seafront, this neo-medieval château and park are the lifework of an American couple, Henry and Marie Clews, who bought the ruins in 1919 and turned them into a fairy-tale castle ("Once Upon a Time" is carved above the main door). The Clews created hundreds of bronze, marble, and wooden sculptures of monsters that appear throughout the château. Works are monogrammed *H* or *M* (for Henry or Marie). They loved animals and owned seven white peacocks, swans, gray cranes, and a white ibis, all of which roamed freely in the gardens. Henry died in 1937 and is buried in a tomb in the castle's La Mancha tower, also designed by the couple (Henry was greatly inspired by Miguel Cervantes's *Don Quixote*). Marie was also laid to rest there when she died in 1959. The **gardens** are a terrific place to wander around, with faux-medieval monuments, fantastical stone creatures, and mature trees to sit under.

Riverside Footpath
path begins at the junction of avenue Jean Rostand and avenue Marcel Pagnol

A signpost for Les Bords de la Siagne indicates a pleasant 3.5-km (2-mi) round-trip walk from the center of Mandelieu down to the sea, following the Siagne river. It's a flat, easy walk on a gravel pathway with lots of places to stop for a picnic, waterside activities, and a children's play area, and it passes a mini-golf course on its way to the mouth of the river.

Beaches and Water Sports
Mandelieu-la-Napoule has seven beaches across its 3 km (1.8 mi) of coastline. Of the public ones, **Plage de la Rague et des Mineurs** is sandy with rocks to climb. It also has a volleyball court and toilets. **Plage de la Raguette** is sandy with a beach snack bar. **Plage de Robinson** is golden sand with a concrete jetty, snack bar, and toilets. **Plage des Dauphins** is sandy with snack bars and toilets. The **Plage du Château,** donated by the Clews family, is the most protected and has a Wi-Fi hot spot, toilets, and lifeguards during the high season. Dogs are not allowed on any of the resort's beaches.

1: Île Sainte-Marguerite **2:** cormorants on Île Saint-Honorat **3:** paddleboarders off the Îles de Lérins

La Route
du Mimosa

BORMES les MIMOSAS
Km 0

le RAYOL CANADEL
15 Km

SAINT MAXIME
42 Km

SAINT-RAPHAEL
59 Km

MANDELIEU la NAPOULE
108 Km

TANNERON
112 Km

PEGOMAS
115 Km

GRASSE
130 Km

Saint – Raphaël

Mandelieu Watersports

Plage de la Rague; tel. 06 65 48 06 06; daily 8am-8pm in high season; 1-hour sunset Jet Ski €130, 2-hour morning Jet Ski €200

Working off the Plage de la Rague, Mandelieu has all kinds of water sports on offer including Jet Skis (€90 for 30 minutes), wakeboarding and waterskiing (adults €55/lesson, children €40), and towed inflatables and parachute rides (adults €55, children €40).

La Cigale

Port La Napoule, avenue Henry Clews; tel. 04 93 90 98 66; www.lacigale-plongee.com; Mon.-Sat. 9am-7pm, Sun. 9am-1pm; lessons €75

La Cigale's 12-person aluminum boat, *La Fourmi,* leaves from La Napoule port and takes divers along the Estérel coastline and into the waters off the Îles de Lérins. Their Diving for Beginners course is €75 per session. Two "discovery" dives (equipment supplied) go for €130. Snorkeling equipment hire costs €35, or €40 with guide.

★ Cycling

Mandelieu is a popular destination for cyclists, as there is a large selection of seafront and inland terrain to explore. The tourist office has a range of cycling route maps for road and mountain biking.

Mandelieu2roues

542 avenue de la Mer; tel. 04 92 97 27 37; www. mandelieu2roues.com; €24-35/day

Mandelieu2roues has a large range of bikes, scooters, and motorbikes for rent at its shop or online. Mountain bikes are €30 per day (€20/ day for a week's rental); racing bikes and tandems are €50 per day (€30/day for a week's rental), and e-bikes run €35 per day (€22/day for a week's rental). Preferential rates are given for booking a month in advance. Repair kits and helmets are included in the price.

1: mimosa forest **2:** Route du Mimosa **3:** Château de la Napoule

Festivals and Events
Mandelieu Mimosa Festival

www.mandelieu-tourisme.com; Feb.

The Mandelieu Mimosa Festival, launched in 1931, takes place over eight days in February, during which a carnival atmosphere takes over the town. A mimosa queen is elected on the first Saturday and joins the parades of yellow flowers. Brass bands play through the streets, floats stream past, tourists throw the prickly yellow flowers, and locals stay at home cursing the traffic jams. Every day there are guided hikes up to the Pays du Mimosa, a 2-hour walk leaving from the tourist office and venturing into the hills of Europe's largest mimosa forest. Tickets for the parades go on sale in December at the Mandelieu tourist office and online, but it is still possible to watch the parade from afar without a ticket.

The festival ties in with the **Route du Mimosa** (https://routedumimosa.com), a 130-km (80-mi) eight-stage tour of the Riviera's main mimosa locations. It is best done January-March when the yellow flowers are blooming and abundant.

La Napoule Boat Show

Port La Napoule; www.lanapouleboatshow.com; Apr.

La Napoule boat show takes place in mid-April and celebrated its 34th edition in 2024, with over 400 boats on display in the marina. It is one of the biggest boat shows in France, attracting around 20,000 visitors for the four-day event.

Food
La Rotonde

391 avenue du 23 Août; tel. 04 93 49 82 60; www. larotondemandelieu.fr; Thurs.-Mon. noon-2pm and 7pm-9:30pm; €27

Perfectly placed for a visit to the château, La Rotonde serves creative French cuisine straight from the market and is a popular destination for foodies. Four- and five-course menus might include zucchini flowers in a basil and ricotta sauce, followed by thyme-infused rack of lamb with Provençal vegetables, and roasted apricots with lavender mascarpone and grilled almonds to finish.

Les Bartavelles

1 place du Château; tel. 04 93 49 95 15; www. restaurantlesbartavelles.com; Thurs.-Mon. noon-1:30pm and 7pm-9pm, Wed. 7pm-9pm; €38

Opposite the château entrance, Les Bartavelles is a popular haunt doing a good range of surf and turf. For the surf, there's mussels and fish carpaccio, scallops and creamy asparagus risotto, and a filet of turbot with lime and butter sauce. For the turf, try tenderloin beef filet in pepper and Armagnac sauce.

Accommodations

Hotel La Corniche d'Or

place de la Fontaine; tel. 04 93 49 92 51; www. cornichedor.com; €60 d

A modest hotel near Mandelieu-la-Napoule railway station, La Corniche d'Or has 12 rooms with terraces or balconies, with breakfast served on a sun terrace. Rooms are clean and tasteful and have air-conditioning and Wi-Fi, making this a good budget stay on the Riviera.

Camping Les Cigales

505 avenue de la Mer; tel. 04 93 49 23 53; www. lescigales.com; open year-round; camping from €15 per person per night, mobile homes €382-1,195 per week

One of the few local campgrounds to stay open all year, the four-star Les Cigales is located on the banks of the river Siagne. Guests can arrive by boat and moor in the private port. The campsite has its own pool and solarium, and it accommodates 43 car camping and tent pitches and 22 mobile homes. It's a 15-minute walk to the sea from the property.

Information and Services

Office de Tourisme de Mandelieu-la-Napoule

806 avenue de Cannes; tel. 04 93 93 64 64; www. mandelieu.com; Mon.-Sat. 9:30am-12:30pm and 2pm-6pm

The tourist office has free maps and brochures on current events and activities in and around Mandelieu-la-Napoule.

Getting There and Around

Car

Take exit 40, Mandelieu-la-Napoule, off the A8 autoroute, which leads directly into Mandelieu center. The resort runs along the coastal road, the boulevard du Midi Louise Moreau between Cannes la Bocca and Théoule-sur-Mer. It's 8 km (5 mi) and a 15-minute drive from **Cannes** along the coast road, and 30 km (19 mi; 25 min) from **Saint-Raphaël** using the A8 autoroute. There are numerous **parking lots** in the center and along the main routes.

Train

Mandelieu-la-Napoule **railway station** (www.sncf-connect.com) is alongside place de la Fontaine, on the line from Les Arcs-Draguignon to Ventimiglia via Nice and Menton. Journey time to Cannes via the local TER trains is 9 minutes (€2.70).

Bus

Palm Bus route 22 (www.palmbus.fr) runs through La Napoule between Théoule-sur-Mer and Cannes railway station. Fares are €1.50 single (€12.50 for a 10-journey pass; journey time around 30 min). **Palm Express Bus A** runs between Mandelieu-la-Napoule and Cannes railway station (30 minutes).

THÉOULE-SUR-MER

Théoule takes its name from *Théou-oule,* meaning the "wood of the gods." Its dense woodland and craggy, red-rock coastline make it one of the most attractive places on the Riviera. It has an excellent coastal walk where the Pointe de l'Aiguille nature park and l'Estérel forest meet at the Mediterranean, with a combined area of 600 hectares (1,482 acres).

To the west of Théoule port is Théoule Plage, backed by the promenade Pradayrol, the beginning of a seaside walk with beach restaurants and a large car park. Farther west toward Saint-Raphaël, the road follows the magnificently jagged coastline where, at the Pointe de l'Esquillon and Cap Roux, the rust-colored rocks and green vegetation tumble into the blue ocean.

For those interested in war monuments, the **Croix de Lorraine** on the avenue de la Côte d'Azur (just to the east of the Pointe de l'Esquillon) commemorates the Allied landing in 1944, the only site in the Alpes-Maritimes where troops landed.

Sights
Palais Bulles
33 boulevard de l'Estérel; www.palaisbulles.com
A 5-minute drive west on the avenue de la Côte d'Azur from Théoule-sur-Mer center, the Palais Bulles (Bubble Palace) looks like something out of a 1960s science-fiction movie set. Designed by Hungarian architect Antii Lovag for French industrialist Pierre Bernard in 1975, it is an amazing sight: organic terra-cotta-colored domes with portholes clinging to the rocks of the Estérel. The narrow, winding road beneath makes it hard to stop for a photograph, so it's best to park a few hundred meters away and walk to the nearest curve in the road. The palais has 29 rooms, a couple of swimming pools, and an outdoor amphitheater and was owned by fashion designer Pierre Cardin from 1991 until his death in 2020.

Beaches and Water Sports
Plage du Suveret
quai Edouard Blondy; parking on promenade Pradayrol or near restaurant Marco Polo
Access to Plage du Suveret is just east of the marina, 10 minutes' walk from the tourist office. It's a sandy beach, and the clear water is great for swimming. It's also a popular place for an evening picnic. Arrive early because the beach is small and gets busy during summer. Amenities include showers and a beach restaurant, **Chez Philippe** (www.restaurantchezphilippe.com; Feb.-Oct.), that has a good reputation for its seafood and friendly ambience.

Plage de l'Aiguille
Follow signs from promenade Pradayrol, where there is also a car park
A 20-minute walk from the promenade André Pradayrol in Théoule, L'Aiguille is a white-sand beach, far enough away from the center to be relatively serene during the summer. It is backed by dense forest that rises steeply up the hillside, providing some shade in July and

Plage de l'Aiguille

August. Bring water, picnic items, and beach inflatables for a relaxing afternoon away from the crowds. There's a beach restaurant-bar, **l'Aiguille Restaurant Plage** (www.laiguille-restaurant-plage.com; Apr.-Sept.).

Bat'ski

Plage de la Figuerette; tel. 06 24 23 33 37; www.batski. fr; Apr.-Sept.; €20/hour kayak rental and 5-person pedalo

Bat'ski offers a huge range of water sports on the Plage de la Figuerette. It conducts five-day sailing and introduction to wakeboarding, kayak, windsurfing, and paddleboarding courses. Wingsurfs can be rented for €70/hour; wingfoils are €90. They also run water-skiing courses and rent out catamarans, optimists, and paddleboats.

Hiking
Pointe de l'Aiguille Nature Park

boulevard de la Corniche d'Or; park open Apr.-Oct. daily 7am-8pm, Nov.-Mar. daily 8am-6pm; parking along promenade du Pradayrol in town center

Mediterranean oaks and maritime pines cover the peninsula surrounding Théoule-sur-Mer, where a 7-hectare (17-acre) nature park was created in 1961. The park, filled with eucalyptus and mimosa trees, is great for walking and biking and is split in two by the RD6098 road. The upper section rises to 200 m (656 ft) and the lower section goes down to the pebble beaches, where the red rocks plunging into the blue water are a picturesque sight.

Coastal Footpath

Distance: *2.5 km (1.5 mi) one-way*
Time: *45 minutes*
Information and Maps: *Théoule tourist office*
Trailhead: *promenade Pradayrol parking lot*

This great walk along the coastal footpath up into the Pointe de l'Aiguille park leads to picnic tables and spectacular views of the Golfe de la Napoule. The local Terre Association organizes guided walks departing every Thursday from the esplanade Charles de Gaulle to discover the delights of the Estérel.

Full details on these and other walks are available at the **tourist office.**

Corniche de l'Esterel

Distance: *12.6 km (7.8 mi) round-trip*
Time: *5 hours*
Information and Maps: *Théoule tourist office*
Trailhead: *Théoule-sur-Mer train station*

The hike begins in the center of Théoule-sur-Mer. Take the avenue du Midi and the piste des Mineurs, which continues up the Rocher des Monges to the piste des Trois Cols. The hiking pathways, steep and irregular, are marked along the **Crête des Grosses Grues, Petites Grues** to the **Col Notre-Dame** at 324 m (1,062 ft) elevation, where all trails are marked in red and white stripes. A short loop takes hikers to the **pic de l'Ours** summit (492 m/1,614 ft) before winding back down the mountain to Le Trayas railway station.

Food and Accommodations
★ La Cabane du Pêcheur

promenade Pradayrol, Plage de la Petite Fontaine; tel. 06 81 47 06 11; Apr.-Sept. daily 9am-6:30pm; €13-35

Great for grilled sardines, the "fisherman's cabin" is about 300 m (984 ft) along the promenade Pradayrol heading toward the beach at the Pointe de l'Aiguille (where there's another great restaurant for salads and fresh fish called l'Aiguille). La Cabane has a takeaway bar, a few tables on the sand, and more tables directly underneath the cliff—which provides some welcome cool in the summer.

Marco Polo

Bord de Mer, 45 avenue de Lérins; tel. 04 93 49 96 59; www.marcopolo-plage.com; daily 9am-midnight; €20-43

Marco Polo has been run by the same family on the same beachfront since 1949, seating diners in a raised room with panoramic views of the sea or at wooden tables on the beach. It has a glamorous Riviera feel with lots of glass and chrome, cacti on the deck, and a wood-slatted facade. Live bands play every night from June through August.

Hotel Le Patio

48 avenue de Miramar; tel. 04 93 75 00 23; www. lepatio.fr; €105 d

There's not much budget accommodation on this part of the Riviera, but Le Patio has 14 cheerfully decorated rooms at reasonable rates. The patio restaurant serves pizzas and local fare, plus there's a pool. The hotel is near the sailing schools and diving center and is a good base for any activity-focused holiday. Apartments are also available.

Tiara Yaktsa

6 boulevard de l'Esquillon; tel. 04 92 28 60 30; https:// yaktsa.tiara-hotels.com; €300 d

With views across the red rocks of the Estérel Massif and out to sea, the Yaktsa (part of the Tiara group, which also owns the Miramar hotel nearby) is an exclusive five-star residence. It has 21 rooms and suites, an infinity pool, a hot tub, a spa, manicured lawns, and a private beach.

Information and Services
Théoule-sur-Mer Office de Tourisme

2 Corniche d'Or; tel. 04 93 49 28 28; www.theoule-sur-mer.org; Oct.-Apr. Mon.-Sat. 10am-5:30pm, May-Sept. Mon.-Sat. 9am-7pm, Sun. 10am-5:30pm

The tourist office has free maps and brochures on current events and activities in and around Théoule-sur-Mer.

Getting There and Around
Car

Take exit 40, Mandelieu-la-Napoule, off the A8 autoroute, which leads directly into Mandelieu center, from where Théoule is a 10-minute drive along the twisting coast road. It takes about 20 minutes (9 km/6 mi) to drive to Théoule-sur-Mer from **Cannes** and 40 minutes from **Saint-Raphaël** (30 km/19 mi) along the coast road. The most central place to park is in the **parking lot** on the promenade Pradayrol.

Train

Théoule-sur-Mer railway station (www. sncf-connect.com) is on the avenue des Lérins, 1 km (0.6 mi) to the east of Théoule-sur-Mer center, on the line from Marseille to Ventimiglia. Journey time to Cannes via the local TER trains is 15 minutes (€2.80).

Bus

Palm Bus number 22 runs between Cannes and Théoule-sur-Mer via Port La Napoule (www.palmbus.fr; 35 min; €1.50 single, €12.50 for a 10-journey pass).

Inland Day Trips

Molinard perfumery mansion

SAINT-PAUL-DE-VENCE

The picturesque perched village of Saint-Paul-de-Vence is one of the most popular destinations on the Riviera. Artists Matisse, Braque, Picasso, and Modigliani all stayed there, exchanging works of art for lodging at the Colombe d'Or inn. The village is still a magnet for painters and sculptors. Its narrow medieval lanes are lined with art galleries as well as perfume shops, interior design boutiques, and walls of jasmine and bougainvillea.

Artist Marc Chagall is buried in the local cemetery, from where there are breathtaking views of the coast and surrounding countryside. Visitors can dine under the art at the **Colombe d'Or** (Place du Général de Gaulle; tel. 04 93 3 80 02; www.la-colombe-dor.com), and there are plenty of cafés, restaurants, and hotels in and around the village. A 5-minute drive away is the **Fondation Maeght** (623 chemin des Gardettes; tel. 04 93. 32 81 63; www.fondation-maeght.com), a gallery set in a pinewood forest that houses one of France's greatest collections of modern art.

Getting There

It's 20 km (12 mi) from Nice, a 30-minute drive past the Polygone Riviera shopping mall, and 28 km (17 mi) from Cannes, a 40-minute drive via the A8 autoroute. To take public transportation, catch the train to Cagnes-sur-Mer and Zou bus 622 (https://zou.maregionsud.fr) to the village.

GRASSE

With its charming old town, trickling fountains, and cobbled streets, hilltop Grasse makes for an ideal day trip from the coast. Visitors can spend a day immersed in a whole industry based on perfume and visit the grand parfumeries located around the town.

The three major perfume houses—**Galimard** (www.galimard.com), **Fragonard** (www.fragonard.com), and **Molinard** (www.molinard.com)—all have large visitor and production centers offering free tours as well as the chance to make your own exclusive scent or attend a perfume-making workshop. The Musée International de la Parfumerie (www.museesdegrasse.com) is a must-visit, but it's just one of a number of museums on fashion, history, and art. These include the Villa Fragonard and a multisensory tour of the Jardins du MIP, 2 hectares (5 acres) of rare roses, jasmine bushes, and aromatic plants.

Getting There

Grasse is 40 km (25 mi) from Nice, a 50-minute drive on the A8 autoroute, and 20 km (12 mi; 30 ||||||| /||||| ▮||| ‖‖ The train (www.sncf-connect.com) takes 70 minutes from Nice Ville station and 30 minutes from Cannes.

MOUGINS

High in the hills between Cannes and Grasse, Mougins is a perfect getaway from the noise and bustle of the coast. That's not to say it doesn't get crowded in the summer, but the village is a pastel-toned, art-inspired location with some of the top restaurants on the Riviera.

Mougins is also a thriving cultural hub with an internationally renowned **photography museum** (43 rue de l'Église; tel. 04 22 21 52 12; https://centrephotographiemougins.com) and the privately owned **FAMM** (32 rue du Commandeur; tel. 04 93 75 18 22; www.famm.com), a museum dedicated solely to women artists that opened in summer 2024.

Getting There

It's 30 km (19 mi) from Nice, a 40-minute drive on the A8 autoroute, and 8 km (5 mi; 10 min) from Cannes. Zou buses 660 and 663 (https://zou.maregionsud.fr) leave from just north of Cannes railway station and go directly to Mougins village.

ROQUEBRUNE-SUR-ARGENS

Named after its magnificent brown rock, Roquebrune feels a million miles from the Riviera. Steep, narrow lanes twist through intimate squares and past arcaded dwellings to reach the 12th-century, newly restored Église Saint-Pierre-et-Saint-Paul at the very top of the village.

The **Maison du Chocolat** (rue de l'Hospice; Tues.-Sat. 9am-12:30pm and 2:30pm-6pm; free) displays the 5,000-piece collection of chocolatier Gérard Courreau, including old enamel signs, chocolate boxes, cooking trays, and molds. It also has a popular chocolate boutique.

Roquebrune's Le Rocher, a mass of rust-red, heavily eroded sandstone covering 6 sq km (2.3 sq mi) and reaching a height of 376 m (1,200 ft) above sea level, is the site of one of France's Grande Randonnée hikes, the GR51. The most popular trek is to the **Summit of Three Crosses** (Le Sommet des Trois-Croix; 9.5-km/6-mi round-trip; 4 hours) created by French sculptor Bernar Venet and dedicated to Crucifixion paintings by the three G's: painters El Greco, Grunewald, and Giotto. The hike is a tricky walk over steep, rough ground requiring hiking boots and plenty of water. Views of the coastline are dazzling, though, as are the colors of the rocks—bright orange, red, amber, and rust—which appear to change throughout the day.

Getting There

The village is 72 km (44 mi) from Nice, a 75-minute drive on the A8 autoroute, and 44 km (27 mi; 50 min) from Cannes. By public transport, bus 60 leaves from Cannes to Fréjus, from where bus 04 goes directly to Roquebrune-sur-Argens.

Golfe de Fréjus and the Estérel

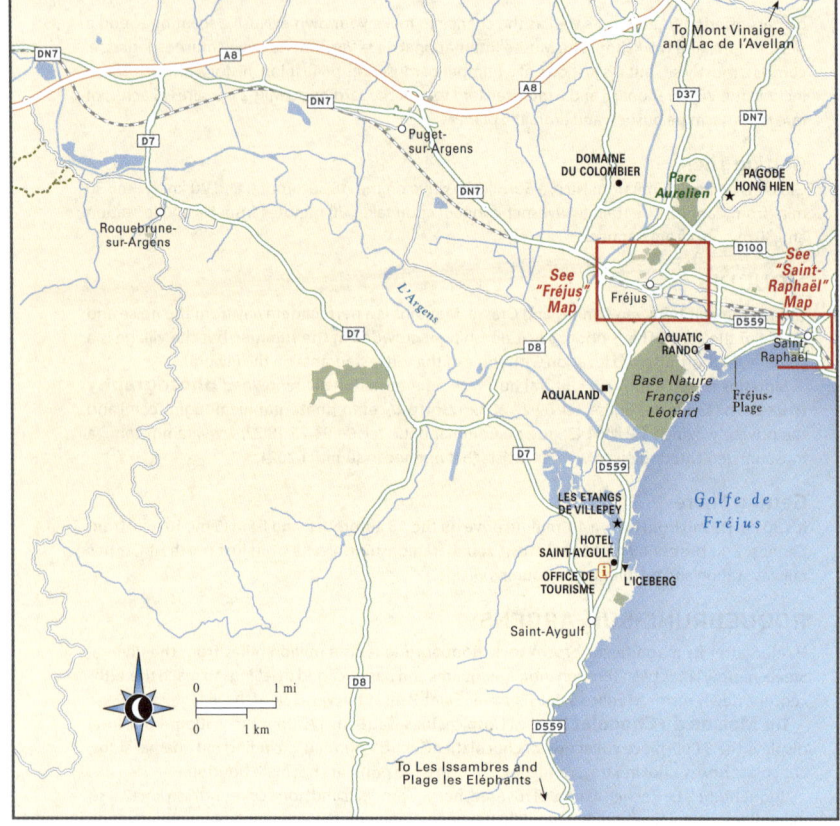

Golfe de Fréjus and the Estérel

The coastline between Théoule-sur-Mer and Sainte-Maxime, with Fréjus at its center, is a fascinating area in terms of geology and wild-life. It's largely free of crowds and lacks the ostentation of Cannes or Saint-Tropez.

ORIENTATION

Saint-Raphaël, at the western end of the Corniche d'Or coast road, is a great family resort with lively pleasure ports and a thriving water sports industry. It may lack a certain cachet, but this means staying and eating here is a lot cheaper than life on the rest of the Riviera. Inland, tourism in Saint-Raphaël's older brother, **Fréjus,** is based on a rich heritage of Roman ruins and some family-friendly water parks. The two resorts merge into each other halfway along the seafront, where the Pédégal and Garonne rivers reach the sea beside the place Kennedy.

The beaches in Saint-Raphaël and Fréjus-Plage, along the Golfe de Fréjus, are sandy and

family-friendly, but the real charm of the area lies in the jagged creeks and calanques, colored deep rust red from the majestic mountain range, the **Massif de l'Estérel,** that dominates this part of the coast and makes snorkeling and exploring rock pools a unique experience. To the southwest of Fréjus is **Saint-Aygulf,** which has the Étangs de Villepey lagoons, a breeding ground for over 200 species of birds and an interesting location for an ornithological tour.

FRÉJUS

Fréjus, founded as Forum Julii by Julius Caesar in 49 BCE, was once the most important Roman naval base in the Mediterranean. The city still has 30 listed historic buildings, including an impressive Roman amphitheater and medieval quarter, but has been overtaken by its more glamorous neighbors—Sainte-Maxime, Saint-Tropez, and Cannes—in the quest for summer visitors. What draws people to Fréjus today is its proximity to two mountain ranges, the Massif des Maures and Massif de l'Estérel, perfect locations for hikers, bikers, and nature lovers.

The heart of Fréjus's old town, or Vieille Ville, is its Cité Episcopal, made up of the cathedral, cloisters, and adjoining religious buildings. They are surrounded by squares,

Fréjus

Reyran

DN7

RUE ARMAND DUVIVIER

RUE JÉACQUES PINELLI

RUE GUSTAVE BRET

AVE DE L'AGACHON

Cimetière

Cimetière

RAMPARTS ★

ROMAN
AQUEDUCT ★

ROMAN
THEATER ★

*Clos de
la Tour*

AVE DU THÉÂTRE ROMAIN

AVE DU XVÈME CORPS

RUE JEAN BACCHI

RUE DU

*Jardin
Public
du Clos
de la Tour*

FRÉJUS
BUS STATION

POUVADOU

RUE GIRARDIN

OFFICE
DE TOURISME
ⓘ

RUE DU DOCTEUR
LOUIS TURCAN

To Roman
Amphitheater ←

RUE JOSEPH AUBENAS

*Cimetière
Saint-Léonce*

RUE PIE BERTAGNA

MUSÉE
ARCHÉOLOGIQUE

RUE JEAN JAURÈS

RUE REYNAUDE

PORTE DES
GAULES ★

GRAND CAFÉ
DE L'ESTEREL ▼

PLACE FORMIGÉ
MARKET ▼

CATHÉDRALE
SAINT-LÉONCE
AND CLOISTERS

RUE ARISTIDE BRIAND

RUE HENRI VADON

AVE DE VERDUN

RUE DU GÉNÉRAL DE GAULLE

LA CAVE DES
CARIATIDES ▼

▼ L'AMANDIER

▼ L'ENTRÉE
DES ARTISTES

D98B

R DE LÀ
JUIVERIE

● HOTEL
LE FLORE

RUE GRISOLLE

Fréjus TER
Rail Station

RUE DE CAMELIN

*Place de
Jésuites*

RUE DES SAUGES

RUE EDMOND
POUPE

*Square
des Moulins*

RUE DES MOULINS

PLATEFORME ★

BLVD SÉVERIN DECUERS

To Lanterne
d'Auguste →

CHEMIN DE LA LANTERNE D'AUGUSTE

AVE DE VILLENEUVE

RUE JEAN CARRARA

D98B

AVE DE PROVENCE

D559

© MOON.COM

0   100 yds

0   100 m

most with fountains at their centers and filled with restaurant tables in the summer. Visitors can find the occasional extravagant doorway along its medieval streets. Other interesting sites in the old town are the place Agricola, with its Gothic chapel dedicated to Saint-François de Paule, and the 16th-century town ramparts, which run behind the tourist office in place Clémenceau. Fréjus's marina is a lively area of arched bridges, hundreds of modern holiday flats, and the kilometer-long beach of Fréjus-Plage. The marina extends east to Saint-Raphaël, with a line of bars, cafés, beach shops, and water sports agencies, and to the west, the Base Nature François Léotard.

Sights

Cathédrale Saint-Léonce and Cloisters

place Formigé; tel. 04 94 51 26 30; www. cathedralefrejus.fr; June-Sept. daily 10am-12:30pm and 1:45pm-6pm, Jan.-May and Oct.-Dec. Tues.-Sun. 10am-1pm and 2pm-5pm; cathedral free, cloisters €3, EU citizens under 26 and children free

Magnificent 16th-century walnut doors give a slightly misleading entrée to this somber cathedral. Erected in the 5th century, the baptistry to the left of the entrance is one of the few remaining examples of Paleo-Christian architecture in France, and was originally designed for total immersion of those being baptized. Two interior naves are dedicated to Notre-Dame and Saint-Étienne, the latter of which has an 11th-century stone altar. Of note are the 16th-century bell tower, whose steeple is covered in yellow and green tiles, and the staircase that gives access to the upper gallery, made from stone seats taken from the Roman amphitheater.

The cathedral's cloisters, accessible through a separate entrance 20 m (65 ft) away on rue de Fleury, are exceptional due to their painted wooden ceiling. The faithful would pass through the cloisters on their way into the church and look up to see a fantastic bestiary of creatures painted within the rectangles of a larchwood ceiling (which is claimed never to rot). The 14th-century paintings depict cannons, priests, troubadours, and prominent citizens, as well as dragons and mythical beasts with human extremities and inanimate objects grafted onto their behinds. Guided tours are available in French and English.

Roman Amphitheater

rue Henri Vadon; tel. 04 94 51 34 31; www.ville-frejus. fr; Oct.-Mar. Tues.-Sat. 9:30am-noon and 2pm-4:30pm, Apr.-Sept. Tues.-Sun. 9:30am-noon and 2pm-6pm; adults €3, under 12 free

Fréjus's amphitheater was built at the entrance to the Roman city at the end of the first century CE and is one of the best-preserved and largest of its kind in France. It originally had three tiers of arches reaching 21 m (69 ft) high, with the top level supporting wooden masts that held a vellum canvas when part of the amphitheater was covered—a Roman-style temporary roof. An estimated 10,000 people could watch gladiators, wild animals, and slaves in the sandy arena below. After Emperor Constantine banned gladiatorial combat in the 4th century, the amphitheater was used as a quarry; many of the medieval buildings in Fréjus are built from its sandstone blocks. This process continued until the mid-18th century, when attempts were made to restore antique architecture in France. The amphitheater has since been repaired, with new stones brought in to replace those stolen, and concrete used to stabilize the tiers, corridors, and passageways. Today, the amphitheater is open for visits during the day, and in the evenings it is a venue for musical concerts, boxing matches, and reenacted gladiator spectacles.

Roman Aqueduct

avenue du XV Corps d'Armée

On the north side of the main road into Fréjus from Cannes and running through the Parc Aurelien is a line of giant brick pillars that once supported the city's Roman aqueduct. Fresh water was brought to Fréjus from streams at La Foux de Montauroux and Siagnole de Mons, some 42 km (26 mi) away, a stunning feat of engineering that originally included underground channels and 86 brick

Roman Holiday in Fréjus

Roman walls and modern walls at the amphitheater in Fréjus

On the Roman via Aurelia between Arles and Rome, Fréjus has some of France's most significant Roman remains. Its role was predominantly as a naval base and port, shipping out the region's wine and olive oil production. This Roman legacy can still be seen throughout Fréjus today.

ROMAN SIGHTS

The Roman buildings, ramparts, and vestiges of roads form the city's open-air museum and take a day to explore fully:

- One of the towers that guarded the port is the 10-m-high (32-ft) hexagonal **Lanterne d'Auguste** (673 chemin de la Lanterne d'Auguste), which was used by sailors as a landmark to help them navigate the coast (before the invention of lighthouses). Today it is in the center of town, the sea having receded over the centuries. At the end of the chemin de la Lanterne d'Auguste is a stone bollard carved out of blue-tinged esterellite, which was quarried nearby. You can still make out the rubbing marks from Roman ships' ropes on the base.

- Most impressive is the **amphitheater,** now open for visits during the day and concerts and boxing matches in the evening (page 351).

- The **Roman theater,** built before the amphitheater, is also open as a summer venue. A steel-framed semicircle is built within the original Roman supports, little of which remain.

- The **Porte des Gaules** (rue Henri Vadon), part of the original ramparts, is the largest Roman gateway in France. However, it was filled in and today has no obvious opening to allow people access through it. The "gateway" was sealed when the place Agricola was constructed, but the stone road (via Aurelia) is still visible underneath it.

- The **Platforme** (l'esplanade de la Butte) was probably used as a military headquarters or residence of the fleet commander, erected at the time of Caesar Augustus just before the Christian era.

- To the northeast along the avenue du XV Corps d'Armée and into the Parc Aurelien is a line of giant brick supports for the **aqueduct,** which brought water into the city from over 40 km (25 mi) away (page 351).

- The most interesting artifacts dug up in excavations from the beginning of the 19th century are displayed inside the **Musée Archéologique** (page 353).

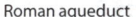

arches over the 700-m (2,300-ft) plain leading into Fréjus. The columns at the roadside are in ruins, but there are nine complete arches within the park that have been partially restored.

Musée Archéologique

place Calvini; tel. 04 94 52 15 78; www.ville-frejus.fr; Oct.-Mar. Tues.-Sat. 9:30am-noon and 2pm-4:30pm, Apr.-Sept. Tues.-Sun. 9:30am-noon and 2pm-6pm; €3, under 12 free

The museum's collection of archaeological artifacts is divided into four sections. By far the most interesting is on the right of the entrance and is devoted to Roman sculptures and, as its centerpiece, the polychrome mosaic of a panther. A marble, two-headed Hermes, roughly the size of a human head and unearthed in 1970 in Fréjus, has become the symbol of the city. Downstairs are the other three sections: a quadrangular Roman house, dug up under the nearby place Formigé, which shows what luxury homes would have looked like during the time of Augustus (around 27 BCE-14 CE); a collection of graves, urns, and funerary monuments (audiculas) from the local necropolis; and finally, a set of Roman ceramic and glassware discoveries from the outskirts of Fréjus.

Based on private collections of dug-up curios, the archaeological museum was founded in Fréjus in 1880, almost 80 years after the first archaeological digs had taken place.

Pagode Hong Hien

726 avenue du Général d'Armée Jean Calliès; tel. 04 94 53 25 29; https://pagodehonghienfrej.wixsite.com/ pagodefrejus; Apr.-Oct. daily 10am-7pm, Nov.-Mar. daily 10am-noon and 2pm-5pm; €4, under 7 free

At the side of a roundabout, 2 km (1.2 mi) from the old town on the DN7, is a Buddhist temple built by Vietnamese soldiers who were stationed at a military camp in Fréjus in 1917. The soldiers had joined France's mainland armed forces to fight on the battlefields in the north of France during the First World War, but were stationed in the south (where the climate was more similar to their own more humid colonies). Visitors are welcome to walk along the alleyways of the temple's grounds, where the pagoda is set within a stepped garden. The site includes a 2.5-m-high (8-ft-high) bell, a 10-m (33-ft) reclining Buddha (the largest in Europe), and brightly painted statues of giant frogs, dragons, a half-buried crocodile, and other colorful figures from Buddhist mythology.

Roman aqueduct

Parks

Parc Aurelien

avenue du Général d'Armée Jean Calliès; daily 8am-6pm Oct.-Apr., daily 8am-8pm May-Sept.

An entrance in front of the arches of the Roman aqueduct leads to a car park and miles of orienteering paths and play areas within this 24-hectare (59-acre) park. It's a popular place for young lovers and dog walkers, and it has bird- and insect-identification panels along its many footpaths.

Deep within the park is the **Villa Aurélienne,** a fabulous Palladian-style mansion built in 1889 by James Crossman, who hailed from a family of British brewers. Taking its name from the nearby Roman road, via Aurelia, the property changed hands many times among the upper bourgeoisie before falling into ruin and being acquired by Fréjus city council in 1988. It's now used to host concerts, receptions, and art exhibitions. The front facade of the rectangular villa has two levels of narrow marble columns, while the interior rooms are constructed around a patio with a stained glass roof. Access is through the park, and admission to exhibitions is free.

Base Nature François Léotard

199 boulevard de la Mer

In what used to be a French naval airbase, the Base Nature is a huge expanse (135 hectares/333 acres) of grassland, cycle tracks, endless paved former runways, and wooded areas Highlights include a children's play area; a skateboard park; sports pitches; pathways for walking, jogging, and Rollerblading; plus picnic areas and a walkway along the beach. Bikes and four-person cycles are available to rent, but most people bring their own equipment. The sandy beach has lifeguards, a couple of snack bars, and showers. There is a 640-space parking lot that is free September-June. In July and August, parking is free for the first 30 minutes, then €0.50 per 15 minutes.

Beaches and Water Sports

Fréjus-Plage

Boulevard d'Alger

With the beach bars Les Sablettes and La Plage at either end, the beach at Fréjus-Plage is a long expanse of golden sand and an active base for water sports and swimming. Parking can be found along the seafront or in the Base Nature parking lot.

Aquatic Rando

quai Cléopatre; tel. 04 83 09 90 35; www.aquatic-rando.fr; open year-round; courses from €75

Aquatic Rando's diving school runs certified PE12 courses for level 1 to level 4 divers and includes all equipment. They also offer diving treasure hunt sessions for children over six, and rent out underwater equipment.

Aqualand

462 RD559, Camp de l'Abbé; tel. 04 94 51 82 51; www. aqualand.fr/frejus; mid-June-early-Sept. daily 10am-6pm or 7pm (see website for hours); €32, children and pensioners €25, family ticket €106

The Côte d'Azur's largest water park has 19 different rides on its 22-hectare (54-acre) site, including the 90-degree free fall, rapid rafting slides called Niagara and Vertigo, Europe's largest waterslide, King Cobra, and rides for couples on a giant inflatable ring through the tunnels of the Flying Boat. It has two spaces for small children, 10 areas for families, sunbathing zones, and plenty of lifeguards in bright yellow T-shirts monitoring the rides. If you purchase tickets online, there's a 10 percent discount at the cafeteria.

Shopping

The old town is mainly pedestrianized, so it's a pleasant place to stroll, with plenty of arts and crafts boutiques along the rue Saint-François-de-Paule and more everyday shops, bakers, and grocers along the rue Jean Jaurès.

La Cave des Cariatides

53 rue Sieyès; tel. 04 94 53 99 67; Mon.-Sat. 9:30am-7:30pm, Sun. 9:30am-1pm

In the heart of the old town, it's worth

1: reclining Buddha at Pagode Hong Hien **2:** swimming pool at L'Aréna hotel **3:** doorway of La Cave des Cariatides

stopping at this wine and spirits shop just to appreciate the 17th-century carved wooden door and magnificent Atlantes. The shop's name translates as "the wine shop of the caryatids," but because the carved figures holding up the door frame are girdled men (caryatids are women), they are known as Atlantes (after Atlas). The wine shop has been around since 1953 and has a good selection of local Provence wines, champagnes, whiskeys, and Armagnacs, and they sell gift boxes for special occasions.

Place Formigé Market

Place Formigé; Wed. and Sat. 8:30am-12:45pm

Fréjus's fresh produce market takes place on the place Formigé and rue Général de Gaulle on Wednesday and Saturday mornings. The market is a good source for olive oil, locally grown melons, charcuterie, cheeses, and fresh vegetables, with stalls lined with barrels of olives and pickled garlic.

Food
Grand Café de l'Estérel

10 place Agricola; tel. 04 94 51 50 50; https://grand-cafe-esterel-frejus.eatbu.com; Mon.-Sat. 8am-10pm; €10-12

Beside a monument dedicated to Agricola, the "big café" with bright green awnings is a favorite with locals, serving everything from pizzas and foie gras to burgers, roast duck, and seafood salad. Portions are big, the staff are friendly, and it's well located between the rail station, old town, and amphitheater. The three-course lunch is €15.90.

★ L'Entrée des Artistes

63 place Saint-François de Paule; tel. 04 94 40 11 60; Tues.-Sun. noon-2pm and 7:30pm-9pm; €18-22

In the delightful place Saint-François de Paule, where bollards are hand-painted and the lampposts have knitted scarves, L'Entrée des Artistes offers an intriguing menu of Provençal and Asian combinations. The restaurant was started by two brothers, Sébastien and Christophe Terrier, in 2009 and has been gaining in popularity

ever since. The two-course lunchtime formule is €17.50. With just a dozen tables, meals are served in a cool dining room or on a wooden-decked terrace.

L'Amandier

19 rue Desaugiers; tel. 04 94 53 48 77; www.restaurant-lamandier-frejus.com; Tues., Thurs., and Sat. noon-1:15pm and 7:30pm-9:15pm, Mon., Wed., and Fri. noon-1:15pm; €19-23

L'Amandier ("the almond tree") is a refined gastronomic restaurant on a side street opposite the city hall. Decorated in autumnal tones, the restaurant has three-course gourmet menus for €35 or €47, with highlights such as roast cod with paella and saffron emulsion or roast duck with honeyed baby carrots and muscat gravy. The weekday two-course lunch is €24.

Accommodations
Hotel Le Flore

35 rue Grisolle; tel. 04 94 51 38 35; www.hotelleflore.fr; €60 d

A salmon-pink facade and mint-green shutters greet guests at this good-value, centrally located hotel. Tiled floors and air-conditioning keep the rooms cool in the summer. Twin rooms, doubles, triples, and quadruples make up the family-run hotel's 11 rooms, which are all on the 2nd and 3rd floors of the 19th-century town house. Note that there is no elevator.

★ L'Aréna

139 rue du Général de Gaulle; tel. 04 94 52 77 20; www.hotel-frejus-arena.com; Dec.-Oct.; €120 d

The three-star L'Aréna is an upmarket option in a city not overly blessed with top-notch hotels. The hotel is made up of three buildings around a central courtyard, with its restaurant, La Table de Guillôme, an attractive place to eat with a creative menu and a large garden terrace for dining beside the pool.

★ Domaine du Colombier

1052 rue des combattants en Afrique du Nord; tel. 04 94 51 56 01; www.frejus-campsite-colombier.co.uk;

Apr.-mid-Sept.; from €50/day for a tent site to €800/ day for a deluxe mobile home

Calling itself an "open-air hotel," the five-star Domaine du Colombier is a prestige campground 4.5 km (2.7 mi) from the sea. It has restaurants, tropical pools, a spa, a fitness center, evening shows, and a 3,600-sq-m (38,750-sq-ft) lagoon waterpark. The 10-hectare (25-acre) property offers 18 different types of rooms and mobile homes to rent, decorated in art deco, Aztec, African, or contemporary styles, plus pitches for tents and camper vans, protected within a forest of olive trees, conifers, and pines.

Information and Services
Fréjus Office de Tourisme
249 rue Jean Jaurès; tel. 04 94 51 83 83; www.frejus. fr; Sept.-June Mon.-Sat. 9:30am-noon and 2pm-6pm, July-Aug. daily 9am-7pm

The office has information about all the city's tourist sites and the film and theater festivals, plus a booklet on the **Circuit des Métiers d'Art** that gives details of the city's 25 arts and crafts workshops and studios.

Getting There and Around
Car
Fréjus is 4 km (2.5 mi) from exit 38 of the A8 autoroute, signposted for Fréjus-Saint-Raphaël. It takes around 1 hour to drive from **Nice** (64 km/40 mi) and 40 minutes from **Cannes** (36 km/22 mi).

Train
Fréjus's **railway station** (rue Martin-Bidouré; www.sncf-connect.com) is on the TER railway from Marseille to Ventimiglia in Italy. The train to Cannes takes 40 minutes (€8.80); to Nice, 1 hour 15 minutes (€14.50); and just under 2 hours to Marseille (€30).

Bus
Fréjus's **gare routière** (tel. 04 94 53 78 46) for buses is at 97 rue Gustave Bret. **Zou network bus 876** (tel. 04 13 94 30 50; https://zou. maregionsud.fr) runs between Fréjus and Saint-Tropez (70 min; €2.10); **bus 90** runs between Fréjus and Nice airport (70 min; €18 single).

SAINT-AYGULF

Five km (3 mi) west of Fréjus, Saint-Aygulf is a seafront resort adjacent to the **Étangs de Villepey,** lagoons that attract thousands of migrating birds. The resort is one of the more built-up along the sea, but it is also the start of a terrific coastal walk heading southwest toward Les Issambres, a streak of rocky inlets, tiny gravelly beaches, and a few parks.

Sights
Les Étangs de Villepey
Stop at the sign for the Esclamandes on the D559 to the east of Saint-Aygulf

Where the Reyran and Argens rivers join just before reaching the sea, a 260-hectare (642-acre) lagoon of brackish water has become a popular migration stopover ground for over 200 bird species. Ducks, pink flamingos, swans, herons, and storks can all be seen in the waters here. With the banks of reeds and surrounding forest of pines and mimosas, it's a great place to while away a few hours, either by walking the 7-km (4-mi) track around the lagoon (closed in parts from June to September) or along one of the lagoon's sandy beaches. Parking costs €4.50 per day or €3 per half day.

Food and Accommodations
★ L'Iceberg
21 place de la Galiote; tel. 04 94 81 95 21; https:// liceberg-restaurant.business.site; Apr.-Oct. Wed.-Sun. 9am-10pm; €20

On a large wooden deck overlooking the water, L'Iceberg specializes in seafood brochettes, barbecues, and a Marmiton de la Mer, a large bowl of fish soup, cod steaks, wild king prawns, and vegetables for €19. Seating is both indoors, with panoramic windows, and out, on a large terrace with wooden barrels and straw sunshades. The restaurant has a great selection of organic wines and craft beers.

Hotel Saint-Aygulf

214 rue d'Alsace; tel. 04 94 52 74 84; www.hotelstaygulf.fr; €120 d

Where the freshwater meets the sea, this hotel is smart, reasonably priced, and well located for a few days in Saint-Aygulf. Owned by the Van de Valk group, the hotel has its own beach club and restaurant and is also equipped for business meetings. Most rooms have balconies and sea or lagoon views, and there are also family suites.

Information and Services

The **Office de Tourisme de Saint-Aygulf** is located on the place de la Poste (tel. 04 94 81 22 09; Mon.-Sat. 9am-6pm). It has brochures and information about activities and events in town, as well as information on accommodations.

Getting There and Around

Saint-Aygulf is 5 km (3 mi) from **Fréjus** on the RN98 and 35 km (21 mi) from both **Cannes** and **Saint-Tropez. Parking** is straightforward at Saint-Aygulf P1 and P2 (420 spaces, signposted) on the RN98. Parking is free for the first 30 minutes and subsequently €1.80 per hour.

Buses **20** and **876** run from Fréjus Ronde-Point Aeronautique to Saint-Aygulf La Poste (7 minutes).

SAINT-RAPHAËL

Saint-Raphaël's old quarter is centered on the **place de la République** and contains the archaeology museum and Romanesque church. A huge iron anchor sits in front of the old port, which today harbors a mixture of pleasure boats, speedboats, launches for diving schools, pedal boats, and the occasional painted pointu (traditional wooden fishing boat). Opposite the tourist office and beneath the **Bonaparte gardens** is the location for Saint-Raphaël's ferry terminus, where boats leave for various destinations including Saint-Tropez, Île Sainte-Marguerite, and Porquerolles. Another trip takes you along the Calanques de l'Estérel, the dramatic

inlets where the Massif de l'Estérel meets the sea. You can also catch a ride on the glass-bottomed *Capitaine Némo.*

Just off the coast, to the south of **Port Santa Lucia,** the new port, are two volcanic outcrops known as the Lion de Terre and Lion de Mer because their rocky formations are said to resemble lions. The new port itself has moorings for 1,630 boats (the third largest on the Côte d'Azur), a conference hall, hotels, and scores of restaurants and bars, creating a new nightlife zone in Saint-Raphaël.

Sights

Musée de Préhistoire et d'Archéologie and Église Romane

Parvis de la Vieille Église; tel. 04 94 19 25 75; www.ville-saintraphael.fr; Nov.-Feb. Tues. 2pm-5pm, Wed.-Fri. 9am-12:30pm and 2pm-5pm, Sat. 10am-12:30pm, Mar.-June and Oct. Tues. 2pm-5pm, Wed.-Sat. 9am-12:30pm and 2pm-5pm, July-Sept. Tues.-Sat. 10am-6pm; free

Objects in this museum come mainly from prehistoric times—Neolithic and Paleolithic tools and human skulls. Nearby underwater digs have unearthed Roman amphorae, canons, hoards of coins, glassware, and pottery. QR codes throughout the museum guide tablet users through the displays and into the church crypt (access in the museum). The church's 22-m-high (72-ft-high) bell tower offers great views of the Estérel, Roquebrune, and Fréjus coastline, but hold on tight if you are inside when the bells ring on the hour.

Basilica Notre-Dame de la Victoire de Lépante

boulevard Félix Martin; tel. 04 94 19 81 29

Consecrated in 1888, Saint-Raphaël's impressive pink sandstone basilica was built in the Romano-Byzantine style popular at the time, like Marseille's Notre-Dame de la Garde, and named after the victory of Christian forces over the Ottoman navy in 1571. It's a stunning building from the outside, standing 35 m (115 ft) high, but hemmed in by hotels and municipal buildings. Of note are the modern stained glass windows depicting the lives of Mary and Joseph and Jesus's miracles, and

Saint-Raphaël

the gilt statues of the Archangel Raphaël and Tobias behind the main altar.

Beaches
Plage du Veillat
promenade René Coty

To the east of the Vieux Port, this pleasant, safe, urban beach is a gentle curve of 400 m (1,300 ft) of fine white sand, with lifeguards, volleyball courts, and water sports available in the summer. Parking is available at Bonaparte parking lot on the seafront (€4 for 2 hours).

Plage de la Peguière
route de la Corniche

A 10-minute walk along the coastal footpath east from the Port Santa Lucia takes sunbathers to the protected and picturesque Plage de la Peguière, where it's a bit calmer and less crowded than the beaches in Saint-Raphaël. Lifeguards are on post in the summer, and showers and toilets are available. Pine trees and overhanging rocks provide a bit of shade. Limited parking is available along the roadside.

Water Sports
Aventure Sous Marine
*155 quai Amiral Nomy; tel. 06 09 58 43 52; www.
aventuresousmarine.fr; Mon.-Sat. 9am-7pm, Sun. 9am-
noon; introductory class €80*

Saint-Raphaël's principal diving school runs a "baptism" class with an introduction to scuba diving to include a 20-minute, 4-6-m (13-17-ft) dive at the Lion de Mer, just off the shore. For more experienced divers, six Level 1 classes are €490 and 12 lessons for Level 2 are €640. ASM has a 40-person boat and a shop on the harbor.

Centre Nautique Municipal
*Port Santa Lucia, 90 place du Club Nautique; tel. 04
94 83 84 50; www.cnsr.fr; courses from €100, rentals
from €8/hour*

Saint-Raphaël's sailing school is open all year but specializes in summer sailing courses based in the town's modern marina. They have catamarans, optimists, kayaks, and windsurf sailboards. Non-members can rent windsurf boards for €22 per hour, paddleboards for €15 per hour, and catamarans for €48 per hour.

Hiking
Sentier du Littoral
Distance: *10 km (6 mi) one-way*
Time: *4 hours*
Information and Maps: *Saint-Raphaël tourist office*
Trailhead: *place du Lion de Terre, Port Santa Lucia*

Starting from opposite the Lion de Terre outcrop at Port Santa Lucia heading east, the Sentier du Littoral (coastal footpath) snakes along the seafront for around 10 km (6 mi), finishing at Agay. Doing the entire route will take around 4 hours, and it's quite up and down—prepare to clamber over rocks and walk along narrow paths and small, sandy coves—but it is an enjoyable walk (note that it's not suitable for wheelchairs or young children). There's a local railway station at Agay to do the return leg by train (10 min; €2.80).

Food
Since most of the tourist areas in Saint-Raphaël run along its port and seafront, fish and seafood restaurants dominate the town's food scene. Fresh produce **markets** take place every morning except Monday on place Victor Hugo and place de la République. The **fish market** is held every morning in the old port.

Le Coelacanthe
*2 place Amiral Ortoli, Port Santa Lucia; tel. 04 94 83 61
04; Fri.-Tues. noon-3pm and 7pm-10pm; €26-39*

This is a popular portside restaurant serving big portions of excellent seafood, with a creative selection of oyster, scallop, and mussel dishes and freshly caught grilled fish. Booking ahead is recommended, especially on weekends.

Le Sirocco
*35 quai Albert 1er; tel. 04 94 95 39 99; www.lesirocco.
fr; Tues.-Sun. noon-2pm and 7pm-10pm; €27-43*

A fish-inspired restaurant overlooking the old port, Le Sirocco serves a lot of oysters, but they also do a nice plate of grilled giant prawns with pesto and parmesan, and a slightly more eccentric snail stew with mushrooms or Saint-Raphaël-style bourride (fish stew with aioli).

Accommodations
Hotel du Soleil
*47 boulevard Domaine du Soleil; tel. 04 94 83 10 00;
www.hotel-dusoleil.com; €105 d*

Friends Linnea and Susanne moved from Sweden to take over this belle epoque villa. There is a good selection of singles, doubles, triples, and studio bungalows. A garden restaurant is on the grounds (breakfast €10), and there are bikes to rent for the short cycle to the beach (towels provided).

★ Le Touring
*1 quai Albert 1er; tel. 04 94 55 01 50; www.letouring.
fr; €260 d*

One of the town's best-known hotels, Le Touring's 10 rooms and suites overlook the old port. The decor is art deco with a modern twist, a stylish mix of stripes and chrome.

There's a 1st-floor Japanese restaurant, Nami, a ground-floor bistro, and two cocktail bars. All rooms have air-conditioning and a mini-bar, and breakfast is free if booked directly with the hotel.

La Villa Mauresque

1792 route de la Corniche, Bolouris; tel. 04 94 83 02 42; www.villa-mauresque.com; Mar.-mid-Nov.; €480 d

The white, Moorish-style villa on the coast road east of Saint-Raphaël is an exclusive residence, having Chakira, Pamela Anderson, and Karlie Kloss among its guest book signatories. The hotel has a gastronomic restaurant, seafront gardens, hot tubs, pools, a patio-terrace set among the red rocks, and direct access onto a private beach and boathouse.

Information and Services

The **Office de Tourisme de Saint-Raphaël** is located at 99 quai Albert 1er (tel. 04 94 19 52 52; www.saint-raphael.com; Mon.-Sat. 9am-1pm and 2pm-7pm, Sun. 9:30am-12:30pm and 2:30pm-6:30pm).

Getting There

Take exit 38 on the A8 autoroute signposted for Fréjus and Saint-Raphaël. The drive takes 1 hour from **Nice** (66 km/41 mi) and 45 minutes from **Saint-Tropez** (37 km/23 mi) on the coast road.

Train

The main train station, **TGV St-Raphaël Valescure** (rue Waldeck Rousseau; tel. 36 36; www.sncf-connect.com), is on the Marseille-Nice line. Fast trains take 1 hour 45 minutes from Marseille and 52 minutes from Nice.

Bus

Saint-Raphaël's bus **gare routière** (tel. 04 94 44 52 70) is 100 rue Victor Hugo. **Zou network bus 876** (tel. tel. 04 13 94 30 50; https://zou.maregionsud.fr) runs between Saint-Raphaël and Saint-Tropez (60 min; €2.10); **bus 90** runs between Saint-Raphaël and Nice airport (80 min; €18 single).

Getting Around

Saint-Raphaël is centered on its old port, and with a tiny old town and new marina, it's easy to get your bearings on foot.

Bus

Saint-Raphaël shares the **Estérel-Côte d'Azur bus network** (tel. 04 94 44 52 70; www.esterelcotedazur-lebus.fr) with Fréjus. Buses 2, 4, A, and B go from Fréjus bus station to Saint-Raphaël bus station. Bus 21 travels east along the Corniche d'Or coast road to Le Dramont and Agay (single fares €1.50).

Ferry

Since traffic can be very heavy during the summer, taking the ferry along the coast can be a pleasant alternative. **Les Bateaux de Saint-Raphaël** (tel. 04 94 95 17 46; https://bateauxsaintraphael.com) leave from the **gare maritime** on quai Amiral Nomy for Saint-Tropez, Sainte-Marguerite, Porquerolles, and points along the coast to the Estérel coves.

Bicycle

Cote eBike-Proxy-Cycle (34 avenue Général Leclerc; tel. 09 80 90 26 67; https://cote-ebike. com; Tues.-Sat. 9:30am-12:30pm and 2pm-6pm) has a large range of city, road, mountain, and electric bikes to rent from their large store in the center of Saint-Raphaël. Electric bikes are €25 for a half day and €35 a full day. Electric mountain bikes are €49 for a full day.

LE DRAMONT

Le Dramont, 7 km (4 mi) east of Saint-Raphaël, has two beaches, the sandy Plage du Camp Long and the pebbly Plage du Débarquement, which sit on either side of the Cap Dramont, a squat peninsula jutting out into the sea just before the rust-colored mountain range of the Massif de l'Estérel. There's a huge free car park at the latter, decorated with several armored vehicles and monuments, leftovers of the Allied landings on the beaches in August 1944. It's a 3-minute walk down to the sea, where there's a café-restaurant, water

1

2

3

4

sports center, and large, shady areas for family picnics.

Sights
L'Île d'Or
Le Dramont

The tiny island and its tower, 200 m (650 ft) from the shore, were the inspiration for Hergé's Tintin adventure *The Black Island*. Like many small islets off the coast, the Île d'Or spent much of the last 150 years in private hands. One Dr. Auguste Lustaud won the island during a game of whist in 1909 from his friend Léon Sergent, who had bought it a decade earlier at an auction from the French state for 280 francs (US\$1,000 in today's money), and built a square Saracen tower on the golden-red rock island. He declared himself king of the island under the name Auguste Le Premier and minted commemorative coins. In 1962, the island was bought by a naval officer named François Bureau, but he drowned while swimming off the island.

One of the most photographed places on this stretch of the Riviera, it's a popular destination for kayakers, who paddle across the **Haddock Strait** (Détroit Haddock) between the coast and the island (named after Tintin's Captain Haddock).

Beaches and Water Sports
Plage du Débarquement
Route de la Corniche

One of the major landing sites of the Allied troops on August 15, 1944, this 500-m-long (1,640-ft-long) beach is one of the most interesting in the region, with a mixture of sand, pebbles, and rocks, and a shaded woody backdrop, perfect for picnics and family games. The rocky seabed means the water is clear and ideal for snorkeling, and there's a nice café with a water sports rental center next door. Public toilets and showers are available. The entrance for the 200-space parking lot is opposite the railway station.

Plage du Camp Long
2400 boulevard de la 36ème Division du Texas

Also known as Tiki Beach, Camp Long is on the other side of the Cap Dramont headland from the Plage du Débarquement and protected by the rough, rubiginous rocks that almost encircle the bay. The near-perfect semicircle of rocks provides a great, safe swimming area for young families, with gritty sand below. Free parking is available on the road above the beach.

Food
La Plage de l'Île d'Or
Plage du Débarquement, 904 boulevard de la 36ème Division du Texas; tel. 04 98 12 43 01; www. laplagedeliledor.com; Apr.-Nov. daily 9am-midnight; €17-28

Right on the beach facing the Île d'Or, the café-restaurant has roped-off wooden decks under the shade of some pine trees and serves a good selection of salads, burgers, pizzas, and fresh fish. Popular dishes include the vegetarian food board (€24) and guacamole with spicy nachos (€9). The restaurant also rents out sunbeds from a cabin next door, and water sports equipment (tel. 07 66 51 16 74; 1-person kayak €15/1 hour, 8-person paddleboard €50/1 hour).

Getting There and Around

Le Dramont is on the D559 coast road, 7 km (4 mi) from **Saint-Raphaël** (15 min) and 31 km (19 mi) from **Cannes** (50 min) on the twisting D1098 road.

Le Dramont railway station (D559, Agay) is on the TER Marseille-Ventimiglia line. Journey time to Saint-Raphaël-Valescure is 6 minutes (€2.40) and to Cannes 30 minutes (€6.70).

Estérel Côte d'Azur Le Bus number 21 travels along the coast road from Saint-Raphaël bus station to Le Dramont (17 min; €1.50; www.esterelcotedazur-lebus.fr).

1: Saint-Aygulf **2:** grand belle epoque buildings in Saint-Raphaël **3:** view from Plage du Débarquement in Le Dramont **4:** mountain biker in the Massif de l'Estérel

THE MASSIF DE L'ESTÉREL

The Massif de l'Estérel is a dramatic mountain range that runs parallel to the coast from Agay to Mandelieu-la-Napoule for about 30 km (19 mi). The entire range is a mass of rust-colored volcanic peaks, created from igneous rock, that plunges into the sea, forming rough-edged creeks and photogenic, dramatic scenery. The mountains were sparsely inhabited by cave dwellers up to the end of the 18th century, and even now there are areas where few have trod. However, the range is a terrific place for exploring, hiking, horseback riding, and mountain biking.

High temperatures, strong winds, and long periods of dry weather mean forest fires are common in the summer, some destroying huge swaths of the green oak, pine, and cork tree forests. Access for hiking and biking is highly regulated; check the status on the government website the day prior to your visit (www.risque-prevention-incendie.fr/var).

Maps, hiking guides, and mountain biking guides and details are available from all the region's **tourist offices** as well as from rental outlets. The massif also provides over 100 km (62 mi) of tracks suitable for horses and pony trekking.

Sights
Mont Vinaigre

At 641 m (2,103 ft), this is the highest point in the Massif de l'Estérel. It takes an hour to climb from the Maison forastière de Malpay car park. It is not a difficult walk and can easily be done by healthy children (but not strollers).

Lac de l'Avellan

This lake and half-ruined dam at 311 m (1,021 ft) are ideal for walking around and fishing, and are popular with mountain bikers. Swimming is not allowed, but there are lovely white water lilies and shady areas for picnics. There is a parking area nearby (signed off the DN7 between Fréjus and Les Adrets).

Grotte de la Sainte-Baume Cave

This cave is accessible as a detour from the Cap Roux hike. Leave the car in the Parking Sainte-Baume and follow signs past the Chapelle de Sainte-Baume and water fountain. Bring a jacket, as it can be a lot cooler in the cave at high altitude than at sea level, even in the hot summer.

Lac de l'Ecureuil

From Agay, follow signs for the Massif de l'Estérel and then the car park signed Col de Barbe Belle. The Lac de l'Ecureuil (squirrel's lake) is signed in the car park. Walk for half an hour along the left-hand bank of the Grenouillet river, which leads up to the red rocks of the Canyon du Mal Infernet, so named because it was believed victims could cast the plague off down to the bottom of the ravine in medieval times. After another half-hour's walk, you arrive at the lake, a great place for a picnic.

Cycling
Sobike 83

1440 avenue du Gratadis, Agay; tel. 04 89 78 12 84; www.sobike-esterel.com; high season daily 9am-noon and 2pm-6pm, low season Tues.-Sat. 9am-noon and 2pm-6pm; €33-39/half day, €49-63/full day

Among other routes, Sobike organizes rides to the Massif de l'Estérel's most challenging locations—**Piste de Castelli** (half day) the **Ferme du Roussiveau** (full day)—and **Mont Vinaigre** (641 m/1,023 ft) and **Pic de l'Ours** (492 m/1,614 ft), the range's two highest peaks.

Hiking
Cap Roux

Distance: *7 km (4 mi) round-trip*
Time: *3-4 hours*
Information and Maps: *Local tourist offices*

Trailhead: *Parking Sainte-Baume*

Cap Roux (red peninsula) is one of the most dramatic parts of the Massif de l'Estérel and a favorite location for hiking, as it's a manageable route for families and takes half a day to complete. Depart from the car park (Parking Sainte-Baume), heading toward a fountain, and follow the yellow markers up to the **Col du Saint Pilon** (281 m/922 ft), which has a great view of the orange-red rocks plunging into the sea. Before continuing toward the Cap Roux summit, take a tour of the headland, enjoying the scents of the wild herbs along the path.

From the summit, you can see Saint-Raphaël to the west and Cannes to the east. It's nice to be there for sunset, but you must be out of the car park by 9pm. The walk is around 7 km (4 mi) and takes 3-4 hours round-trip, although there is another worthwhile detour of 30 minutes to the **Grotte de Saint-Honorat** (Saint-Honorat's cave), a stone chapel built into the rocks.

Horseback Riding and Pony Trekking
Les 3 Fers

chemin des Sangliers; tel. 06 85 42 51 50; www.les3fers. com; open year-round

The Les 3 Fers riding school runs holiday courses and individual lessons as well as trips (from half a day to 2 full days) into the Estérel range. The 90-minute Panoramer trek is along the coastline and costs €52 while the 2-hour Les Ferrièrs trek into the wilds costs €62. Riders must be over 11 years old to go trekking.

Information and Services

The Saint-Raphaël, Fréjus, and Roquebrune-sur-Argens **tourist offices** all have hiking guides and maps available. There are around 100 km (62 mi) of marked horseback riding tracks, 100 km (62 mi) of signed mountain biking tracks, and 40 km (25 mi) of hiking trails through the mountains. Check out www.esterel-cotedazur.com/decouvrir/sites-naturels/massifs/massif-de-lesterel for more information.

★ Saint-Tropez

Saint-Tropez is an intoxicating place to be. Its famous sandal shops and fashion boutiques are crammed with visitors, the marinas are full of superyachts, and it's almost impossible to find a restaurant table. Visitors should prepare themselves for a hot, crowded experience and the knowledge that someone with an orange tan and white linen suit will be waved past them in the queue no matter how long they've been waiting. In the low season, many of the hotels, shops, and restaurants close, but enough are open to reveal a more authentic, simple side of this once-humble fishing village. It may be easier to wander through its narrow peach-colored lanes in the autumn or spring, but if you want to see the Riviera in full, irrepressible, hedonistic swing, then go in July and August and stay late into the evening, when the marina is lit up by the superyachts and champagne is flowing in the surrounding bars, with live music and dancing in the streets accompanied by the growl of vintage cabriolets and supercars queuing for the parking lots.

Bouillabaisse Plage is Saint-Tropez's most accessible beach, but serious sunbathers head to the eastern side of the peninsula, where buses (or taxis) take them to Tahiti Plage and Plage de Pampelonne. It's worth following the coastal path past the Baie des Canebiers (also known as Canoubiers) to the Cap Saint-Pierre and Pointe de la Rabiou for some calmer sections of the seafront. Saint-Tropez has four museums, of which the portside Annonciade art museum is the most prominent, but the real fun is enjoying the spectacle of the boats, the people, the daily fish market, the lively restaurants, and the suntans, all in a setting

Saint-Tropez

Golfe de Saint-Tropez

© MOON.COM

so intimate you could walk around the entire village in 1 hour.

ORIENTATION

Most of Saint-Tropez's sights are arrayed around the **Vieux Port** (or Old Port, which begins at the place aux Herbes), its sable-, peach-, and apricot-colored facades forming a picturesque backdrop to the superyachts and pleasure boats in the harbor; and the **place des Lices,** a square lined with hotels, restaurants, ice cream wagons, and snack bars. The marina has space to moor 734 boats, most of which are larger than the town houses behind them. Port business is

done at the **Capitainerie** (harbormaster's office) on the **quai de l'Epi,** near the tourist office and opposite the restaurants of the **quai Jean Jaurès.** Meanwhile, the place des Lices, formerly a place for animals to graze, hosts a **market** on Tuesday and Saturday mornings, and games of pétanque under the plane trees.

SIGHTS
Musée de l'Annonciade

place Georges Grammont; tel. 04 94 17 84 10; www. saint-tropez.fr; Nov.-Mar. Tues.-Sun. 10am-5pm, Apr.-June and Oct. Tues.-Sun. 10am-6pm, July-Sept. daily 10am-7pm; €4-6, under 12 free

The former Chapel of the White Penitents

was built in 1568 and converted into an art museum by local collector Georges Grammont in 1955 to house his private stash of 56 paintings, as well as works owned by the town. L'Annonciade has some exceptional examples of pointillism by Paul Signac (who had a home in Saint-Tropez), Henri-Edmond Cross, and Maximilien Luce, as well as works from Postimpressionists Raoul Dufy and Henri Manguin and some nudes by Pierre Bonnard and Henri Matisse. The ground and upper floors house the permanent collection and the lower-ground annex has temporary exhibitions.

La Citadelle

1 montée de la Citadelle; tel. 04 94 9/ 59 43; www.
saint-tropez.fr; Apr.-Sept. daily 10am-6:30pm, Oct.-
Mar. daily 10am-5:30pm; €4, under 12 free

Forming part of Provence's defense fortifications against potential Spanish invaders, French king Henry IV's engineer Raymond de Bonnefons undertook the construction of a tower overlooking the bay of Saint-Tropez in 1602. The Citadelle was completed in 1608 and subsequently rebuilt many times to include soldiers' garrisons and repair damage to its ramparts. Abandoned as a military base in 1875, it was bought by the town council in 1993.

Visitors can enter the Citadelle's hexagonal dungeon, which houses the **Musée d'Histoire Maritime.** The ground floor, accessed over a drawbridge, has displays on the local fishing industry, coastal trading, deep-sea diving for coral and sponges, and the town's former torpedo factory. Exhibits on the 2nd floor look at the age of steamships and Tropeziens in naval wars, and the 3rd floor, an open terrace, covers powerboating, yachting, and regattas. The hilltop around the Citadelle is a large space for picnics, walks, and great views over the entire peninsula, with a resident peacock and some impressive cannons from the 17th century.

La Maison des Papillons

17 rue Étienne Berny; tel. 04 94 97 63 45; www. sainttropeztourisme.com; May-mid-July and Sept.-early-Nov. Sat.-Wed. 2pm-5pm, mid-July-Aug. and late-Dec.-early Jan. daily 2pm-5pm; €2, under 12 free

This museum features the collection of the Tropezien artist and amateur lepidopterist Dany Lartigue, who collected more than 35,000 butterflies that are now on display in hundreds of drawers, cases, and frames in his former house. Lartigue also pinned butterflies over paintings of where they were caught, adding another dimension to the three-dimensional art. It's an interesting spectacle from a time when animal welfare was less important than visual significance, especially since many of the butterflies he collected are now under threat of extinction.

BEACHES

Saint-Tropez's most famous beach, **Plage de Pampelonne,** is actually on the other side of the headland, in **Ramatuelle.** The waters around the resort's port are so busy with yachts and pleasure boats that it's worth traveling away from Saint-Tropez for a day on the beach.

Plage de la Ponche and Plage de la Glaye

rue de la Rampe and rue Saint-Pierre; no parking

On the northern edge of Saint-Tropez's old town are two tiny beaches, La Ponche and La Glaye, nice places to visit for a quick dip, to cool off, or to spend an hour sunbathing and watching boats buzz back and forth across the gulf. Plage de la Glaye is rocky but has beautiful clear-green waters. Plage de la Ponche is more sandy, hidden away, and clear of bathers.

cannons at La Citadelle

Saint-Tropez on Celluloid

sculpture of Brigitte Bardot by artist Milo Manara

ET DIEU... CRÉA LA FEMME

Perhaps the most famous movie filmed in Saint-Tropez is *Et Dieu... Créa la Femme,* the 1965 film that launched Brigitte Bardot's career. Bardot is so entwined into the life of Saint-Tropez that black-and-white photographs and giant postcards of her image are still all over the village. A bronze statue of a young Bardot by sculptor Milo Manara, posing coquettishly in a shell, was unveiled in 2017 outside the Musée de la Gendarmerie, and her house, La Madrague, is not far from the center.

OTHER FILMS

The children's classic **Chitty Chitty Bang Bang** (1965) was filmed at Cap Taillat in Ramatuelle. In **Le Viager** (1972), a doctor's brother buys an old man's Saint-Tropez house en viager (while he's still living in it). Highly successful in France, it was one of Gérard Depardieu's first feature films. The TV soap opera **Sous le Soleil** was filmed in the town until 2008. However, most famous (at least for the French) are the five films starring Le Gendarme de Saint-Tropez, whose popularity is celebrated at the **Musée de la Gendarmerie et du Cinéma.**

Other movies filmed in Saint-Tropez include:

- **Bonjour Tristesse** (1956), a tragic family drama based on Françoise Sagan's novel of the same name.

- **The Vintage** (1957), a story of jealousy and fugitives set in a vineyard.

- **The Collector** (1967), the third part of Eric Rohmer's *Six Contes Moraux* (Six Moral Tales), which won the Silver Bear at the Cannes Film Festival.

- **Bad Girls** (1968), a story about erotic parties and wild relationships at a Saint-Tropez villa.

- **La Piscine** (1969), in which Alain Delon, Jane Birkin, and Romy Schneider star in a troubling melodrama set in a villa with a pool.

- **La Cage aux Folles** (1978), a farce about a gay nightclub owner, his boyfriend, and their encounter with an ultraconservative couple.

- **The Transporter** (2002), a violent but spectacular film with car chases all along the Riviera starring Jason Statham and Shu Qi.

Bouillabaisse Plage

route Départementale 98A, Quartier de la Bouillabaisse; parking lot is 300 m (0.2 mi) away toward the town center

Just before reaching Saint-Tropez, to the west of the Pointe de la Pinède, is La Bouillabaisse Plage, a long sand-and-pebble beach that looks across the gulf at the Maures mountain range. It's a short walk past the heat-radiating parking lot, worth it for a few less-crowded beach restaurants, like **Pearl Beach** (www.thepearlbeach.com).

Plage les Graniers

chemin des Graniers; limited parking on chemin des Graniers

A 20-minute walk from Saint-Tropez's old port on the Sentier du Littoral coastal footpath and through the woods is this great little cove, free from the crowds and with a fresh-fish beach restaurant, **La Plage les Graniers** (https://lesgraniers-sainttropez.com).

Plage les Salins

chemin des Salins; limited parking on chemin des Salins

Three km (1.8 mi) from the old port of Saint-Tropez, on the other side of the headland, is Plage les Salins, a soft-sand beach with a well-regarded family restaurant, **Les Salins** (http://lessalins.com), serving pizzas and seafood for lunch and dinner beside the waves.

SPORTS AND RECREATION

Boat Trips

Boat trips (promenades en mer) are advertised all along the harbor, with the most popular being an hour-long cruise (with commentary) past the old fishing harbor, La Citadelle, the sailors' cemetery, villas of the rich and famous, film locations, La Madrague (Brigitte Bardot's house), the Baie des Canebiers, and the Cap Saint-Pierre. *La Pouncho* (quai du Président Meiffret, parking at Nouveau Port; tel. 04 94 97 09 58; www.lapouncho.com) and *Brigantin II* (tel. 06 07 09 21 27; www.lebrigantin.com) offer tours and trips along the coast.

FESTIVALS AND EVENTS

Les Bravades de Saint-Tropez

place de l'Hôtel de Ville; May 16-18; free admission

For three days in mid-May, the citizens of Saint-Tropez celebrate the day Caius Torpetius's (Saint-Tropez's) body was washed ashore in the year 68 CE. While it's technically a religious festival, in reality it's a chance for hundreds of people to dress up in military uniform and red-pom-pommed sailor hats, holding pikes, waving flags, and setting off smoke-filled blunderbusses. The main events take place in the place de l'Hôtel de Ville (town hall square), beginning May 16 with an artillery salvo at 8am and finishing on May 18 when a young musketeer (usually the grandson of one of the bravadeurs) ties a red scarf around the neck of a bust of Saint-Tropez, which has been paraded round the town, and kisses it in front of the crowd.

Les Voiles de Saint-Tropez

Vieux Port; www.societe-nautique-saint-tropez.fr; Sept.-Oct.; free standing room on the side of the port

Saint-Tropez's most prestigious regatta takes place at the end of September and beginning of October with a week of sailing, partying, and plenty of posing. Organized by the Société Nautique de Saint-Tropez, the weeklong event attracts more than 300 classic sailing yachts and ultramodern racing boats up to 50 m (164 ft) long. On land, there are hundreds of exhibition and merchandise stalls in the temporarily erected village des Voiles, pop-up bars, a blessing of the boats on the first Monday by the parish priest of Saint-Tropez, live bands, boat parades, and boules tournaments in the place des Lices. **Les Bateaux Verts** (www.bateauxverts.com) takes trips out to sea to follow the yachts for close-up photos.

SHOPPING

Saint-Tropez is one of the premier shopping destinations on the Riviera, catering to day-trippers, wealthy holidaymakers, and those stepping off their superyachts straight into the portside boutiques. It has its own fashion trends, a Saint-Trop "look" combining Bohemian chic, nautical dash, and pale beachwear: spotless espadrilles or gladiator sandals, tight-fitting white or ivory-colored tunics over skimpy swim costumes, and wide-brimmed hats. (And always lots of chilled rosé wine and raffia baskets.)

Rue Général Allard has mainly clothes boutiques, sunglasses shops, and art galleries; **rue Sibilli,** which runs from behind the tourist office to the place des Lices, has many of the big fashion names in some spectacular buildings: Versace, Louis Vuitton, Dior, Gucci, Fred, and Dolce & Gabbana. In the old town, **rue des Commerçants** and **rue de la Citadelle** are the best for souvenirs, off-the-peg clothes, and Saint-Tropez trinkets.

Clothing and Fashion
Les Tropéziennes Sandales Rondini
18 rue Georges Clémenceau; tel. 04 94 97 19 55; www. rondini.fr; Mon.-Sat. 9:30am-1pm and 2pm-6:30pm, Sun. 10:30am-1pm and 3pm-6:30pm
Saint-Tropez's legendary handmade sandal boutique was founded in 1927 and has supplied the gladiator-style, Salome, Saharan, and, most famously, Tropéziennes strappy shoes to stars and visitors from across the globe. Founder Dominique Rondini was the first sandal-maker in Saint-Tropez and passed his skills on to his son, Serge, who expanded the range of designs and inspired his own son, Alain, to continue the family tradition. Today, the next generation of Rondinis is in charge, and around a dozen people manufacture Rondini sandals in the workshop at the back of the shop. They are available to order online, but it's more fun to have a fitting in the shop.

Wine
Da Vini Code
4 bis avenue Augustin Grangeon; tel. 04 94 56 27 90; www.davinicode.com; daily 9:30am-1pm and 3:30pm-8pm
Unquestionably the flashiest wine shop in town, Da Vini Code has over 35,000 bottles to choose from, ranging from €7 local wines to the finest of Grand Cru vintages—Romanée

Conti, Lafite-Rothschild, Petrus, and Cheval Blanc. Just off the place des Lices, the shop also stocks a large selection of rosé wines, spirits, special editions, and champagne with free delivery in the Golfe de Saint-Tropez. Sommelier Mathias Biscot speaks English and is happy to offer advice to customers.

Markets
For those on a budget, there's a well-stocked **Utile Supermarket** (avenue Général Leclerc; tel. 04 94 97 03 20; daily 8am-8:50pm) next to the bus station, which is a good place to buy drinks and food supplies and which also has a bakery and chicken rotisserie.

Place des Lices Market
place des Lices; Sat. and Tues. 6:30am-1:30pm
The place des Lices hosts a big market on Tuesday and Saturday mornings among the plane trees, with around 150 stalls selling locally grown fruit and vegetables, plus cheese, charcuterie, plants, clothing, and ice cream.

Place aux Herbes Market
Place aux Herbes
The tiny square behind the tourist office, the place aux Herbes, was the town's original fruit and vegetable market, and there are still a few stalls every morning covered in produce and local ceramics mixing in with the café and restaurant tables.

Fresh Fish Market
Tour du Port
Under the arches of the Tour du Port, one of the town's fortified gates, the photogenic fresh fish market is still in full swing, with fishmongers displaying trays of wrapped-up lobsters, cuttlefish, scallops, and giant tuna under a fish-themed mosaic.

FOOD
Mediterranean
Le Petit Pointu
12 rue des Remparts; tel. 04 94 54 36 78; daily noon-2pm and 7pm-9pm; lunch €17-23, dinner €17-34
Opposite La Ponche hotel in the old town is

LOTTE
1

2

LE PETIT
POINTU
BAR - RESTAURANT
LE PETIT
POINTU
LE PETIT POINTU
MENU DU MIDI
3

a down-to-earth, good-value seafood restaurant. Le Petit Pointu (the Little Fishing Boat) has a tapas menu—plates of oysters, sardines in olive oil, fish-and-chips, squid a la plancha—or a full seafood menu in the dining room or the outdoor terrace, which spills out onto the pedestrianized street. Everything is very seafaring—tables have traditional lanterns and model boats on them.

★ Zinzin

18 rue Henri Seillon; tel. 04 83 68 02 88; www. zinzinsainttropez.com; Mon.-Sat. 9am-1am, Sun. 4pm-1am; €24-43

Zinzin (*loopy* in colloquial French), run by the Di Benedetti family, is a new edition to Saint-Tropez's restaurant scene. Mediterranean specials include chickpea salad with feta and spices, grilled octopus with rouille sauce or sea-urchin pasta, and, for dessert, a limoncello tiramisu or spiced strawberry soup. The ambience is fun and friendly.

★ La Petite Plage

9 quai Jean Jaurès; tel. 04 94 17 01 23; www. sainttropeztourisme.com; Apr.-Oct. daily noon-3am; €34-79

More laid-back than its Senequier neighbor, La Petite Plage (the Little Beach) has white sand on the floor, raffia light-shades and furnishings, and driftwood tables and chairs. They do a black truffle croque monsieur, salade Tropézienne, and a Parma ham with cherry tomatoes and basil Burrata. It's a great place for people-watching with a cool beach vibe, serving mainly fish and seafood dishes.

Senequier

29 quai Jean Jaurès; tel. 04 94 97 20 20; www. senequier.com; daily 7:30am-3am; €34-135

The first thing that many people see as they step off their superyacht are the bright red awnings and matching red furniture of Senequier. The café-restaurant is a landmark that has appeared in many films and documentaries about Saint-Tropez and where waiters in white jackets serve a glass of sparkly water for €11, a 50-cl (16-oz) beer for €22, or a bottle of pink champagne for €190. In the place des Herbes behind the restaurant is the **Senequier boutique,** which sells red signs and mementos. It was the company's original cake shop, dating from 1887.

Italian
CasaCri

30 rue Philippe Tilliye; tel. 04 07 42 52; www. casacrilamaisondesjumeaux.com; Oct.-June Tues.-Sat. noon-3pm and 7:30pm-11pm, Sun. 7:30pm-11pm, July-Sept. daily noon-2pm and 7pm-11pm; €18-34

A few yards from the butterfly museum, CasaCri serves home-cooked Italian food in a light, bright dining room with a garden terrace down a long, narrow alleyway. Specials include seafood risotto, seabream with capers and tomatoes, and beef tagliata. With a terra-cotta floor and white cotton tablecloths, the restaurant has an airy feel, a nice alternative to the tightly packed restaurants of Saint-Tropez's quayside.

NIGHTLIFE

In the summer, many of the restaurants stay open until well after midnight; there's singing in the street, dancing on the tables, and plenty of people walking around the old port clasping a bottle of rosé wine trying to find their boats. Most clubs stay open until dawn.

Gaïo

4 rue du 11 Novembre 1918; tel. 04 94 97 89 98; http:// gaio.club; daily 8pm-6am

Gaïo is open for evening meals, where diners can enjoy the fusion of Japanese and Peruvian flavors with a plate of sweet potato fries for €8, or a Casparian caviar served on a blini with cream, quail's egg, and herbs for €620! It's always full, and it's hugely stylish with black Oriental lacquer decor and touches of art deco.

1: fish market **2:** place des Lices **3:** Le Petit Pointu

La Tarte Tropézienne

La Tarte Tropézienne—a brioche-sponge cake filled with a lemon-vanilla cream and sprinkled with sugar crystals—is the most delicious and notorious cake in the South of France. It was originally created by Polish baker Alexandre Micka, who opened a shop in Saint-Tropez in the early 1950s and developed a cream sponge based on his grandmother's recipe. Micka was in charge of local catering when the film crew arrived in 1956 for *Et Dieu… Créa la Femme,* and everyone began making special requests for his cake. According to legend, it was Brigitte Bardot who told him to name it after the town.

Micka patented the name Tarte Tropézienne in the 1970s and the cake took off, with new bakeries opening in Saint-Raphaël, Toulon, Aix-en-Provence, and Marseille. There are now more than 20 shops in France—including four in Saint-Tropez alone. The cake is available at pop-up stores in ski resorts and airports around France, and the original flagship patisserie on the **place des Lices** (tel. 04 94 97 94 25; www.latartetropezienne.fr; daily 6:30am-9pm) has been expanded to include a 1st-floor restaurant. Their website allows eager customers to click and collect from a TT boutique or receive a home delivery. Bite-size Baby Trops and raspberry tartes are also now available.

La Tarte Tropézienne

Les Caves du Roy

27 avenue Marechal Foch; tel. 04 94 56 68 00; www. lescavesduroy.com; Apr.-Oct. Fri.-Sat. midnight-6am, July-Aug. daily midnight-6am

Approaching its 60th anniversary, the most legendary of Saint-Tropez's night-clubs is still going strong. It's famous for its dance floor surrounded by illuminated palm trees, a ceiling of stars, and carpeted floor (changed every few weeks), not to mention the film stars, sports stars, and Saint-Tropez's "in" crowd who frequent the place.

VIP Room

Allée du quai d'Épi; tel. 06 77 07 77 07; June-Sept. daily 8pm-5am

VIP Room is spread over two floors, with a dance floor upstairs. Popular among clubbers and groupies, it plays on its image of being an exclusive venue for pop stars, rappers, DJs, and designers, but is open to "normal" people too. Just dress the part—it's easier to get in if

you visit the shop and buy a VIP Room logo baseball cap and bag.

ACCOMMODATIONS
Under €200
Hotel Playa

99 rue Général Allard; tel. 04 98 12 94 44; www. playahotelsttropez.com; Apr.-Oct.; €140 d

Centrally located on one of the town's main shopping streets, Hotel Playa is a downright bargain for Saint-Tropez. Rooms are simple yet comfortable in a seaside-Provençale style, all with safes, air-conditioning, and private bathrooms. The breakfast buffet is €12 per person and is served in the interior court-yard, which remains cool even in the summer.

€200-400
★ Hotel des Lices

10 avenue Auguste Grangeon; tel. 04 94 97 28 28; www.hoteldeslices.com; end-Mar.-early Nov. and last week of Dec.; €220 d

A Saint-Tropez favorite for 50 years, the Hotel

des Lices is just off the square of the same name and epitomizes the charm and color schemes of Provence. The hotel has 40 rooms, decked out in blond wood and pastel shades, plus three apartments and three nearby villas that it rents out even when the hotel is closed. Sun loungers surround the large swimming pool, and there's also an elevated whirlpool. Breakfast is beside the pool, and there's also a boutique selling locally produced bags and Saint-Trop beachwear.

Hotel Sube

23 quai Suffren; tel. 04 94 97 30 04; www. hotelsubesainttropez.com; mid-Mar.-mid-Nov.; €220 d

Access to this three-star hotel is down an arcade behind a statue of 18th-century vice admiral of the French navy Pierre-André de Suffren (known as the Bailli de Suffren) on the old port. Definitely a hotel for yachting aficionados, the decor is a mishmash of naval artifacts, models of boats, old brass compasses, maps, and leather club armchairs. The most expensive rooms have views over the port with a balcony; others have garden views, patio views, or village views (and the cheapest rooms have no views). There's a huge bar on the 1st floor, and breakfast is also open to nonguests.

Over €400
★ La Ponche

5 rue des Ramparts; tel. 04 94 97 02 53; www. laponche.com; late Mar.-late Oct.; €420 d

The discreetly luxurious five-star hotel built into the ramparts of the old town started life as a simple bistro, serving fishermen in the old port. It was bought in 1938 by the Armando family, who added a few rooms that attracted artists and writers from Paris. Jean-Paul Sartre, Simone de Beauvoir, and Françoise Sagan all stayed there in the 1950s, and the hotel has been a film set for the French productions *Princesse Marie* and *Hors de Prix*. La Ponche has 22 rooms, including four luxury suites and a couple of apartments with sea views, street views, or rooftop terraces.

The original bistro is now a bistronomic restaurant.

INFORMATION AND SERVICES
Office de Tourisme

11 Allée du Quai de l'Epi; tel. 04 94 97 45 21; www. sainttropeztourisme.com; Apr.-Oct. daily 9:30am-7pm, Nov.-Mar. Mon.-Sat. 9:30am-1pm and 2pm-5:30pm

The Office de Tourisme is the best place to find out what's going on and details of any cultural and sporting events. They also organize accommodations and tours around Saint-Tropez.

GETTING THERE
Ferry

The least stressful way to reach Saint-Tropez is via the passenger ferry from **Sainte-Maxime. Bateaux Verts** (14 quai Léon Condroyer; tel. 04 94 49 29 39; www.bateauxverts.com) runs a navette (ferry) service every 15 minutes to Saint-Tropez 24/7 in the summer (less frequently in the low season). Adults are €8.40 single (return €15.20), ages 4-11 €4.50 single (return €8.40).

Ferries arrive on the **quai Jean Jaurès** in Saint-Tropez's Vieux Port, 50 m (164 ft) from the tourist office.

Car

From **Fréjus,** it's a 38-km (23-mi) drive (1 hour) along the coast roads (D8 and D559) to Saint-Tropez, or a 54-km (33-mi) drive (1 hour) via the A8 motorway. The D98A is the only main road in and out of the resort, so it can be extremely slow in July and August—it can take anywhere from 20-90 minutes from Sainte-Maxime, 15 km (9 mi) away.

There are plenty of small **car parks** in Saint-Tropez. The largest and easiest to access is the **Parking des Lices** (41 avenue Paul Roussel; 2 hours/€7).

Bus

The **gare routière** is on avenue Charles de Gaulle. **Zou network bus 876** (https:// services-zou.maregionsud.fr) connects

Abbaye du Thoronet

Thought to be the oldest of the three Cistercian abbeys in Provence (the other two are Sénanque and Silvacane), the **Abbaye du Thoronet** (Le Thoronet; tel. 04 94 60 43 90; www.le-thoronet.fr; Apr.-Sept. daily 10am-6:30pm, Oct.-Mar. 10am-1pm and 2pm-5pm; €8, reduced €6.50, under 18 free) is in the heart of La Daboussière Forest. Construction began around 1176 and the church and the rest of the abbey buildings were finished early the following century, meaning the entire construction boasts a rare architectural unity. The monks used only basic and pure elements in the design, concentrating on rock, light, and water to achieve a simplicity envied by modern architects. When famed French-Swiss architect Le Corbusier was asked to build a convent near Lyon in the 1950s, he came to Le Thoronet for inspiration. After his visit he commented, "The light and shadows are the loudspeakers of this architecture of truth."

Abbaye du Thoronet

Visitors today tour the church and the monks' buildings, including the dormitory, the sacristy, the library, and the chapter house, as well as the cloisters. In the summer, there are themed days with a special focus on subjects such as the music of the Middle Ages and Romanesque symbols.

GETTING THERE

Le Thoronet makes an ideal day trip from Saint-Tropez. It's only a 1-hour drive away, heading directly north on the D558. There are no public transport links to the abbey.

Saint-Tropez with Saint-Raphaël, Fréjus, and Sainte-Maxime (single €2.10).

The minibus service **Var Express** (www.varexpress.eu; single €34.90, return €59.90) links Saint-Tropez and Nice airport with stops at Sainte-Maxime, Port-Grimaud, and Marines de Cogolin. There are three journeys a day each way.

There is no train station in Saint-Tropez; the easiest to reach is in **Saint-Raphaël-Valescure,** a 45-minute drive away on the D8, D559, and D98A coast road.

GETTING AROUND

Saint-Tropez is a tiny village; walking distances are small, so the best way to travel is on foot.

Bus

From June to the end of September, air-conditioned **shuttle buses (navettes)** do a

circuit from the place des Lices to the Plage des Canebiers, Plage des Salins, and Plage de Pampelonne (Pomme de Pin stop) from Monday to Saturday for €0.50 single fare (five per day).

Bike

Saint-Tropez village is not an easy place for cycling, as much of the center is cobbled, the streets are narrow and crowded, and the overall "look" is more superyacht and suntan than mudguards, paniers, and cycle clips. However, there is a cycle path from Sainte-Maxime to Saint-Tropez, and the roads that cross the peninsula to Ramatuelle (D93 and D61) are a pleasure to cycle, gently undulating and passing through pine forests and vineyards.

A 5-minute walk from the Bouillabaisse Plage roundabout, **Blue Bikes** (86 route des

Plages, 43 ZA Saint-Claude; tel. 04 94 96 34 39; https://bluebikes.fr; Apr.-Oct. Mon.-Sat. 9am-12:30pm and 3pm-7pm, Sun. 5pm-7pm) rents city bikes for €20 per 24 hours (€80/week), electric mountain bikes for €40 per 24 hours (€190/week), and racing bikes for €65 per 24 hours (€290/week).

Taxi

Taxis can be eye-wateringly expensive, but are sometimes the only option, since local buses along the coast are few and far between. Three recommended taxi services are **Services Azur** (tel. 06 07 47 47 45 75), **Taxi Philippe Saint-Tropez** (tel. 06 09 10 52 54), and **Taxi Golfe de Saint-Tropez** (tel. 06 24 85 23 66).

Around Saint-Tropez

Directly across the Golfe de Saint-Tropez is Sainte-Maxime, a pleasant family resort with excellent sandy beaches, plenty of accommodations, café-restaurants, and a night market in the summer. Halfway on the coast road between Saint-Tropez and Sainte-Maxime is Port-Grimaud, a late-1960s construction made to look like Venice with canals, arched bridges, and canal-side brasseries, with an older brother, Grimaud, perched 6 km (3.7 mi) inland, overlooking, sometimes perhaps disapprovingly, the yachts and nightlife below.

SAINTE-MAXIME

Known as Saint-Tropez's little sister, Sainte-Maxime is a very lively resort, popular with families and far removed from the chichi pretentions and often exorbitant prices of its prettier neighbor across the bay. Sainte-Maxime's part-cobbled old town has been tastefully renovated and is a pleasant place to shop and eat. The center of town has scores of restaurants and bars, a large boules court, and a palm tree-lined promenade, with water sports on offer along the Plage de la Nartelle. Sainte-Maxime also has a botanical garden and a casino. The resort is a good base for visits to Saint-Tropez; a Bateaux Verts ferry leaves every 15 minutes in the summer (less frequently in low season) from the quai Léon Condroyer.

Sights
Casino Barrière

23 avenue Charles de Gaulle; tel. 04 94 55 07 00; www.casinosbarriere.com; daily 9am-3am

Sainte-Maxime's casino entrance is an extravagant art deco masterpiece. Inside the casino are roulette, blackjack, and 123 slot machines in a room with a retractable roof, a beach restaurant, and a bar and disco for evening entertainment.

Jardin Botanique des Myrtes

boulevard Jean Moulin; www.sainte-maxime.com; Apr.-Sept. daily 8am-8pm, Oct.-Mar. daily 8am-5pm; free

Southwest of the city center on the coast road toward Saint-Tropez, Sainte-Maxime's beautifully maintained botanical gardens have over 60 varieties of trees, including dwarf palms, Chilean coconut trees, parasol pines, and local myrtles. A path runs through the gardens from the entrance behind Plage de la Croisette, where there's a fish pond and a children's play area. No dogs, bikes, or picnics are allowed.

Musée du Phonographe et de la Musique Mécanique

Parc Saint-Donat, route du Muy; tel. 04 94 96 50 52; www.sainte-maxime.com; May-Sept. Wed.-Sun. 10am-noon and 4pm-6pm; €4, students €3, children €1.50

This terrific museum, a 15-minute drive north of Sainte-Maxime on the D25, is dedicated to mechanical instruments and looks

Around Saint-Tropez

Sainte-Maxime

HOTEL LE PETIT PRINCE
RUE DE VERDUN
BLVD A. BRIAND
BLVD MISTRAL
LE BISTRO DE LOUIS
COVERED MARKET
NIGHT MARKET
LE CAFÉ DE FRANCE
D559
D25
CASINO BARRIÈRE
BATEAUX VERTS
Quai Leon Condroyer
HOTEL ROYAL BON REPOS
OFFICE DE TOURISME
LES PALMIERS

0   1 mi
0   1 km

D8
D559

Plage des Elephants

Plage de la Nartelle
PLAGE LA VOILE
D559

D25
Sainte-Maxime
See Detail

D44

JARDINS BOTANIQUES DES MYRTES
Plage de la Croisette

Golfe de Saint-Tropez

LA TABLE DU MAREYEUR
D559

See "Saint-Tropez" Map

Sentier du Littoral

LE SUFFREN
D14   Port Grimaud
ÉGLISE SAINT-FRANÇOIS-D'ASSISE
☆ SAINT-TROPEZ
Grimaud
OFFICE DE TOURISME
D61
Bouillabaisse Plage
D98

Plage Les Salins

D558
D98

BLUE BIKES

Cogolin
D559   C1
CHATEAU MINUTY
D93

CHÂTEAU BARBEYROLLES
DOMAINE DE LA ROUILLÈRE

D98
D61
TOISON D'OR
CLUB 55
✚ PLAGE DE PAMPELONNE

Gassin

Ramatuelle
DOMAINE FONDUGUES-PRADUGUES
INDIE BEACH

La Croix-Valmer
D559
D93

Baie de Cavalaire

Cavalaire-sur-Mer
D559

Sentier du Littoral

0   1 mi
0   1 km

© MOON.COM

like a giant white barrel organ. Inside, some 350 mechanical devices retrace the origins of the phonograph and the beginnings of radio transmission and television. There are also early sewing machines, a mélophone (a precursor to the accordion from the late 18th century), pianolas, wind-up music boxes, typewriters, cameras, telephones, and pipe organs. Early versions of the jukebox, karaoke machine, and record player complete a fascinating look at the past. Note access is only by car, and the final 100 m (328 ft) is up a steep, stony slope.

Beaches
Plage de la Croisette
boulevard Jean Moulin; street parking and Magali car park nearby

A short walk from the tourist office westward toward Saint-Tropez is the Plage de la Croisette. Offering good views of Sainte-Maxime port and the boats out to Saint-Tropez, the bay is split into three sections; the nearest to Sainte-Maxime is pebbly and the other two are sandy. It is a wide, flat beach, excellent for swimming and paddling, but can get busy in the summer. Showers and toilets are on-site, and a few beach bars along with a water sports station are at its westward end.

Plage de la Nartelle
avenue Général Touzet du Vigier; parking on roadside and in car park where D559 joins avenue du Débarquement

Heading east along the coast road toward Fréjus, past gated villas and giant parasol pines, the D559 arrives at Sainte-Maxime's best beach, the Plage de la Nartelle, a 2-km (1.2-mi) stretch of uninterrupted sand and a hub for water sports (not for children's paddling, as the ledge into the water is quite steep). Beside the road at Nartelle are the rusty remains of an amphibious Sherman M4 DD tank that was used in the Allied military landings on August 15, 1944. Damaged by a German land mine, the tank remained buried beneath the sand until 2011, when it was dug up and partly restored.

Plage des Eléphants
avenue du Croisier léger le malin

One km (0.6 mi) east from Plage de la Nartelle on the D559 toward the resort of Les Issambres is the Plage des Eléphants, a thin strip of sand that takes its name from Jean and Cécile de Brunhoff's Babar the elephant. The couple had a house beside the beach in the 1930s, and the bay inspired Jean to illustrate Babar's balloon-trip honeymoon with Celeste in *The Travels of Babar*. The beach is narrow and not for toddlers, as the ledge dips steeply into the sea. The beach club, **Plage les Eléphants** (tel. 04 94 49 11 49; Apr.-Sept. daily 11am-midnight), serves grilled fish and seafood cocktails.

Festivals and Events
Corso de Mimosa
Sainte-Maxime seafront; www.sainte-maxime.com; Feb.; free, €7, under 7 €3

On the first weekend in February, Sainte-Maxime hosts its Corso de Mimosa, a parade through the streets celebrating the region's ubiquitous yellow mimosa flowers. There is a night parade on Saturday, and the corso begins at 2:30pm on Sunday.

Fête de Saint-Pierre
Église Sainte-Maxime and port; www.sainte-maxime. com; June 29; free

On June 29, the resort celebrates the Fête de Saint-Pierre, patron saint of fishermen, with a Mass of Thanksgiving at 6pm followed by a procession through the streets, a blessing of the boats in the harbor, the sacrificial burning of a pointu fishing boat, and finally a sardinade (grilled sardines barbecue) beside the port.

Shopping
Night Market
place du Marché and old town; www.sainte-maxime. com; mid-June-mid-Sept. daily 4pm-midnight

From mid-June to mid-September there's a night market in the streets around the place du Marché, with stalls selling handmade

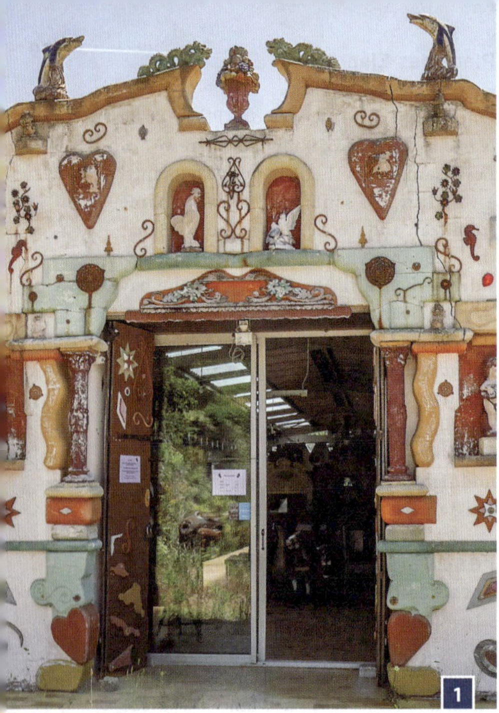
1

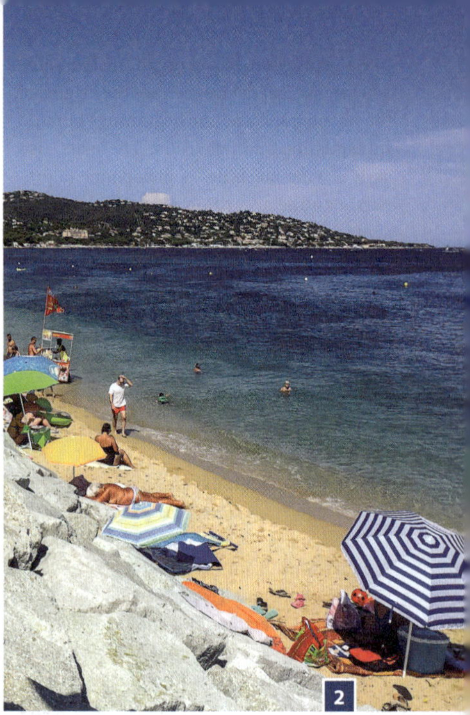
2

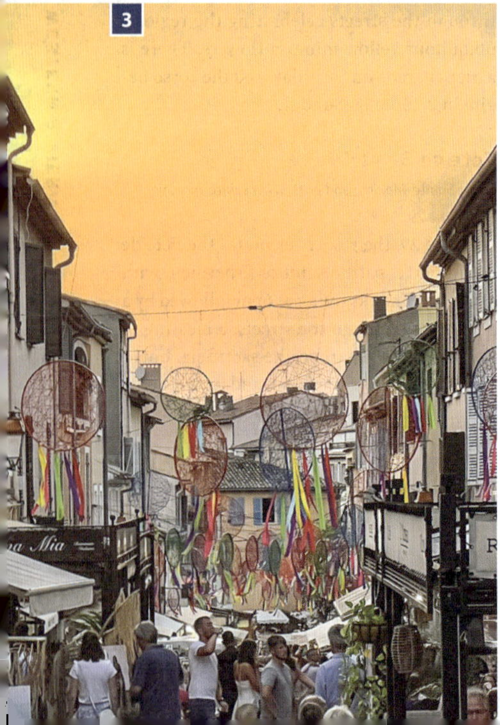
3

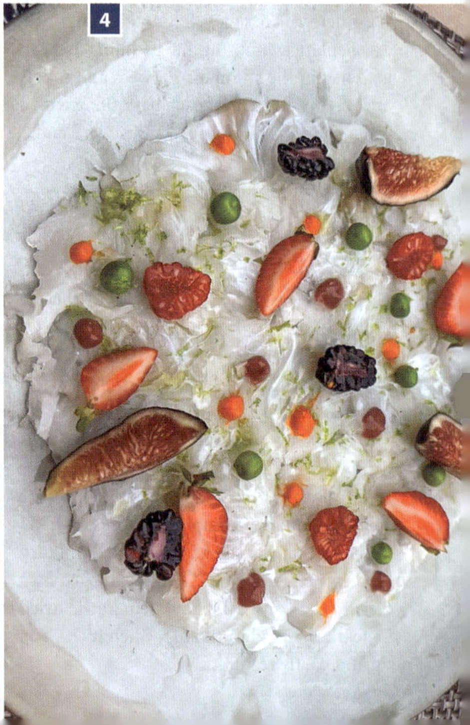
4

jewelry, straw hats, garlands, beachwear, original paintings, and lavender soaps.

Food
Covered Market
rue Fernand Bessy; tel. 04 94 79 42 42; www.sainte-maxime.com; Jan.-Mar. and Oct.-Dec. Tues.-Sun. 7am-2pm, Apr.-mid-July and Sept. daily 7am-2pm, mid-July-Aug. Mon.-Sat. 7am-2pm and 5pm-7:30pm, Sun. 7am-2pm

The covered market on rue Fernand Bessy sells fresh produce as well as roast chickens, honey, spices, and flowers.

Le Café de France
2 place Victor Hugo; tel. 04 94 96 18 16; www. lecafedefrance.fr; June-Sept. daily 6:30am-11pm, Oct.-May Sun.-Wed. noon-2pm, Thurs.-Sat. noon-11pm; €19-34

On the edge of the cobbled old town, the café opened in 1852, when Sainte-Maxime was still a fishing village. It has a huge outdoor terrace and air-conditioned dining room and serves mainly Mediterranean cuisine. It is a nice place for an afternoon coffee, and in the summer it hosts jazz nights.

Le Bistrot de Louis
9 place Colbert; tel. 04 94 43 88 27; https://le-bistrot-de-louis.metro.bar; late Mar.-early Nov. daily noon-2pm and 7pm-10:30pm; €19-42

Protected by a canopy of plane trees on a pedestrianized square, this smart bistro serves a Mediterranean menu including ceviche, steak tartare, rockfish soup, and sea bream filet. It is a popular haunt for locals as well as tourists, so booking ahead is recommended.

Plage La Voile
55 avenue Général Touzet du Vigier; tel. 04 94 49 19 12; www.lavoile-plage.com; Mar.-Nov. daily 10am-11pm; €24-68

One of the best restaurants along La Nartelle beach, La Voile specializes in seafood—lobster,

giant prawns, a grilled catch of the day, and a marmite du pêcheur (fish stew). The main restaurant is within a glass-fronted salon but spills out onto the sand under large parasols. Sun loungers and mattresses in lines on the seashore are for rent.

Accommodations
Hotel Royal Bon Repos
11 rue Jean Aicard; tel. 04 13 51 02 74; www.sainte-maxime-hotel.fr; €128 d

Ideally located in the heart of the old town, this comfortable hotel has 22 elegant rooms in mustard, pink, and cream tones with spacious bathrooms and views of the bay of Saint-Tropez or the old town. All rooms have TVs, minibars, and safes, and there's a shaded cocktail terrace, a billiards room, and a bar.

Les Palmiers
28 rue Gabriel Péri; tel. 04 94 96 00 41; http://hotel-les-palmiers-sainte-maxime.fr; €131 d

Located at the edge of the old town, opposite the embarkation quay for boats to Saint-Tropez, the three-star Les Palmiers has a range of room sizes, including family rooms and suites. There's also a ground-floor restaurant, Le First, open from April to September, with an outdoor terrace. Rooms all have flat-screen TVs and air-conditioning.

Hotel le Petit Prince
11 avenue Saint-Exupéry; tel. 04 94 96 44 47; www. hotellepetitprince.com; €153 d

A top-floor solarium with panoramic views is an attractive addition to this substantial family hotel just 50 m (164 ft) from the sea. Breakfast is served on the terrace in the summer, and there's free parking for guests. Supérieur and Privilège rooms have balconies overlooking the sea, and family rooms with connecting doors are available.

Information and Services
The **Office de Tourisme de Sainte-Maxime** is located at 21 place Louis Blanc (tel. 04 94 55 75 55; www.sainte-maxime.com; Oct.-Mar. daily 9am-noon and 2pm-6pm,

---

1: Musée du Phonographe et de la Musique Mécanique **2:** Plage de la Nartelle **3:** night market in Sainte-Maxime's rue Hoche **4:** ceviche at Le Bistrot de Louis

Apr.-June and Sept. daily 9am-12:30pm and 2pm-6:30pm, July-Aug. daily 9am-7pm).

Getting There
Car
From **Fréjus,** take the D1098 and D98b coast road for 20 km (12 mi; 30 min).

From **Cannes,** take the A8 autoroute and leave at junction 36 for the D25 down to the coast (56 km/34 mi; 1 hour), or the D1098 via Fréjus (60 km/37 mi; 75 min).

From **Saint-Tropez,** it's a 15-km (9-mi), 20-minute drive along the D98A and D559.

The easiest-to-locate and most convenient place to park is the 300-space **Parking du Port** (avenue du Général Leclerc), which has direct access to the departure port. There is also **Parking Louis Blanc** (boulevard Frédéric Mistral), which has 200 spaces. Parking fee is €3.30 per hour April-October and €1.20 per hour November-March.

Bus
Zou bus network Line 876 (tel. 04 13 94 30 50; https://zou.maregionsud.fr) connects Sainte-Maxime to Saint-Raphaël-Valescure, the closest rail station (one per hour; 45 min; €3). **Line 876** continues to Saint-Tropez (1.5 hours, €2.10).

Ferry
Bateaux Verts (14 quai Léon Condroyer; tel. 04 94 49 29 39; www.bateauxverts.com) runs a navette (ferry) service every 15 minutes to and from Saint-Tropez 24/7 in the summer, less frequently in the low season. Boats arrive and depart from the **quai Léon Condroyer** where there is a ticket office, a few restaurants, chandlers, and tourist shops, just in front of the **Capitainerie** (adults €8.40/€15.20 single/return, ages 4-11 €4.50/€8.40, under 4 free).

Getting Around
Sainte-Maxime's cobbled, gently sloping old town is a very pleasant place to walk, as is the lively seafront promenade. For any longer journeys, the resort is served by the **SimpliBus** network (tel. 04 94 54 86 64; www.simplibus.fr), a fleet of green-and-blue shuttle buses with two routes. SimpliBus route 1 runs inland to the out-of-town shopping centers; route 2 runs toward Fréjus, stopping at the Plage de la Nartelle on the seafront.

In July and August, the **Navette Plage (NP)** shuttle bus also runs to Plage de la Nartelle from the tourist office (10 min). Tickets cost €1 per journey (all journeys must be made within 60 min of ticket purchase; day pass is €2.50).

PORT-GRIMAUD

Port-Grimaud occupies the western end of the Golfe de Saint-Tropez, exactly halfway between Saint-Tropez and Sainte-Maxime. The "Venice of Provence," so called because of its canals, gently arching bridges, striking buildings, and touristic charm, is a great resort, unlike anything else on the south coast of France.

Sights
Église Saint-François-d'Assise
place de l'Église; daily 9am-7pm

Port-Grimaud's imposing church took four years to build (1969-1973) and was inspired by the Camargue churches of Saintes-Maries-de-la-Mer, complete with fortifications and gargoyles. Inside, the nave is relatively plain, but made interesting by the 25 arched stained glass windows designed by Hungarian op artist Victor Vasarély, representing the trajectory and reflections of the sun over the sea. You can access the roof of the church via an exterior staircase (96 steps) for great views of the port.

Food and Accommodations
★ La Table du Mareyeur
10 place des Artisans; tel. 04 94 56 06 77; www. mareyeur.com; late Mar.-mid-June and mid-Sept.-Oct. Wed.-Sun. noon-2pm and 7pm-10pm (Fri. dinner only), mid-June-mid-Sept. daily noon-2pm and 7pm-10pm; €31-70

Extravagant and very photogenic platters of

glistening seafood are the specialty at "the fishmonger's table." The restaurant is right on the water's edge and specializes in fish and shellfish dishes with smart service. Le grand mareyeur, a dish to share, is a tray of crab, lobster, oysters, prawns, clams, shrimp, cockles, and whelks for €182, and is certainly a last-night-of-the-holiday treat.

Le Suffren

16 place du Marché; tel. 04 94 55 15 05; www.hotel-suffren.com; Apr.-Oct.; €300 d

On the water's edge of the market square, Le Suffren replicates the turquoise, blues, and silvers of the canal waters in its decor. Rooms are stylish, well-equipped, and have excellent views of the marina. The hotel also has five studios and three apartments with kitchens available for self-catering stays in a separate résidence.

Information and Services
Office de Tourisme de Port-Grimaud

Les Terrasses de Port-Grimaud, rue l'Amarrage; tel. 04 94 55 43 83; www.grimaud-provence.com; Oct.-Mar. Mon.-Sat. 9am-12:30pm and 2pm-5:30pm, Apr.-June and Sept. 9am-12:30pm and 2:30pm-6pm, July-Aug. daily 9:30am-12:30pm and 2:30pm-6:30pm

The tourist office has free maps and brochures on current events and activities to do in and around Port-Grimaud. They can also book accommodations.

Getting There and Around

Port-Grimaud is 25 minutes by car from the A8 autoroute, leaving at junction 36 (Le Muy/Saint-Tropez) and following signs for Sainte-Maxime and then toward Saint-Tropez on the D559 and D98. Port-Grimaud is 8 km (5 mi) before **Saint-Tropez** on the D559.

Zou bus network Line 874 (tel. 04 13 94 30 50; https://zou.maregionsud.fr) connects Sainte-Maxime to Port-Grimaud (10 min; €2.10).

RAMATUELLE

Ramatuelle is a small, arty, medieval village tucked into the side of a hill. Its narrow lanes twist around into each other, keeping the houses cool in the hot summer. The flagstone **rue des Amoureaux** and **rue Rompe Cuou** are always covered in flowers; look out for the intricate door knockers and carved stonework. Ramatuelle is located at the edge of the Ramatuelle plain, which is covered in vineyards and parasol pines, and is a short drive from the most famous beach resort in France, the Plage de Pampelonne, a 5-km-long (3-mi-long) strip of white sand.

Beaches

TOP EXPERIENCE

★ Plage de Pampelonne

route des Plages; park at one of the six entrances spread out off the route des Plages behind the beach; fixed fee of €5.50 payable at the entrance

Saint-Tropez's most famous beach is actually in neighboring Ramatuelle, a huge expanse of white sand backed by gentle dunes and scores of trendy beach clubs, pine forests, and the occasional luxury campsite. The sand is cleaned early every morning, the water is clear and shallow, and it's a wonderful walk end to end, from **Cap du Pinet** to **Cap Camarat,** watching the yachts drop anchor and strolling past beach restaurants, volleyball courts, swaths of sun loungers and parasols (€20-50/day), and the occasional nudist area (**Neptune Plage,** approximately halfway along Pampelonne, is for naturists). If you have your own transport, the best time to visit is after 5pm when the beaches clear of tourists, the water sports stop, and there is plenty of space at the bars.

Food and Accommodations

Most of the beach clubs serve the same type of food: warm octopus salad, ceviche, grilled fish, salmon tartare, and crème brûlée or homemade ice cream for dessert, with iced rosé wine and sparkling water. The style is smart and relaxed, but prices can be astronomical.

Club 55

43 boulevard Patch; tel. 04 94 55 55 55; www.club55. fr; Apr.-early Nov. and Dec. 20-Jan. 5 daily 11am-7pm; €30-70

With its own jetty, a history going back 70 years, and bookings taken months in advance, there's nowhere quite like Club 55. The food can be quite ordinary, and a popular hors d'oeuvre is just a huge bowl of raw vegetables, but the ambience is permanently buzzing with yacht captains, château-owners, and mega-rich entrepreneurs rubbing shoulders with celebrities (it's a favorite with Bono, Kate Moss, Leonardo DiCaprio, and Rihanna). The car park is full of Bentleys and Maseratis (with sand in their tires).

On the beach, Club 55 has a self-conscious but authentic simplicity; furniture is made from reclaimed wood, and there's a rainwater storage system and recycling bins hidden in the sand for guests. A shuttle boat service picks up guests from their yachts, who are guided along the pontoon by staff before being seated in the crammed restaurant, which has room for 400 diners.

Indie Beach

route de Bonne Terrasse; tel. 04 94 79 81 04; www. indiegroup.fr/indiebeach; May-Oct. daily 10am-1am; €34-195

Three local friends, Vincent, Raphaël, and Tobias, constructed this cool beach bar in washed-out wood, creams, and mushroom tones. Sun loungers can be rented by the day, with lunch served on comfortable banquettes in the shade, or you can sit on a pouf with an aperitif later in the day. It's one of the few beach bars to stay open at night.

★ Toison d'Or

Plage de Pampelonne, chemin des Tamaris; tel. 04 94 79 83 54; www.riviera-villages.com/toison-d-or; Apr.-Oct.; from €310

One of only a handful of campgrounds behind the Plage de Pampelonne, Toison d'Or is part of the group that runs Kon Tiki (also on Pampelonne) and the Prairies de la Mer (near Port-Grimaud). There are over 100 beach cabins spread out across the site, with super-expensive ones looking directly out at the ocean, and 1-, 2-, and 3-bedroom cabins (some with two decks and a hot tub) gathered around palm trees in collective gardens. The site has a grocery shop, a pool, a pizzeria, a fitness spa, a hairdresser, a restaurant, a crêperie, and direct access onto the beach.

Information and Services

Office de Tourisme de Ramatuelle

place de l'Ormeau; tel. 04 98 12 64 00; www. ramatuelle-tourisme.com; Apr.-June and Sept. Mon.-Fri. 9am-1pm and 2pm-6pm, Sat. 10am-1pm and 2pm-6pm, July-Aug. Mon.-Sat. 10am-1pm and 2pm-6pm, Sun. 9am-1pm, Oct.-Mar. Mon.-Fri. 9am-12:30pm and 2pm-5pm

Located in the old village, the tourist office has free maps and brochures on current events, activities, and suggested walks to do in Ramatuelle.

Getting There and Around

Car

Ramatuelle is 10 km (6 mi; 15 min) from **Saint-Tropez** on the D93 and D61. It's a lovely drive through the vineyards on a gently undulating road. The center is pedestrianized, but there are several car parks on the roads leading up to the village (free of charge).

The easiest way to reach the beach is to drive or take a **taxi** (€25 from Saint-Tropez or €15 from Ramatuelle village).

Bus

Zou bus network Line 875 (tel. 04 13 94 30 50; https://zou.maregionsud.fr) connects Saint-Tropez to Ramatuelle (40 min; €2.10) via half a dozen stops just inland of Plage de Pampelonne.

1: Plage de Pampelonne **2:** Port-Grimaud **3:** Club 55 **4:** Toison d'Or beach huts

Vineyards of the Saint-Tropez Peninsula

When the Greeks arrived on the southern coast of France 3,000 years ago, it didn't take them long to begin planting vines. They produced something that looked very similar to today's rosé wine—a rust-colored alcoholic liquid. The vineyards around Saint-Tropez are therefore some of the oldest in France, with the Var département producing around 40 percent of the country's rosé wine and comprising 6 percent of global rosé production.

Made primarily with Granache, Tibouren, Mourvèdre, and Cinsault grape varieties, rosé wine has been a longtime favorite in Saint-Tropez, but has become more popular in the last two decades because of its less-formal appeal, the public's greater interest in ethnic cuisine, and the fact that rosé wine can be opened and drunk immediately. Rosé goes with everything, and the French tend to drink it with pizzas and barbecues as a carefree picnic wine, served with a couple of ice cubes to keep it light, summery, and refreshing.

All the vineyards are open for visits and some for tastings, usually by appointment. Bottles are always a better value when purchased directly at the vineyard and range in price from €4-62. Here are some of the peninsula's best vineyards:

CHÂTEAU MINUTY

2491 route de la Berle, Gassin; tel. 04 94 56 12 09; www.minuty.com; Mon.-Fri. 9am-7pm (7:30pm July-Aug.), Sat.-Sun. 10am-1pm and 2pm-6pm (6:30pm July-Aug.)

This is one of the most prestigious estates and one of the few Côtes de Provence vineyards where the grapes are still picked by hand. Its slopes were first planted in 1936, and the estate now covers 110 hectares (271 acres). Around 90 percent of Minuty's production is dedicated to rosé wine, all managed by the third- and fourth-generation Matton family.

CHÂTEAU BARBEYROLLES

2065 route de la Berle, Gassin; tel. 04 94 56 33 58; www.barbeyrolles.com; Apr.-Sept. Mon.-Sat. 9am-7pm, Oct. Mon.-Sat. 9am-6pm, Nov.-Mar. Mon.-Fri. 9am-5pm

The Western Riviera

The farther away from Saint-Tropez, the more "normal" the coastal resorts become: reasonable prices, ordinary families driving average cars with bikes on roof racks, inflatable sharks hanging out of windows, and unsophisticated beach restaurants without a voiturier (car valet) in sight. The D559 highway hugs the coastline, with sandy bays, pine forests, and lots of water sports on offer.

★ BORMES-LES-MIMOSAS

If there's one village worth leaving the coast for, it's Bormes-les-Mimosas. Perched on a hill with great views out to sea and over the Îles d'Or, Bormes added the "Mimosa" label over 50 years ago, claiming to have the largest density of the yellow flowers on the south coast. It's a pretty village filled with flowers even outside of mimosa season (December-March), its pink-tiled houses covered in bougainvillea, flowering creepers, and trays of carnations. In the summer it can be very crowded, especially walking up and down the 83 steep steps of the **rue Rompi Cuou** (Provençal for "ass-breaker"), which is lined with tourist stores and workshops and joins the lower village to the upper village.

Bormes-les-Mimosas has a huge boules court and an Australian-inspired garden, the **Parc González,** at the far end of the **place Saint-François.**

A little farther along route de la Berle, this 12-hectare (29-acre) organic estate was acquired by Régine Sumeire in 1977. Her grapes are also hand-picked, and plowing is still done by horses. Sumeire was one of the first female vintners in Provence, and in 1985 she created the Pétale de Rose, her flagship wine.

vineyards of the Saint-Tropez Peninsula

DOMAINE DE LA ROUILLÈRE

route de Ramatuelle, Gassin; tel. 04 94 55 72 60; www. domainelarouillere.com; Apr.-Oct. Mon.-Fri. 9am-8pm, Sat., Sun. 10am-8pm

Another of the prestigious wine estates on the D61, halfway between Saint-Tropez and Ramatuelle, the Domaine de la Rouillère has been growing grapes since 1900. Approximately 75 percent of production is dedicated to rosé, 15 percent to red, and 10 percent to white wine over the 120-hectare (300-acre) estate.

DOMAINE FONDUGUES-PRADUGUES

7677 route des Plages, Ramatuelle; tel. 04 94 79 09 77; www. fondugues.fr; May-Sept. Tues.-Sun. 11am-2pm and 5pm-9pm; free visits and tastings

This domaine has launched L'Éphémère, a pop-up wine-tasting truck in a vintage Citroën van where clients can have a game of boules, lunch, and a choice of organic rosé wines.

Sights
Chapelle Saint-François

place Saint-François; daily 9am-6pm

The Romanesque-Provençal-style chapel was built in 1560 to honor Saint-François de Paule, who delivered the village from the plague in 1481. Having been restored in 1989, it has an altar dating from the 17th century and paintings of Saint-François. Its contemporary stained glass windows by Georges Pescadère, dating from 1993, depict Saint-François on his journey to save the village and, in the background, vegetation and animals of Provence.

Musée d'Art et d'Histoire

103 rue Carnot; tel. 04 94 71 56 60; http://musee-bormes.com; Nov.-Apr. Wed.-Sun. 11am-6pm, Apr.-Nov. Tues.-Sun. 10am-7pm; €8.50

This museum is inside a building dating from 1650, which has been, at various times, a boys' school, a courthouse, and a prison, but since 1985 has been the village's main gallery and exhibition space. Bormes has a large collection of art from the 19th and 20th centuries, which is exhibited alongside contemporary paintings on the two floors of the stone building.

Food and Accommodations
★ La Tonnelle

23 place Gambetta; tel. 04 94 71 34 84; https:// restaurant-la-tonnelle.com; Wed.-Sat. noon-2:30pm and 7pm-11:30pm, Tues. 7pm-11:30pm; €25-29

Inspirational chef Gil Renard does all the food shopping himself from local sources. His Gourmandise menu at La Tonnelle (two starters, two main courses, and two desserts) is €75, but would be twice that in a more prestigious location. Dishes include guinea fowl with vegetables or a navarin of lamb, served while diners sit alongside the bay windows of the restaurant's veranda. There's also a good range of vegetarian, vegan, and gluten-free dishes.

Hotel Belle Vue

14 place Gambetta; tel. 04 94 71 15 15; www. bellevuebormes.com; mid-Jan.-Nov.; €130 d

Across the street from the tourist office is the family-friendly Hotel Belle Vue, which, as its name suggests, has fantastic views out to sea from its elevated location in the heart of the village. The hotel has 17 rooms, decorated in soft tones with patterned carpets, some with baths, others just showers. It's worth asking for a balcony (rooms 1, 5, 7, 10, and 15) and a sea view, although the view over the square is also great. The attached restaurant has a panoramic terrace.

Information and Services

The **Office de Tourisme Bormes-les-Mimosas** is located at 1 place Gambetta (tel. 04 94 01 38 38; www.bormeslesmimosas.com; Oct.-June Mon.-Sat. 9:30am-12:30pm and 2pm-6pm, July-Sept. daily 9:30am-7:30pm).

Getting There and Around

Bormes-les-Mimosas is a 40-minute drive (38 km/23 mi) from **Sainte-Maxime** along the D559 and D98. It's a picturesque route of tight bends from the valley below.

Once in the village, the most convenient place to leave the car is the 370-place **car park** under the **place Saint-François.**

Zou network bus 878 (tel. 04 13 94 30 50; https://zou.maregionsud.fr) runs from Saint-Tropez (90 min) to Hyères (35 min) via Bormes-les-Mimosas (€3 single).

HYÈRES

A completely different ambience from Saint-Tropez greets visitors in Hyères, an unpretentious town whose fame as the first winter resort on the Riviera has now faded, but whose charm lingers in its aristocratic buildings and thousands of palm trees. The former public washhouse (lavoir) has been redesigned and is now a meeting place for artists and romantics,

and the old town has been revitalized with the **Parcours des Arts,** a walking route comprising 30 art and craft workshops and galleries.

Sights

The square in the heart of the old town, the **place Massillon,** is accessed through the **Porte Massillon,** a 14th-century gate, which still has one of its original watchtowers. Dominated by the huge, curved **Saint-Blaise tower,** the last remnant from the 12th-century Templars, the square is lined with restaurants, bars, and tourist shops, and is a great stop for an afternoon drink.

Tour des Templiers

Place Massillon; tel. 09 63 53 22 36; Sept.-June Tues.-Sat. 10am-1pm and 2pm-5pm, July-Aug. Tues.-Sat. 10am-1pm and 3pm-7pm; free

Hyères was a strategic site for the Knights Templar, for whom this building housed a chapel on the ground floor and a guard room on the 1st floor. It is still possible to see the arrow-slits cut into the facade, as well as a stained glass red Templars' cross high above the modern stairwell at the back of the building. Today, the building is open as a gallery space.

Villa Noailles

47 Montée de Noailles; tel. 04 98 08 01 98; www. villanoailles-hyeres.com; early Sept.-early June Wed.-Thurs. and Sat.-Sun. 1pm-6pm, Fri. 3pm-8pm, mid-June-early Sept. Tues.-Sun. 2pm-8pm, Thurs. 3pm-9pm; free

A steep 10-minute walk up through the old town leads to the Villa Noailles, a modernist masterpiece of concrete and glass with amazing views of the coastline. Inside are white-walled rooms now dedicated to art and design exhibitions.

Parisian art patrons Charles and Marie-Laure de Noailles commissioned the construction of the villa to Robert Mallet-Stevens in 1923. It took four years to build and contained the first indoor swimming pool in Europe. There is a thick Perspex lid to it today, so visitors can walk over the top of the white-tiled pool and inspect the art exhibits. Guests in

Hyères and Les Îles d'Or

the 1930s were given swimming costumes and encouraged to do exercises around the pool. Salvador Dalí, Luis Buñuel, Jean Cocteau, and Man Ray all spent time there. Having been inspired by Mallet-Stevens's designs in 1929, Ray even made a short film about the place titled *Les Mystères du Château du Dé,* in which the de Noailles couple and the pool exercises feature prominently.

The villa is enclosed within the **Parc Saint-Bernard gardens** (winter daily 8am-5pm, summer daily 8am-7:30pm), where visitors can wander along the pathways, observe the Îles d'Or through many square-framed gaps in the walls, and smell the cultivated lavender, rosemary, and silverberries. There's also a cubist garden of square bushes in a triangular frame designed by Gabriel Guevrekian.

La Banque, Musée des Cultures et du Paysage d'Hyères

14 avenue Joseph Clotis; tel. 04 83 69 19 40; July-Aug. Tues.-Fri. 10am-1pm and 3pm-7pm, Sat. 3pm-7pm, Sun. 10am-1pm, Sept.-June Tues., Thurs., Fri., and Sun. 2pm-6pm, Wed. and Sat. 10am-1pm and 2pm-6pm; free

Opened in 2021, Hyères's local history museum comprises a former bank building, the apartment of the bank manager, gardens, and an underground bank vault. The museum houses historical artifacts, fine arts, sculpture, and contemporary art.

Les Vieux Salins d'Hyères

Espace Nature des Salins d'Hyères, rue de Saint-Nicolas, Village des Salins; tel. 04 94 01 09 77; Jan.-Feb. Wed.-Sun. 9:30am-4:30pm, Mar.-June Wed.-Sun. 9am-noon and 2pm-5:30pm, July-Aug. Wed.-Sun. 9am-noon and 4pm-8pm, Sept.-Oct. Wed.-Sun. 9am-noon and 2pm-5:30pm, Nov.-Dec. Wed.-Sun. 9:30am-4:30pm; free

Hyères's former salt marshes are today a 141-hectare (350-acre) biological sanctuary with a nature center offering information on bird and animal life, flora, the hydraulic management of the marshes, and the history of the site. The center has permanent and temporary exhibitions and offers 2-hour guided tours of some of the salt marshes (binoculars supplied), which focus on the history of the salt industry and the migrating and mating birds that fill the marshes.

Beaches
Almanarre Plage

route de l'Almanarre; parking in the large parking lot beside the roundabout or along the road

A 5-km (3-mi) stretch of sand, this is Hyères's wildest, windiest beach. It's popular with windsurfers and kite-surfers, and every October hosts the Grand Prix de l'Almanarre, a windsurfing competition. However, when the mistral wind drops it's a great beach for families, with fine, pale sand and a gentle slope leading into clear, turquoise waters.

Food and Accommodations
Le Jardin

19 avenue Joseph Clotis; tel. 04 94 35 24 12; http:// restolejardin.flashenligne.com; summer daily 11am-3pm and 7pm-10pm, hours vary seasonally; €19-24

Through an iron gate and up a few stone steps, Le Jardin is set in a garden romantically lit with fairy lights at night, a green oasis in the center of town. A large palm tree beside the entrance provides most of the shade, but wooden tables on a gravel terrace all have parasols or straw beach umbrellas, and there are flowers everywhere and a

swordfish at Chez Soi

vegetation wall. There are plenty of choices for vegetarians.

★ Chez Soi

9 place de la République; tel. 04 94 31 51 21; Tues.-Sat. noon-2pm and 7:30pm-10:30pm, Sun. noon-1:30pm; €52

Opposite the early-Gothic Église Saint-Louis on the place de la République, Chez Soi is one of Hyères's finest restaurants serving imaginatively presented dishes such as lamb with smoked rosemary and Riesling wine or Blanquette of scallops and monkfish. There is seating for 50 inside and an outdoor terrace for 30 more diners in the summer. It's popular with locals so booking is recommended.

★ La Reine Jane

Port de l'Ayguade, 1 quai des Cormorans; tel. 04 94 66 32 64; www.lareinejane.fr; €220 d

Having appeared in Jean-Luc Goddard's 1965 film *Pierrot le Fou,* this hotel has since been completely revamped by Jean-Pierre Blanc, manager of the Villa Noailles, and his friend, local restaurateur David Pirone, who commissioned 14 different designers to create 14 rooms, all inspired by the sea. Some are completely blue or completely white; others are based on an aquarium, underwater cartoons, or seaside changing cabins. The former guesthouse still has a 1950s feel, with blue and white nautical stripes decorating the common parts, but the pièce de résistance is the modernist-style rooftop with 360-degree views of the sea and surrounding landscape. Yoga classes are held there in the summer.

Information and Services
Office de Tourisme

Rotonde du Park Hôtel, 16 avenue de Bélgique; tel. 04 94 01 84 50; www.hyeres-tourisme.com; Sept. Mon.-Fri. 9am-5pm, Sat. 9am-4pm, Oct.-Mar. and Apr.-June Mon.-Fri. 9am-6pm, Sat. 9am-4pm, Sat. 9am-4pm, July-Aug. Mon.-Sat. 9am-6pm, Sun. 9am-1pm

The Office de Tourisme has maps for two 1.5-hour self-guided walking tours around central Hyères: one for the old town, which follows a steep, uneven path to the Villa Noailles, and the other a tour around Hyères's grand 19th-century buildings. The office has details of all current events happening in the town as well as museums, galleries, and art studios, part of the Parcours des Arts that was created as a way of rejuvenating the artistic heritage of Hyères.

Getting There and Around
Air

Toulon-Hyères airport (Aeroport Toulon/ Hyères; tel. 08 25 01 83 87; www.toulon-hyeres.aeroport.fr) is 5 km (3 mi) southeast of Hyères. **Shuttle bus 102** (navette) goes directly from the airport to Hyères bus station and Hyères port in 10-15 minutes (www.reseaumistral.com; €1.40 single).

Flights within France include Paris Orly, Paris Roissy-CDG, Brest, Ajaccio, Lille, Nantes, and Strasbourg. International flights include Brussels, Antwerp, Geneva, Rotterdam, and London Gatwick.

Car

From **Saint-Tropez,** the D98 heads directly west to Hyères, but can be very slow in the summer (1-1.5 hours; 50 km/31 mi). The main **parking lot** for visiting the old town is underneath the **place Clémenceau.**

Bus

Zou network bus 878 (tel. 04 13 94 30 50; https://zou.maregionsud.fr) runs from Saint-Tropez to Hyères (2 hours; €3 single). The station is on place du Maréchal Joffre.

Train

The **TGV InOui fast train** connects to Paris (tel. 3635; www.sncf-connect.com; 4 hours 40 minutes; €50). The local **TER** trains are on the Marseille-Toulon-Hyères line, which takes around 1.25 hours from Marseille Saint-Charles station (€20).

★ LES ÎLES D'OR

The golden islands, also known as the Îles d'Hyères, have some of the region's best beaches and have become havens for birdlife, hiking, biking, and, on one of the islands, nudity.

In Greek legend, Prince Olbianus had four daughters who loved swimming. One day, as the four were out bathing in the sea, a pirate ship appeared on the horizon, and the prince, in despair, begged the gods to save his daughters from capture. Taking pity on him, the gods turned his four daughters to stone. The three who were the farthest away became Porquerolles, Port-Cros, and Le Levant, and the fourth sister, who was closest to shore, was turned into the Giens peninsula.

Sights
Porquerolles

Porquerolles, regularly sacked by Saracen pirates, was fortified by French king François I and turned into a sodium plant in the 19th century. It was bought as a wedding gift by Belgian gold and silver mine owner François Fournier for his young wife in 1912. They introduced the fabulously exotic flora to the island, and its 60 km (37 mi) of footpaths make it a popular destination for walkers, cyclists, and nature lovers.

The **Fondation Carmignac** (La Courtade; tel. 04 65 65 25 50; www.fondationcarmignac. com; end-Apr.-early Nov. Tues.-Sun. 10am-6pm, to 7pm July-Aug.; adults €15, reduced €11, ages 12-26 €6, under 12 free) is a villa and gardens exhibiting over 300 works of contemporary art. The island has several vineyards, fortifications to explore, and bike hire outlets.

Port-Cros

Port-Cros, the smallest of the three islands at just 3.2-by-2.2 km (2-by-1.4 mi), is also the most mountainous. It was the first national maritime park in Europe and is a wonderful place for hiking and swimming. No vehicles are allowed on the island.

Le Levant

Over 90 percent of the third island, Le Levant, is owned by the French military and used as a launching base for submarines. The remainder is reserved for a harmonious-living naturist colony called Heliopolis, where nudists are welcome to wander everywhere except the harbor and village square.

Food and Accommodations
Héliotel

Montée du val des Moines, Île du Levant; tel. 04 94 00 44 88; www.heliotel.net; late Apr.-Sept.; €135 d

Part of the naturist complex created by

Porquerolles Island

the Durville brothers in 1931, the 14-room Héliotel overlooks the beach and has a pool, Mediterranean garden, and restaurant. It's a 15-minute walk to the hotel from the port, but a minibus meets every boat and drops guests outside the hotel. Nudity is obligatory on the beach and seafront.

Hostellerie Provençale

Île de Port-Cros; tel. 04 94 05 90 43; www. hostellerieportcros.fr; mid-Apr.-early Nov.; €175 d
This seafront hotel has six pleasant double rooms and a nice fish restaurant, and it organizes boat hire and picnics for guests.

Auberge des Glycines

22 place d'Armes, Porquerolles; tel. 04 94 58 30 36; www.auberge-glycines.com; €220 d
This characterful inn on the island's central square has a Mediterranean restaurant and rooms decorated in a Provençal style.

Le Mas du Langoustier

Le Langoustier, Porquerolles; tel. 04 94 58 30 09; www. langoustier.com; late Apr.-Oct.; €540 d with half board for two
Le Mas du Langoustier is a luxury hotel and restaurant where guests are picked up in a golf cart from the harbor and whisked off to a deluxe holiday experience. Amenities include a spa, heated outdoor pool, tennis courts, and fine dining.

Getting There

TLV-TVM (www.tlv-tvm.com/reservations) runs regular ferry service to the three islands all year, with greatly increased service in July and August. The journey to Porquerolles takes 20 minutes departing from **La Tour Fondue** in Giens (10 km/6 mi south of Hyères). There are 7 trips per day November-March; 14 in September, April, and May; and 20 per day in July and August. Adult fare is €24 return ticket, reduced fares and those under 26 are €21, and those under 4 are free. Online booking is available via the company's website.

The ferry to Port-Cros (1 per day Sept.-Apr. rising to 4 per day July-Aug.) and Le Levant (1 per day Sept.-Apr. rising to 3 per day July-Aug.) departs from **Hyères Le Port.** Fares are to Port-Cros (1 hour) and Le Levant (1.5 hours). Adult fare is €29 return ticket, reduced fares and those under 26 are €25, and those under 4 are free.

Monaco and Menton

Seen from above, Monaco is a huge blot of concrete: skyscrapers and towers of tiny apartments soaring ever upward, traffic jams of luxury sedans, pavements and elevated walkways teeming with tourists and grumpy billionaires, and the marina, buzzing with boat engines and roadworks. It's not for everyone, but the experience is unique, and no visit to the Riviera is complete without a trip to Monaco.

The second-smallest independent state in the world after the Vatican, Monaco has a surface area of just 2 sq km (0.8 sq mi) and more millionaires per square kilometer than anywhere else in the world. A further 6 hectares (15 acres) of land is being reclaimed at Portier Cove (Mareterra), a development preparing to accommodate another raft

Highlights

Look for ★ to find recommended sights, activities, dining, and lodging.

★ **Casino de Monte-Carlo:** The legendary belle epoque building made Monaco the playground of the rich it is today (page 403).

★ **Musée Océanographique:** One of the most sumptuous buildings on the Riviera houses the voluminous pools of Monaco's aquarium and an entertaining collection of marine artifacts donated by Monaco's former ruler Albert I. It's worth a half-day's visit (page 408).

★ **Formula One Grand Prix:** The world's greatest car race has been a feature of Monaco

life since 1929. It's a noisy, thrilling experience (page 413).

★ **Roquebrune Village:** Perched on a rocky crag, the château ruins here are a reminder of the region's turbulent history. The tiny village looks and feels as if little has changed since the Middle Ages (page 423).

★ **Cap Moderne:** The remarkable beach homes of the modernist designers Eileen Gray and Le Corbusier can only be visited via a guided tour (page 424).

Monaco and Menton

To La Defi
de la Madone

Roquebrune-Cap-Martin

LA BOMBONIÈRE

AU GRAND
INQUISITEUR

CHÂTEAU DE
ROQUEBRUNE

CASARELLA

LA GROTTE
& L'OLIVIER

LA BELLE VUE

*Train
Station*

Plage de
Buse

VILLA
E-1027

LE
CABANON

CAP
MODERNE

D2564

*Baie de
Roquebrune*

0   0.25 mi

D53

D22

D23

D50

See Detail

D2564

A8

Roquebrune-
Cap-Martin

*Train
Station*

CAP
MODERNE

D2564

MAYBOURNE
RIVIERA

D6007

D6098

Plage
du Golfe
Bleu

*Baie de
Roquebrune*

GRANDE CORNICHE

A8

D2564   La Turbie

D6007

MOYENNE CORNICHE

D6007

See
"Monaco"
Map

CASINO DE MONTE-CARLO

Monte
Carlo

See
"Monaco
Center"
Map

D37

MUSÉE
OCÉANOGRAPHIQUE

FRANCE

M6007

M6098

Cap-d'Ail

MONACO

of millionaires, all eager to avoid paying income tax. The lingua franca is English, despite proximity to France, but there are also plenty of Russian and Italian voices. It's a world of yachting and driving, large jewels and big wallets. But it is also incredibly safe. There is one police officer for every 60 residents, hundreds of security cameras, and no paparazzi, and the microstate is spotlessly clean. Around 40,000 people travel there each morning to work from the surrounding towns and villages, doubling the local population and filling the trains, buses, and car parks.

Besides its fantastic oceanography museum, Monaco has some excellent cultural centers, concert halls, galleries, and gardens. Head of state Prince Albert II is very active in ecological issues, and his foundation is at the forefront of marine conservation in the Mediterranean. Though the constant building work is a bothersome and noisy distraction, and it can be a tiring place to walk around, it's worth spending time here just to see the traffic directed by white-gloved police officers, the marina's superyachts, the casino, and the guests coming through the swinging doors at the Hotel de Paris.

Menton, 14 km (9 mi) east and the closest town to Italy, was made popular by Britain's Queen Victoria during her winter visits there at the end of the 19th century. It claims to have the most days of sunshine in France (320 average), and it certainly has one of the oldest populations. Despite that, it is a vibrant seaside resort with peach- and tan-colored buildings, an annual music festival, the long-running

Fête du Citron, and one of the "best restaurants in the world."

Monaco and Menton are separated by Roquebrune and its rocky cape. The perched village feels almost unchanged since the Middle Ages. It's popular with cyclists, and its peninsular beaches are great places to swim, explore, and visit the modernist holiday homes on the coastal footpath, a world away from flashy Monaco.

PLANNING YOUR TIME

Just 20 km (12 mi) east of Nice, Monaco will take a **full day** to explore—the **Musée Océanographique** being a highlight—and is definitely worth a return after dark to see the lights and nightlife around the **Port Hercule** and **casino.** However, hotels there are **extremely expensive,** and Menton, which allows easy access to Monaco, may be a better place to stay (the train journey takes only 10 min; driving takes 20 min along the coast road). Since Monaco is extremely crowded at rush hours (7am-9am and 4pm-6pm), traveling there is best done during the middle of the day or on the weekend. Four to five days would be enough time to explore the region: Monaco, Menton, Roquebrune, and across the border into **Italy,** with Menton as a good home base. For beach lovers, **Larvotto** to the east of Monaco is a protected sandy bay with plenty of beach clubs, and the **Plage du Buse** close to Roquebrune-Cap-Martin railway station is good for swimming and exploring the coast, and is rarely crowded even in the **summer.**

Itinerary Ideas

TWO DAYS IN MONACO AND MENTON

Day One

1 Aim to arrive in Monaco in the late morning by train, skipping the rush hour traffic, and enjoy a drink at one of the cafés on the **place d'Armes.**

2 Head up the slope of Le Rocher to Monaco-Ville and take a tour of the **Musée Océanographique,** which provides a history of underwater exploration and features a modern aquarium in the basement.

3 Have a Mediterranean lunch under the arches of the **Port Hercule** while watching the superyachts in the harbor.

4 Swim in the large outdoor pool on the seafront, the **Stade Nautique Rainier III.**

5 Visit the **Collection de Voitures,** one of the world's great automobile collections.

6 Have supper at restaurant **Les Perles de Monte-Carlo** overlooking Fontvieille marina.

7 Head to the **Casino de Monte-Carlo** for a few hands of blackjack and an hour at the roulette table.

Day Two

1 Enjoy breakfast and people-watch at the art nouveau **Café de Paris** on the place du Casino.

2 Take the train to Menton and enjoy an assiette Mentonnaise at **La Mandragore** restaurant under the arches of the place des Herbes.

3 Wander through the old town and pay a visit to **Basilique Saint-Michel Archange,** a magnificent baroque cathedral.

4 Eat dinner at the fabulous **Mirazur,** just a stone's throw from Italy.

Monaco

Monaco was once Europe's poorest state, a country built on a large rock with steep slopes and no agriculture. It had lost the revenues from the lucrative tax on lemons when it separated from Menton in 1848. However, 15 years later, it was granted a gambling franchise, and immediately wealthy guests from Nice and Cannes began to flock to what France's then-largest newspaper *Le Figaro* described as "un paradis sur terre"—paradise on earth. (The paper's owner was rewarded with a villa in Monaco.) The railway arrived, and profits from the casino were so high that taxation was abolished for its citizens and a company was set up to manage the gambling, the fantastically misleading Société des Bains de Mer

Itinerary Ideas

FRANCE

ITALY

Menton

See Detail

Baie du Soleil

TOURIST OFFICE

Menton

Plage des Sablettes

QUAI BONAPARTE

RUE IGUYAU

RUE DE LA RÉPUBLIQUE

RUE SAINT-MICHEL

PROMENADE DU SOLEIL

Plage du Marché

| 0 | | 200 yds |
| 0 | | 200 m |

DAY ONE	DAY TWO
1 Place d'Armes	1 Café de Paris
2 Musée Océanographique	2 La Mandragore
3 Port Hercule	3 Basilique Saint-Michel Archange
4 Stade Nautique Rainier III	4 Mirazur
5 Collection de Voitures	
6 Les Perles de Monte-Carlo	
7 Casino de Monte-Carlo	

Monaco

A8   D2564   D6007

GRANDE CORNICHE

CHEMIN ROMAIN SUPÉRIEUR

BLVD GUYNEMER

MONTE CARLO COUNTRY CLUB

ROUTE DE BEAUSOLEIL

AVE PRINCE RAINIER III MONACO

CHEMIN DE LA CRÉMAILLÈRE

BASSE CORNICHE

BLVD D'ITALIE

HOTEL MONTE-CARLO BAY

BEAUSOLEIL

FRANCE

AVE DES ANCIENS COMBATTANTS DE L'AFRIQUE DU NORD

MOYENNE CORNICHE

BLVD DE LA RÉPUBLIQUE

BLVD DU LARVOTTO

GRACE

AVENUE 31

Larvotto Beach

AVE PRINCESSE GRACE

NOUVEAU MUSÉE NATIONAL DE MONACO VILLA SAUBER

★ TWIGA MONTE-CARLO

MONACO

HOTEL OLYMPIA

SASS CAFÉ

Jardin Japonais

ADAGIO APART-HOTEL MONACO PALAIS JOSÉPHINE

SONG QI

Beausoleil

PHARMACIE DE MONTE CARLO

IL TERRAZZINO

HOTEL METROPOLE MONTE-CARLO

MÉTROPOLE MALL

TOURIST OFFICE

★ CASINO DE MONTE-CARLO

D6007

BLVD PRINCESSE CHARLOTTE

BLVD DE SUISSE

MONTE CARLO

See "Monaco Center" Map

AVE PRINCE RAINIER III MONACO

MONEGHETTI

BLVD DU JARDIN EXOTIQUE

RUE GRIMALDI

BLVD ALBERT 1ER

AVE JOHN F. KENNEDY

Port Hercule

LA CONDAMINE

AVE DE LA PORTE NEUVE

MONACO-VILLE

NOUVEAU MUSÉE NATIONAL MONACO VILLA PALOMA

★ MUSÉE OCÉANOGRAPHIQUE

PRINCESS GRACE HOSPITAL

MUSÉE DES TIMBRES ET DES MONNAIES

■ LES JARDINS SAINT-MARTIN

CATHÉDRALE DE MONACO

Port Fontvieille

LES PERLES DE MONTE-CARLO

FONTVIEILLE

M6098

STADE LOUIS II

BOUTIQUE HOTEL MIRAMAR

Parc Paysager de Fontvieille

Mediterranean

Sea

0   0.5 mi

0   0.5 km

© MOON.COM

(SBM, sea bathing company). The Crown took 10 percent of SBM's profits and the dynasty began to flourish. Even today, the SBM owns most of the leading hotels and casinos in Monaco.

ORIENTATION

Monaco is divided into six main districts. The best known, **Monte-Carlo,** is the area around the casino and the opera house, with the grandest hotels and glitzy shopping centers. Down the hill, **La Condamine** is the zone around the harbor that includes the outdoor swimming pool, artists' studios on the **quai Antoine I,** and the daily market in the **place d'Armes. Monaco-Ville** is on top of **Le Rocher,** the medieval town on a huge outcrop of rock jutting out into the Mediterranean, dominated by the prince's palace, the cathedral, and the oceanographic museum. Monaco-Ville overlooks **Fontvieille** to the west, a **marina** full of yachts and sports cars, and the Princess Grace rose garden just in front of the **heliport.** Fontvieille is built on land reclaimed in the 1970s and is characterized by its hundreds of confectionary-looking apartment blocks.

To the far east of the principality is **Larvotto,** dominated by its beach, great for people-watching in the many restaurants and bars along the seafront esplanade and **avenue Princesse Grace. Moneghetti** is the residential zone on the steep, twisting lanes where the landmass rises up toward **Beausoleil,** a dormitory town for some of Monaco's 40,000 daily commuters, and a cheaper place to stay for the night if you want to sleep nearby. Its pastel-colored belle epoque villas are now dwarfed by residential tower blocks, full of millionaires living in tiny studio apartments.

Elevators and moving walkways throughout the principality help mitigate long stretches of Tarmac and Monaco's hills. The **Mareterra** development, which runs along the seafront from Port Hercule to **Larvotto,** is due to be finished by 2025 and will include 110 apartments and six villas as well as recreational spaces, shops, a small marina, a cycle track, and a footpath.

SIGHTS
Monte-Carlo

TOP EXPERIENCE

★ Casino de Monte-Carlo
place du Casino; tel. +377 98 06 21 21; www. montecarlosbm.com; daily 2pm-early hours of the

Casino de Monte-Carlo

morning when the games rooms close; identification required; atrium free, game rooms 18+ only €19, tours daily 10am-1pm €18, ages 13-17 €12, ages 6-12 €8

The world's most famous casino opened in 1863, and it is the single most important institution in the history of Monaco. Its capacity for earning money made Monaco what it is today, though Monegasque citizens (who make up only around 15 percent of the local population) are not allowed to gamble.

The original idea for a casino in Monaco is attributed to Princess Marie Caroline, business-minded wife of Prince Florestan I, whose House of Grimaldi was facing bankruptcy in the mid-19th century. Having had several unsuccessful incarnations around Monaco, the princess handed the reins to businessman François Blanc, who was responsible for the Bad Homburg casino in Germany and who set about founding the Société des Bains de Mer (SBM). He asked Charles Garnier to design an ornate building, and the casino was born. In 1962, the Monaco state became the SBM's principal shareholder. It is a place of myths and legends, far too glamorous to be a tourist trap, and, like all casinos, more likely to be a place of demise than glorious triumph.

The casino offers games of English roulette, European roulette (one of the few casinos still to offer the game), Trente-et-Quarante, Punto Blanco, and Baccara. Card games, dice games, Texas Hold'em poker, blackjack, craps, and slot machines alone provide €120 million in jackpots each year. The high-roller "real" business gets done in the Salons Super Privés, where the tailor-made games and pro tournaments take place. Facing the swinging doors in the lobby of the Hotel de Paris across the square is a bronze statue of a mounted King Louis XIV. The king's horse's knee is shiny from where gamblers leaving the hotel have touched it to bring them a bit of luck. In the main casino hall, the minimum bet is €5 and the maximum €2,000. In the private gaming rooms, the minimum is €10 and there is no maximum!

The dress code has been relaxed in recent years, but gentlemen still require a jacket after 7pm, and no shorts or T-shirts are allowed. It's hard to define what is acceptable attire, since many of the high-rollers come straight off their boats, but generally, the smarter the better.

Place du Casino

place du Casino

The place du Casino is the centerpiece of Monte-Carlo, and it's where crowds gather for family photo shoots and selfies alongside expensive sports cars and vintage convertibles on Monte-Carlo's most famous roundabout. The enormous terrace of the **Café de Paris** is a front-row seat for watching the glamorous wander up the steps of the Hotel de Paris across the square. To the south are the casino and opera house. Running up the middle of the square are the Allée des Boulingrins (a French version of the word "bowling green," which testifies to what the gardens were originally designed for), modern sculptures, and some extraordinary botanical specimens.

Nouveau Musée National de Monaco Villa Sauber

17 avenue Princesse Grace; tel. +377 98 98 91 26; www. nmnm.mc; Sept.-June daily 10am-6pm, July-Aug. 11am-7pm; €6, under 26 free, free for all on Sun.

One of Monaco's two contemporary art museums, the belle epoque Villa Sauber houses half of the principality's contemporary art collection. With high ceilings and huge French windows, the gallery opens out onto a sloping garden with views over Grimaldi Forum and Larvotto Beach. Hosting two exhibitions a year, Villa Sauber's general theme is art and performance, and the selection of modern works, which includes sculptures, neon signs, and garden installations, is designed to promote contemporary art as well as give support to "creators, thinkers, and researchers."

1: place du Casino **2:** Le Palais Princier **3:** Princess Grace Irish Library **4:** Musée Océanographique

Monaco Center

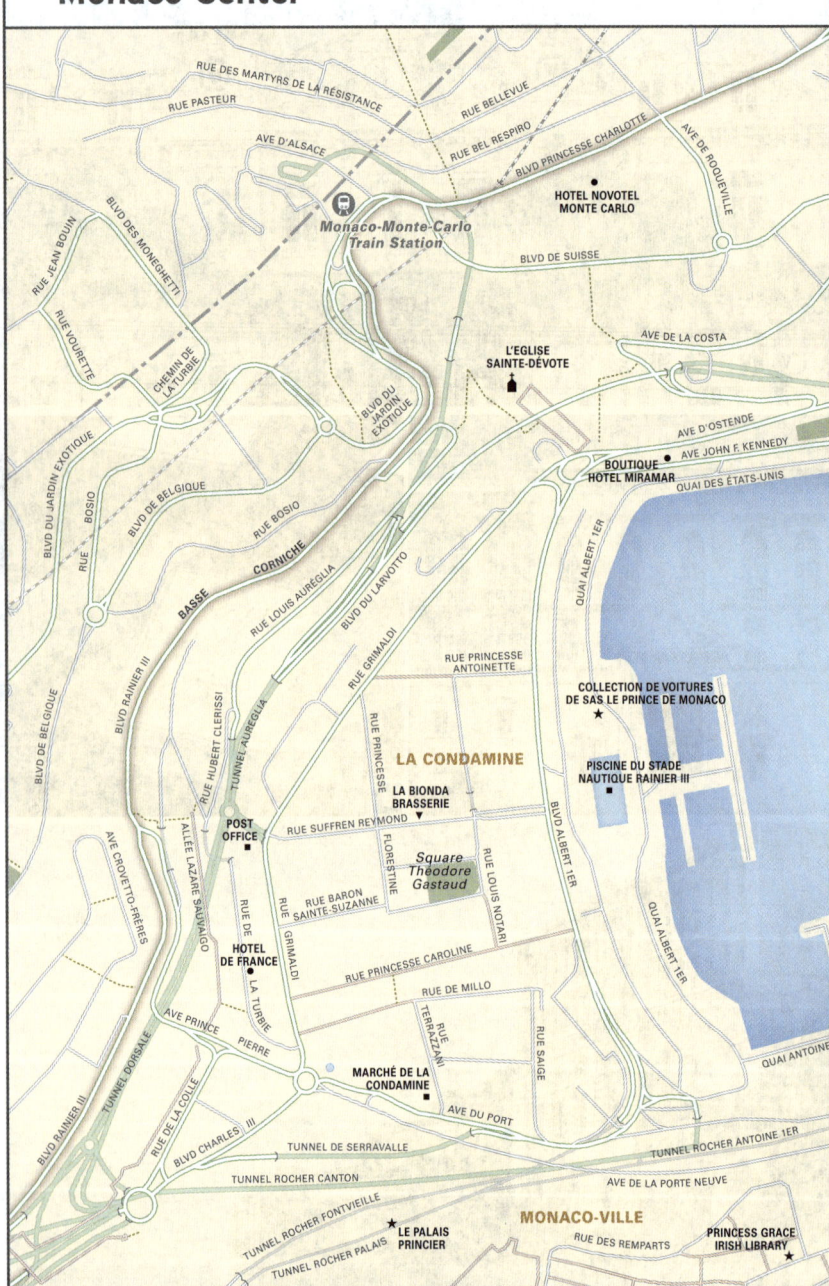

RUE DES MARTYRS DE LA RÉSISTANCE

RUE PASTEUR

RUE BELLEVUE

AVE D'ALSACE

RUE BEL RESPIRO

BLVD PRINCESSE CHARLOTTE

AVE DE ROQUEVILLE

HOTEL NOVOTEL MONTE CARLO

🚉 Monaco-Monte-Carlo
Train Station

BLVD DE SUISSE

RUE JEAN BOUIN

BLVD DES MONEGHETTI

RUE VOIRETTE

CHEMIN DE LA TURBIE

BLVD DU JARDIN EXOTIQUE

AVE DE LA COSTA

L'EGLISE SAINTE-DÉVOTE

BLVD DU JARDIN EXOTIQUE

RUE BOSIO

BLVD DE BELGIQUE

RUE BOSIO

CORNICHE

BASSE

RUE RAINIER III

RUE LOUIS AURÉGLIA

BLVD DU LARVOTTO

AVE D'OSTENDE

AVE JOHN F. KENNEDY

BOUTIQUE HOTEL MIRAMAR

QUAI DES ÉTATS-UNIS

QUAI ALBERT 1ER

RUE PRINCESSE ANTOINETTE

COLLECTION DE VOITURES DE SAS LE PRINCE DE MONACO ★

RUE PRINCESSE

RUE HUBERT CLERISSI

TUNNEL AURÉGLIA

LA CONDAMINE

PISCINE DU STADE NAUTIQUE RAINIER III ◼

BLVD DE BELGIQUE

LA BIONDA BRASSERIE ▼

RUE SUFFREN REYMOND

FLORESTINE

Square Théodore Gastaud

RUE LOUIS NOTARI

BLVD ALBERT 1ER

QUAI ALBERT 1ER

POST OFFICE ◼

ALLÉE LAZARE SAUVAIGO

RUE DE

RUE DE LA TURBIE

RUE BARON SAINTE-SUZANNE

RUE GRIMALDI

HOTEL DE FRANCE

RUE PRINCESSE CAROLINE

AVE CROVETTO-FRÈRES

AVE PRINCE PIERRE

RUE DE MILLO

RUE TERRAZZANI

RUE SAIGE

QUAI ANTOINE

TUNNEL DORSALE

MARCHÉ DE LA CONDAMINE ◼

AVE DU PORT

BLVD RAINIER III

RUE DE LA COLLE

BLVD CHARLES III

TUNNEL DE SERRAVALLE

TUNNEL ROCHER ANTOINE 1ER

TUNNEL ROCHER CANTON

AVE DE LA PORTE NEUVE

TUNNEL ROCHER FONTVIEILLE

LE PALAIS PRINCIER ★

MONACO-VILLE

RUE DES REMPARTS

PRINCESS GRACE IRISH LIBRARY ★

TUNNEL ROCHER PALAIS

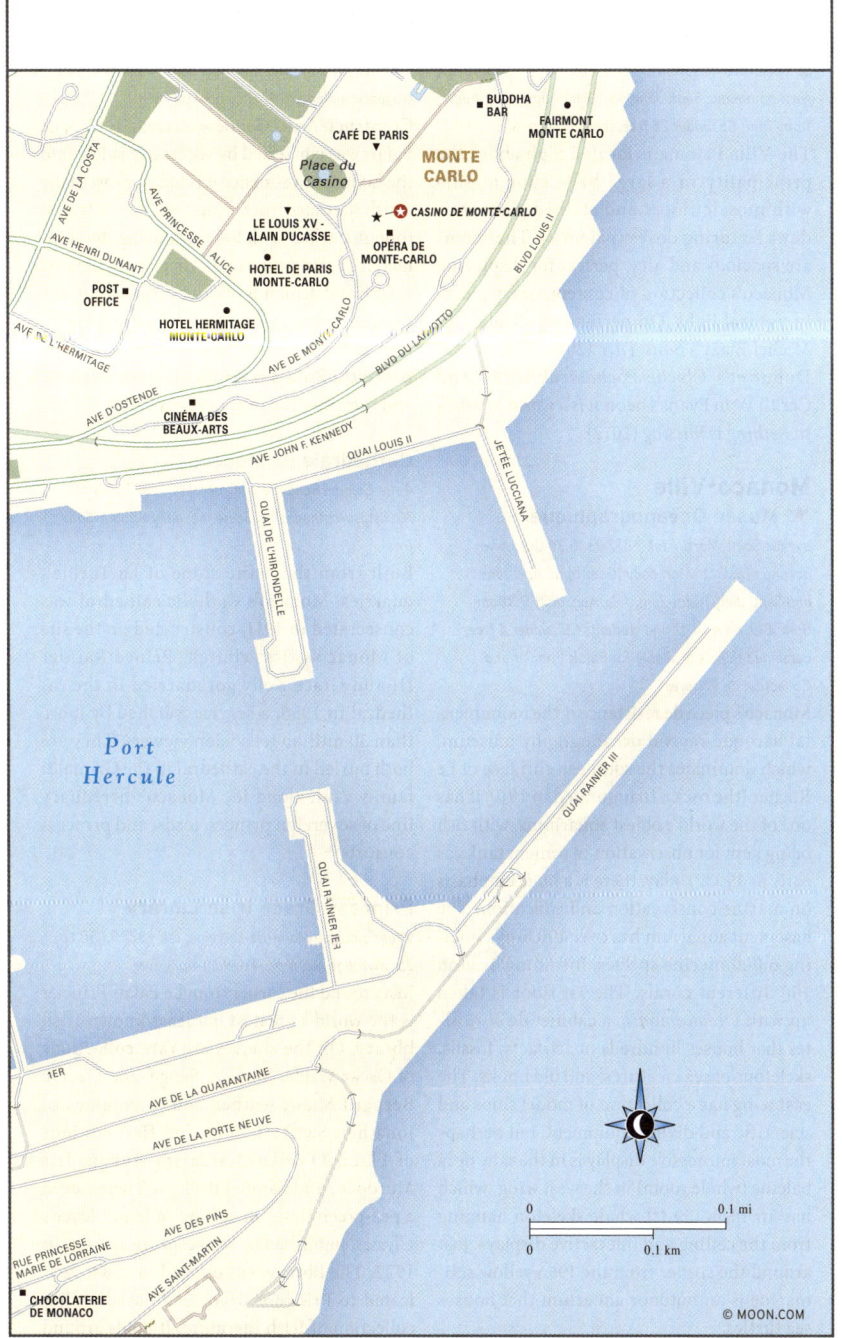

Nouveau Musée National de Monaco Villa Paloma

56 boulevard du Jardin Exotique; tel. +377 98 98 48 60; www.nmnm.mc; Sept.-June daily 10am-6pm, July-Aug. 11am-7pm; €6, under 26 free, free for all on Sun.

The Villa Paloma is located high above the principality in a large belle epoque villa with mosaic floors and stained glass windows featuring doves (palomas). The rooms are spacious and airy, perfect for displaying Monaco's collection of contemporary pieces in natural light. On permanent display are Michel Blazy's *Sans Titre* (2014), one of Jean Dubuffet's *Cloche-Poche* sculptures, and Cerith Wyn Evans's neon *It is a world in which something is missing* (2012).

Monaco-Ville

★ Musée Océanographique

avenue Saint-Martin; tel. +377 93 15 36 00; www. oceano.org; Oct.-Mar. daily 10am-6pm, Apr.-June and Sept. daily 10am-7pm, July-Aug. daily 9:30am-8pm; €19, ages 14-17 and students €12, under 4 free, combined tickets available for Palais Princier and Collection de Voitures

Monaco's pièce de résistance is the monumental baroque-revival oceanography museum, which dominates the southern cliff face of Le Rocher (the rock). Inaugurated in 1910, it has one of the world's oldest aquariums, with fish being kept for observation in cement tanks as early as 1903. Today, there is a huge emphasis on marine conservation and education. The basement aquarium has over 100 tanks holding 6,000 marine species, including around 100 different corals. The 1st floor is taken up with *Océanomania,* a cabinet de curiosités that houses hundreds of artifacts, fossils, skeletons of sea creatures, and old books. The east wing has a collection of model ships and scientific and diving equipment, but perhaps the most impressive display is in the salle de la baleine (whale room) in the west wing, which has an 18-m (59-ft) whale skeleton hanging from the ceiling and interactive displays. Just around the corner from the 1966 yellow submarine is an outdoor aquarium that houses sea turtles.

Le Palais Princier

Le Rocher; tel. +377 93 25 18 31; www.palais.mc; Sept.-June daily 10am-6pm, July-Aug. daily 10am-7pm; €10, students and ages 6-17 €5, under 6 free

Constructed as a Genoese fortress in 1215, Le Palais was converted by successive rulers into the luxurious residence it is today. Most of the lavish state apartments are open for visits, and though packed with tourists during the summer, it's a sumptuous tour displaying the vast wealth and demands of a succession of royalty. The most photogenic of all the natural balconies in Monaco, one side of the immaculate **place du Palais** overlooks the Port Hercule and the other the port of Fontvieille.

Cathédrale de Monaco

4 rue Colonel Bellando de Castro; tel. +377 93 30 87 70; https://cathedrale.diocese.mc; daily 8:15am-6pm; free

Built from the white stone of La Turbie's quarries, Monaco's Catholic cathedral was consecrated in 1911, constructed on the site of Monaco's first church. Prince Rainier III and Grace Kelly got married in the cathedral in 1956, a service watched by more than 30 million television viewers. They are both buried in the cathedral in the Grimaldi family vault alongside Monaco's hereditary line of sovereign princes, lords, and princess consorts.

Princess Grace Irish Library

9 rue Princesse Marie-de-Lorraine; tel. +377 93 50 12 25; www.pgil.mc; Mon.-Fri. 9am-4pm; free

Just around the corner from Le Palais Princier is the world's greatest but least-known Irish library. On the shelves are rare collections of Oscar Wilde, John M. Synge, and George Bernard Shaw; leather-bound volumes of Jonathan Swift's letters; and first editions of Flann O'Brien, Katharine Tynan, Iris Murdoch, and Samuel Beckett. There's even a pea-green first full edition of James Joyce's *Ulysses,* published by Shakespeare and Co. in 1922. The library was opened in 1984, dedicated to Princess Grace and based on her collection of Irish literature. It holds around

9,000 volumes, many donated by the late fashion designer Karl Lagerfeld.

Fontvieille
Musée des Timbres et des Monnaies

Terrasses de Fontvieille; tel. +377 98 98 41 50; www.
mtm-monaco.mc; daily 9:30am-5pm; €3, students and
ages 12-18 €1.50, under 12 free

For a principality built around money, the museum of stamps and currency is a relatively low-key affair. Definitely one for traveling numismatists and philatelists, the prince's collection of rare stamps and coins includes original sketches for coinage, mint designs, and artist's proofs, as well as commemorative medallions and printing machines. Opened in 1996, the museum displays the original stamp collection of Prince Albert I, which was added to by Prince Rainier III (who described the Monaco stamp as "the first ambassador of the country") and has been enlarged further by the current Prince Albert II.

La Condamine
Collection de Voitures de SAS le Prince de Monaco

54 route de la Piscine; tel. +377 92 05 28 56; www.
mtcc.mc; Sept.-June Mon.-Sat. 10am-6pm, July-Aug.
Mon.-Sat. 10am-7pm; €10, ages 6-17 €5, under 6 free

Monaco's premier car museum displays the private collection of Prince Rainier III, who amassed over 100 classic, vintage, city, and muscle cars over a 40-year period. Some of the collection was sold off at auction in 2012, but it has been reorganized and added to by the reigning prince, including a silver Bugatti Chiron from 2016 and Sébastien Loeb's Citroën DS3, which won the Monte-Carlo rally in 2013. The collection moved from the palace garage into the Terrasses de Fontvieille in 1993 and, since 2022, has been housed alongside the swimming pool in the Port Hercule. On display is the Bugatti Type 35C driven by Marcel Lehoux in the first Monaco Grand Prix in 1929 and a jet-black Bugatti W16 Mistral from 2024. Among the most elegant cars in the collection is the Renault Floride, presented to Princess Grace in 1959, and an identical version of the 1954 Sunbeam Alpine MK1 roadster she drove in Alfred Hitchcock's film *To Catch a Thief,* which is set on the French Riviera.

L'Église Sainte-Dévote

1 rue Sainte-Dévote; tel. +377 93 50 52 60; https://
saintedevote.diocese.mc; daily 10am-noon and 4pm-
6pm; free

According to the legend, Devota, a young Christian girl from Corsica, was martyred around 303 CE under Emperor Diocletian. Her body was saved by the faithful and placed on a boat headed for Africa, but a wind carried the craft to Monaco, where it arrived on January 27, led by a dove. That date, now a public holiday in Monaco, is when a boat is symbolically burned outside the church. Legends aside, the courtyard of this cream-pink church is a tranquil haven among the high-rise blocks and superyachts. It dates from the early 16th century, but a chapel is recorded on the same site as early as 1070.

SPORTS AND RECREATION

Despite the high-rise tower blocks, a convoluted road network, and constant construction, Monaco has a good number of parks, squares, gardens, and outdoor sports facilities.

Parks and Squares
Les Jardins Saint-Martin

avenue Saint-Martin; Apr.-Oct. daily 9am-6:45pm,
Nov.-Mar. daily 9am-5:45pm; free

Next door to the Musée Océanographique and opened in 1816, these were the first public gardens in the principality. There are almost as many sculptures as plants, including a life-size bronze of Prince Albert I, which has the monarch at the helm of a ship in waterproof hat and capes, staring out to sea. There are too many tourists to make this a place of tranquil contemplation, but the views are fantastic— and there are plenty of benches, interesting trees, and a public toilet nearby.

Jardin Japonais

avenue Princesse Grace; tel. +377 98 98 83 36; Apr.-Oct. daily 9am-6:45pm, Nov.-Mar. daily 9am-5:45pm; free

The Japanese garden's air of tranquility and sea view will no doubt vanish when the Portier Cove urban development is completed in 2026, but until then, it remains one of Monaco's most restful green spaces. The gardens were created in 1994 in accordance with the principles of Zen design by Yasuo Beppu—a pond with koi carp swimming under water lilies and lotus flowers, islands, decked walkways, bamboo water features and hedges, a waterfall, a teahouse, stone lanterns, red bridges, and a Zen garden.

Beaches
Larvotto Beach

avenue Princesse Grace

Monaco's largest stretch of beach is Larvotto, divided into several sections by its private beach clubs that fill the gravelly sand with their parasols and sunbeds. Behind the beach is a two-deck esplanade lined with beach-wear shops, a gymnasium, and plenty of restaurant-bars. Swimming is great for families, as the T-shaped concrete quays protect the bay. There's a floating pontoon to swim to, an anti-jellyfish mesh, lots of children's play areas, and a cycle path.

Swimming
Piscine du Stade Nautique Rainier III

quai Albert I; tel. +377 93 30 64 83; www.guide-piscine. fr/alpes-maritimes/piscine-du-stade-nautique-rainier-iii-a-monaco-2991_P; late Apr.-June and mid-Sept.-mid-Nov. Fri.-Sun. and Tues.-Wed. 9am-6pm, Mon. and Thurs. 7am-6pm, July-early Sept. Fri.-Sun. and Tues.-Wed. 9am-6pm, Mon. 7am-6pm, Thurs. 7am-6:30pm, closed during Grand Prix week; €7.50, ages 3-11 €6, under 3 free

Symmetrically placed in the middle of the Port Hercule, the swimming pool has diving boards and a 45-m (148-ft) waterslide. It's a popular place to sit and watch swimmers, divers, and the superyachts beyond. Swimming lessons and sunbeds are also available. From December to March, the site becomes an **ice-skating rink** (Mon.-Thurs. 11am-6pm, Fri.-Sat. 11am-11pm, Sun. 11am-9pm, daily during school winter break 11am-11pm; admission and skate rental €7, under 5 free).

Hiking
Monte-Carlo to La Turbie

Distance: *6.5 km (4 mi) one-way*
Time: *3.5 hours*
Information and Maps: *Monaco tourist office*
Trailhead: *Monaco Monte-Carlo railway station*

This hike begins at the Monaco Monte-Carlo railway station, following signs for Beausoleil and taking the avenue d'Alsace to the place Moneghetti and the start of the chemin de la Turbie. Follow a zigzag path behind les Hauts de Monte-Carlo up toward the village of **La Turbie** and to the Roman **Trophée des Alpes** monument. La Turbie has a drinking-water fountain and lots of cafés for a meal or refreshments. From there, it's a short walk to the **Tête de Chien** (dog's head) rock, visible from the coast. Descend along the **old Roman road,** which is signposted back toward Monaco.

Cycling
Bike Trip

tel. 06 88 06 13 62; www.rent-bike.fr; bicycle and motorbike delivery Apr.-Oct. Mon.-Fri. 9:30am-12:30pm and 2:30pm-6:30pm, Sun. delivery available July-Aug.; city bikes €14/24 hours, road bikes €35-74/24 hours

Bike Trip specializes in online bicycle rentals, which the company delivers to customers in offices, at hotels, on boats, or at the Monaco Monte-Carlo railway station. Payment and reservation can be made via the website. The company rents city, touring, racing (carbon or aluminum frames), and electric bikes, plus scooters and motorbikes. It also has agencies in Nice, Cannes, and Menton.

1: Jardin Japonais **2:** Stade Louis II **3:** Larvotto beach

Moneco Bike

tel. +377 99 90 24 04; www.monecobike.com; deliveries any time 9am-6pm; electric bike rental €45/day (1-2 days), €38/day (3-4 days), €32/day (4-6 days), or €28/day (7 or more days), including lock, helmet, and basket

Moneco Bike delivers bikes anywhere in Monaco to homes, hotels, boats, or the railway station. The company specializes in foldable electric Fat Bikes, but also sells and rents Hoverboards, pavement scooters (trottinettes), and electric, mountain, city, and vintage-look bikes.

Spectator Sports
Stade Louis II

3 avenue des Castelans; tel. +377 92 05 40 21; www.stadelouis2.mc; Sept.-June Mon.-Sat. 7am-11pm, Sun. 8am-7pm, July-Aug. Mon.-Fri. 7am-9:30pm, Sat. 8am-6pm, Sun. 8am-3pm, stadium tours in English, French, and Italian Sept.-June Mon.-Fri. and July-Aug. Tues.-Sat. 10:30am, 11:45am, 2:30pm, 3:45pm, and 5pm.; €5.60, under 12 and over 65 €2.80

Football club AS Monaco plays its home games at the 18,250-seat Stade Louis II in Fontvieille. The stadium is so close to the border that the practice pitch next door is actually in France. It includes a gas station, shops, offices, a huge four-level underground car park, and Monaco University. There are also squash courts, a weight-training room, a gymnasium, and a swimming pool, and a total of 2,100 doors in the building. Besides the football and occasional music concerts, the stadium is best known for hosting the annual athletics meeting Herculis every July.

ENTERTAINMENT AND EVENTS
The Arts
Opéra de Monte-Carlo

place du Casino; tel. +377 98 06 28 28; www.opera.mc and www.balletsdemontecarlo.com; tickets from €21, under 25 from €12

Monaco's opera house, which is situated at the back of the casino, was designed by Charles Garnier, who had built the Paris opera house in 1874. He was given only six months by the Société des Bains de Mer to create something spectacular, so over 1,000 laborers worked day and night on the project, and a steam engine was brought down from Paris to provide the power to light up the building site after dark. On January 25, 1879, the opera house was inaugurated with a performance by Sarah Bernhardt in front of 800 guests, mainly members of the European aristocracy. The principal hall, known as the Salle Garnier, is one of Europe's grandest theaters for opera and ballet.

Cinéma des Beaux-Arts

12 avenue d'Ostende; tel. +377 93 25 36 81; www.cinemas2monaco.com; €12, students and over 60 €9.50, under 12 €7

The belle epoque Théâtre Princesse Grace houses the excellent Cinéma des Beaux-Arts on its lower-ground floor, which shows current movies in their original version (with French subtitles). In June, July, and August, the Monaco open-air cinema launches its season. Films begin at 10pm in the Terrasses du Parking des Pêcheurs on Le Rocher, which has 500 plastic seats (with free cushions) and a snack bar for soda, beer, and popcorn.

Festivals and Events
Printemps des Arts

various arts venues in Monaco; tel. +377 93 25 58 04; www.printempsdesarts.mc; mid-Mar.-Apr.; average ticket price €26

The Printemps des Arts festival takes place in March and April at a selection of venues throughout the principality, including the Grimaldi Forum and Théâtre Princesse Grace. The festival includes classical music concerts, master classes, roundtable discussions, workshops, dance performances, and small-scale concerts given in private apartments in Monaco.

Monte-Carlo Masters

Monte-Carlo Country Club, 155 avenue Princesse Grace; tel. +377 97 98 70 00; www.montecarlotennismasters.com; mid-Apr.; tickets €20-340

Part of the ATP World Tour Masters 1000 series, the Monte-Carlo Masters is one of the most prestigious men's tournaments in the tennis calendar. It is played on clay courts and runs for nine days in mid-April each year. The main Rainier III court has a capacity of 10,200 seats, and during the tournament, trains stop at the rarely used Monte-Carlo Country Club station (between Monaco and Roquebrune-Cap-Martin rail stations), 500 m (1,640 ft) from the entrance. Buses 1, 2, 4, 5, and 6 are also free for tournament ticket-holders, and there are free shuttle buses from the Portier roundabout to the club. The best way to watch a match is to reserve tickets online and travel to the courts by train.

★ Formula One Grand Prix

www.monaco-grand-prix.com; late May; tickets from €120 for a seat in Le Rocher to €6,000 for a two-day VIP pass, official online ticket sales at www.formula1monaco.com

Monaco's famous 2-mi (3.2-km) circuit has changed little since the Grand Prix began there in 1929. It runs along the harbor front, up the hill to the casino, under the Fairmont hotel, and down Portier, into a tunnel and alongside the swimming pool to the very tight La Rascasse corner (named after the bar-restaurant), and back along the home straight—78 times. There is little chance to overtake on the course, which Brazilian driver Nelson Piquet compared to "riding a bicycle round your living room." While the race is an integral part of the Grand Prix Formula 1 World Championships, the Grand Prix weekend here is as much about the glitz and glamour of the event as it is about the racing.

A buzz builds in Monaco in the weeks leading up to the Grand Prix. Some six weeks before the event, fencing, safety barriers, temporary grandstands, and the mobile Formula One village are slowly constructed around the circuit. Spectators line the entire route, except through the tunnel. Tickets are sold for the practice day (Thursday before the race), qualifying (Saturday), and the big race on Sunday.

Do not attempt to drive anywhere near the center a fortnight before. On race day, get there a couple of hours before the start to enjoy the ambience and watch the parades. Bring sunglasses, sunscreen, and a sunhat, but most importantly, bring ear protectors, as the noise can be astonishing. Dismantling the grandstands and safety barriers takes another three weeks after the race, a welcome relief for many residents for whom traffic delays and noise affect over two months of their Monaco lives.

The Formula One Grand Prix motor race is held through the streets of Monaco.

Monaco's Love Affair with Cars

Monaco's streets are filled with the gentle purr of luxury sedans and the over-revved cackle of hired-for-the-day sports coupés. Whether it's a line of Ferraris parked outside the casino, a vintage Bugatti in Fontvieille, or a Bentley Continental driven at top speed for 30 m (98 ft) alongside the harbor, Monaco has been obsessed with automobiles since the **Automobile Club de Monaco** was inaugurated in 1909. What better way to show off your wealth than with a gleaming sports car?

THE GRAND PRIX

Today, the Formula One Grand Prix is the most famous and obvious evidence of this love affair. The original idea for a Grand Prix in Monaco came from Antony Noghès, president of the Automobile Club de Monaco and friend of Prince Louis II. The first Grand Prix de Monaco took place in 1929, won by British driver William Grover-Williams in a Bugatti. A life-size bronze statue of him and his car is opposite place Sainte-Dévote, just meters from where racers speed up the avenue d'Ostende on their way to the casino. Racing through the narrow streets with tight turns and steep inclines, there is little room for error. No deaths have ever occurred, but in 1955, Alberto Ascari's Lancia and in 1965, Paul Hawkins's Lotus both famously ended up in the harbor.

GETTING AWAY FROM THE CROWDS

If you don't want to deal with the crowds, high prices, and street closures, there are plenty of other ways to get your car fix while in Monaco. Prince Rainier III spent 40 years gathering one of the world's great automobile collections—the **Collection de Voitures de SAS Le Prince de Monaco** is open to the public. Several other shows and races take place in the principality throughout the year:

- The **Top Marques show** (www.topmarquesmonaco.com) is a prestigious annual car extravaganza that takes place over five days in June and is the only event in the world where visitors can test-drive some of the world's most sought-after cars on a section of the Grand Prix circuit.

- **Salon International de l'Automobile de Monaco** (SIAM) (www.salonautomonaco. com), inaugurated in 2017, fills the tarmac next to Port Hercule with vintage sports cars and dealers exhibiting their prestigious wheels.

- The **e-rally Monte-Carlo** is also part of the FIA calendar, with points counting toward the Electric and New Energies Championship, dedicated to 100 percent electric and 100 percent hydrogen cars. The race lasts for four days in late October, over 12 legs around Monaco and into the Alpes-Maritimes, Ardèche, Drôme, and Hautes-Alpes départements, with the finish line in Monaco at around 1:30am on the final day.

- Founded in 1997 by the Automobile Club of Monaco, the **Historic Grand Prix** (www. monacograndprixticket.com/grand-prix-historic) takes place a fortnight before the Formula One Grand Prix on even-numbered years. More than 220 classic cars and sports models from the 1950s, '60s, and early '70s race around the Monaco circuit, with special races for older F1 and F2 Grand Prix cars. It's a good-humored, though still competitive, spectacle of gleaming nostalgia.

- Keeping up with the times and Monaco's reputation for pioneering ecological thinking, the **Formula E Monaco ePrix** (https://acm.mc) has taken place on the Saturday a fortnight before the F1 Grand Prix on odd-numbered years since 2015. Due to limited battery lives, the much quieter races last 45 minutes. Tickets are fixed at €30, and are much easier to come by than those for the real Grand Prix.

Herculis

Stade Louis II, 3 avenue des Castelans; ticket desk tel.
+377 92 05 42 60; https://monaco.diamondleague.
com; July

Herculis is Monaco's annual athletics meet at the Stade Louis II. It is one of the most important track and field competitions in the world and is part of World Athletics' Diamond League events, where athletes compete in a league of 14 global meetings. Herculis is usually held in mid-July, with tickets available online and at the stadium on the day.

Rallye Automobile Monte-Carlo

https://acm.mc; Jan.

Monaco's first competitive car race started over a decade before the Grand Prix, in 1911. Prince Albert I's idea was for cars to start from different points across Europe and converge in Monte-Carlo. Judging was done not just on speed but on elegance, comfort, and the condition the car arrived in. Drivers still have to tackle the hairpins of the Col de Turini, usually covered in snow, during the night. There is no enclosed circuit or grandstands in the rally. Spectators must stand in the designated spectator zones, and park their cars in the official parking lots at least 30 minutes before the first car is due to pass through. Full details of the course, maps, and spectator areas are available on the Automobile Club of Monaco website. Immediately following the Monte-Carlo rally, it's the turn of the classic cars, which follow some of the same route but without any timekeeping, and are able to drive to Monaco for the final stage on whatever roads they wish before gathering in Monaco for a final concours d'élégance.

SHOPPING

With more millionaires per square meter than anywhere else in the world, it's little wonder that Monaco has shops to cater to their tastes. In the super-swanky (and well-guarded) boutiques near the casino, jewels and watches can cost an annual salary, and the haute-couture stores and shopping malls nearby are full of luxury goods. Even if you can't afford to buy

anything, it's fun just to look through the windows. Down by the port are the yacht brokers, and toward Fontvieille or Larvotto, the luxury automobile showrooms, where the super-rich sell their cars after a few spins along the coast road. There are, perhaps surprisingly, a few normal shops, but as the food writer and critic AA Gill wrote, "Monaco is a money puddle, a cash delta"—no one goes there to buy anything useful.

Chocolaterie de Monaco

place de la Visitation; tel. +377 97 97 88 88; www.
chocolateriedemonaco.com; Thurs.-Sat. and Mon.-Tues.
9:30am-6pm, Sun. 11am-6pm (tearoom closes at 5pm)

A wall of dripping chocolate greets visitors in the air-conditioned entrance to the principality's official chocolate supplier to Albert II. Inside, glass-fronted cabinets are full of boxes of chocolate crowns (€30 for 16), marbled chocolate sticks (€15 for 250g), and the Coffret Splendide, a showcase selection of milk, dark, and white chocolate for €95. There's also a tearoom at the back of the shop.

Marché de la Condamine

place d'Armes; tel. +377 93 30 63 94; daily 7am-3pm

The Monegasques come here to buy their fresh fruit, vegetables, flowers, and fish. The square has a large selection of market stalls and plenty of places to have a very affordable, freshly prepared lunch—sushi, pasta, pizzas, Asian food, organic options, bakeries, cheeses, and local specialities like fougasse and barbajuans for a couple of euros. From Tuesday-Saturday, the covered Halle Gourmande on the same site is also open 6pm-9:30pm and has become a popular place for an early-evening aperitif.

FOOD
Monte-Carlo
Il Terrazzino

rue des Iris; tel. +377 93 50 24 27; www.il-terrazzino.
com; Mon.-Sat. noon-2:30pm and 7:30pm-10:30pm;
€15-25

One of the most popular Italian restaurants in Monaco, Il Terrazzino is only 2 minutes' walk from the tourist office and casino square. It's a fun place for lunch, with Neapolitan specialties including Degustazione di antipasti napoleatani (€17), a spread of Neapolitan starters. There's even a table dedicated to the late Argentinian footballer Diego Maradona.

Café de Paris

place du Casino; tel. +377 98 06 76 23; www. montecarlosbm.com; daily 8am-2am; €29-54

The Café de Paris is the people-watching mecca of Monaco. With a huge pavement terrace across from the casino and Hotel de Paris, the SBM-owned brasserie has been on the same site for 150 years. It has a huge choice of snacks and meals, from a plate of beef tartare with french fries for €30 to a dish of caviar for €200. Sit down, order quickly (as the waiters don't wait for long), and enjoy looking at the luxury cars parked on the place du Casino and the art nouveau styling in the café's 1st-floor salon.

★ Avenue 31

31 avenue Princesse Grace; tel. +377 97 70 31 31; www. avenue31.mc; Sun.-Thurs. 12:30pm-2:15pm and 7:30pm-10:45pm, Fri.-Sat. 12:30pm-11:45pm; €28-110

This extensive, stylish, family-run Italian restaurant halfway to Larvotto has four dining rooms, with the plant-covered "veranda" being the most romantic. Mediterranean specialties include linguini with lobster, ceviche, tenderloin on the bone (cooked over a wood-fired grill), and their vaunted vanilla ice cream, which is prepared on request (it takes 20 min).

★ Song Qi

7 avenue Princesse Grace; tel. +377 99 99 33 33; www. song-qi.mc; daily noon-2:30pm and 7:30-11pm; €29-120

A Chinese restaurant opposite the Japanese garden, Song Qi specializes in fine dining yet offers a good-value lunchtime set menu. Emerald and deep-red velvet furniture contributes to a restful setting, disturbed occasionally by the sound of sports cars heading down the avenue Princesse Grace. The lunchtime noodle set menu is €26, and the dim sum set menu is €29.

★ Le Louis XV–Alain Ducasse

Hotel de Paris, place du Casino; tel. +377 98 06 88 64; www.montecarlosbm.com; Thurs.-Mon. 7:30pm-9:15pm (July-Aug. Wed.-Mon. 7:30pm-9:15pm), Sat.-Sun. 12:15pm-1:30pm; €90-150

An opulent, sumptuous setting with tones so hushed you can hear the napkins being folded, the Michelin guide describes this restaurant as the place where chef Alain Ducasse "forged his style, imposing his new exacting and masterful culinary classicism, always guided by the true character of the ingredients." It is the summit of fine dining in Monaco, with a menu of mesmerizing combinations. The Riviera lunch is €210, the Agape set menu €420, and the vegetarian Jardin menu €280.

Fontvieille
Les Perles de Monte-Carlo

quai Jean-Charles Rey; tel. +377 97 77 84 31; www. perlesdemontecarlo.com; Mon.-Sat. noon-2:30pm and Wed.-Fri. 7pm-10pm; €20-30

Facing the dramatic rock face of Le Rocher at the far end of Fontvieille's harbor wall, Les Perles de Monte-Carlo specializes in creative seafood dishes. One Sunday per month they host Les Perles du Dimanche, which includes a live DJ and a seafood brunch menu with wine and dessert for €50.

La Condamine
La Bionda Brasserie

7 rue Suffren Reymond; tel. +377 97 98 71 90; Mon.-Sat. noon-2:30pm and 7:30pm-11:30pm; €11-25

Always packed with local Italians, La Bionda is a popular choice for meat and pasta dishes one block from the Port Hercule in La Condamine. The grilled picanha steak is their signature dish, but they also do Italian sausages, salad, and fries. The building looks refreshingly scruffy from the outside, but it is one of the best reasonably priced Italian grills in Monaco.

Larvotto
Giacomo's Spiaggia
Plage du Larvotto; tel. +377 99 92 12 55; www. bigmammagroup.com; daily noon-2:15pm and 7pm-10:45pm; €19-36

On the far east of Larvotto beach, the Big Mamma group's Giacomo's Spiaggia has a huge parasol-covered terrace and a Capri-style coral-and-mint-toned dining room with an open kitchen. Dishes include lobster risotto, octopus brochettes, and a big selection of pizzas.

BARS AND NIGHTLIFE
Buddha Bar
place du Casino; tel. +377 98 06 19 19; www.buddhabar. com; Tues.-Sun. 6pm-2am

Buddha would probably have given Monte-Carlo a wide berth, but his bar is one of the top locations for a pre- and post-casino drink. This mixture of high-end Asian restaurant and relaxed lounge-bar is in the former cabaret at the eastern end of the casino, with two open-air terraces serving cocktails and shisha, a fruit-flavored tobacco smoked through a water pipe. It's plush with Asian reds and purples mingling with gentle, tasseled amber lighting and the luxurious discretion characterized by venues run by the Société des Bains de Mer.

Sass Café
11 avenue Princesse Grace; tel. +377 93 25 52 00; www. sasscafe.com; daily 8:30pm-3am

Set among the sports car showrooms and arcaded walkways of the avenue Princesse Grace, the Sass Café is a favorite late-night hot spot for Monaco's jet set and wealthy visitors. Opened in 1993, it's worth going to see the diners packed into tables along the outdoor terrace or in the piano bar and later dancing in the nightclub. Named after owner Sassa, it's a Monaco institution with an exclusive clientele of Formula One drivers, tennis and basketball players, and French and Hollywood film stars.

Twiga Monte-Carlo
10 avenue Princesse Grace; tel. +377 99 99 25 50; https://twigaworld.com; Tues.-Sun. 7pm-3am (until 5am Fri.-Sat.)

On the 2nd floor above the Grimaldi Forum, Twiga has a spectacular terrace operating as a lounge and shisha bar during the evening with a restaurant serving both Asian and Italian dishes. After 1am, Twiga, which means "giraffe" in Swahili, turns into an exclusive nightclub with celebrity DJs, amber lighting, and millionaires dancing until dawn.

ACCOMMODATIONS
Besides staying at the Hotel de France in La Condamine, a night in Monaco for a couple will top €200 even in low season. However, the hotels are, across the board, welcoming and regularly refurbished, with huge lobbies offering a generous amount of space to often very demanding guests.

€100-200
Hotel de France
6 rue de la Turbie; tel. +377 93 30 24 64; www. hoteldefrance.mc; €135 d

Near the shops and restaurants of La Condamine, the Hotel de France has been under the same ownership for the last 30 years and is unashamedly the cheapest hotel in Monaco. It has 26 air-conditioned rooms and is simple, unpretentious, and clean. Breakfast is €10 per person, although there are lots of cafés for croissants and coffee within easy walking distance.

€200-300
Hotel Novotel Monte Carlo
16 boulevard Princesse Charlotte; tel. +377 99 99 83 00; www.accorhotels.com/fr/hotel-5275-novotel-monte-carlo/index.shtml; €230 d

Halfway between the train station and the casino, the well-located Novotel is very reasonably priced for Monte-Carlo. It has a swimming pool and lots of airy lounges and places to hang around. The hotel is particularly

Monaco's Most Glamorous Hotels

Hotel de Paris Monte-Carlo

Many of Monaco's accommodations are out of the price range of the normal traveler, even for a once-in-a-lifetime splurge. However, many of these instutitions deserve a peek into the lobby, or a meal at one of their world-class restaurants. Here's a list of some of the principality's most outrageous—and luxurious—places to lay one's head for an evening:

HOTEL DE PARIS MONTE-CARLO

place du Casino; tel. +377 98 06 30 00; www.hoteldeparismontecarlo.com
One of the world's most prestigious (and expensive—rooms from €680 per night) hotels, the Hotel de Paris hopes to fulfill its founder François Blanc's dream of "a hotel that surpasses everything that has been created until now." For eating, there's **Le Grill** restaurant on the top floor, the three-star Alain Ducasse **Le Louis XV** restaurant on the ground floor, and **Em Sherif,** a Mediterranean-Lebanese-inspired lounge-restaurant with an outdoor terrace.

The hotel's **wine cellars,** where 20,000 wine bottles were hidden from the Nazis during World War II, are worthy of a visit on their own. The cellars were reopened in 1945 by Winston Churchill, who drank a bottle of 1811 rum to celebrate the occasion. For private views, contact the Hotel Operations department (tel. +377 98 06 89 01).

popular with families, and has a restaurant, a breakfast bar open for outside guests, and modern, well-equipped rooms.

Boutique Hotel Miramar

1 avenue John F Kennedy; tel. +377 92 00 21 00; www. hotelmiramar.mc; €250 d
This nautical-themed boutique hotel overlooks Port Hercule with a great view of Le Rocher. Guests who reserve a room for the Grand Prix will have a trackside view from their balconies. Each of the 14 rooms has a seafaring name including Noah's Ark, Potemkin, and Yellow Submarine, and there's a great rooftop bar on the sundeck.

INFORMATION AND SERVICES

Monaco has a separate international dialing code to France. The country code is +377, and numbers have eight digits.

HOTEL HERMITAGE MONTE-CARLO

square Beaumarchais; tel. +377 98 06 40 00; http://fr.hotelhermitagemontecarlo.com
This super-luxury hotel managed by the Société des Bains de Mer in Monte-Carlo is just a casino-chip's throw from the Hotel de Paris. General manager Pascal Camia describes it as "the most romantic hotel in Europe." A carved belle epoque facade, a huge winter garden atrium designed by Gustave Eiffel, and a drink at the Crystal Bar make it more fairy-tale than bling. The hotel's duplex "diamond suites" have a glass staircase leading to a hot tub on their private roof terrace.

HOTEL METROPOLE MONTE-CARLO

4 avenue de la Madone; tel. +377 93 15 15 15; www.metropole.com
This spectacular belle epoque building hidden behind gates and lawns on the east side of place du Casino opened in 1886 on land originally owned by the pope. There's a Givenchy-styled spa and a Karl Lagerfeld-designed pool fresco. The hotel employs a sound designer for the different salons and lobby bar.

HOTEL MONTE-CARLO BAY

40 avenue Princesse Grace; tel. +377 98 06 02 00; www.montecarlobay.com
More a resort than a hotel, this property has the feel of a film set of a '60s-style beach party. There's a huge sand-bottomed lagoon, an indoor pool, a snooker room, Cinq Mondes spa, a children's club, a swish casino, and a Michelin two-starred restaurant, the Martinique-inspired **Blue Bay.**

FAIRMONT MONTE CARLO

12 avenue des Spélugues; tel. +377 93 50 65 00; www.fairmont.com/montecarlo
Built into the rock behind Monaco's casino, the Fairmont is a glamorous monster of a hotel. Constructed in the early 1970s, the hotel has great views of the bay and houses the **Nobu** Japanese restaurant.

MAYBOURNE RIVIERA

1551 route de la Turbie, Roquebrune-Cap-Martin; tel. +33 4 93 37 50 00; www.maybourneriviera.com
The address may be in Roquebrune, but the experience and views are all about Monaco. The spectacular-looking hotel is perched on a cliff high above the coast with an uninterrupted panorama of the Mediterranean all the way to Italy. There's an infinity pool, spa, a patisserie serving bespoke cakes, and the hotel's own beach club at the tip of the Cap-Martin.

Tourist Information

The **tourist office** (tel. +377 92 16 61 16; www.visitmonaco.com; Mon.-Sat. 9am-7pm, Sun. 11am-1pm) is located on the boulevard des Moulins at the northeast end of the Jardins du Casino. They provide free maps, brochures, transport timetables, and listings of what is going on in Monaco, and can also help you book accommodations. Visitors can download the office's Monapass application (www.visitmonaco.com/en/39061/monapass), which gives opening hours and travel information in the principality.

Post Offices

There are six post offices in Monaco. The two main offices are at **Palais de la Scala** (near the Hermitage hotel in Monte-Carlo) and at **17 rue Grimaldi** in La Condamine (both Mon.-Fri. 8am-7pm, Sat. 8am-1pm). Monaco produces its own stamps, but French stamps are valid there too.

Beausoleil: Monaco's More Affordable Neighbor

Beausoleil

Beausoleil, formerly known as Monte-Carlo-Supérieur, is a French commune wrapped around the top of Monaco. Roads that begin in the principality can end in Beausoleil, and in some cases, one side of the road is in Monaco while the other side is in France (and its residents therefore pay income tax).

Perhaps unfairly, Beausoleil has become known as a dormitory town to Monaco, with hundreds of workers living there and descending to Monaco every day. It has substantial Filipino, Portuguese, and British communities. It's an active place with a covered market designed by Gustave Eiffel, food shops, bars, doctors and dentists, pretty villas, and children (there aren't many children in Monaco).

For anyone wanting to visit Monaco but not wishing to pay to stay the night there, Beausoleil is a much cheaper option and is just a 5-10-minute walk from the place du Casino, along a pavement covered in pretty suns, a reminder you are in *beau soleil*.

WHERE TO STAY

- **Adagio Apart-hotel Monaco Palais Joséphine** (2A avenue du Général de Gaulle; tel. 04 92 41 20 00; www.adagio-city.com; €160 d): Run by the Adagio group, Beausoleil's apart-hotel has a rooftop pool and 100 fully air-conditioned rooms with kitchens for 2-6 people. Decor is functional and clean, with red-and-white fittings and tiled floors. Breakfast is €12.90.

- **Hotel Olympia** (17 bis boulevard Général Leclerc; tel. 04 93 78 12 70; www.olympiahotel. fr; €130 d): The Olympia is in a belle epoque residence in the center of Beausoleil. Art deco-style rooms are clean and functional and include a safe, a hair-dryer, and electronic blinds. Balcony rooms are a little more expensive but worth the extra cost for a view of Monaco. Breakfast is €10.

Money

There are many **banks** with ATMs around the place du Casino and branches in La Condamine and Fontvieille, and there is a **Bureau de Change** at 20 rue Comte Félix Gastaldi (tel. +377 97 77 54 82; www. rivierachangemonaco.mc; Mon. 10am-5pm, Tues.-Fri. 10am-6pm, Sat. by appointment).

Medical and Emergency Services

Monaco's **emergency telephone numbers** are the same as they are in France: 15 for SAMU (medical emergency), 17 (police), 18 (firefighters who are also paramedics), and 196 (coast guard).

Centre Hospitalier Princesse Grace (1 avenue Pasteur; emergencies tel. +377 97 98 97 69, pediatric emergencies tel. +377 97 98 95 33; www.chpg.mc/the-hospital): Monaco's only public hospital is located in the hills above the principality. It is open 24/7 and has an accident and emergency unit, with a separate pediatric emergency department. It has an underground car park for private cars. Bus routes 3 or 5 pass the entrance; alight at bus stop "Hôpital."

Pharmacie de Monte-Carlo (4 boulevard des Moulins; tel. +377 93 30 83 10; https://pharmaciedemontecarlo-monaco. pharmavie.fr; Mon.-Sat. 8:30am-8pm): Most pharmacies will have a sign in their window to show which local pharmacy is open on Sundays; otherwise, the Princess Grace hospital also has an online list.

GETTING THERE
Car

Monaco is 20 km (12 mi) from **Nice.** There's a scenic route along the **Basse Corniche** via Beaulieu-sur-Mer that takes around 40 minutes, but the traffic can be slow. The fastest way is via the A8 autoroute. With a toll (€3.90 cash, contactless or by credit card) and occasional bottlenecks approaching Monaco, the journey time can range from 25 minutes to an hour from Nice airport. **Menton** is 14 km (19 mi) from Monaco and the journey takes

25 minutes along the coast road or 20 minutes using the A8 autoroute.

Train

Monaco Monte-Carlo railway station (place Sainte-Dévote; tel. +377 93 10 60 05; www.gares-sncf.com/fr/gare/frxmm/monaco-monte-carlo; daily 5am-2am) is on the TER line between Marseille and Ventimiglia in Italy and the TGV route from Paris-Marseille to Ventimiglia. You can access the station from the harbor (at Sainte-Dévote), La Condamine and Fontvieille via moving walkways. The entrance nearest the place du Casino (10-min walk) is on avenue du Prince Pierre.

The journey from **Paris** takes just over 6 hours (average cost €80), from **Marseille** 3 hours (average cost €35), and from **Nice** 25 minutes (average cost €5).

Bus

Zou network (https://zou.maregionsud.fr) bus numbers **602** (Nice Vauban to Monaco casino), **603** (Nice Vauban to Monaco place d'Armes via La Turbie), and **607** (Nice Square Normandie Niemen to Monaco, place des Moulins) cost €2.10 for a single. The journey takes around an hour depending on the time of day and travel via Villefranche, Eze-sur-Mer, and Cap d'Ail.

There is also a direct bus from **Nice airport** to Menton via Monaco (Line 80 Express; normally 1 per hour), which costs €19 single (ages 4-11 €9.50, under 4 free, discounts for group of three or more). Tickets can be purchased from the airport ticket office or from the driver (departing from Terminal 2). The ride takes 45 minutes.

Helicopter

Monacair (tel. +377 97 97 39 00; www. monacair.mc) has flights every 15 minutes between **Nice airport** (Terminals 1 and 2) and the principality. Flight time for moneyed tourists and executive business passengers is 7 minutes. Fares start from €140, including chauffeur-driven pickup and drop-off.

GETTING AROUND

The best way to visit Monaco is on foot, since there are elevators all over the principality to move from level to level and moving walkways (flat escalators) to help with the long stretches.

Car

Driving in Monaco can be great fun for a few weeks every May when drivers can speed around the Grand Prix circuit. However, for most of the rest of the year, roadwork can cause considerable congestion. Streets and landmarks are well signposted, but the road network is complicated and it's not an easy place to navigate.

For Le Rocher and Monaco-Ville, follow signs for the **Parking Pêcheurs** (547 spaces); for the casino, **Parking Casino** (411 spaces); and for Monte-Carlo, follow signs for **Parking Grimaldi Forum** (442 spaces). Most of the hotels have private car parks and valets to park diners' cars outside the major restaurants. The first hour at public parking lots is usually free, after which it costs approximately €2.40 per hour. Parking after 7pm is only €0.40 per hour.

Bus

The **Compagnie Autobus de Monaco** runs six bus lines in the principality. Potentially useful to tourists are line 1 from Monaco-Ville to Saint Roman; line 2 from Monaco-Ville to the Jardin Exotique; line 5 from Hôpital to Larvotto; and line 6 from Fontvieille to Larvotto.

Tickets can be purchased directly on the bus or from the **bus station** (22/24 rue du Gabian; tel. +377 97 70 22 22; www.cam.mc). Bus tickets cost €2 per journey (which lasts 30 min from first use on any route), €11 for six journeys, and €5.50 for a day pass. Children under 7 travel free. Buses run from around 7am to 9:30pm, after which there is a reduced nighttime service.

Bike

MonaBike (https://monabike.mc) is a local electric bike-rental service There are currently 49 bike stations around Monaco and 390 red bikes available. Renting a bike for a single journey of up to 30 minutes costs €1. Renting a bike (unlimited journeys in Monaco up to 30 min each) for a 24-hour period (Pass24) costs €3, or €8 for a seven-day pass (of 30-min maximum journeys). Registration can be done online and bikes are released via a CAM card or QR code via a mobile phone.

Taxi

Since Monaco is smaller than New York's Central Park, walking and buses are the most common ways to travel around the principality. However, there are **taxi ranks** on the port and outside the railway station. Taxis charge €18 for a journey within Monaco and the fare to Nice airport is €100, baggage included. If traveling in a group of more than four, there are also **Taxis-Buses** (smart minibuses) with two pickup areas: outside the railway station (débarcadère) and the Port de Monaco (Nouvelle Digue). **Uber** is not available in Monaco but can be used in Beausoleil, a short walk from Monaco's place du Casino.

Roquebrune-Cap-Martin

Heading east from Monaco means a sudden return to France, income tax, and the commune of Roquebrune-Cap-Martin, a perched medieval village and rugged peninsula. The village of Roquebrune feels like stepping several centuries back into the past. It was controlled by the Grimaldi family for five centuries, until 1793, when it became French. Although all part of the same commune, the village feels very separate from the beaches of Cap-Martin beneath. The seaside **Carnolès** district separates Roquebrune-Cap-Martin from Menton farther east and has its own train station and pebble beaches.

Not merely popular with tourists, Cap-Martin has a storied celebrity history: Its maze of roads is lined with belle epoque villas that were popular haunts for England's King Edward VII, Winston Churchill, Emperor Franz-Josef of Austria, and Irish poet William Butler Yeats, who took residence for a year. The village was also home to the American-born entertainer Josephine Baker from 1969-1975. Fashion designer Coco Chanel had her seaside manor, La Pausa, built above the village in the 1930s.

SIGHTS
★ Roquebrune Village

https://roquebrune-cap-martin.fr; parking lot on avenue Raymond Poincaré

Despite the influx of art galleries, craft shops, and other visitors, Roquebrune's tiny village seems to have remained unchanged since medieval times. It has panoramic views of the coast and some excellent restaurants. A 200-m (656-ft) walk from the edge of the village on the chemin de Menton leads to a 2,000-year-old olive tree, claimed to be one of the oldest in the world.

Château de Roquebrune

place William Ingram; tel. 04 93 35 07 22; https://roquebrune-cap-martin.fr; Oct.-Jan. Sat.-Thurs. 10am-5pm, Feb.-Apr. daily 10am-5pm, May-Sept. daily 10:30am-6:30pm; €5, children and students €3, under 6 free

One of France's oldest castles, this Carolingian château was built at the end of

Roquebrune village

the 10th century by Conrad I, the Count of Ventimiglia, to stop invading Saracens. From the 15th century on, the fortress belonged to the Grimaldis, who tried to make it less austere. The original donjon (keep) became the castle, and the walled fortress below became the village. Visitors can walk through to the archers' dormitory, a kitchen, the roofless Ceremonial Hall, and the Common Room, where a primitive toilet is still visible. The castle was bought by an Englishman, Sir William Ingram, in 1911. He tried to do up the brutal-looking structure, which had been burned and savaged by cannonballs over many centuries, but he eventually sold it a decade later to the local commune.

★ Cap Moderne

esplanade de la Gare SNCF de Cap-Martin Roquebrune; tel. 07 62 53 30 82; https://capmoderne.monuments-nationaux.fr; May-Sept. daily, Oct. Tues.-Sun.; 2-4 tours per day; €18, ages 7-17 €7

The bright white Cap Moderne building was first a railway carriage depot and then a storage unit for Roquebrune's festival floats before being inaugurated in 2014 as an architecture foundation dedicated to the remarkable collection of buildings on the cape: Le Corbusier's **Le Cabanon** and its associated holiday cottages, Eileen Gray and Jean Badovici's **Villa E-1027,** and Thomas Rebutato's **L'Etoile de Mer,** a former restaurant and bar.

The excellent **tour,** which must be booked via the website, gives visitors controlled glimpses into these remarkable structures. Built for his wife, Le Corbusier's Le Cabanon is fundamentally a square cabin, filled with modernist furniture and everything one could need. In exchange for the plot of land on which he built Le Cabanon, Le Corbusier had five holiday cabins, also visible on the tour, built for Thomas Rebutato, owner of L'Etoile de Mer restaurant. L'Etoile de Mer itself is a perfect example of how to enjoy the seaside, with its panoramic terrace, decorated bar, and art brut gardens. Villa E-1027 is Irish-born furniture designer and architect Eileen Gray's "house by the sea." It's an amazingly modern building for the 1920s, a bilevel, L-shaped concrete structure built on pillars. The tour lasts 2 hours and leaves from the Cap Moderne wagon in the railway station car park. Private tours are also available in French and English on request.

SPORTS AND RECREATION
Beaches

Roquebrune has two beaches to the west of the Cap-Martin peninsula, both easily accessible by foot from the railway station. In the Carnolès district, there are also narrow stony beaches along the promenade du Cap-Martin. Swimming is possible off the cape, although it can take a bit of doing to descend the rocks.

Plage du Buse

9026A avenue Le Corbusier

The main beach in Roquebrune is accessible from the western end via a 100-step staircase. It's a shingly, gravelly beach, great for a dip, though there are quite a few offshore rocks lurking just below the surface. The eastern end of the bay is the start of Promenade Le Corbusier, a delightful walking path around the peninsula. Parking for the beach is in the railway station car park (a 5-min walk).

Plage du Golfe Bleu

chemin du Golfe Bleu

Farther from the railway station and car park, the Plage du Golfe Bleu is less crowded than the Plage du Buse, separated from the latter by the Cabbé peninsula. The western end is a popular landing site for hang gliders (**RoqueBrun'Ailes;** www.roquebrunailes. com/en/paragliding-roquebrune.htm) descending from Mont Gros, but otherwise it's a peaceful, pebbly beach excellent for swimming.

Hiking
Promenade Le Corbusier

Distance: *4.6 km (2.8 mi) one-way*
Time: *2 hours*

Information and Maps: *Roquebrune's tourist office*

Trailhead: *Plage du Buse, directly under Roquebrune railway station*

From the eastern end of Plage du Buse directly underneath Roquebrune railway station, it is possible to walk around the entire peninsula to join the avenue Winston Churchill. The footpath is officially called the Sentier Massolin, but it is signposted as the Promenade Le Corbusier. It's an easy walk; Eileen Gray's villa and Le Corbusier's Le Cabanon beach house are just 10 minutes from the beach, but walking round the entire cape, past Cap-Martin at the tip, takes about 2 hours.

Cycling

Road cycling has become a huge part of the tourist industry around Roquebrune-Cap-Martin and Menton. However, because the terrain is steep, routes around Roquebrune tend to be for serious cyclists. The **Via Sportiva de Mer en Forts** starts at the car park in the village and includes La Turbie, the medieval village of Peille, the Col de Braus, Saint-Roch fort, and Caramel viaduct, and ends at the Palais Carnolès in Menton. The route is 84 km (52 mi) and takes around 5.5 hours.

The Alpes-Maritimes département has created guides for three types of **cycling tours:** for experienced cyclists (www.departement06.fr/choisir-une-boucle-cyclable-selon-vos-envies/echappee-sportive-1953.html); for "tourists," or fair to competent cyclists (www.departement06.fr/choisir-une-boucle-cyclable-selon-vos-envies/decouverte-touristique-2288.html); and for families (https://fr.calameo.com/read/000334644435f03f23a3d).

FOOD

Casarella

15 rue Grimaldi; tel. 04 93 35 03 57; Wed.-Sun. noon-2:30pm and 7pm-10pm, Mon.-Tues. 7pm-10pm; €14-18

An authentic Neapolitan restaurant in the heart of the medieval village, Casarella has

a small dining room but plenty of tables in the street under canopies. The building dates from 1885, and fare includes pizzas, pastas, and handmade Italian ice cream.

La Grotte & l'Olivier

2 place des Deux Frères; tel. 04 93 35 00 04; www.lagrotte-lolivier.fr; Thurs.-Mon. 10am-11pm, Wed. 5pm-11pm; €15-26

When the original road was being constructed through the village (over 100 years ago), the solid mass of brown puddingstone beneath the château proved too hard to dig, so the workers stopped and moved 20 m (66 ft) to the right. The cave left behind is now the interior of this excellent and friendly Mediterranean restaurant, "The Cave and the Olive Tree." It specializes in creative pasta dishes, wooden platters to share (€15), and pizzas (€10-16).

Au Grand Inquisiteur

15 et 18 rue du Château; tel. 04 93 35 05 37; www.augrandinquisiteur.com; daily 7pm-9:30pm; €20-25

The metal sign outside says this restaurant was established in 1965, but it could easily be 1365. Hidden behind a wall of ivy on the road up to the château, the intimate restaurant seats only 22 diners under the cloak-and-dagger decor of whitewashed walls and rough-hewn stones. They also run a tiny wine bar next door, La Petite Cave, which opens a few afternoons a week.

ACCOMMODATIONS
€100-200
La Bombonière

24 rue du Château; tel. 06 80 86 99 43; www.bomboniere.fr; €120 d

Right opposite the entrance to the château steps, La Bombonière is a little gem of a guesthouse, already popular with an international crowd. They have a spa with a massage area carved out of the rock, and rooms for larger groups, including the roof terrace suite over two floors with views of the château and Monaco, and a 60-sq-m (645-sq-ft) duplex apartment.

★ La Belle Vue

45-47 avenue Gabriel Hanotaux; tel. 04 93 51 32 15;
www.labellevuercm.com; €195 d

This is an attractive stone villa with, as its name suggests, "a beautiful view" of a garden of orange, lemon, and grapefruit trees, and rooms full of antiques and tasteful paintings. The guesthouse is located on the Grande Corniche, which runs between Nice and Menton, a 5-minute walk above Roquebrune village. It has three rooms, all with direct access to the garden.

INFORMATION AND SERVICES

The **main tourist office** is situated in Carnolès (218 avenue Aristide Briand; tel. 04 93 35 62 87; www.rcm-tourisme.com; Sept.-June Mon.-Fri. 9am-12:30pm and 2pm-5:30pm, July-Aug. Mon.-Sat. 9:30am-12:30pm and 2pm-6pm).

GETTING THERE AND AROUND

The perched village of Roquebrune-Cap-Martin is steep with uneven, cobbled streets, so it's not easy to get around for those with limited mobility. The footpath from the railway station to the village is a breathless 25-minute walk uphill.

Car

Roquebrune-Cap-Martin is accessible from the A8 autoroute (exit for La Turbie), approximately 22 km (13.5 mi) from **Nice** (35 min) and 12 km (7.5 mi) from **Ventimiglia** in Italy (20 min). It is also on the more scenic coastal roads, the D6007 and D2564 (Grande Corniche) from Nice (45 min). It takes approximately 25 minutes to drive the 9 km (6 mi) from **Monaco.**

The streets in the village are too narrow for cars, so there's a **parking lot** at the bottom of the village.

Train

It is possible to reach Roquebrune-Cap-Martin by train, but confusingly, when purchasing tickets, the station is called **Cap-Martin-Roquebrune** (listed under C). TER trains stop at the station on the Ventimiglia-Grasse lines between Monaco and Nice. The walk to the perched village takes 25 minutes up a series of steep staircases, or a **taxi** can be ordered (Les Taxis de Roquebrune; 185 avenue Aristide Briand; tel. 04 93 35 15 00); there is no taxi stand at the train station. It's easier to catch the train to Carnolès, one stop to the east on the same line, where Zestbus line 24 goes to the old village.

Bus

Roquebrune is served by the **Zou network** (https://zou.maregionsud.fr) bus numbers 602 and 608 (Nice to Menton via Monaco; €2.10). The local bus network **Zestbus** (www.zestbus.fr) line 18 travels between Menton and Roquebrune village.

1: Plage du Buse **2:** La Grotte & l'Olivier
3: Cap Moderne

Menton

Famed for its annual lemon festival, Menton, only 12 km (7.5 mi) from Monaco and 7 km (4 mi) from Roquebrune, is the last town on the coast before the Italian border and feels as much Italian as it does French. It was acquired by the Grimaldis in 1346 from the wealthy Genoese Vento dynasty, and remained under their influence until it became permanently attached to France in 1861. By this time its climate (one of the sunniest in France) and mild winters were attracting European aristocracy and a host of writers and artists. Today, despite its aged population (one of the oldest in France), it is a vibrant, attractive resort town with an excellent covered market, majestic gardens, and a thriving arts and music scene.

ORIENTATION

Modern Menton runs behind the pebble beaches and seafront **promenade du Soleil,** which ends at the old port and **Bastion fortress.** The area immediately north, **Les Sablettes,** has sandy beaches and a marina, behind which sits the **old town (Vieux Menton).** A kilometer east is

the Italian border and the **Garavan quarter,** now mainly residential and once home to the Spanish writer Vicente Blasco-Ibáñez and New Zealand author Katherine Mansfield, who stayed at Villa Isola-Bella in 1920 (now on avenue Katherine Mansfield). Other writers—Gustave Flaubert, Guy de Maupassant, Anton Chekhov, Thomas Carlyle, Laurence Sterne, Aubrey Beardsley, Robert Louis Stevenson, and, later, Alan Sillitoe and William Butler Yeats—were drawn to Menton, and some of them even died here. The town's two **cemeteries** are lapidary registers of European fine art and high society.

SIGHTS
Le Vieux Menton
rue Saint-Michel, rue de Bréa, and rue Longue
Menton's old town begins in the pedestrianized area around Les Halles market and ascends the steep hill to the Basilique Saint-Michel Archange, which can be reached by walking through the gateways of Saint-Julién and Saint-Antoine along the rue Longue, or via the steep Rampes Saint-Michel from the

Basilique Saint-Michel Archange

Menton

© MOON.COM

quai Bonaparte. Cobbled pathways often turn into staircases so steep they require handrails, or narrow lanes of brick and smooth cobbles that lead up to the old château cemetery. The old town facades are painted in lemon and orange tones, reminiscent of the citrus fruits that made the town so famous.

Basilique Saint-Michel Archange

rue Saint-Michel; Mon.-Fri. 10am-noon and 3pm-5pm

The highlight of the old town is the Basilique Saint-Michel, a baroque masterpiece built 1640-1653 on the remains of two former churches (also dedicated to Saint Michael). Its 53-m (173-ft) clock tower is a symbol of the city. The parvis—a pebbled square in front of the church—depicts the Grimaldi coat of arms.

Musée des Beaux-Arts– Palais de Carnolès

3 avenue de la Madone; tel. 04 93 35 49 71; www. menton.fr; Wed.-Mon. 10am-noon and 2pm-6pm

In 1717, Prince Antoine I Grimaldi of Monaco bought land to construct the Palais de Carnolès, an ornate pink residence that was turned into a luxury casino in 1863 before the Menton municipality made it a fine-arts

museum in 1994. It houses paintings from the 13th century to the modern day. The palace gardens were replanted in 1970 with the objective of creating one of Europe's most remarkable collections of citrus trees. There are 140 different citrus varieties including Marsh pomelos, Moro oranges, Persian limes, and huge Japanese grapefruits. Most fruits ripen toward the end of October but can remain dangling on the trees until the spring. The best time to visit the park is November to March, when the fruit is gleaming and the Potager des Princes is producing the best herbs and vegetables.

Musée de Préhistoire Régionale

rue Lorédan Larchey; tel. 04 93 35 84 64; www. menton.fr; Wed.-Mon. 10am-noon and 2pm-6pm; €3, students, teachers, and over 65 €2.25, under 18 free, free admission first Sun. of the month

Menton's museum of prehistory, built at the start of the 20th century, is art nouveau and classically inspired, with a stunning arched entrance surrounded by orange trees. Inside are many artifacts, from *Homo erectus* tools found in the Vallonnet caves (dating from one million years ago) to the Bronze Age (700 BCE) in Liguria and the Alpes-Maritimes.

Musée de Préhistoire Régionale

Displays include animal bones, early chopping tools, and items found in many other caves at Lazaret, Terra Amata, and Grimaldi.

Cimetière du Vieux Château

Montée du Souvenir; Apr.-Oct. daily 7am-8pm, Nov.-Mar. daily 8am-5pm; free

On land once occupied by Menton's 13th-century château, the tombstones of the Cimetière de Vieux Château have fantastic views of the coast. A life-size sculpture of William Webb Ellis, the schoolboy "inventor" of rugby, who eventually became Anglican vicar of Menton, greets visitors; his grave is adorned with rugby balls, sports flags, and engraved plaques from rugby associations around the world. Dozens of other British and Russian "visitors" are buried in the cemetery.

Musée du Bastion

quai Napoléon III; tel. 04 93 18 82 61; www.menton.fr; Wed.-Mon. 10am-12:30pm and 2pm-6pm; adults €3, students and over 65 €2.25, under 18 free

During his time in Menton in the 1950s, artist, poet, and filmmaker Jean Cocteau became fascinated by the abandoned 17th-century fort at the end of the promenade du Soleil. The city's mayor, Francis Palmero, let him display his work there and, following the 2018 flood that destroyed much of the purpose-built Cocteau Museum on the quai de Monléon, the Bastion now exhibits some of Belgian-American businessman Séverin Wunderman's huge collection of Cocteau's works. The ground floor includes a large pebble mosaic of a lizard—the artist's symbol for Mediterranean laziness.

SPORTS AND RECREATION
Beaches
Plage des Sablettes

quai Gordon Bennett; parking on roadside or in Parking Vieille-Ville Sablettes

The sand-and-pebble beach in front of the newly developed esplanade is popular with families for its calm waters. It's protected by two stone-built jetties at either end. There are showers, toilets, and lifeguards in July and August.

Plage du Casino

promenade du Soleil; parking on place d'Armes

This pebbly beach runs beneath Menton's famous seafront promenade. The pebbles are big and it's not a beach to sunbathe on unless you have a lounger (it's a public beach so you'll have to bring your own), but the water is clear, there are lifeguards in July and August, and it has a special area for dogs. Valuables can be left in lockers at the entrance to the beach.

Cycling
La Route des Grandes Alpes

www.route-grandes-alpes.com

The roads above Menton, through Sainte-Agnès, to the Col de la Madone, Castellar, or into the Alps provide some of Europe's greatest cycling challenges. La Route des Grandes Alpes, whose summits and climbs are part of the Tour de France, departs from Menton and reaches Lake Geneva, 700 km (435 mi) away. The RGA-organized route includes 17 passes, with accommodations suitable for cyclists, food advice, climb details, and nature observations. The RGA website suggests tours lasting 3-7 days.

RBike Menton

quai Nord Port de Garavan; tel. 06 26 03 31 37; www.rbikementon.com; Tues.-Sat. 8am-noon and 3pm-7pm; €20 50/day

Being in RBike feels like being in Italy. Husband-and-wife team Luciano and Paula offer advice and do bike repairs and bike rental, and they also have a huge selection of new and secondhand bikes for sale. Their shop has become a center for top-of-the-range triathlon equipment and triathlon chat. Visitors can rent racing, mountain, or city bikes, and also e-bikes to ride up to the mountain pass, the Col de la Madone, or along the coast to the Italian seaside resort of Bordighera for lunch.

Menton's Gardens

Menton has been celebrated for its gardens ever since British botanists began introducing rare plant species into their scientific gardens in the early 19th century. When the railway arrived in 1869, it brought with it scores of aristocrats looking for a winter palace who adored the idea of a tropical or subtropical garden to wander around. A dozen of these gardens and villas still exist, and many are open to the public. Several that are not, such as the **Jardin Colombières, Jardin Clos de Peyronnet,** and **La Citronneraie,** do occasionally open to the public on tours organized by the tourist office, and the annual garden festival, **Le Mois des Jardins,** opens most of the town's exotic and Mediterranean gardens to the public for tours. Here are some of Menton's top gardens:

JARDIN DU VAL RAHMEH

avenue Saint-Jacques; tel. 04 93 35 86 72; Wed.-Mon. 9:30am-12:30pm and 2pm-5pm; €7, guided visits Sept.-Apr. Mon. 3pm, May-Aug. Mon. 3:30pm, €11

Perhaps the most notable of Menton's gardens is the Jardin du Val Rahmeh. British army general Sir Percy Radcliffe and his wife, Rahmeh Theodore Swinburn, introduced the first exotic plants into the garden in the 1920s. The property was bought by the Natural History Museum in 1966, which opened the grounds to the public a year later. Today the garden covers 1.5 hectares (3.7 acres) and has 1,500 species, including dozens of cacti and an olive grove behind the house.

JARDIN SERRE DE LA MADONE

74 route de Gorbio; Jan.-Mar. Tues.-Sun. 10am-5pm, Apr.-Oct. Tues.-Sun. 10am-6pm; €8, under 18 free, guided visits 3pm €8

The Jardin Serre de la Madone on the road to Gorbio, a commune to the north of Menton, was the brainchild of British-American soldier Lawrence Johnston, who was injured in the First World War and wanted a place in a warm climate where he could nurture his botanical interests. The gardens, developed in 1924, spread out over 6 hectares (15 acres) of olives, cypress, figs, peach trees, vines, Mexican nolina, and Yunnan magnolia, as well as Johnston's extensive collection of sculptures from antiquity.

FESTIVALS AND EVENTS

Fête du Citron

tel. 04 92 41 76 95; www.fete-du-citron.com; Feb.; €25/€12 (seated/standing), ages 6-14 €10/€6, under 6 free

Menton's Lemon Festival attracts around 240,000 visitors a year, second only to Nice Carnaval in its size and popularity on the Riviera. Usually held in the last two weeks of February, it uses 140,000 tonnes (154,000 tons) of citrus fruit, although most is now imported from Spain, Menton's own lemon crop having been dramatically reduced in size over the last few decades. The city's first carnival took place in 1877, but it wasn't until February 1929—when a winter exhibition of citrus-tree flowers was held in the gardens of the Riviera Hotel—that the festival took on more of a citrusy tinge. The following year, donkeys pulled carriages full of locals along the rue de la République, with the floats decorated with the colorful fruits from citrus trees. Booking rooms during the festival at least three months in advance is recommended, as visitors come from all over the world to see the carnival.

Le Mois des Jardins

tel. 04 92 41 76 93; https://rendezvousauxjardins. culture.gouv.fr; June

Menton holds its annual garden festival over the entire month of June, when the town's exotic and Mediterranean gardens are open to the public for educational and pleasure tours.

JARDIN MARIA SERENA

21 promenade Reine-Astrid; tel. 04 92 10 97 10; guided tours only, Tues. 10am and Fri. 2:30pm; €6, under 18 free

The Jardin Maria Serena, at the foot of the cliffs near the Italian border, was designed by Charles Garnier (of Paris and Monaco opera house fame) for the Foucher de Careil family, but it was bought in 1922 by wealthy English banker Henry Konig, who developed the garden with hundreds of palm trees and cycas trees. Complete with a dragon fountain and jacarandas beside the railway track, it's one of Menton's most striking gardens.

JARDIN FONTANA ROSA

avenue Blasco-Ibáñez; tel. 04 92 10 97 10; guided tours only, Mon.-Fri. 10am; €6, under 18 free

Perhaps the most poetic garden is the Jardin Fontana Rosa, which is in a near-abandoned state. It was classified as a national monument in 1990, but more than 30 years on it still requires rebuilding. It was the masterwork of exiled Spanish writer Vicente Blasco-Ibáñez, who founded

Square des Etats-Unis

Spain's first socialist newspaper and decided to settle in Menton in 1921. He built a garden full of Spanish shrubs and plants dedicated to the world's great novelists, with busts of Dickens, Hugo, Flaubert, Dostoyevsky, and Balzac on display. Much of his land is now taken up with apartment blocks, but what remains of the garden is still a delight.

SQUARE DES ETATS-UNIS

Menton also has secret gardens dotted around the town. The contemporary-looking Square des Etats-Unis off avenue Carnot is surrounded by apartment blocks but has some huge, draping ficus trees, a bridge, and a fishpond.

Besides visits to the gardens, there are painting and photography exhibitions. The first weekend of the festival coincides with the nationally organized "Rendez-vous aux jardins" event, which has encouraged around two million people to visit 2,000 French public and private gardens every year since 2003.

Festival de Musique

Basilique Saint-Michel Archange; tel. 04 92 41 76 76; www.festival-musique-menton.fr; July-Aug.; tickets €24-54

Founded by the Hungarian musician André Böröcz, Menton's Festival de Musique has been held annually since 1950. The classical music event takes place over two weeks in July and August. Besides the nightly concerts, there are master classes, a mass in the Basilique Saint-Michel Archange, guided tours, exhibitions, workshops, outdoor recitals, and a fringe festival. A shuttle bus ferries audience members from the tourist office to the Basilica, where most of the concerts take place.

SHOPPING
Allo Robert

3 rue Galliéni; tel. 04 93 28 88 21; Mon.-Tues. and Thurs.-Sat. 10am-12:30pm and 3:30pm-6:30pm

Menton's giant bric-a-brac warehouse is better than any museum for curios, ancient crockery, glasses, and shop signs. Owner Robert is interested in everything to do with old bars, clubs, restaurants, furniture, paintings, and garden decoration. The entrance looks like a small shop of piled-up antiques, but it opens onto huge salons full of glass cabinets packed with liqueur glasses, landscape paintings, old bicycles, and huge enamel signs. You could stay there for hours, despite the dust.

FOOD

Marché des Halles

5 quai de Monléon; daily 6am-1pm

Menton's food market is housed in a beautiful mustard-colored hall between the seafront and old town. The building was inaugurated in 1898 and refurbished in 2015, retaining all the original art nouveau fittings, painting, and ceramic arches. It's always full of locals and restaurateurs, and it has a strong Italian feel, with sausages, fresh pasta, and fruits and vegetables from across the border. Saturday mornings are the busiest and most colorful time to visit.

Sini

7 rue des Marins; tel. 04 89 98 71 77; July-Aug. daily 9am-5pm, Sept.-June Tues.-Sun. 9am-5pm; from €2.50

Behind the Marché des Halles food market, Sini serves snack-size local pizza specialties in a relaxed, fast-food ambience. The chef describes the food as "a mixture between Rome and Menton," with pick of the lunchtime alternatives, the planche de dégustation, with six different pizza slices for €18.50.

★ La Mandragore

place aux Herbes; tel. 04 93 35 43 19; Wed.-Sun. 10am-7pm; €12-16

Under the arches of the place aux Herbes, La Mandragore is a great little restaurant specializing in local dishes—the assiette Mentonnaise (€17) is a platter of deep-fried beignets, stuffed tomatoes, salad, and quiches. The square, 100 m (328 ft) from the sea, is full of places to eat, and in the summer buskers entertain diners beside the central water fountain.

★ Mirazur

30 avenue Aristide Briand; tel. 04 92 41 86 86; www. mirazur.fr; Tues.-Sun. 12:15pm-2pm and 7:15pm-10pm, phone reservations can be made 9am-11:30am and 3pm-6:30pm; €80-260

Mirazur, a super-exclusive dining room close to the Italian border, has three Michelin stars, only 20 seats, and has held the title Best Restaurant in the World by *The World's 50 Best Restaurants* panel (www.theworlds50best. com). Argentinian chef Mauro Colagreco prepares exquisite dishes with herbs coming from the restaurant's own garden, vegetables from the surrounding countryside, saffron from nearby Sospel, and prawns from Sanremo across the border. The restaurant is housed in a spectacular 1930s rotunda with panoramic views over the Mediterranean and across to Menton. Downstairs is the bar and a glass-fronted kitchen.

ACCOMMODATIONS
€100-200
Hotel Pavillon Imperial

9 avenue de la Madone; tel. 04 93 35 75 69; http:// pavillonimperial.com; €100 d

This friendly hotel is a good value in a great location, at the calmer western end of Menton, one block from the beach. There is on-site parking, and most of the 17 rooms overlook a courtyard garden where guests can enjoy breakfast or lounge around in the afternoon.

Sous l'Olivier

406 chemin de la Colle Supérieur; tel. 06 51 88 19 15; www.chambre-hotes-menton.com; €160 d, two-night minimum

A charming bed-and-breakfast in the hills above Menton, this three-suite, soft-pink manor house is in a tropical garden overlooking the Mediterranean. In the garden, guests are encouraged to sit and read about the locality from the hosts' collection of books while enjoying homemade jams with breakfast.

1

2

€200-300
★ Hotel Napoléon

29 porte de France; tel. 04 93 35 89 50; www.
napoleon-menton.com; €200 d

The Hotel Napoléon's blue and white rooms, inspired by artist Jean Cocteau, overlook the sea or the hotel's tropical garden, complete with banana palms and mountains behind. The ground floor has a large bar, a breakfast room with orange- and lemon-colored seating, a pool, a gymnasium, and a solarium. The hotel is well located for visits to Italy or to the gardens in the Garavan quarter of Menton. Most of the 44 rooms have balconies. Buffet breakfast is €14, parking €12.

INFORMATION AND SERVICES
Tourist Office

8 avenue Boyer; tel. 04 83 93 70 20; www.menton-
riviera-merveilles.fr; mid-June-mid-Sept. Mon.-Sat.
9am-noon and 2pm-6pm, mid-Sept.-mid-June daily
9am-7pm

The tourist office is in the Palais de l'Europe, a belle epoque building that was previously a casino, next door to the town's concert hall and children's library. The office has a wall of brochures regarding art, cultural, and sporting activities, as well as free beach ashtrays and dog-waste bags on the counter—catering to two of Menton's bêtes noires.

GETTING THERE AND AROUND
Car

Menton is easily accessible from the A8 autoroute approximately 25 km (16 mi) from **Nice** and 10 km (6 mi) from **Ventimiglia** in Italy. The drive takes 35 minutes from

Nice on the A8 and 1 hour on the coast road; it is 20 minutes' drive from **Monaco** and 15 minutes from Ventimiglia. There are several underground **parking lots** around the town; the most convenient for the old town and seafront is the esplanade Palmero.

Train

Menton station (place de la Gare; www. sncf-connect.com; daily 4:55am-1:15am) is on the TER line between Nice and Ventimiglia and the TGV route from Paris-Marseille to Ventimiglia. A second station, **Menton-Garavan** (place Gare; Mon.-Fri. 8am-1pm and 2pm-4:10pm), receives only local TER trains; it is less than 2 km (1.2 mi) from the Italian border and serves eastern Menton. The journey from Paris takes just over 6 hours (average cost €78), from Marseille 3 hours (average cost €33), and from Nice 25 minutes (average cost €6).

Bus

Zou network bus number 607 (https:// zou.maregionsud.fr) leaves Nice Square Normandie Niemen to Monaco, place des Moulins (€2.10 for a single) with a free connection on number 608 to Menton bus station. The journey takes around 1 hour 40 minutes depending on the time of day.

There is also a direct bus from **Nice airport** to Menton (Line 80 Express; normally 1 per hour), which costs €19 single (ages 4-11 €9.50, under 4 free, discounts for group of three or more). Tickets can be purchased from the airport ticket office or from the driver (departing from Terminal 2). The ride takes 1 hour.

Background

The Landscape

GEOGRAPHY

The geography of Southern France is varied, encompassing mountains, river deltas, fertile planes, and coastal areas.

Provence is bordered to the west by the Rhône river and to the east by the Côte d'Azur (blue coast). The term *Côte d'Azur* can also be used synonymously with "French Riviera," which is essentially a literary term and, besides the Italian frontier, has no formal boundaries. For the purposes of this guidebook, it stretches for around 180 km (110 mi) from the Italian border in the east to Hyères in the west.

The mountainous areas (Mont Ventoux, the Luberon Nature Park, Les Alpilles Nature Park, the Dentelles de Montmirail, Mont Sainte-Victoire, and the Gorges du Verdon) were created by the same tectonic lifting that gave rise to the Alps and the Pyrenees. The highest peak is Mont Ventoux at 1,912 m (6,273 ft).

The region's main rivers are the Rhône, the Durance (a tributary of the Rhône), the Verdon (a tributary of the Durance), the Var, and the Argens.

CLIMATE

Provence has three climatic zones: continental, Mediterranean, and alpine.

The continental zone, referring broadly to inland Provence below the tree line, experiences significant annual variations in temperature, with temperatures in July and August regularly nudging 40°C (104°F) and in winter frequently falling below zero (32°F).

The alpine climate is concentrated in the northwestern Vaucluse and in Haute Provence above 1,000 m (3,280 ft). Once again, there are large variations in temperature. Warm Mediterranean air means summer temperatures reach the upper 20s to mid-30s Celsius (low 80s to +90°F) during the day. However, in the winter the temperature can easily plummet to -20°C (-4°F). Snow is common.

Finally, there's the Mediterranean zone, which is situated along the Côte d'Azur, the Var Coast, and the inland areas bordering the coast. Here, winter temperatures remain mild with frosts uncommon.

The good news for visitors is that Provence is blessed with an inordinate amount of sunshine. The bad news is that it rains as much, if not more, in Provence than it does in Paris or London. Luckily, the rain usually comes in heavy bursts rather than a dull drizzle. The skies soon clear again.

No discussion of the climate of Provence is complete without a mention of the mistral wind. Sweeping down the Rhône valley from the Alps, the wind brings cold mountain air to Provence. Gusts frequently surpass 80 km (50 mi) an hour, and the mistral can blow for up to 15 consecutive days. More commonly, it lasts one or two days. It can blow at all times of year and is notoriously unpredictable. Months can pass without a whiff of wind, and then suddenly the branches of trees stir and then their trunks bend. The mistral is at its strongest in the Rhône delta around Arles, Avignon, and the Camargue. It is also a fearsome force in Les Alpilles and the Luberon, but it fades away as you enter the Var.

ENVIRONMENTAL ISSUES

Fires have devastated large areas of forest and scrubland in the Var during periods of excessive drought. Many of the fires are caused by humans, intensified by the strong mistral wind, dry pine needles, and flammable secretions from some vegetation. Public access to the Massifs des Maures and de l'Estérel is limited between mid-June and the end of September in an attempt to reduce the risk of forest fires; the local prefecture announces whether access will be restricted the following day at 7pm each evening during this period. Even if access is allowed, rules prohibit smoking, barbecues, and camping in the wild and require informing the fire brigade if any smoke is seen.

Severe flooding was experienced on the Riviera on October 3, 2015, when 200 mm (7.8 in) of rain fell in under 3 hours, killing 20 people, carrying off 17,000 cars, destroying four campsites, flooding railway stations, and causing over €600 million in damages. Flooding also occurred in the Alpes-Maritimes in 2017 and 2018, with long-term solutions, such as run-off reservoirs along the coast, under investigation.

The population has grown by 73 percent in the Provence-Alpes-Côte d'Azur (PACA) region since 1962. This is the largest growth in France and has had a major impact on the natural environment. More roads, increased presence of domestic cats and dogs in natural areas, and global warming are all having a negative impact on the region's fauna, particularly the number of reptiles and amphibians. See www.paca.developpement-durable.gouv.fr/IMG/pdf/cenpaca_listerougeamphibiens_def.pdf for a list of all threatened species.

Plants and Animals

PLANTS

Much of the landscape of Provence is taken up by the garrigue, a scrubland typically found on limestone soils where wild herbs such as lavender, thyme, sage, rosemary, and artemisia thrive.

Pine trees flourish throughout Provence. The main varieties are aleppo, maritime, stone, and northern. Oak trees are also prevalent in three main varieties: holm, kermes, and cork oak. The kermes oak is the principal tree of the garrigue. Its leaves have a special protective layer that slows evaporation. Other species of trees include chestnut and olive trees. The latter grow wild as well as in farmed plantations.

Trees also play an important part in the landscape of the Riviera. Dense forests cover much of the hills behind the coast, while in the towns, palm trees line the seafront promenades and rare species fill the many botanical gardens along the coast.

Village squares are traditionally protected by plane trees, while Nice's and Cannes's seafront promenades are lined with palms. Hyères, once the plant capital of France, claims to have over 7,000 palm trees. Menton has its lemon, lime, citron, and grapefruit trees, although most lemons for its annual festival now come from Spain. The region's forests are made up of chestnut, oak (Holm, kermes cork), and parasol pines, which stick out like spindly umbrellas in the martial landscape around the Massif des Maures.

ANIMALS
Mammals

The most frequently spotted large wild mammals are wild boars. They are nocturnal, so sightings are most common at dusk and dawn. Boars travel in family troops and can be dangerous to humans, particularly if there are young. It's best to steer clear if possible. There are also rare wolf sightings, largely in Haute Provence, and as recently as 2024, pairs attacked livestock around Aix-en-Provence. Less threatening is an encounter with the largest species of beaver in Europe, which weighs between 20 and 40 kg (44-88 lb) and lives in the Luberon. You'll know these animals have been present if you spot claw markings on trees and branches sharpened to a point like pencils. Wild hares, chamois, and deer are also commonly seen throughout Provence.

Sea Life

Striped dolphins *(Stenella coeruleoalba)* are commonly seen in the Golfe de Saint-Tropez and around the Îles de Lérins off Cannes, and fin whales have also been spied in the waters off the Cap d'Antibes. In terms of fish, tuna *(Thunnus)*, wreckfish groupers *(Polyprion americanus)*, mahi mahi *(Coryphaena hippurus)*, and devil rays *(Mobula mobula)* can all be seen in the Mediterranean off the Côte d'Azur. Underwater Mediterranean scenes from the revolutionary documentary *Le Monde du Silence* (The Silent World), filmed by Jacques Cousteau and Louis Malle, which won the Palme d'Or at the Cannes film festival in 1956 and an Academy Award the following year, were famously filmed in the waters off the Riviera.

Anti-jellyfish boats with nets patrol the most popular beaches during the summer, but swimmers, especially those more than

50 m (164 ft) from the coast, need to watch out for jellyfish, which can produce painful stings. The current advice if stung is to wash the affected area with seawater or vinegar (not fresh water, alcohol, or urine) and to remove any filaments stuck to the skin, possibly by gentle rubbing with sand or scraping with a credit card. Once all filaments have been removed, apply a disinfectant and keep away from the sun. Lifeguard posts on surveyed beaches will be able to offer treatment and advice, and it's worth remembering that even washed-up, dead jellyfish can sting.

Birds

Birdlife is abundant in Provence. The Luberon and the Gorges du Verdon are home to rare birds of prey, including three species of vulture and three eagle species. Pheasants and guinea fowl roam wild throughout Provence. Nightingales are also common. The Camargue delta is a center for migratory birds. Some species end their migration in the nature reserve, with large colonies of herons and egrets nesting in the trees. During the winter, resident species include ducks, geese, cranes, birds of prey, and several rarer species, such as the tiny penduline tit. Winter is also the best time to watch flamingos in the Camargue, when the population shows off their colorful new plumage during courtship displays.

Reptiles, Arachnids, and Insects

The diverse flora and fauna of the garrigue host nearly 300 species of insect. Lizards are common, including, in the Luberon, the ocelle lizard, the largest lizard in Europe. Turning over too many rocks is not advisable, because Provence has two species of scorpion. The sting of the dark brown, almost black one (*Euscorpius flavicaudis*) is relatively harmless, but if you get stung by the lighter, yellow-orange scorpion (*Buthus occitanus*) it's best to hurry to the hospital.

History

PREHISTORY

Proof that this region has been inhabited for over a million years comes from prehistoric tools discovered in the Vallonnet caves near Roquebrune-Cap-Martin, found in 1958 by a teenage girl. This became one of the earliest sites of known human settlement in Europe. The earliest sign of man in inland Provence is found in the Grotte Baume Bonne near Quinson in the Gorges du Verdon. Here, marks on the walls suggest the presence of men some 400,000 years ago. In the Cosquer Cave, in Les Calanques just outside Marseille, the walls of the cave (now accessed by an underwater passage) contain drawings dating between 27,000 BCE and 19,000 BCE. An archaeological dig at Châteauneuf-les-Martigues suggests that around 6,500 BCE the Castelnovian people who lived on the wetlands outside what is now Marseille were one of the first groups in Europe to abandon a migratory existence in favor of the domestication of sheep. From 10 BCE on there is evidence of the arrival of the Ligurian people eastward into Provence. It is thought the Ligurians were responsible for the earliest bories (conical-shaped stone buildings that you can still see all over Provence today). Shortly afterward, the Celts arrived from the north and the combined Celto-Ligurian civilization constructed fortified towns, or oppida, an example of which is the Oppidum d'Entremont built around 200 BCE outside Aix-en-Provence.

EARLY HISTORY

The Greek trading post of Massalia (Marseille) was established around 600 BCE by traders from Phoccea. They also developed outposts (emporia) along the coast at Olbia (Hyères),

Antipolis (Antibes), and Nikaia (Nice). When Phoccea was destroyed by the Persians around 550 BCE, Massalia became a refuge for its citizens. Massalia became an independent republic governed by its 600 wealthiest citizens. The settling Greeks were not conquerors but traders. The Greeks introduced money, wine, the olive tree, and fruit trees such as cherry and fig into Provence. They were responsible for its first inland towns at Glanum near Saint-Rémy-de Provence and Mastrabala (Saint-Blaise) on the Étang de Berre. By 3 BCE the Roman Empire was becoming an increasing force. Its initial efforts were concentrated on the conquest of Spain. The Greeks supported this effort and were allies of the Romans in the Roman Empire's war against the Carthaginian Empire. In return, the Romans supported Greek efforts to suppress the Celto-Ligurian tribes of Provence.

After the defeat of Spain, the republic of Massalia in 125 BCE invited Roman forces into Provence to help with the fight against the Celto-Ligurians. The Romans won notable victories, including a battle in 123 BCE at the Oppidum d'Entremont under general Gaius Sextius, which led to the establishment of Aix (Aqua Sextius). The Romans remained in Provence to keep the land route open between Rome and the empire in Spain, but they held back from full-scale conquest.

Things changed in 49 BCE when Massalia chose to support Pompey in his dispute with Caesar. Caesar laid siege to and defeated Massalia, and over the next few centuries the Romanization of Provence began in earnest. Cities such as Apt, Arles, Avignon, Fréjus, and Vaison were established to give homes to former Roman legionaries. A major program of public works was established, resulting in the construction of theaters, arenas, baths, and aqueducts. Just as the power of the Roman Empire was fading in the 3rd and 4th centuries, Christianity began to take hold in Provence. Early churches were built on the sites of Roman forums and churches. A synod of bishops took place in Arles in 314. However, around the same time German tribes attacked

and began to make progress into Provence. These invasions continued, and by the 5th century and the fall of Rome, the Visigoths had captured Arles.

THE FRANKS

Between the 5th and the 9th centuries Provence was not the safest place to live. Facing the dual threat of warring Germanic tribes from the north and the Muslim Saracens from the south, hill villages sprang up to provide at least some protection to inhabitants. By the 8th century Provence at least in theory, was part of the Frankish empire. There were, however, many virtually autonomous local rulers; the Saracens controlled much of the Mediterranean coast and frequently sent armies inland. In 732 the Saracens challenged Frankish authority at the Battle of Tours near Poitiers in the center of France. Subsequently, Frankish power was consolidated under Charlemagne and the Carolingian dynasty. However, dynastic infighting in the 9th century once again opened the door to the Saracens. Marseille was pillaged and Arles sacked. In 879 Charles Boso (known as Charles the Bald) broke away from the Carolingian kingdom of his brother-in-law Louis III and was elected king of Provence and Burgundy. The title was a hollow one, with Provence being controlled by rival local counts.

THE HOUSE OF GRIMALDI

In Monaco, Genoese nobleman François Grimaldi, having been expelled from his native city in 1297, hoodwinked his way into Monaco's castle at the top of the hill dressed as a monk, killed the guards, and founded the Grimaldi dynasty, which has ruled Monaco, with one or two brief intervals, ever since. Different branches of the Grimaldis controlled Monaco and built castles (all of which still stand today) in Antibes, Haut-de-Cagnes, and Grimaud. Charles Grimaldi acquired Menton in 1346, and neighboring Roquebrune in 1355.

BACKGROUND
HISTORY

Historical Timeline

1 million years BCE	Tools and animal bones found in Vallonnet cave near Roquebrune-Cap-Martin prove presence of early man on the Riviera.
About 400,000 years ago	The first evidence of humans in inland Provence, some 400,000 years ago, is in the Grotte Baume Bonne.
27,000 BCE to 19,000 BCE	Cave drawings in the Cosquer Cave provide further evidence of Stone Age humans in Provence.
6,500 BCE	Castelnovian people abandon their migratory existence and settle near Châteauneuf-les-Martigues.
600 BCE	Trading posts in Massalia (Marseille) and along the coast at Hyères, Antibes, and Nice are established.
125 BCE	Massalia invites Roman armies into Provence to help fight the Celto-Ligurians.
49 BCE to 3rd century CE	Julius Caesar lays siege to and enters Massalia. Provence is "Romanized" with the construction of arenas, theaters, and forums.
10 BCE onward	Provence sees the arrival of Ligurians from the east and Celts from the north.
6 BCE	Emperor Augustus builds la Trophée des Alpes to celebrate his victory over 45 tribes from the Alps.
5th century	The Roman Empire falls. Abbey of Lérins founded on an island off Cannes.
8th century	The Frankish Carolingian dynasty rules Provence.
879	Charles Boso breaks from the Carolingian dynasty and becomes king of Burgundy and Provence.
11th to 13th century	Provence is ruled by competing counts. This is the construction period for many Romanesque ecclesiastical buildings.

THE COUNTS OF PROVENCE

By the 12th century, control of Provence had passed to the Counts of Toulouse and the Counts of Barcelona. To the northeast, the land around Forcalquier formed an independent fiefdom. During this period, Provence enjoyed relative prosperity. Major Romanesque ecclesiastical buildings were constructed, including the Cistercian Abbeys of Thoronet, Silvacane, and Sénanque.

The Counts of Toulouse lost much of their land in the 13th century as a consequence of supporting the Cathars during their war with the French king. During the same century the Catalan count Raymond Bérenger made his capital Aix-en-Provence and succeeded in creating a Provençal ministate that stretched as far north as Barcelonnette. After Bérenger's death, control of Provence passed to the House of Anjou.

The 14th century was less than pleasant, as Provence faced twin menaces: The Black Death decimated the population, and returning mercenaries from the 100 Years' War pillaged what they could find.

1229	The last Count of Barcelona, Raymond Bérenger V, conquers Nice.
1297	Grimaldi dynasty begins in Monaco.
1346-1355	Grimaldis take over Menton and Roquebrune-Cap-Martin. In 1348, the Black Death strikes Provence.
1388	Nice becomes part of the Kingdom of Savoy.
1483	Provence passes to the French crown.
1536	Army of Holy Roman Emperor Charles V enters Provence.
1543-1545	The Siege of Nice leads to the fall of the town to France and Saracen pirates. Massacre of Vaudois Protestants in Merindol.
1789	Rioting in Marseille in spring signals the beginning of the French Revolution in Provence. The following year, Provence is divided into four départements.
1793-1814	Nice briefly becomes part of France under Napoleon.
1849	The Avignon-Marseille railway opens.
1860	Inhabitants of Nice vote overwhelmingly to join France. Railway constructed along the coast from Marseille to Menton.
1942-1944	Italian and German troops occupy Provence and the Riviera.
August 1944	Allied landings along the coast, part of Operation Dragoon, begin the liberation of Provence and the Riviera from Nazi occupation.
1962	Algerian independence starts a wave of mass immigration to Provence.
1995	National Front wins control of Orange, Toulon, and Marignane.
2013	Marseille is declared the year's European Capital of Culture.

Prosperity returned in the 15th century under Good King René (1434-1480), who established his court in Aix-en-Provence and was a notable patron of the arts. Control of Provence passed to the French crown when René's son died just one year after his succession to the crown.

THE HOUSE OF SAVOY

Nice had begun to gain importance as a trading station, first under the Counts of Provence and then, from 1388, under the House of Savoy, which further developed

trade. Fortifications were built to protect the settlement against France, as the locals preferred to look east for their fortune into Italy and Sardinia, rather than westward toward Provence.

Nice had its own "free" port at Villefranche-sur-Mer and became, after 1526, the Comté de Nice, which enjoyed different customs, culture, and a different language, Niçois. When the French king François I tried to conquer Nice in 1543 (known as the Siege of Nice), allied with the Turkish forces of Barbarossa, the inhabitants of Nice repelled the invaders, aided by heroine Catherine Ségurane, the teenage washerwoman who stole the Turkish flag and beat the invading soldiers with her washing board. Nice did, however, eventually fall to Barbarossa, who carried off 2,500 citizens to become slaves.

THE FRENCH KINGS

Any feeling that the Provençaux had of lingering independence was swiftly crushed by French king Louis XI and his successor Charles VIII. Local officials were removed from their posts and replaced by royal appointees. The castles at Les Baux and Toulon were deemed too threatening and promptly destroyed. Jews, who until now had enjoyed religious freedom in Provence, were massacred in Arles and Manosque, and a policy of convert or leave was instigated.

In the early 16th century Provence was drawn into the conflict between King François I of France and the Holy Roman Emperor Charles V. The French succeeded in taking but not keeping Milan; in response Charles sent armies into Provence; they arrived at the gates of Aix-en-Provence and Marseille but never succeeded in entering the cities. The French army eventually drove the emperors out, earning the king at least a modicum of loyalty from the Provençaux.

The second half of the 16th century was just as difficult. The Wars of Religion tore French society apart. Provence was split between the two faiths. For example, Avignon was staunchly Catholic, whereas Orange allowed Protestants to practice freely. Tension turned to action when in 1545 the parliament of Aix ordered the massacre of Vaudois Protestants in the Luberon town of Merindol.

Religious confusion reached its peak in 1584 when the European Catholic League seized Paris and drove out the king. The new Parisian government appointed their own alternative governor to Provence and named their own alternative capital of the region. The mess was only resolved when Henri IV converted to Protestantism.

Over the following centuries, royal power in Provence, as in the rest of France, became ever more firmly entrenched. The almost obsessive bureaucracy that still characterizes France today was born under Louis XIV, France's "Sun King" (1614-1715). Plague, famine, and taxes kept the countryside and small towns impoverished. Only Aix, Avignon, Arles, and Marseille prospered. In the run-up to the 1789 revolution, Provence was hit by twin economic shocks: a fall in the price of wine, and a severe winter that killed off thousands of olive trees.

REVOLUTIONS

The French Revolution has been so popularized in literature and film that anglophones sometimes believe the dramatic version of events. It's as if, in one swoop in 1789, France rejected monarchy and became a republic. In fact there were successive revolutions and restorations. The French Revolution, the first one, began in 1789 and ended in 1799 with a coup d'état that installed Napoleon Bonaparte as First Consul. In 1814 with Napoleon supposedly defeated and exiled on Elba, Louis XVIII was restored to the throne of France. His reign was briefly interrupted by the return and subsequent defeat of Napoleon in 1815 at the Battle of Waterloo. Royal authority did not last long. In 1830 rioting over three days in Paris led to the second revolution, the abdication of Louis XVIII's successor, Charles X, and the creation of a new constitutional monarchy under Louis Philippe, Duke of Orleans. The big idea was to substitute the principle of hereditary right to throne with popular sovereignty. This proved about as successful as the previous regime, and Louis was overthrown in 1848 by more rioting in Paris. The second revolution installed Louis Napoleon Bonaparte as president of the Second French Republic.

Events in Provence need to be set against this broader context. During this 50-year period there were revolutionaries, royalists, and forces with local rather than national loyalties, and the allegiances of cities, towns, and villages shifted between the three positions. The beginning of the first revolution in Provence can be traced to the spring of 1789 when the population of Marseille revolted against the nobility and magistrates. From Marseille, Republicanism spread throughout Provence, and it was in Marseille that the new French national anthem, "La Marseillaise," was born and taken north to Paris by the city's Fédérés guard. In inland Provence, counterrevolutionary forces organized and Royalism remained a dominant force up until the 1830s and the restitution of Louis Philippe to the throne. As with the rest of France there are horror stories from Provence of excessive revolutionary zeal, of arrest without charge, and even cannons being shot into cells as an expeditious method of execution. For a time, chaos reigned. Retrospectively, one of the notable consequences of the revolution in Provence was the creation of four cohesive government départements (Var, Vaucluse, Bouches-du-Rhône, and Basses Alpes) and the incorporation of papal territories of Avignon and the Comtat Venaissin.

Just as significant as the political upheavals was the Industrial Revolution, with a railway opening between Marseille and Avignon in 1849. Agricultural areas found it easier to get their produce to market, particularly with the expansion of the rail network to Paris. The same rail network gave birth to the tourist industry, with the beaches of Provence and the Côte d'Azur becoming popular meeting spots for high society.

In 1860, the local inhabitants of Nice voted overwhelmingly (by 25,743 to 160) to join France. Napoleon III signed the Treaty of Turin with the king of Sardinia, Victor Emmanuel II, and was handed the keys to the city alongside his wife, Empress Eugênia, who arrived on a large yacht. Giuseppe Garibaldi, who was born in Nice and is regarded as one of the fathers of Italian unification, claimed the vote was rigged and fled, along with many other "Italians," across the border.

WORLD WAR I

The battlefields of the First World War were a long way from Provence, and so the countryside and towns of the region were spared destruction. However, villages in particular suffered depopulation, losing entire generations of their young, who were sent to fight. To provide soldiers with their ration of wine, field upon field was planted with vines. The low-quality crop proved unprofitable after the war, and many vines were ultimately ripped up to make way for housing.

WORLD WAR II

France fell to the Germans in June 1940 and was divided into occupied and unoccupied zones. Provence was in the unoccupied zone until November 1942, when the Germans annexed the rest of France. The French Mediterranean fleet at Toulon destroyed its ships rather than let them fall into Nazi hands. Resistance fighters found refuge in the mountainous areas of Provence. Urban resistance in Marseille, in particular, was particularly difficult for the Germans do deal with. Parts of the city were no-go areas for German troops. To solve this problem, the Germans demolished Le Panier in 1943. Further destruction was caused by Axis bombing ahead of the planned Allied landings in 1944. Marseille, Toulon, and Avignon were badly damaged. On August 15, 1944, Operation Dragoon was launched with landings across the South of France. The Allied forces were assisted by local resistance fighters, and a demoralized German army quickly fled north.

THE GOLDEN AGE

The Riviera's Golden Age began even before the First World War, when the arrival of the railway caused a surge in tourism, sea bathing became more popular, European royalty arrived, and the seafront "season" began to change from winter to summer. The casino in Monaco had been attracting rich clients to the coast since the end of the 19th century, but the wealthy also wanted entertainment on which to spend their money. The first Monte-Carlo rally took place in 1911, and in 1929 the first Monaco Grand Prix raced around the streets of the principality, won by Anglo-French

The Provençal Language

Provençal is an Occitan dialect that was widely spoken throughout the South of France in medieval times. It was the base language of the lyric medieval poetry used by troubadours, who traveled from town to town performing their works. The troubadour tradition spread from the South of France throughout much of Europe. The beginning of the decline of the Provençal language was a royal decree of 1539 requiring all new laws made in Provence to be translated into French, not Provençal. At the end of the 19th century, the language was revived by a group of poets led by Frederic Mistral. Intent on preserving traditional Provençal culture, Mistral created a dictionary and transcribed the works of the troubadours. He also organized a traditional dress parade that still takes place in Arles. Today it is rare to hear Provençal spoken, although place names and business names are sometimes given in Provençal. For example, *lou pebre d'ai* is Provençal for the summer savory herb sarriette, a favorite of local donkeys, and is also the name of a restaurant in the Luberon village of Lauris.

driver William Grover-Williams in a green Bugatti 35B. American tourists, buoyed by the strong American dollar, loved the water sports, the house parties, and the gambling. Railroad magnate Frank Jay Gould built the art deco Palais de la Mediterranée casino in Nice, while Gerald and Sara Murphy opened their Villa America on the Cap d'Antibes to a host of summer visitors, including F. Scott Fitzgerald, Pablo Picasso, Ernest Hemingway, Cole Porter, Dorothy Parker, and Jean Cocteau. The Murphys convinced the Hotel du Cap near their villa to stay open during the summer and started sunbathing on the Plage de la Garoupe nearby, while Coco Chanel, who had a house in Roquebrune-Cap-Martin, made the suntan fashionable.

By 1936, all the hotels on the coast were staying open for the summer, encouraged by the French Front Populaire government's decision to allow paid holidays for all workers.

CONTEMPORARY TIMES

Cannes organized its first film festival in 1946 in part to compete with the annual Venice film festival. When the film *Et Dieu Créa la Femme* (And God Created Woman) was released in 1956, making a star of Brigitte Bardot, it also made a star of Saint-Tropez, the location of the film, and the newly styled international jet set began to arrive, appreciating the sunbathing

opportunities, the yachts, and the relaxed style of the Riviera. Months before the film's release, American actress Grace Kelly married Prince Rainier III of Monaco, cementing the Côte d'Azur as a place of romance, the high life, and unfathomable luxury.

Algerian independence in 1962 led to an influx of hundreds of thousands of so-called pieds noir refugees. The pieds noir were mostly ethnic French, born in Algeria during the period of French rule from the 1830s onward. A second wave of immigration from North Africa was encouraged by the government, stoking anti-immigration sentiment. By the late 20th century, the far-right Front National had become a major force in Provence. In the 1995 elections, the party under Jean-Marie Le Pen won control of Toulon, Orange, and Marignane. In 2002 Le Pen entered the presidential race and enjoyed victories in five out of the six départements that constituted the Provence-Alpes-Côte d'Azur (PACA) region. The onward march of a more sanitized version of the Front National continued in the 21st century. Le Pen's granddaughter Marion Maréchal Le Pen was elected from the Vaucluse to the National Assembly in 2012, while his daughter Marine Le Pen made it to final round of the presidential campaign in 2017 and 2022, losing to Emmanuel Macron in both contests.

Government and Economy

ORGANIZATION

The head of state in France is the president. He is directly elected for a five-year term. At the same time the electorate votes for candidates in the National Assembly, the legislative arm of the French government. The president appoints a prime minister and additional ministers who can call on the support of the majority of the members of the National Assembly. Because the presidential election and the National Assembly election take place at the same time, the president is usually able to control policy making. A second less powerful political force is the Senate. Members of the Senate are elected by local officials, and in the event of a dispute between the Senate and the National Assembly, the National Assembly takes precedence.

On a local level, France is split into 96 départements, with each département controlled by a prefect appointed by the central government. Since 1986 Regional Councils have been elected to discuss and pass laws relevant to each individual département. The Regional Councils take responsibility for matters such as housing, transportation, and schools. This book covers five départements: the Vaucluse, the Bouches-du-Rhône, the Var, part of the Alpes-de-Haute-Provence and the Alpes-Maritimes.

POLITICAL PARTIES

The main political parties in France are as follows:

- **Renaissance:** the centrist party formed by President Emmanuel Macron to support his candidacy in the 2017 election. Originally called En Marche, then La République En Marche, it adopted its current name in September 2022 after Macron was elected to a second term.

- **The Republicans:** a right-of-center, socially conservative, pro-European party

that holds the second-largest number of seats in the National Assembly.

- **The Socialist Party:** France's main left-wing party.

- **La France Insoumise:** radical left-wing party.

- **Rassemblement National (National Rally):** formerly the Front National, the party has a strong presence at the local level in Provence.

ELECTIONS

During the final round of the 2022 presidential campaign, Emmanuel Macron of Renaissance faced off against Marine Le Pen, the daughter of the founder of the Front National. The overall share of the vote was 58.5 percent to Macron and 41.5 percent to Marine Le Pen. The départements of Provence and the Riviera voted as follows: Var: 44.9 percent to Macron and 55.1 percent to Le Pen; Vaucluse: 48 percent to Macron and 52 percent to Le Pen; Bouches-du-Rhône: 52.1 percent to Macron and 47.9 percent to Le Pen; Alpes-de-Haute-Provence: 48.5 percent to Macron and 51.5 percent to Le Pen; Alpes-Maritimes: 50.13 percent to Macron and 49.87 percent to Le Pen. Therefore, compared to the national average, the far right performed much better in the Southeast than in the rest of France. Such results are attributable to concerns over immigration and a loss of identity. The next French presidential election will take place in June 2027. Emmanuel Macron is not eligible to stand again because he has already served two terms. All French citizens over age 18 are eligible to vote.

AGRICULTURE

Provence is a rich agricultural region, thanks to a varied landscape including both coastal and mountainous areas. Wine dominates employment with 20,340 vignerons in the PACA region. Fruit is the next biggest employer with 4,530 farmers, and then livestock (sheep and goats) with 1,550 farmers. Taken together, Provence and the Côte d'Azur account for 4.5 percent of France's total farm produce.

Geographically, production is spread as follows: sheep and goats tend to be raised in the mountainous areas and their foothills. On the plateau of Valensole, mixed grains and lavender are grown. Fruit cultivation is centered on a small area in the Rhône valley and the river valley of the Durance. Vineyards can be found throughout Provence and the Riviera, although they are particularly predominant in the Var.

INDUSTRY

Industry tends to be concentrated around the major cities. Important sectors are alimentary products for agriculture, steel, energy, and chemicals. Major industrial sites include the Eurocopter helicopter-manufacturing business located adjacent to Marseille Marignane airport; petrochemical refineries at Fos; the Agroparc outside Avignon; and the ITER experimental fusion reactor near Manosque.

The PACA region is one of the least industrialized areas of France. The two leading employers are the aeronautical and scent industries, the latter focusing on perfumes and medical or food flavorings. The Riviera remains the number one region for cut flower production in France, supplying almost all of France's mimosas, 42 percent of all roses, and three-quarters of the country's carnations.

DISTRIBUTION OF WEALTH

The PACA region displays the second-highest level of inequality in France after the Île de France (Paris). The top 10 percent have a disposable income of more than €3,000 per month, while the bottom 10 percent have less than €800. Urban areas fare the worst, with an estimated 31 percent of the population of Avignon and 26 percent of the population of Marseille living below the poverty line. Carpentras (30 percent), Cavaillon (28

percent), Orange (23 percent), and Nice (21 percent) also have notably high levels of poverty. Part of the problem is lack of jobs; Provence has a high unemployment rate compared to the rest of France: 3.4 percent compared with the national rate of 3.2 percent.

TOURISM

Tourism is an essential part of the economy in Provence. International visitors account for approximately 6 million out of 18 million tourist arrivals. Major recent tourist infrastructure projects have included a reconstruction of the Cosquer Cave in the Villa Mediterranée in Marseille and the blockbusting Luma Arles project. At the same time there is a growing awareness of the dangers of overtourism. Villages such as Les Baux-de-Provence have become overrun by tourists and are no longer as enjoyable to visit as they once were. Directors of tourism across the regions covered in this book indicate that there is a real desire to promote the new and the undiscovered, whether this be a sight, village, or experience. This broadening of what Provence has to offer has successfully driven visitor numbers higher.

Meanwhile, the French Riviera has always considered itself the most attractive destination in Europe. Tourism contributes almost a fifth of the region's gross domestic product (GDP). In terms of numbers, the French Riviera is the second most popular tourist destination in France after Paris, producing an annual turnover of €12 billion. The Alpes-Maritimes, Var, and Monaco combined receive around 20 million tourists per year, with Nice-Côte d'Azur being the second busiest airport in France. More than 800,000 cruise passengers arrive each year, filling many of the deep-harbor resorts, while there are 35 permanent pleasure-boat ports along the coast. The average number of tourists on the Côte d'Azur at any given moment is 200,000, ranging from 50,000 in mid-January to 650,000 during mid-August weekends. French tourists spend on average €62 per day on the Riviera, while foreign tourists spend on average €110 and conference attendees €223. North Americans spend on average €140 per day, the third-highest spenders after Middle Eastern and Turkish visitors.

The Riviera has 16 casinos, 4 of which are in Monaco; 17 golf courses; 28 spa and thalassotherapy centers; and, excluding Monaco, 5,782 restaurants. During the winter season (December to April) Riviera visitors also have access to 15 ski resorts providing 700 km (435 mi) of slopes, the three closest being Isola 2000, Valberg, and Auron.

People and Culture

DEMOGRAPHY AND DIVERSITY

The Provençaux are descended from the Celt, Ligurian, Italic, Greek, and Germanic peoples, as well as Moors and Saracens. Modern data on the ethnicity of the population is hard to come by. A law dating back to 1872 prevents the national census from collecting this information. However, surveys and polls are free to ask such questions, so a demographic pattern can be built up. Italians make up the most numerous non-indigenous population; around 8 percent of residents are at least partly of Italian origin if their lineage is traced back over three generations. Provence also has a disproportionately large North African population due to waves of migration after the end of World War II.

RELIGION

Provence, like the rest of France, is predominantly Roman Catholic. Secularism is spreading, and church attendance figures are not high. A survey of the whole of France in 2003 found that 65.3 percent of people considered themselves Roman Catholic, 27 percent atheist, and 12.7 percent belonged to a religion other than Catholicism. It is estimated that 5 million Muslims live in France. Precise figures are not available for Provence or the whole of France because of a law preventing the national census from collecting religious data.

LANGUAGE

The official language of Provence and the French Riviera is French. It is a Romance language and belongs to the Indo-European grouping of languages. Very rarely in Provence, you may hear Provençal being spoken. It was codified by poet Frederic Mistral at the turn of the 20th century. It is an Occitan language once used by medieval troubadours in their lyric poetry. Some place- and business names are still given in Provençal form.

Niçois (or Nissart), a subdialect of the Occitan language, is also heard in Nice and its surroundings, and adding a few Niçois expressions into conversation has become a way for many young people to assert their identity. It is taught in some schools and seen written on street signs in the old town, in restaurant names, and in local dishes. Attempts have been made to revive its use on local television news, and the song "Nissa La Bella," sung at home football matches of OGC Nice by its ardent supporters, has become the city's anthem.

LITERATURE

Although he was not a native Provençal writer or adoptive resident, Alexandre Dumas wrote some of the most memorable accounts of Provence in the middle of the 19th century. In *The Count of Monte Cristo* (1844) the description of Château d'If prison off the coast of Marseille is so evocative that it has transcended the boundary of fiction. Today many visitors to the island believe that Edmond Dantès, the hero of the book, was actually imprisoned there. Another classic of French literature, Victor Hugo's *Les Misérables* (1862), features Toulon prison, where the character

Jean Valjean is incarcerated for stealing a loaf of bread.

Perhaps Provence's most famous literary son is Jean Giono. Born in 1895 in Manosque, Giono is the writer of works including *Colline* (1929), *The Horseman on the Roof* (1951), and *The Man who Planted Trees* (1953). After military service during World War I, Giono became a pacifist. In his work he returned again and again to the theme of man returning to nature in search of a more basic way of life. He even tried it for real, creating a Utopian community in the wilds of Haute Provence during the interwar years. As an introduction to Giono, *The Man who Planted Trees* is an inspiring short story that has achieved worldwide fame.

Henri Bosco's *Mas Théotime* (1945) is another deliberation on man's relationship with nature. An epic family story, it concerns the daily lives of peasant farmers. The narrative is centered on a farmhouse just outside the Luberon village of Lourmarin. It's a gentle, slow, subtle book.

The Riviera has been the scenic setting to scores of novels and short stories from authors who have appreciated its hedonistic pleasures and the complicated, sybaritic characters who inhabit it. Perhaps the most celebrated novel set on the Riviera is F. Scott Fitzgerald's *Tender Is the Night,* first published in serial form in 1934. It tells the story of an expatriate American couple, Dick and Nicole Diver, who were based on Fitzgerald's hosts in Cap d'Antibes, the Murphys, and his own marriage to Zelda. The Riviera also features in the second half of Edith Wharton's *The House of Mirth,* set on a yacht off the Riviera. Wharton moved to Hyères 14 years after the publication of *The House of Mirth.*

Nice-born Jean-Marie Gustave Le Clézio was awarded the Nobel Prize for literature in 2008, for his life's work of poetry, essays, short stories, and novels. He worked outside of France for much of his life and was seen as something of a cult writer, little appreciated internationally before his Nobel Prize, not least because few of his books were translated.

JMG Le Clézio, as he is known, divides his time between New Mexico and Mauritius, and his work is not associated with the French Riviera. His first novel, *Le Procès-Verbale* ("The Interrogation"), won the prestigious French Prix Renaudot when he was only 23.

Rarely has an author provoked as much unjustified criticism and jealousy as Peter Mayle. Mayle was an advertising executive in London when he ditched the day job and started a new life in Provence. His weekly musings about the idiosyncrasies of the locals became first a newspaper column, and then a multimillion-selling series of books, beginning with *A Year in Provence* in 1989. The success of the series was such that it created a whole new literary genre, the "good life abroad" book. The books attracted tens of thousands of wannabe Mayles to the Luberon. Expats who already lived in the region turned on Mayle for ruining the area they loved. Without a hint of self-awareness, they complained that farmhouses were being converted into second homes at such a rate that the traditional character of the region was being lost. Literary critics attacked Mayle's portrayal of the locals as patronizing and disparaged his books. The French, however, took Mayle to their heart, recognizing what a great service he had done promoting Provence. He was awarded the government's most prestigious award and made a knight of the Legion d'Honneur in 2002. He was still writing right up until his death in 2018. Reading his work today, it's clear that Mayle was a great storyteller with a gentle and generous sense of humor.

A full and anecdotal account of these and many other writers' lives in the South of France can be found in Ted Jones's *The French Riviera: A Literary Guide for Travelers,* while a good source of tales of glamour and excess is Jonathan Miles's *Once Upon a Time World,* which chronicles the people and events of the French Riviera over the last 200 years.

VISUAL ARTS

In the late 19th century, an artistic revolution began in the South of France. In Arles

Film Locations in Provence and the French Riviera

Cinema's origins were in Provence. In 1895 the Lumière brothers directed a short black-and-white film titled **A Train Arriving at La Ciotat Station.** It was one of the first moving picture films, and in 1896 when it was shown, audience members were reportedly so horrified at the image of the train steaming toward them that they screamed and ran to the back of the room. Fast-forward a century and Provence and the Riviera are still regularly portrayed in cinema. Famous films include:

- **To Catch a Thief** (1955), starring Cary Grant and Grace Kelly, was filmed all along the Riviera seafront
- **La Baie des Anges (Day of Angels)** (1963) has a peroxide-blonde Jeanne Moreau and Claude Mann addicted to roulette in Nice's casino
- **The Day of the Jackal** (1973) has some of its most memorable scenes on the Riviera with the professional assassin, played by Edward Fox, seen driving across the Italian-French border and checking into the Negresco Hotel in Nice
- **Never Say Never Again** (1983), the Bond film, was shot around Nice and Monaco
- **Jean de Florette** (1986) and **Manon of the Spring** (1986), starring Gérard Depardieu, Daniel Auteuil, Yves Montand, and Emmanuel Béart, were shot mainly in the southern Luberon and in the countryside outside Aubagne
- **Dirty Rotten Scoundrels** (1988), starring Steve Martin, Glenne Headly, and Michael Caine, was shot around Beaulieu-sur-Mer
- **Horseman on the Roof** (1995) starred Juliette Binoche and was shot on location throughout Provence
- **French Kiss** (1995), starring Meg Ryan and Kevin Kline, was shot in the southern Luberon
- **Love Actually** (2003) shot its final scenes at Bar de La Marine on Marseille's old port, where Colin Firth's character proposes to his cleaning lady
- **Magic in the Moonlight** (2014), Woody Allen's storybook of the Riviera in the 1920s, has memorable scenes shot at Nice's observatory and the Villa Eilenroc on the Cap d'Antibes
- **At Eternity's Gate** (2018), starring Willem Dafoe as Vincent van Gogh, was filmed on location outside Arles and Saint-Rémy-de-Provence

and Saint-Rémy-de-Provence, Vincent van Gogh (1853-1890) was painting with aggressive, wild swirls of color. For a stormy nine-week period of intense painting, he was joined by Paul Gauguin, before the two men fell out. Meanwhile in Aix-en-Provence and L'Estaque, Paul Cézanne (1839-1906) was throwing artistic norms out of the window. Painting in a wide variety of different styles, he played with perspective—first by using color to create it, and then by using geometry (at Bibémus quarry outside Aix) to flatten it.

Both van Gogh and Cézanne were outliers who struggled for success during their careers. Van Gogh famously died by suicide and Cézanne only achieved a sense of financial security at the end of his life, when the death of his father left him with enough money to build his own atelier. Both artists broke with the prevailing impressionist school that was associated with painters such as Monet, Renoir, and Sisley. Their use of color and distortion of the people and places before them paved the way for artistic movements to follow: fauvism, expressionism, and cubism.

In the early 1900s, Georges Braque (1882-1963) and Pablo Picasso (1881-1973) were working together in the South of France developing ideas around the multiple uses of perspective. Braque was heavily influenced by Cézanne and in 1906 followed in his footsteps to L'Estaque, where Cézanne had marveled at the unique quality of the light. Braque and Picasso deconstructed objects and turned them into geometric pieces before putting them back together again. Unlike their predecessors, they produced work that was very much of their time and commercially successful. Picasso, who had a showman's personality to match his artistic skill, made a fortune. In 1958 he bought Château de Vauvenargues, outside Aix-en-Provence, where he was buried in 1973.

A friend of Picasso, Henri Matisse (1869-1954) spent the last 37 years of his life in Nice, first in an apartment overlooking the busy cours Saleya and then in a 17th-century hilltop villa in Cimiez beside the Roman ruins. Matisse was known for his still lifes and portraits, which he depicts illuminated by the intense light of the Riviera. After becoming ill, he found painting and sculpture too much of a physical challenge and set about developing a new medium based on cut-outs. Matisse is buried in Cimiez cemetery, a few hundred meters from his villa, which is open to the public.

Born near Paris, Jean Cocteau (1889-1963) became one of the most influential figures of the Riviera's art and culture scene. He knew everyone from Marcel Proust to Edith Piaf and was heavily involved in jazz, opera, poetry, and theater, inspiring artistic movements and developing a highly personal decorative style, which can be seen in his murals on the walls of the Salle des Marriages in Menton town hall, the Chapelle Saint-Pierre in Villefranche-sur-Mer, and the Chapelle Notre-Dame de Jérusalem in Fréjus.

Victor Vasarely (1906-1997) is a unique figure in the history of 20th-century art, the creator of an entirely new movement: optical art. While painting at Gordes, 1 hour to the north of Aix, Vasarely had a revelation. In the intense summer light, he noticed a contradictory perspective to the linear one he had been using: "Never can the eye identify to what a given shadow or strip of wall belongs, solids and voids merge into one another, forms and backgrounds alternate. Thus, identifiable things are transmuted into abstractions," he said of Gordes. A center dedicated to his

Nice Jazz Festival

work was constructed outside Aix in 1973, with Vasarely writing a personal message in the cement foundations. He shared only the first few words: "From Cézanne to Vasarely, we will be worthy."

The more modern movement known as the École de Nice is an art collective that ran from the mid-1960s until the mid-1970s, with some of its major figures still working today. The epigrams and axioms of artist Ben can be seen on school bags and notepads all over France, and his cursive writing style appears on Nice's tram stops. Sacha Sosno, Yves Klein, and sculptors Bernar Venet and Arman are internationally known, as is Niki de Saint-Phalle, who joined the movement in its later stages.

MUSIC AND DANCE

The traditional dance of Provence is the farandole. It is a line dance performed to a flute or a drum. The dancers skip with every beat, alternating lead feet. To see it performed in traditional dress, head to the Roman Arena in Arles on the first Sunday in July.

The Riviera's summer music scene begins with the nationwide Fête de la Musique on June 21, when local music groups play all day and night in almost every town in France, but the "season" is dominated by the Riviera's two big jazz festivals. Taking place in mid-July, Nice Jazz Festival has been running since 1948, although it has morphed into a more universal-music event. It used to take place in the Roman amphitheater and gardens on the hill above Nice in Cimiez but now takes place off the place Masséna in central Nice.

The other big jazz event is Jazz à Juan, a 10-day festival of jazz in July, set in a pine grove above the beach in Juan-les-Pins, called Le Pinède Gould. The event, which began in 1960, is more focused on traditional jazz. Acts in the past have included Ray Charles, Chick Corea, Sonny Rollins, Wynton Marsalis, and Ella Fitzgerald, many of whom have their handprints embedded in the pavement walkway outside.

Essentials

Transportation

GETTING THERE
From North America

North Americans can fly from New York direct to **Nice-Côte d'Azur Airport** (www.nice.aeroport.fr). **United** offers direct flights from New York's JFK Airport, while **La Compagnie** offers direct flights from Newark Airport. **American Airlines** flies direct from Philadelphia every day May-October and **Delta** flies daily from Atlanta May-September, while **Air Transat** (www.airtransat.com) has direct Nice-Montreal flights. **Marseille Provence Airport**

(www.marseille-airport.com) also offers direct flights via Air Transat from Montreal. Visitors coming from other North American destinations must fly first to major hubs such as Paris, Amsterdam, and London, and then take a connecting flight to Marseille or Nice.

From Europe and North Africa
Air
Nice-Côte d'Azur Airport has direct flights from all major European cities including Rome, Dublin, Amsterdam, Barcelona, Madrid, London, and Frankfurt. It also has direct connections with Morocco, Algeria, Tunisia, and 17 airports in France.

Marseille Provence Airport has low-cost and charter flights to 62 different European destinations and 20 airports in France.

Toulon-Hyères Airport has seasonal flights to and from Southampton in the UK with Flybe, Geneva with Swiss International Airlines, and Rotterdam with Transavia. From Toulon-Hyères Airport, it's a 1.5-hour drive to Marseille or 2-hour drive to Nice.

Train
The easiest way to get to Southern France by train is via the **TGV** (high-speed train) network (www.sncf-connect.com). The main connection is out of Paris. There are TGVs from the Gare de Lyon every hour to Marseille, which takes about 3.5 hours; most stop both at Avignon and Aix-en-Provence. From Marseille, the TGV stops at Saint-Raphaël, Cannes, Antibes, Nice, and Menton. The train from Marseille to Nice takes about 2.5 hours. There are also now daily trains (spring-autumn) from Barcelona to Avignon, Aix-en-Provence, and Marseille.

Interrail (www.interrail.eu) offers a one-month train pass for €528 for travel throughout Europe, a 7-days-within-one-month global pass for €264, and an 8-day one-country pass for France for €235. There are reductions for those under age 28, children, and seniors, and accompanied children under 11 travel free, making this a potential option for travelers.

If coming from the UK, **Eurostar** (www.eurostar.com) operates a daily service from London to Paris and London to Lille. From these stations passengers can change for a train to Avignon, Aix-en-Provence, Marseille, or Nice.

Bus
BlaBlaCar (www.blablacar.fr) and **Flixbus** (www.flixbus.fr) operate long-distance bus routes between all major towns in France. They are usually cheaper than traveling by train, but can be very slow, with lots of stops and waiting time. They do have Wi-Fi, toilets, and reclining seats.

Journey time from Paris to Nice is around 13 hours.

Ferry
Corsica Ferries (www.corsica-ferries.fr) operates services from Corsica, Sardinia, and Sicily to Nice. **Corsica Linea** and **La Meridionale** operate services from Corsica to Marseille. La Meridionale also operates a service from Sardinia to Marseille.

Car
The main route from the north to Provence is the **A7 autoroute,** which runs between Lyon, Avignon, and Aix-en-Provence. It is nicknamed the Autoroute du Soleil. On weekends in July and August it is advisable to find alternative routes because the stretch between Avignon and Lyon becomes one long traffic jam.

The **A8 autoroute** intersects with the A7 just west of Aix-en-Provence and runs all the way to the Italian border. It serves all the main towns and cities along the Riviera, passing alongside the Massif de l'Estérel between Saint-Raphaël and Cannes and the Massif des Maures west of Fréjus.

The **A9** runs from just outside Avignon to the Spanish border near Barcelona.

An alternative to driving from Paris yourself is to use **Hiflow.** In conjunction with **SNCF,** Hiflow (https://hiflow.sncf-connect.com) offers a private chauffeur-driven service from Paris to any address in the South of France from €233 one-way. Your car can also be driven by an independent driver (from €98) or loaded onto a lorry for a pickup at fixed locations in the south (from €186) or delivered to your door (from €245).

From Australia and New Zealand

The simplest way to get to the South of France from Australia and New Zealand is to fly to Dubai and then to Nice with **Emirates** airlines. Transavia offers a limited service from Dubai to Marseille. There are also direct flights to Nice from Riyadh, Doha, Hong Kong, and Beijing, along with multiple flights a day to major European hubs such as London, Frankfurt, Amsterdam, and Paris, from where it's easy to catch a connecting flight to Nice.

From South Africa

There are no direct flights to the South of France from South Africa. South Africans should fly to either London, Amsterdam, or Paris and pick up a connecting flight to Marseille or Nice.

GETTING AROUND
Car

The most convenient way to travel around Southern France is by renting a car; it makes the perched villages or more remote parts of the coastline easily accessible. There are public car parks in most towns and villages, so parking is rarely a problem.

Car Rental

All the major car rental companies, including **Hertz** (www.hertz.com), **Europcar** (www.europcar.com), **Avis** (www.avis.com), and **Sixt** (www.sixt.com) have offices at the airports, main railway stations, and downtown areas of major cities. In high season, a small rental car will cost around €400 per week. Non-EU residents can take advantage of low rates on long leases.

Routes and Tolls

French roads are divided into **autoroutes** (highways marked with an A); **main national routes** (marked with an N or RN), which often run parallel to the autoroutes; and **local roads** (marked with a D). Least scenic, though fastest, are the autoroutes, which make use of a toll system (péage). Drivers take a ticket when they enter through a péage gate and pay as they progress or leave the autoroute at an exit. The cost varies from €0.03 to €0.16 per km (0.6 mi), depending on the stretch of road, and péages accept coins, cards, and contactless payment. Autoroutes have regular aires de répos (rest stops with toilets and places to sit) and aires de services (petrol stations with restaurants and toilet/shower facilities).

If you are not in a hurry, the coastal roads provide the best visual experience of the Riviera. The Corniches between Nice and Menton are also fun to drive, but they are not for inexperienced drivers because they are narrow with sharp bends, blind corners, hair-raising views, and steep inclines. There can also be stones or mudslides on the roadway.

Rules of the Road

The **speed limit** on autoroutes is 130 kph (80 mph) and 110 kph (68 mph) in the rain, although some sections are fixed at 110 kph in all weather. On Routes Nationales, the speed limit is 80 kph (50 mph). The limit on local roads is between 20 kph (near schools) and 80 kph (12-50 mph), depending on the section. There are regular **speed cameras** (fixed radars) on all French roads, but warning signs are given. In France, it is illegal to use radar identification software that is present in some satellite-navigation systems.

Drivers and passengers must always wear **seat belts.** Children under 10 must travel in an approved child seat. Licenses from all

EU countries as well as the United States, Canada, South Africa, and New Zealand are valid in France for up to a year. The car must have a warning triangle and fluorescent, high-visibility jackets. It is illegal to use a mobile phone while driving, with on-the-spot fines of up to €135, and the allowable alcohol level for driving is 0.5mg/l of alcohol (one glass of wine).

The government website (www.bison-fute.gouv.fr) gives up-to-the-minute traffic information, and the motorway routes. Roads are often busy on weekends around 5pm when visitors are returning from the beach. Beware of the weekend following the end of school for the summer holidays (early July) and again just before the return to school (usually the last weekend in August) when traffic density is highest. The middle weekend of August is also busy.

Train

The comfortable, efficient, but often crowded French **TER rail network** runs along the Riviera between Marseille and Ventimiglia just across the Italian border. It services all the major towns, some of which are also on the **TGV fast train network.** Tickets can be purchased at all stations or online (www.sncf-connect.com). Train fares are cheaper if traveling during off-peak times (période bleue) and can be booked three months in advance. A typical journey from Nice to Cannes on the local TER train takes 38 minutes and costs €7.90, or 26 minutes on the faster TGV InOui train (€9.80).

It is worth noting that there is no train line to Saint-Tropez, the closest stations being Saint-Raphaël or Les Arcs, from where buses run to Saint-Tropez.

Bus

The bus network across Provence is patchy. The various regional bus companies are grouped together under one brand: **Zou** (https://zou.maregionsud.fr).

The Riviera's main towns are also connected by coaches with bus stations (gares routières) usually centrally located (except in Nice, whose bus station is in Riquier, 2 km/1.2 mi from the main place Masséna, though connected by tram line). **Zou** offers large discounts for multi-journey tickets and regular travelers. Go online for more information on intercity bus routes (https://zou.maregionsud.fr).

Bike

Bicycle and e-bike are popular means of travel around the Riviera. Most of the way along the coast from Nice to Antibes, there is a well-maintained cycle track. Most bike shops rent high-quality road and mountain bikes for hire, but since budget airlines and the French railway are very bike-friendly, it's also possible to bring your own if you are a serious cyclist.

Nice has two privately run bike hire operators, Lime and Pony. American-owned **Lime** (www.li.me) has 1,000 bikes available in Nice and surrounding areas, all electric with baskets at the front. French-owned **Pony** (https://getapony.com) has 800 electric bikes available and 200 pedal bikes, all of which can transport a second person behind the cyclist on a padded seat. Bicycles can be picked up from 250 designated zones around the city and use a GPS locating system via the companies' downloadable apps. Many large hotels and the SNCF rail network also have bikes for hire. **Holiday Bikes** (www.loca-bike.fr) has rental stations in Nice, Villeneuve-Loubet, Antibes, Juan-les-Pins, and Cannes, and rents scooters and bikes from €14 per day for a city bike to €35 for a carbon-frame racing bike. Lime also operates in Marseille. Station Bees e-bike rental points are located in many of the larger and more popular Provençal villages (www.stationsbees.com).

Taxi and Ride Share

Taxis are available throughout the region, but they tend to be expensive and usually have to be booked in advance over the phone. In Nice, Cannes, Marseille, and Aix-en-Provence there are designated taxi stands near large hotels.

Uber cars are present in most towns on the Riviera, but not Provence, where they only operate out of Marseille. A popular budget way to travel from town to town (usually over long distances) is via the carpooling scheme BlaBlaCar (www.blablacar.fr).

Visas and Officialdom

PASSPORTS AND VISAS

The very latest visa requirements can be quickly checked at www.diplomatie.gouv.fr/en/coming-to-france. There is a "visa wizard" that will quickly tell you your requirements. A summary of the current regulations is found below.

United States, Canada, United Kingdom, Australia, and New Zealand

Beginning in 2024, the European Union, of which France is a member, is implementing the European Travel Information and Authorization System (www.etiasfrance.com), which will require visitors from visa-free countries to obtain authorization before they arrive in France. Travelers will need to fill out a form online and receive approval by email. The cost is €8 and the authorization will be good for multiple entries over three years. Passport holders can stay for 90 days in a 180-day period without a visa. For longer stays, a special visa is required.

European Union/Schengen

Nationals of EU-member countries who have a valid passport and national identity card can travel freely to France.

South Africa

South African nationals require a short-stay visa for stays of up to 90 days, and a long-stay visa for more than 90 days.

VACCINATIONS

There are no specific requirements, although health bodies recommend that visitors be vaccinated for hepatitis A and B and yellow fever. In addition, a rabies shot is recommended for those likely to come into contact with animals.

Recreation

There's so much to do in the South of France. Here's a round-up of ways you might spend your time.

BEACHES

Beach life is a big part of most people's holidays on the French Riviera. Every seafront will have a public beach (meaning there are no sun loungers for rent), but rules can vary regarding smoking, ballgames, and access for dogs. Regulations for each beach will be posted on a board at the access point, along with dates for when the beach has lifeguards present and flags depicting water quality and safety.

Seafronts in large resorts such as Nice, Cannes, and Saint-Tropez have a mixture of public beaches and private beaches (plages privées). The private beaches are open to the public, but there is a charge for hiring a sun lounger, umbrella, and sometimes towels. Anyone can walk across a "private beach," but you can't use the facilities unless you pay. Prices range €15-50 for a lounger and umbrella, depending on the location of the beach. If you eat in the beach restaurant or have a drink in the beach bar, it does not automatically mean you can use the loungers or umbrellas at the water's edge.

There are also handiplages along the coast (always signposted), which provide access for those in wheelchairs or with restricted mobility.

Finally, certain sections of Riviera beaches, or, in places such as the island of Le Levant, the entire beach is naturist, where sunbathers are nude. Going topless on a Côte d'Azur beach is regarded as very normal, but it's not as popular today as it once was.

WATER SPORTS
Sailing and Regattas

Two of the Mediterranean's most prestigious sailing events take place on the Riviera: the **Voiles d'Antibes** in June, (www.voilesdantibes.com) and the **Voiles de Saint-Tropez** in September and October (www.lesvoilesdesaint-tropez.fr), and there are scores of **sailing schools** and practice centers all along the coast (www.voilepaca. fr). Sailboats can be hired out privately or chartered with a skipper from all ports, although Antibes and Saint-Tropez are definitely the places to go for sailing fans.

Scuba and Snorkeling

For scuba-diving enthusiasts, the Côte d'Azur offers some of the best locations in Europe, with diving schools divided into two categories: those affiliated with the Fédération Française de Plongée and those linked to the American PADI system. Some instructors will have both qualifications, but those wishing to dive must bring their qualification certificates and a certified medical certificate proving a clean bill of health for diving. Summer is the most popular season for scuba diving, but most schools are open all year round. They offer "baptism" introductory courses and take experienced divers to natural sites, special reserves, and shipwrecks. The most popular diving sites on the Riviera can be found off **Porquerolles** and **Port-Cros,** near Hyères and Cap Bénat, **Ramatuelle** near Saint-Tropez, and **Golfe-Juan** near Antibes.

Other Water Sports

Other water sports, such as Jet Skiing, wakeboarding, kitesurfing, paddleboarding, and being dragged behind a boat on an inflatable banana are all possible on the beaches of the Riviera; waterskiing was invented in Juanles-Pins. Besides kayaking and canoeing in the rivers of the Var and Alpes-Maritimes départements, sea kayaking (www.canoepaca.fr) has become a popular pastime in recent years, and is a great way to explore the coastline and creeks inaccessible by car.

OUTDOOR ACTIVITIES

Outdoor enthusiasts tend to visit in spring or autumn when average temperatures are lower. France's huge network of long-distance hiking trails, the **Grandes Randonnées,** is managed by the Fédération Française de la Randonnée Pédestre (French Hiking Federation; www.ffrandonnee.fr). The trails are marked by red and white stripes, and six pass through or near the Riviera. Details of routes and maps, advice, and accommodations can be found at www.gr-infos.com/gr-fr.htm.

Besides the Grandes Randonnées, Provence and the Riviera have many excellent hikes, notably in the **Massif des Maures,** the **Massif de l'Estérel, Mont Sainte-Victoire,** the **Gorges du Verdon,** and in the hills above Monaco and Menton. Hiking Les Calanques is also a popular activity, and for cyclists there is no greater challenge than the ascent of Mont Ventoux, one of the most punishing climbs on the Tour de France.

The South of France is generally a safe place to cycle, with drivers giving a generous amount of space on the road. There are extensive cycle paths all along the coast and plenty of cycling shops for any repairs and equipment needs. For serious road cyclists and mountain bikers, the great joy is riding the small roads, high up in the hills above the Riviera, and venturing into the massifs in the Var département.

The Alpes-Maritimes département (www.departement06.fr) has published three cycling maps for different levels of ability: the **Cartes des Boucles Cyclosportives** for long-distance, semipro cyclists; the **Cartes des Boucles Cyclotouristes** for medium-level cyclists who want to explore the region by road bike; and the **Cartes des Balades Famille** for those who want a relaxed ride through parkland or along the seafront. Individual tourist offices will also have their own routes and printed brochures.

SHOPPING

Nice, Cannes, Monaco, and Saint-Tropez provide some of Europe's most exclusive shopping, with **designer boutiques** and **watch and jewelry outlets** filling up the seafront, while the old towns of Nice, Fréjus, and Antibes are best for **fashionable clothes** and **locally produced arts and crafts.** Provence is famous for its **weekly markets,** and most tourists will visit at least one of them during their trip. Fresh fruit and vegetables are a major lure, as are artisan-made products. Popular **Provençal goods** include Marseille soap, lavender nosegays, crystallized salt, and packets of herbes de Provence. On the Riviera, there are always raffia bags, sunhats, and designer T-shirts for sale in beach shops—upmarket Roman sandals in Saint-Tropez, nautical wear in Antibes, and pottery from Vallauris, plus perfume, wine, olives, and tapenade.

For those interested in **antiques** and secondhand knickknacks, there are always plenty of **brocante** fairs during the year. Bargain hunters should head for vide-greniers (the garage sales of Provence), where similar items are on sale for half the price. Popular purchases include old irons that are used as doorstops and antique coffee grinders. Tourist offices will have details of regular antiques fairs.

Remember that non-EU residents can claim back VAT (TVA) on purchases that exceed €175. Make sure the shop fills out a form, which you can present at the airport customs desk when departing the country. The goods must accompany you when you leave, and this must be within three months of the purchase date.

WINE-TASTING

Wine-focused tourism is booming in Provence. Private companies offer daylong tours of the most famous wine-producing areas, such as Châteauneuf-du-Pape, Gigondas, and Vacqueyras. On a more basic level, local vineyards and wine shops offer tastings. Here the drink of choice is rosé. There is simply no better place to drink pink wine in the world. The pale rosé produced by Provençal vineyards is a perfect match for the blue sky and is typically drunk in copious amounts.

Festivals and Events

SPRING
Feria d'Arles

Arles; April

This festival is a bullfighting spectacle with street parties and live music (page 90). It is one of the largest events in Provence. Expect crowded streets, dancing, drinking, and plenty of noise. Arles is transformed during the Feria into the party capital of Provence.

Festival de la Mode

Hyères; April

Internationally renowned fashion and photography exhibitions showcase rising designers in the town's modernist Villa Noailles.

Monte-Carlo Masters

Monaco; mid-April

An international men's hard-court tennis tournament in the prestigious Monte-Carlo tennis club (page 412).

Cannes Film Festival

Cannes; two weeks in May

This famed annual film festival celebrates cinema and the business around it, with two weeks of parties, paparazzi, red carpets, film competitions, and beach screenings (page 324).

Formula One Grand Prix

Monaco; late May

The most prestigious and glamorous sporting event of the year, when Formula One racing cars charge around the public roads of Monaco (page 413).

Fête de la Transhumance

Saint-Rémy-de-Provence; late May

On Whit Sunday (the seventh Sunday after Easter), herds of sheep are driven through the streets of Saint-Rémy en route for the mountains where they spend the summer (page 103). It's a crazy affair. The traffic is halted, and for half a day the animals take over the town. Along with the sheep come donkeys, goats, and, of course, shepherds.

SUMMER

Fête de La Musique

Across France; June 21

During this annual festival there is live music in the streets of every town and village. Many restaurants and cafés put on special menus for the evening. Tables and chairs spill into the streets, stages are erected on street corners, and everybody has a party. Best of all, it's all free.

Voiles d'Antibes

Antibes; June

One of the biggest sailing events on the Mediterranean, where classic and modern yachts race against each other over four days with free festivities on the quayside (page 293).

Rencontres d'Arles Photography Festival

Arles; July to September

Now well past its 50th year, this festival includes photography exhibitions in venues across the town (page 92). Major venues host ticketed shows by well-known photographers, but even without a pass you can partake in the atmosphere of the festival. Photos decorate windows and street corners and animate a visit to the city. The opening week is particularly busy with various VIP and press events. The festival attracts an estimated 100,000 people to the city.

Festival d'Aix-en-Provence

Aix-en-Provence; July

Every year, Aix hosts an internationally renowned opera festival with performances throughout the month (page 132). The festival tends to feature five major operas and a host of subsidiary events, including performances of everything from classical music to flamenco guitar.

Avignon Festival

Avignon; July

This annual theater festival presents 60 shows spread across 40 venues, attracting an audience of 155,000 people (page 47). Shows range from dance, theater, and comedy to mime, readings, and films. Many events, such as dance, don't require any language skills and are often performed in unforgettable venues such as the courtyard of the Palais des Papes.

Jazz des Cinq Continents

Marseille; mid-July

Now well into its second decade, this festival continues to attract large crowds and top musicians (page 206). There are large open-air concerts in spectacular venues such as outside the Palais Longchamp, or at the entrance to MuCEM. If you are in Marseille, it's great fun to attend.

Jazz à Juan

Juan-les-Pins; mid-July

The most famous jazz event in the South of France is now over 60 years old and has hosted all the great names in jazz, many of whom have left their handprints in the ceramic tiles beside the pine grove where the festival takes place (page 306).

Nice Jazz Festival

Nice; mid-July

Since 1948, one of the summer's most popular festivals takes place off place Masséna in the center of Nice, and includes world music, rock, and reggae as well as jazz (page 255).

AUTUMN
Richerenches Truffle Festival

Richerenches; November

This festival marks the opening of the town's truffle market. The event includes a truffle-based holy mass and all sorts of delicious truffle-based delicacies to try (page 59). Check the village website for the date of the celebration: www.richerenches.fr.

Monaco Fête Nationale

Monaco; November 18-19

On Monaco's national day, the Monégasques unite to celebrate their identity and fidelity to their princely ruler, Albert II, with fireworks, open-air concerts, and fairs.

WINTER
Rognes Truffle Market

Rognes; last Sunday before Christmas

This village's ever-more-popular truffle market is the perfect occasion to stock the fridge with delicacies for the big day and to taste truffle dishes such as scrambled eggs laced with truffles (page 144).

Rallye Automobile Monte-Carlo

Monaco; January

One of the world's great car rallies now has "historic" and e-sport editions. Drivers finish the race in Monaco, accompanied by car exhibitions and a selection of prizes (page 415).

Festival International du Cirque de Monte-Carlo

Monaco; late January

This long-running festival features world-renowned circus artists performing an array of acts in Monaco's Espace Fontvieille.

Nice Carnaval

Nice; mid-February to March

The biggest carnival on the coast takes place in the month before Lent, when giant floats, bands, and street performers parade through Nice with streamers and flying flowers. Tickets are required for the main events behind closed doors (page 255).

Corso de Mimosa

Sainte-Maxime; February

The advent of spring is celebrated with a parade of flower sculptures, traditional dances, musical acts, and floats. Around 80,000 flowers are displayed around the town (page 379).

Fête du Citron

Menton; February

Menton's lemon festival attracts more than 200,000 visitors each year, with fruit-covered floats and performers parading through the streets (page 432).

Food and Drink

EATING IN

Markets are the best places to shop for fruit, vegetables, cheeses, dried meats, and aperitif snacks. Villages usually have one market each week, although some may have two: a farmers market and a more traditional weekly market. Markets open early around 8am, and the stall holders pack up between 12:30pm and 1pm.

EATING OUT

Most restaurants start serving lunch at noon. The locals tend to get hungry early, and so tables fill up quickly and it's not unusual for a restaurant to sell out of its plat du jour before 1pm. At lunchtime, restaurants will often offer a fixed-price menu in addition to à la carte items. Service usually lasts until 2pm, but try to be seated by 1:45pm. In the summer, serving hours tend to be extended.

In the evening, restaurants open around 7pm. There is not usually a plat du jour, although a fixed-price menu may still be offered. At night, restaurants will often not accept new customers after 10pm.

Besides the posh end of dining and Michelin-starred restaurants, there is no rigid dress code. However, men should always wear a shirt when eating, even in a beach restaurant. French people generally have refined table manners, and children are expected to sit up at the table. Restaurants will have high chairs available but are not very tolerant of noisy children.

When making a reservation make sure you specify where you want to eat—on the terrace or inside. If at all possible, make a reservation in person. By doing this you can select your table and pick one with a view or one that is nicely shaded. In the winter, be aware that a lot of restaurants cover their terraces with a plastic awning. For the purposes of smoking laws, the terrace (with plastic cover) is still considered to be outside and smoking is

therefore permitted. If you object to smoking, it is best to eat inside.

Tips

Restaurant prices include a service charge, so there is no obligation to tip, but especially in tourist areas it is seen as polite to leave an additional cash sum for the waiter or waitress if the service has been good. For a simple coffee, leaving a coin of 10-20 cents is normal; for a meal of €80-100, leaving €5 is appreciated.

WINE

Most vineyards offer free tastings. However, there is an implicit understanding that visitors have an intention of buying at least one bottle of wine. It's considered rude to turn up, taste multiple wines, and leave empty-handed. Larger, more commercial vineyards have started charging for tasting. In this case, it is entirely at your discretion whether you choose to purchase wine or not.

SPECIALTIES

Look for the following regional products while traveling in Provence and on the Riviera.

Liquor and Wine
Pastis

The French drink a mind-boggling 20 million glasses of this aniseed-flavored liquor a day. Depending on whom you believe, it was invented by either a hermit in the Luberon or Jules Pernod, who decided to capitalize on the banning of absinthe manufacture in 1715. Either way, pastis is synonymous with Provence and is a perfect slow drink to enjoy with the setting sun.

Châteauneuf-du-Pape

The dream of many oenophiles is to enjoy a tasting in the cellar of a Châteauneuf-du-Pape vineyard: to pour the deep-ruby liquid into a glass and inhale as the heady scents of ripe fruits are released; to close their eyes and sip, counting the seconds as the flavors reach a crescendo in the mouth before receding gently like a tide. Note, though, that Châteauneuf-du-Pape is a wine that needs to age. A young Châteauneuf (under five years) can be a disappointment. Also, beware of buying Châteauneuf-du-Pape in a supermarket; the quality is unlikely to be good.

Rosé

The world is experiencing a boom in rosé consumption. Pale, coral-colored pink wines that glint in the sun are all the rage. These wines were first made in Provence and now the region offers the finest range of rosés in the world, including oaked rosés, rosés d'une nuit (where the vinification is completed in one night), and a rosé so pale and light in color it is called a gris (gray).

Lunch and Light Bites
Banon Goat's Cheese

This is fast food Provençal style and probably the greatest, simplest takeaway available in the world. Buy a Banon goat's cheese from a market trader, grab a baguette, and you are away. Rather than the greasy paper wrapping of a burger, you gently unfold dried chestnut leaves that are secured around the cheese by a raffia tie. Note, these are not just any chestnut leaves; they have been soaked in wine to impart flavor into the cheese. Eaten at perfect ripeness, a Banon goat's cheese hovers on the dividing line between solid and liquid, and can simply be mopped up with the end of a baguette.

Pan Bagnat

Literally "soaked bread": a large, crusty roll filled with tuna, olives, peppers, onion, and salad. It's a very popular lunchtime sandwich but can be a challenge to eat without the contents spilling out.

Truffles

Between late November and early February, Provence goes truffle mad. Nicknamed "black diamonds," truffles are a fungus that grows underground, attached by a gossamer-thin thread to the roots of oak trees. Prices for truffles commonly top €1,000 a kilo (2.2 lb). The good news is that they are best eaten as a complement to basic ingredients. Eggs, cheese, even the common baked potato are elevated to almost ethereal levels by the judicious addition of truffle shavings.

Socca

A traditional Niçois pancake made from chickpea flour, served hot, sprinkled with olive oil and black pepper, and usually cut up in thin, folded slices from a round pan the size of a dustbin lid.

Pastries and Dessert
Tarte au Citron

A creamy, tangy lemon tart with a sweet pastry base, sprinkled with icing sugar. Popular for a dessert or teatime snack.

Tarte Tropézienne

A specialty of Saint-Tropez, the dessert is a light sponge cake filled with vanilla custard cream.

Soups and Stews
Bouillabaisse

The famous fish soup of Marseille was originally cooked by fishermen using the rockfish they were unable to sell to restaurants. The word *bouillabaisse* refers to the cooking method of reducing the fish to a stock using boiling water. Restaurants now sign an official charter to cook the dish. They use agreed-upon ingredients and serve it in the traditional way. Bouillabaisse is accompanied by a feisty garlic-infused rouille paste that is spread on croutons, adding an extra punch to the already intensely flavored soup. As memorable as the dish is the pomp and ceremony that accompany its serving.

Daube de Boeuf

This hearty winter beef stew became so popular that it is now served year-round. Cooks need to be patient; the meat is marinated for three days, cooked for 4 hours, and then rested for another day. The result is rich, pungent, melt-in-the-mouth, deeply satisfying comfort food.

Ratatouille

A hearty stew of local vegetables including tomatoes, onions, eggplant, zucchini, and red peppers, cooked with olive oil, garlic, and herbs. Originally from Nice, ratatouille is usually served as a side dish.

Soupe au Pistou

This is a traditional vegetable soup finished off with a large dollop of basil-garlic pesto. It's more common to find this on menus during the winter. Beware it can be quite garlicky.

Accommodations

HOTELS AND GUESTHOUSES

Booking accommodations in advance is essential during the summer months. In May, June, and September, it is still advisable to reserve ahead. From October to April, it is possible to wait until the last minute before deciding on accommodations.

Hotels

The French Riviera and Provence have some of the most prestigious and expensive hotels in the world, and even finding out how much a room costs can be challenging. Reservations can be made using the usual booking websites (www.booking.com, www.hotels.com), but it is usually better (and cheaper) to **contact hotels directly,** especially if you have specific requests. Local tourist offices can also book accommodations. Prices for a standard double room in a two-star hotel in high season start at around €85 per night, rising to €150 in the more touristy areas. Almost all rooms will have en suite bathrooms and Wi-Fi, though the latter can be of variable quality even in the large hotels. Cheaper hotels will not usually have a bar or restaurant.

A three-star hotel usually means the rooms will have air-conditioning and there will be a bar/restaurant, a garden, and a pool. Prices for a standard double room in a three-star hotel in high season range from €100-250 per night.

Prices for four- and five-star hotels start around €300 a night and can rise to over €1,000 and considerably more for suites, rooms with top-floor terraces, and penthouses.

Hotels typically charge €10-15 per day for small, well-behaved dogs, but it is always worth checking to see if they accept pets when booking.

Unless otherwise noted, all rooms in hotels listed in this guide have Wi-Fi, air-conditioning, and en suite bathrooms.

Guesthouses

Guesthouses in France are referred to as chambre d'hôte. They all offer bed-and-breakfast and may also offer an evening meal. They tend to be small with no more than five rooms. Some are as luxurious as the top hotels and have prices to match. Typically, they offer a much more intimate experience than a hotel with the host offering advice on what to do and where to eat.

CAMPING

Nearly all villages, towns, and cities in Provence have camping nearby. Between July 14 and August 14, booking is advisable. Outside this period (which coincides with the main French summer holiday), it is not normally necessary to book. Campsites tend to open around May and close at the end of September. Campsites are not generally listed in this book, but information is available on all the tourist office websites under the Accommodations heading.

Conduct and Customs

The French tend to be more soft-spoken than Anglophones. Apart from students, they rarely eat or drink when walking about in the street. They tend to dress smartly and take care with their appearance. Taking a shirt off in public, other than on the beach, is not done. Wine is consumed slowly and in moderation.

Visitors are unlikely to cause great offense by contravening any of these conventions, but they may risk being commented upon. Both French men and French women greet each other with a kiss on both cheeks. In some

Provence villages, two kisses rise to three or even four kisses. Unless visitors become close friends with a local, they are not expected to kiss when saying hello; a handshake suffices.

Most waitstaff, hotel staff, and people working in the tourist industry will speak enough English to enable them to perform their jobs. Speaking just a little bit of French will endear a visitor to locals, even if it is as simple as saying "Merci beaucoup" rather than "Thank you."

Health and Safety

The South of France is a very safe place to travel. Petty crime such as mobile phone or wallet theft occurs infrequently in crowded places such as the weekly markets, though pickpockets are known to operate on crowded trains and trams on the Riviera. When visiting any large city in the world travelers are always advised to keep an eye on the neighborhoods they are exploring and not to wander too far from the established trail. This remains true in Provence, although violent crime is extremely rare.

There are no specific health risks associated with traveling to Provence and the French Riviera. Perhaps the greatest threat is the summer heat. Visitors should take sensible precautions such as wearing a hat, applying plenty of sunscreen, and staying in the shade or indoors during the hottest hours of the day between 2pm and 5pm. Insect bites and stings are common, and it is wise to pack repellent sprays and calming lotions.

The French health care service is excellent. It has a basic principle of universal access. In

emergencies, treatment is given first and questions about insurance asked later. European Union nationals should always travel with their European Health Insurance card. This entitles them to the same level of treatment as French nationals. Non-EU nationals should travel with appropriate medical insurance. It is usually relatively easy and quick to get a doctor's appointment. To get the phone number of the nearest practitioner near you, go to the local pharmacy and ask. If you have problems making an appointment, you can also ask in the pharmacy for help. An appointment with a doctor costs around €30. You will be expected to pay up front and will be given a form that you can use to claim back the money from your insurer.

In the event of an emergency, dial 112, the European emergency number, which is staffed by English-speaking operators. To be connected directly to a French operator, dial 15 for an ambulance, 17 for the gendarmerie (police), or 18 for emergency medical help from the fire service.

Practical Details

WHAT TO PACK
Clothing

In the summer, light dresses, shorts, and T-shirts should make up the majority of what you pack. However, it is always worth packing a long pair of trousers and a warm sweater because the evenings can get cold, particularly if the mistral wind is blowing, or if you are staying at altitude. It's also sensible to pack a light raincoat.

In spring and autumn, shorts, T-shirts, and light summer dresses should be supplemented by more warm clothing, and again a light raincoat.

In the winter, sweaters, jeans, and long trousers should take up most of your suitcase. A warm coat is also advisable. It's still worthwhile packing some shorts/light dresses because temperatures can occasionally rise into the mid-20s Celsius (high 70s Fahrenheit).

If you are planning to engage in outdoor sports, it's best to bring as many appropriate clothes as possible. Items such as bikes and riding helmets can be rented from the activity center. However, buying specialist clothing like hiking boots will entail spending precious holiday time hunting down items in out-of-town shopping centers.

Always bring a smart outfit that you could wear to a decent restaurant; this should give you access to places of worship, hotel lobbies, and casinos. Some places, like Monte-Carlo Casino, have more specific dress codes.

Electronics

Bring telephone and electronic equipment chargers that are compatible with the French **two-pin power sockets (230 voltage).** These can be purchased at the airport.

Medication

Bring your own medication, as, even if the products are available in France, they may have a different name and dosage.

MONEY

The currency of France is the euro. Notes that are commonly in circulation are 500, 100, 50, 10, and 5. Some businesses may not wish to take notes larger than €50.

Cash points (ATMs) are readily available throughout France. Even small villages have them. In general, you should not need to carry much cash, nearly all restaurants and businesses accept credit cards.

Places where you can exchange foreign currency are ever rarer and are limited to the main airports and large train stations. The rates of exchange do not tend to be favorable, and more often than not it is best to take money out from a cash machine rather than to bring cash from home and exchange it for euros. Check with your bank before leaving about charges.

COMMUNICATIONS
Cell Phones

The international dialing code for France is **00 33.** When using a foreign cell phone to dial a number in France, you will need to use this prefix, and drop the first 0 at the beginning of the French number. The dialing code for Monaco is **00 377.**

Roaming charges across the European Economic Area and the EU were restricted in 2016, so most Europeans should be able to use their mobile phones without incurring extra charges. It's a good idea to check with your service provider about using your phone in France, and what the charges will be.

Non-Europeans should consider changing their SIM cards to ones from a French phone service provider. All the main French operators (Orange, Bouygues Telecom, and SFR) offer **prepaid SIM cards** with plans for data, talk, and text that can be purchased at boutiques in all big towns and swapped into your personal phone. Phone chargers must be compatible with French power sockets.

Provence & the French Riviera on a Budget

If you are on a tight budget, Provence and the Riviera are not generally ideal places to visit, but there are ways to balance the big expenses of a foreign trip with more budget-minded behavior. In the first place: Don't come in the summer, during the Monaco Grand Prix, or during the Cannes Film Festival.

GETTING THERE

- If traveling from another European destination, most budget airline flights are more expensive the more the airplane fills up, so **reserve as far as possible in advance** for a good deal.

- Take **local buses** but **longer-distance trains.** From Nice airport, Express buses cost €20-22 to Cannes or Monaco, but walk 10 minutes from the airport to Nice Saint-Augustin train station and the journey will be less than half the price on the train.

ACCOMMODATIONS

- **Book accommodation early ... or very late.** Reserving early generally gets you access to cheaper rooms; the only exceptions are when hotels want to fill up and offer good, last-minute, same-day deals.

- **Camping** is the cheapest option, especially if you have your own tent. Book **budget hostels** early to guarantee the cheaper rooms and beds in a large dormitory.

- Hotels have started fighting back against the third-party booking websites by matching deals and even undercutting them, so **contact the hotel** directly. Sometimes they have budget rooms (beside lifts or fire escapes) that they do not advertise on their website.

FOOD

- Shop for fresh fruit, drinks, and grocery items in **local markets** and **supermarkets.** If you are really on a budget, most large supermarkets reduce the price of perishable goods after 5pm.

- Have your main meal at **lunchtime,** when most restaurants do a cheaper formule or plat du jour.

- Instead of ordering a bottle of water, ask for a **carafe d'eau**—free tap water—and instead of a bottle of wine, order a pichet de vin—a jug of house wine for a quarter of the price of a bottle.

- Don't take the hotel breakfast (€8-18); go out for a **petit-déjeuner** at a nearby café, or just have a coffee and croissant at the local boulangerie (€3-4).

ACTIVITIES AND RECREATION

- If you want to experience the full cultural scene, inquire in the **local tourist offices.** They may have a multi-pass ticket for the museums, which also often includes public transport.

- Take identification to museums and galleries, as most are free or offer **reduced admission** to those under 26 or with proof of student status.

- Use **public beaches.** If you want to rent a sun lounger and parasol, beaches run by the municipality offer much cheaper rates. Loungers and parasols are cheaper to rent (sometimes half-price) on private beaches after 3pm.

- Find out about all the **free events** and activities from local tourist offices. Even during the Cannes Film Festival, there are free screenings on the beach every night, but you have to pick up the limited number of tickets at the tourist office.

Internet Access

Internet access is available in nearly all hotels, restaurants, and cafés. Provided you are a customer, you just need to ask for the password.

Shipping and Postal Service

Even small villages have post offices. The postal service is quick and reliable. Shipping packages abroad is easy, and prepaid boxes can be picked up at the post office. To courier items, the post office offers a service called Chronopost, which guarantees next-day delivery in France. Details can be found online: www.chronopost.fr.

OPENING HOURS

Most businesses are open Tuesday to Saturday from 9am to noon and from 3pm to 6pm. Restaurants and shops in touristy locations will open seven days a week during the busy summer season. Many shopkeepers also choose not to shut for lunch in the summer months.

PUBLIC HOLIDAYS

- January 1: New Year's Day (Jour de l'An)
- Easter Monday (Lundi de Pâques)
- May 1: Labor Day (Fête du Travail)
- May 8: VE Day (Fête de la Victoire 1945)
- Ascension Day: 39 days after Easter Sunday (l'Ascension)
- Whit Monday: 50 days after Easter, 10 days after Ascension Day (Lundi de Pentecôte)
- July 14: Bastille Day (Fête Nationale)
- August 15: Assumption of the Blessed Virgin Mary (l'Assomption)
- November 1: All Saints' Day (La Toussaint)
- November 11: Armistice Day (Armistice 1918)
- December 25: Christmas Day (Noël)

In Monaco, there is also a public holiday to celebrate **Sainte-Dévote** on January 27 and **National Day** on November 19.

When a public holiday falls on a Thursday or Tuesday, it is common for the French to take an extra day off and create a long weekend. This is known as a *pont,* or bridge.

WEIGHTS AND MEASURES

France uses the **metric system.** All distances are given in kilometers and all weights in grams and kilograms. Drinks are served by the centiliter. A beer is 25 centileters (half pint) and a bottle of wine 75 centiliters (a pint and a half).

TOURIST INFORMATION
Tourist Offices

There are tourist offices in most villages, and in all towns and cities in Provence. They usually open from Monday to Friday outside of the main season, and Monday to Saturday in the summer. Main tourist offices in cities will be open seven days a week in the summer season. They are excellent places to pick up information on everything from accommodations to tours, experiences, and hikes. They also offer free local maps.

The main tourist office websites are as follows:

Provence

- **Aix-en-Provence:** www.aixenprovence-tourism.com
- **Arles:** www.arlestourisme.com
- **Avignon:** www.avignon-tourisme.com
- **Gorge du Verdon:** www.verdontourisme.com
- **The Luberon:** www.destinationluberon.com
- **Marseille:** www.marseille-tourisme.com
- **Northern Vaucluse:** www.ventoux-provence.fr
- **The Var:** www.visitvar.fr

The Riviera

- **Antibes:** www.antibesjuanlespins.com
- **Cannes:** www.cannes-destination.fr

- **Fréjus:** https://frejus.fr
- **Hyères:** www.hyeres-tourisme.com
- **Mandelieu-la-Napoule:** www.ot-mandelieu.fr
- **Menton:** www.menton.fr/-L-Office-de-tourisme-de-la-Ville-de-Menton-.html
- **Monaco:** www.monte-carlo.mc/fr/visites/office-tourisme-monaco
- **Nice:** www.nicetourisme.com
- **Sainte-Maxime:** www.sainte-maxime.com
- **Saint-Raphaël:** www.saint-raphael.com
- **Saint-Tropez:** www.sainttropeztourisme.com

Maps

Tourist offices are very generous when it comes to giving out free maps. Besides a €2 charge at Saint-Tropez, other offices have a good range of free one-page local plans and larger, fold-out maps of the region, as well as specialist maps for hikers and cyclists. A road map is a good idea if you are driving around the region; a recommended one is Michelin's Carte de Provence-Alpes-Côte d'Azur. The company's website (www.viamichelin.com) is excellent for working out route alternatives, traveling times, and the locations of restaurants and hotels.

Traveler Advice

OPPORTUNITIES FOR STUDY AND EMPLOYMENT

Nationals of EU-member states are free to move to and work in Provence and on the Riviera. Nationals of other countries will need a visa and authorization to work.

In the summer season, bars and restaurants are always looking for bilingual staff. In cities, there is high demand from businesses for English language lessons. A good resource for vacancies is the Teaching English as a Foreign Language website (www.tefl.com).

French universities welcome applications from foreign students. Opportunities can also be found on the AFS intercultural program website (https://afs.org) and the American Institute for Foreign Study (www.aifsabroad.com).

For more detailed advice on working or studying in France, see *Moon Living Abroad France*.

ACCESS FOR TRAVELERS WITH DISABILITIES

Most public transportation is adapted for passengers with disabilities. Stations have special access points, and trains have carriages with ramps. Buses, particularly in cities, are adapted for wheelchairs.

Museums and art galleries have almost universally been adapted to accommodate disabled access. Historical sights are not always easy to access, and their websites should be checked for information. Hotels will mention on their websites if they have ground-floor rooms accessible to wheelchairs. Generally, it is better to check before booking accommodations that the establishment is suitable for personal disabilities. Establishments can apply for the official **Tourisme et Handicap** label, which certifies them for five years under four disability criteria (visual, motor, audio, and mental). Certain beaches participate in the Handiplage scheme, in which specially adapted wheelchairs take users into the sea under professional supervision.

TRAVELING WITH CHILDREN

Most restaurants welcome children and offer children's menus (menu enfant). Hotels rarely have interconnecting rooms, but they often have large rooms that can accommodate up to six people. They will also put an additional bed or a cot into a room for a small extra charge.

Public transportation is free for children under 4. Museums are usually free for children under 12 and offer reduced rates up to the age of 18 or 26 (if a member of the EU).

Tourist offices have plenty of suggested activities for children. It is also always worth enquiring whether there are any "stage" holiday activities being run in the near vicinity. For example, horse riding and tennis clubs frequently have weeklong courses for children during holidays.

WOMEN TRAVELING ALONE

Women travelers should feel as safe traveling in Provence and on the Riviera as anywhere else in France. There should be no issues with overtly sexist behavior. It is very common for women to go to the beach alone and eat alone in restaurants. However, the Riviera tends to be quite sexually charged in the summer, and women sitting on their own or in small groups may be approached by individuals looking for the chance to flirt and compliment. What begins with the request for a light can advance quite quickly, but a clear, "*Non, merci*" and avoiding eye contact should prevent any unwanted attention. The usual safety precautions should be taken late at night or when traveling on public transport alone—be aware of your surroundings and sit near other people.

SENIOR TRAVELERS

Visitors over the age of 65 have free access to museums and most tourist attractions, although sometimes proof of age is required. Seniors are generally well-catered to on the Riviera, and large towns like Menton and Nice have a substantial populations over the age of 65, with pensioners having retired to the sun. The **Carte Senior +** offers a 25 percent discount on all rail journeys and 40 percent on first-class travel.

LGBTQ+ TRAVELERS

France has a liberal attitude toward sexuality, and LGBTQ+ travelers should not normally encounter any discrimination. The most useful websites for LGBTQ+ travelers in the South of France are www.gayfrenchriviera. com, www.gay-sejour.com, and the Nice tourist office guide: https://en.nicetourisme.com/gay-friendly-nice.

TRAVELERS OF COLOR

While France as a whole has experienced problems associated with racism (mainly against the Arab community), travelers of color should feel safe and welcome in Provence and on the French Riviera. If you encounter any racism, or are the victim of threats or racist insults, contact the police, the consulate (the nearest ones are in Marseille), or SOS Racisme, a Europe-wide organization that fights discrimination and racism (tel. 01 40 35 36 55; www.sos-racisme.org; Tues., Thurs., Fri. 10:30am-1pm).

Resources

Glossary

AROUND TOWN

bastide: fortified town or manor house

boules: France's national sport, the object of which is to throw heavy metal balls (boules) nearest to a smaller one—the jack, known in French as *le cochonnet*

centre-ville: town center

chambre d'hôte: bed-and-breakfast or guesthouse

château: a castle or name given to a wine estate

commune: an administrative division controlled by a village, town, or city, similar to a municipality in the United States or a parish in the UK

département: an administrative division in France, larger than a commune but smaller than a region

domaine: a name given to a wine estate

église: church

fête votive: annual village or town festival with carnival rides

gare: train station

gendarmerie: the French police. Gendarmes belong to a branch of the French army and live in barracks. They are armed.

hôpital: hospital

hotel de ville: the town or city hall

jardin des enfants: playground

mairie: the mayor's office, which administers the local area

marché: market

mas: provençal farmhouse

office de tourisme: tourist office

parc: park

pétanque: a game of boules in which the feet do not move

pharmacie: pharmacy

place: square in a village or town

police (municipal): the level of law enforcement below the gendarmerie. Each village will normally have one police municipal. It is their job to deal with small civic crime and parking/traffic issues. Towns and cities have a larger police municipal force. They are armed.

poste: post office

préfecture: an administrative center responsible for driving licenses, resident permits, etc.

region: a large administrative area. France is divided into 13 separate regions on the European continent and five overseas regions. Provence and the French Riviera are located within the Provence-Alpes-Côte d'Azur region, whose capital is Marseille.

stade: sports stadium

tabac: shop that sells magazines, cigarettes, stamps, and lottery tickets

tomette: red earthenware floor tiles that are present in many villas and old houses in the South of France, usually hexagonal in shape

trompe l'oeil: optical illusion that gives a three-dimensional look to a flat, usually painted surface

Urgence: emergency department of a hospital

vieille ville: old town

FOOD AND DRINK

bouillabaisse: fish soup served in Marseille and the surrounding coastal area

boulangerie: bakery

bistro: a small restaurant serving reasonably priced, home-cooked food. Also spelled *bistrot.*

brasserie: a traditional "French" eatery with printed menus and a more formal style than a bistro. Open long hours and with a large choice of dishes. It can also mean "brewery."

café: a place serving coffee and most other drinks, as well as light meals and snacks

daube: slow-cooked stew, popular in Southern France, usually made with wild boar or beef

formule: fixed-price menu, most often served at lunchtime

plat du jour: daily special at a restaurant, usually reasonably priced

restaurant: a place that has tables and chairs and serves food to customers on the premises

tarte tropézienne: a light sponge cake with a vanilla cream filling, created by pâtissier Alexandre Micka in Saint-Tropez in 1955

vigneron: a winemaker

LANDSCAPE AND NATURE

calanque: steep-sided creek or inlet, sometimes only accessible by boat

sentier: marked footpath or trail

ON THE ROAD

aire de repos: a place to stop the car, usually on autoroutes, with toilets and a rest area

aire de service: a place to stop the car on autoroutes where they sell petrol, and there is often a restaurant, shops, toilets, showers, and a children's play area

autoroute: French highway that operates a toll system; it usually has three lanes in both directions

corniche: a road cut into a mountainside. It is used to suggest a more thrilling and picturesque drive than the simple word "route."

péage: toll on France's highway system

French Phrasebook

PRONUNCIATION

French is a language full of homophones: words with the same sound but different meanings and spellings, so for the casual visitor with a basic knowledge of the language, misunderstandings can be common. The word *vert,* for example, means "green" in French, but its pronunciation is indistinguishable from *vers* ("toward" or "a verse"), *ver* ("worm"), *verre* ("glass"), and *vair* ("squirrel fur"). Most French people will be pleased to hear tourists attempting to speak and understand a foreign language but are also rather impatient and might answer in English anyway. Be prepared not to understand, don't get flustered, and be good-humored.

Vowels

The pronunciation of French vowels is notoriously difficult for Anglophones, but here's a general guide.

a   like *ah* in *father*

à   same as above; the accent is used for spelling purposes to avoid confusion between identical words

au like *oh* in *rose*

e   usually short, as in the word *le* ("the"), the sound is akin to the English *er*. In the middle of a word, such as *raclette,* *e* sounds like the English *ai,* as in *fair.* At the end of a word, *e* is silent.

é   like short *e* in *hey*

è   like the *e* in *bet*

i   like *ee* in *feet*

o   like the short *o* in *not*

ô   like the long *o* in *oh*

œ   like the *u* in *upset*

oi not heard in English, but the *oi* in *moi* or *mademoiselle*

ou like the *ue* in *true*

u   like a long *u* sound, but more emphasized like the *oo* in *moon*

ù   only used in the word *où* ("where") to distinguish it from *ou* ("or")

Consonants

The following French consonants are pronounced the same, or nearly the same, as English ones:

B D F K L M N P T U X V Z

The following have both hard and soft ways of being pronounced, depending on the letters that follow:

C G S

c has a hard *k* sound before a, o, and u; for example, *cadeau;* and a soft *s* sound before e, i, and y; for example, *cerise.*

ç always pronounced *s;* for example, *garçon* (pronounced *garsson*)

g is hard, as in *gold,* before consonants and vowels a, o, and u (*gamine, Gordes, guru*), but soft (like the *g* in *massage*) before e, i, and y (*Géraldine, GiGi,* and *gymnase*)

s mostly pronounced in the same way as it is in English, except for at the end of words, when it is usually not pronounced

The following consonants differ from English:

ch always pronounced *sh,* as in soft *champignon* (mushroom) and the Champs Elysées

h always silent; *hôtel* sounds like *ôtel*

j pronounced like a soft *g,* as in Jean, Jacques, or Julie

ll after the letter i, it's like the *y* in *yes.* After other vowels it's like the *ll* in *mull.*

qu pronounced like a *k* sound

r particularly difficult for foreigners to master: a strongly rolled *r* sound, produced by positioning the tongue in the same place as if saying the word "get."

th pronounced *t,* as in *thé* (tea, pronounced *tey*).

w rare in France. It usually occurs in an imported German or English word. If the word is of German origin, pronounce with a *v* sound. If English, pronounce the English *w.* Wi-Fi is pronounced "wee-fee."

Finally, almost all consonants at the end of words are silent. Paris is pronounced *paree* and *c'est trop tard* (it's too late) is *se tro tar.*

ESSENTIAL PHRASES

Hello Bonjour
Hi Salut

What's your name? Comment vous appelez-vous?
My name is… Je m'appelle …
Mrs./Mr./Miss Madame/Monsieur/ Mademoiselle
Nice to meet you Enchanté
How are you? Comment ça va? (informal) / Comment allez-vous? (formal)
Excuse me Excusez-moi
Please S'il vous plaît
Thank you Merci
You're welcome De rien (informal) / Je vous en prie (formal)
Sorry Pardon
Cheers Santé
See you later A tout à l'heure
See you soon A bientôt
Goodbye Au revoir
Yes Oui
No Non
Good Bon (masculine) / bonne (feminine)
Bad Mauvais (masculine) / mauvaise (feminine)
Beautiful Beau / Belle
Can you help me? Pouvez-vous m'aider?
Do you speak English? Parlez-vous anglais?
I do not speak French Je ne parle pas français
I do not understand Je ne comprends pas
Speak more slowly, please Pouvez-vous parler plus lentement, s'il vous plait?
How do you say… in French? Comment dit-on… en français?
Where are the restrooms/toilets? Où sont les toilettes?
Can you take my/our photo? Pouvez-vous me prendre en photo / Pouvez-vous prendre notre photo?

Transportation/Directions

Where is…? Où se trouve… ?
How far is…? A quelle distance est… ?
Is it far/close? C'est loin / proche?
To the left? C'est à gauche?
To the right? C'est à droite?
Straight ahead? C'est tout droit?

Is there a bus to…? Il y a t-il un bus pour …?

Does this bus go to…? Est-ce que ce bus va à …?

Where do I get off? Où est-ce-que je descends?

Where is the nearest station? Où se trouve la gare la plus proche?

What time does the bus/train leave/ arrive? A quelle heure part/arrive le bus/ train?

I would like to look at the timetable Je voudrais regarder l'horaire

Where can I buy a ticket? Où puis-je acheter un billet?

I would like to reserve a ticket Je voudrais réserver un billet

I would like to purchase a one-way ticket/a return ticket to… Je voudrais acheter un billet aller simple/aller-retour pour…

Where is a good restaurant? Où se trouve un bon restaurant?

Where is the beach/city center? Où se trouve la plage/le centre-ville?

I am looking for the train station/ airport Je cherche la gare/l'aéroport

I am looking for the hotel/hospital/ bank Je cherche l'hôtel/l'hôpital/la banque

Hotels

I would like a double room Je voudrais une chambre pour deux/double

I would like to cancel my reservation Je voudrais annuler ma réservation

At what time should we check out? A quelle heure faudrait-il partir?

Do you accept animals? Acceptez-vous les animaux?

air-conditioning climatisation (la clime)

bathroom salle de bain

balcony balcon

parking parking

breakfast petit déjeuner

sea view vue sur la mer

Shopping

money argent

cash en espèces

Where are the shops? Où se trouvent les magasins?

Can I pay with a credit card? Est-ce que je peux payer avec une carte de crédit?

At what time is it open? A quelle heure ouvre-t-il?

At what time is it closed? A quelle heure ferme-t-il?

I am looking for a supermarket Je cherche un supermarché

How much does it cost? Combien cela coûte?

It's too expensive / cheap C'est trop cher / pas cher

I'm just looking (for now) Je regarde pour l'instant

Restaurants

There are two of us Nous sommes deux

The menu, please La carte, s'il vous plaît

Do you have a menu in English? Avez-vous la carte en anglais?

I'm going to have… Je vais prendre…

I would like a coffee Je voudrais un café

I would like a glass of (red/white/rosé) wine Je voudrais un verre de vin (rouge/ blanc/rosé)

I would like some water Je voudrais de l'eau

What is the daily special? Quel est le plat du jour?

The bill, please L'addition, s'il vous plaît

Waiter / waitress Serveur / serveuse

I'm a vegetarian Je suis végétarien

I'm a vegan Je suis végan

breakfast le petit déjeuner

lunch le déjeuner

dinner le dîner

snack snack / goûter

salad la salade

soup la soupe

beef le bœuf

lamb l'agneau

chicken le poulet

pork le porc

fish le poisson
vegetable le légume
bread pain
pasta les pâtes
fruit le fruit
cake le gâteau
ice cubes les glaçons
ice cream la glace
pie la tarte
glass le verre
still water eau plate
tap water eau du robinet
mineral water eau minérale
sparkling water eau gazeuse
coffee café espresso
coffee with milk café au lait
beer bière
wine vin

Health

drugstore pharmacie
pain douleur
fever fièvre
headache mal de tête
stomachache mal au ventre
toothache mal aux dents
burn la brûlure
cramp la crampe
nausea la nausée
vomiting vomissement
medicine médicament
antibiotic antibiotique
pill/tablet le comprimé
aspirin aspirine
I need to see a doctor J'ai besoin de voir un médecin
I need to go to the hospital J'ai besoin d'aller à l'hôpital
I have a pain here... J'ai mal ici...
She/he has been stung/bitten Il/elle a été piqué(e)
I am diabetic/pregnant Je suis diabétique/enceinte
I am allergic to penicillin/cortisone Je suis allergique à la pénicilline/la cortisone
My blood group is...positive / negative Mon groupe sanguin est... positif / négatif

Numbers

0 zéro
1 un
2 deux
3 trois
4 quatre
5 cinq
6 six
7 sept
8 huit
9 neuf
10 dix
11 onze
12 douze
13 treize
14 quatorze
15 quinze
16 seize
17 dix-sept
18 dix-huit
19 dix-neuf
20 vingt
30 trente
40 quarante
50 cinquante
60 soixante
70 soixante-dix
80 quatre-vingts
90 quatre-vingt-dix
100 cent
101 cent un
200 deux cents
500 cinq cents
1,000 mille
10,000 dix mille
100,000 cent mille
1,000,000 un million

To write numbers from 20 to 69 in French, add the single number to the tens number.
vingt (20) + trois (3) = vingt-trois (23)
trente (30) + sept (7) = trente-sept (37)
quarante (40) + deux (2) = quarante-deux (42)
cinquante (50) + neuf (9) = cinquante-neuf (59)
soixante (60) + six (6) = soixante-six (66)

Time

What time is it? Quelle heure est-il ?

It's three/seven o'clock Il est trois/sept heures

midday midi

midnight minuit

morning matin

afternoon après-midi

evening soir

night nuit

yesterday hier

today aujourd'hui

tomorrow demain

Days and Months

day jour and journée

week semaine

month mois

year an and année

Monday Lundi

Tuesday Mardi

Wednesday Mercredi

Thursday Jeudi

Friday Vendredi

Saturday Samedi

Sunday Dimanche

January Janvier

February Février

March Mars

April Avril

May Mai

June Juin

July Juillet

August Août

September Septembre

October Octobre

November Novembre

December Décembre

Verbs

to be / I am / he/she is / they are être / je suis / il-elle est / ils sont

to go / I go / he/she goes / they go aller / je vais / il-elle va / ils vont

to come / I come / he/she comes / they come venir / je viens / il-elle vient / ils viennent

to stop / I stop / he/she stops / they stop arrêter / j'arrête / il-elle arête / ils arrêtent

to get off / I get off / he/she gets off /

they get off descender / je descends / il-elle descend / ils descendent

to arrive / I arrive / he/she arrives / they arrive arriver / j'arrive / il-elle arrive / ils arrivent

to return / I return / he/she returns / they return revenir / je reviens / il-elle revient / ils reviennent

to stay / I stay / he/she stays / they stay rester / je reste / il-elle reste / ils restent

to leave / I leave / he/she leaves / they leave partir / je pars / il-elle part / ils partent

to look at / I look at / he/she looks at / they look at regarder / je regarde / il-elle regarde / ils regardent

to look for / I look for / he/she looks for / they look for chercher / je cherche / il-elle cherche / ils cherchent

to have / I have / he/she has / they have avoir / j'ai / il-elle a / ils ont

to want / I want / he/she wants / they want vouloir / je veux / il-elle veut / ils veulent

to need / I need / he/she needs / they need avoir besoin de / j'ai besoin de / il-elle a besoin de/ils ont besoin de

to buy / I buy / he/she buys / they buy acheter / j'achète / il-elle achète / ils achètent

to give / I give / he/she gives / they give donner / je donne / il-elle donne / ils donnent

to take / I take / he/she takes / they take prendre / je prends / il-elle prend / ils prennent

to eat / I eat / he/she eats / they eat manger / je mange / il-elle mange / ils mangent

to drink / I drink / he/she drinks / they drink boire / je bois / il-elle boit / ils boivent

to read / I read / he/she reads / they read lire / je lis / il-elle lit / ils lisent

to write / I write / he/she writes / they write écrire / j'écris / il-elle écrit / ils écrivent

Suggested Reading

FICTION

Dumas, Alexandre. *The Count of Monte Cristo*. The story of Edmond Dantès, a man wrongly imprisoned in the Château d'If island prison off the coast of Marseille. While held captive, Dantès plots his brilliant revenge. It's an epic, world-famous story.

Fitzgerald, F. Scott. *Tender Is the Night*. Inspired by his own time on the Riviera in the company of wealthy U.S. socialites, the Murphys, the novel details the high life and mental demise of American couple Dick and Nicole Driver, who rent a villa in the South of France.

Giono, Jean. *The Man who Planted Trees*. An award-winning short story about a shepherd who over the space of four decades reforests a valley in the Alpes de Haute Provence.

Greene, Graham. *Loser Takes All*. Typical of Greene's main characters, the protagonist is an ordinary character who becomes embroiled in an extraordinary world. This time, it's the very English accountant, Bertram, whose boss moves his wedding to Monte-Carlo, where Bertram loses all his money before devising a winning system at the Monte-Carlo Casino, only to lose his wife, only to win her back.

Pagnol, Marcel. *Manon of the Spring* and *Jean de Florette*. These books center on the attempt of Provençal peasant Ugolin to make his fortune from growing carnations. To do so he needs water, which is in abundant supply on his neighbor's land. After a couple of murders, the land in question is eventually inherited by Jean de Florette, a tax clerk. Jean decides to quit work and move to the countryside to make his fortune raising rabbits. Before his arrival Ugolin blocks the spring, rendering Jean's land worthless and unsuitable for the rabbits. Ugolin sits back and waits for what he assumes is Jean's inevitable departure, when he will have an opportunity to buy the land. Jean refuses to leave and a generational saga ensues.

Maugham, W. Somerset. *The Razor's Edge*. Maugham's novel moves from America to Paris to the Riviera (where the author lived for 40 years), describing the trapezing fortunes of troubled World War I pilot Larry Darrell and the corrupt, greedy, and banal society that surrounds him.

Sagan, Françoise. *Bonjour Tristesse*. Published when the author was only 18, this is the ultimately tragic story of the difficult relationship between a teenage girl, her father, and his female companions during one summer on the Riviera.

Sobin, Gustaf. *The Fly Truffler*. A moving, poetic book, the title refers to the practice of looking for truffles by searching for the flies that lay their eggs in the soil next to where the tuber grows. The lead character, a university lecturer called Cabassac, discovers that when he eats a truffle he enters a state of receptivity where he is able to dream clearly about his dead wife, Julieta. This realization drives him to an obsession with the "black diamond," and we follow him through the countryside, stick twitching ahead of him, shadow always behind to avoid disturbing the flies.

NONFICTION

Bromwich, James. *The Roman Remains of Southern France*. A comprehensive guide to the Roman Empire's ruins and restorations in Nice, Fréjus, and La Turbie, among other locations.

Christian, Glynn. *Edible France, a Traveler's Guide.* A region-by-region exploration of France's food and culinary traditions. An indispensable companion for any foodie.

Fotheringham, William. *Put Me Back on My Bike.* Fotheringham tells the story of British cyclist's Tom Simpson's death while racing to the summit of Mont Ventoux.

Garrett, Martin. *Provence, A Cultural History.* This is a fascinating literary history of Provence's main cities and regions.

Gayford, Martin. *The Yellow House.* The Yellow House digs behind the myth and tell the story of the artist's commune that van Gogh hoped to establish in Arles.

Malterre-Barthe, Charlotte, and Zosia Dzierzawska. *Eileen Gray, A House Under the Sun.* An astute examination of the Irish designer and architect's modernist house E-1027 in Roquebrune-Cap-Martin.

Mayle, Peter. *A Year in Provence, Toujours Provence,* and *Encore Provence.* These are the books that popularized Provence as a holiday destination. They consist of collections of humorous anecdotes about Mayle's life in the Luberon.

Ring, Jim. *Riviera: The Rise and Rise of the Côte d'Azur.* An entertaining historico-social account of the Riviera from the arrival of the first English tourists to modern-day celebrities.

Internet Resources and Apps

TRAVEL BLOGS

www.provencepost.com
A popular Saint-Rémy-de-Provence-based blog with updates on daily life, new businesses, and events.

http://shuttersandsunflowers.com
A Lourmarin based blogger writes about her favorite places in Provence.

www.provencewinezine.com
A site focused on food and wine pairings from Provence.

https://curiousprovence.com
Based in Les Alpilles, Ashley Tinker blogs about all things Provençal, particularly house renovation.

www.bestofnice.com
A useful guide about to what to do and how to do it in the capital of the Côte d'Azur.

https://rosajackson.com
A blog about food and travel from the creator of Les Petits Farcis cooking school in Nice.

ACCOMMODATIONS

www.provenceweb.fr
A good overview of what to do and where to stay in Provence and on the Riviera.

www.gites-de-france-alpes-maritimes. com
Official listings of the villas and holiday apartments to rent in the Alpes-Maritimes.

www.onlyprovence.com
A luxury villa rental business with an excellent selection of houses.

www.theluberon.com
Luberon-based villa, cottage, and apartment-rental business.

https://provence.emotional-escapes.com

More luxury villas for rental in Provence.

www.homeaway.com

Offers the largest selection of villas, many of them rented from the owner.

www.clajsud.fr

A good website for cheap accommodations and youth hostels for young people staying on the Riviera.

FOOD

www.france-voyage.com/gastronomie/provence-alpes-cote-d-azur-region.htm

A guide to Provence and the French Riviera's most celebrated recipes and gastronomy.

SPORTS AND RECREATION

www.ffrandonnee.fr

The official site for hiking in France.

www.hikideas.com/walk-provence-alpes-cote-d-azur.html

Has a good selection of hikes with maps, directions, distances, all downloadable in pdf form.

Visorando

A useful app to help keep track of your location during hikes.

www.provence-a-velo.fr

All the information you need to plan a great cycling holiday in Provence is available on this site.

NAVIGATION AND WEATHER

www.lachainemeteo.com/meteo-france/previsions-meteo-france-aujourdhui

The day's weather forecasts for all of France, with annotated maps, graphs, and videos.

www.var.gouv.fr/acces-aux-massifs-forestiers-du-var-a2898.html

Gives information for the following day's access to the Var's mountain ranges for hikers and bikers.

www.autoroutes.fr/fr/trafic-en-temps-reel.htm

The latest live information on traffic on the autoroutes.

Waze

A route-planning app with up-to-date information on traffic conditions.

www.avignon.aeroport.fr

Flight arrival and departure information for Avignon Airport.

www.marseille.aeroport.fr

Flight arrival and departure information for Marseille Airport.

www.nice.aeroport.fr

Flight arrival and departure information for Nice Airport.

www.sudmobilite.fr

A good travel-planning site that gives you access to both train and bus timetables at the same time.

www.sncf-connect.com

For train bookings and timetables.

www.flixbus.com

For long-distance coach journeys across France and Europe. If you're booking on short notice, the prices tend to be much cheaper than the trains.

www.zou.maregionsud.fr

Information on bus service between major destinations as well as between towns and villages in Provence.

www.blablacar.fr

Coach and carpooling website.

www.lametropolemobilite.fr

Bus information in the Bouches-du-Rhone department.

NEWS AND CULTURE

www.cotedazur-tourisme.com/information/applications-mobiles-06_2256.html

Details the best applications to download for planning a trip to the Riviera.

www.cotedazur-en-fetes.com

Details of all the festivals and events taking place on the Côte d'Azur.

www.thelocal.fr

French news website in English.

http://rivieraradio.mc

Website for the English-language radio station based in Monaco.

www.departement06.fr/documents/Import/decouvrir-les-am/catalogue_exposition_premiere_guerre_mondiale.pdf

A look at the impact of the First World War on the Riviera following an exhibition by the Alpes-Maritimes département.

SHOPPING

www.sunshine-riviera-tour.com

Help with luxury shopping trips on the Riviera.

www.leboncoin.fr

A treasure trove for secondhand goods, selling everything from stone fireplaces to cars. There is also an accompanying app.

Index

R

List of Maps

Photo Credits

Title page photo: Veniamin Kraskov | Dreamstime.com; page 5 © (top) Tomas Marek | Dreamstime.com; (left middle) Jon Bryant; (right middle) Sophie Spitéri, Aix en Provence Tourisme; (bottom left) MalaikaCasal/Shutterstock.com; (bottom right) Jon Bryant; page 6 © Darius Dzinnik | Dreamstime.com; page 8 © Faithiecannoise | Dreamstime.com; page 10 © Jon Bryant; page 12 © (top) Romasph | Dreamstime.com; (bottom) Barmalini | Dreamstime.com; page 13 © Ryhor Bruyeu | Dreamstime.com; page 14 © David Evison | Dreamstime.com; page 15 © (top) © Philippe Fritsch - Dreamstime.com; (bottom) Joe Sohm | Dreamstime.com; page 16 © (top) Jon Bryant; (bottom) Brasilnut | Dreamstime.com; page 17 © Vvoevale | Dreamstime.com; page 18 © (bottom) © Madrabothair | Dreamstime.com; page 20 © (top) Abricol | Dreamstime.com; (bottom) Napa735 | Dreamstime.com; page 21 © (bottom) Ryhor Bruyeu | Dreamstime.com; page 24 © Tanya Ivey; page 27 © Jon Bryant; page 28 © (bottom) Ryhor Bruyeu | Dreamstime.com; page 29 © © Barmalini | Dreamstime.com; page 31 © Jon Bryant; page 32 © Kabvisio | Dreamstime.com; page 33 © Jon Bryant; page 34 © (top) VVShots | Dreamstime.com; (bottom) Rrushton6 | Dreamstime.com; page 35 © Barmalini | Dreamstime.com; page 36 © Emicristea | Dreamstime.com; page 37 © (top left) © Gunold | Dreamstime.com; (top right) Robert Zehetmayer |Dreamstime.com; page 46 © (top) Saiko3p | Dreamstime.com; (left middle) Unclejay | Dreamstime.com; (right middle) Photoprofi30 | Dreamstime.com; (bottom) Neil Letson | Dreamstime.com; page 49 © (top) Lucamato | Dreamstime.com; (left middle) Valerie Gillet; (right middle) Alain Hocquel; (bottom) Alain Hocquel; page 50 © © Znm | Dreamstime.com; page 54 © (top left) Valerie Gillet; (top right) © Gunold | Dreamstime. com; (bottom) Gunold | Dreamstime.com; page 56 © Barmalini | Dreamstime.com; page 58 © Robert Zehetmayer | Dreamstime.com; page 59 © Slowmotiongli | Dreamstime.com; page 62 © (top) Lev Levin | Dreamstime.com; (bottom) Barmalini | Dreamstime.com; page 66 © Izanbar | Dreamstime.com; page 67 © Christian Mueringer | Dreamstime.com; page 71 © (top) Robert Paul Van Beets | Dreamstime.com; (left middle) Julia Kuznetsova | Dreamstime.com; (right middle) Jacques Vanni | Dreamstime.com; (bottom) Jacques Vanni | Dreamstime.com; page 75 © Listen900701 | Dreamstime.com; page 76 © (top left) Moreno Soppelsa | Dreamstime.com; (top right) Leonid12sr | Dreamstime.com; page 84 © Sorincolac | Dreamstime.com; page 86 © (top) Arles Tourisme; (left middle) Arles Tourisme; (right middle) Arles Tourisme; (bottom) Arles Tourisme; page 88 © Legacy1995 | Dreamstime.com; page 91 © (top left) Arles Tourisme; (top right) Tanya Ivey; (bottom) Arles Tourisme; page 97 © Arles Tourisme; page 102 © (top left) Office Tourisme Intercommunal Alpilles en Provence; (top right) Office Tourisme Intercommunal Alpilles en Provence; (bottom) Office Tourisme Intercommunal Alpilles en Provence; page 108 © (top) Office Tourisme Intercommunal Alpilles en Provence; (left middle) Office Tourisme Intercommunal Alpilles en Provence; (right middle) Erick Venturelli/Carrières de Lumières; (bottom) Karlo12 | Dreamstime.com; page 111 © Steveheap | Dreamstime.com; page 113 © Xantana | Dreamstime.com; page 114 © (top left) Phbcz | Dreamstime.com; (top right) Hatman12 | Dreamstime.com; page 118 © Sophie Spitéri, Aix en Provence Tourisme; page 119 © Toxawww | Dreamstime.com; page 123 © (top) Sophie Spitéri, Aix en Provence Tourisme; (bottom) Neirfy | Dreamstime.com; page 127 © Sophie Spitéri, Aix en Provence Tourisme; page 129 © Thomas Luppo, Aix en Provence Tourisme; page 131 © Cecoffman | Dreamstime. com; page 135 © Anna Pustynnikova /Shutterstock.com; page 142 © Bbsferrari | Dreamstime.com; page 144 © Andrew Pattman, Aix en Provence Tourisme; page 149 © (top left) Wessel Cirkel | Dreamstime. com; (top right) Beatrice Preve | Dreamstime.com; (bottom) Martin Molcan | Dreamstime.com; page 151 © Barmalini | Dreamstime.com; page 153 © (top) Philophotos | Dreamstime.com; (bottom) Jennifer Barrow | Dreamstime.com; page 155 © Fullempty | Dreamstime.com; page 161 © Archeophoto | Dreamstime. com; page 164 © Amoklv | Dreamstime.com; page 166 © Wirestock | Dreamstime.com; page 168 © Serban Enache | Dreamstime.com; page 170 © Gbphoto27 | Dreamstime.com; page 173 © (top) Jef Wodniack | Dreamstime.com; (left middle) Znm | Dreamstime.com; (right middle) Sarah2 | Dreamstime.com; (bottom) Ermess | Dreamstime.com; page 174 © Antoniogravante | Dreamstime.com; page 176 © Salparadis | Dreamstime.com; page 178 © Rudolf Ernst | Dreamstime.com; page 181 © (top) Ermess | Dreamstime. com; (left middle) Cmfotoworks | Dreamstime.com; (right middle) Phbcz | Dreamstime.com; (bottom) Martin Molcan | Dreamstime.com; page 185 © Sam74100 | Dreamstime.com; page 188 © Aletheia97 | Dreamstime.com; page 189 © (top left) Fouquehdprod | Dreamstime.com; (top right) Marseille Tourisme; page 195 © Sam74100 | Dreamstime.com; page 199 © Saiko3p | Dreamstime.com; page 201 © (top left)

Marseille Tourisme; (top right) Nikolai Sorokin | Dreamstime.com; (bottom) Marseille Tourisme; page 203 © (top) Marseille Tourisme; (left middle) Marseille Tourisme; (right middle) Eric Laudonien | Dreamstime.com; (bottom) Marseille Tourisme; page 207 © Photodynamx | Dreamstime.com; page 209 © Marseille Tourisme; page 221 © (top left) Meinzahn | Dreamstime.com; (top right) Marseille Tourisme; (bottom) Palaine | Dreamstime.com; page 226 © (top left) Tanya Ivey; (top right) Barmalini | Dreamstime.com; (bottom) Tanya Ivey; page 230 © Gevisions | Dreamstime.com; page 231 © (top left) Jon Bryant; (top right) Jon Bryant; page 236 © Rglinsky | Dreamstime.com; page 241 © (top) Miluxian | Dreamstime.com; (left middle) svetlanasf | Dreamstime.com; (right middle) Jon Bryant; (bottom) Jon Bryant; page 247 © (top) Carabiner | Dreamstime.com; (left middle) Jon Bryant; (right middle) Demerzel21 | Dreamstime.com; (bottom) Tashka2000 | Dreamstime.com; page 249 © Jon Bryant; page 250 © Jon Bryant; page 256 © (top) Jon Bryant; (left middle) Jon Bryant; (right middle) Bruno Charvin; (bottom) Jon Bryant; page 262 © Jon Bryant; page 264 © Jon Bryant; page 271 © (top) Jon Bryant; (left middle) Jon Bryant; (right middle) Ruby Soames; (bottom) Jon Bryant; page 276 © (top) Giancarlo Liguori | Dreamstime.com; (left middle) Jon Bryant; (right middle) Jon Bryant; (bottom) Photogolfer | Dreamstime.com; page 278 © Lukaszimilena | Dreamstime.com; page 285 © (top left) Marcello Celli | Dreamstime.com; (top right) Billy Potočnik; (bottom left) Jon Bryant; (bottom right) Jon Bryant; page 287 © Jon Bryant; page 288 © Jon Bryant; page 295 © (top left) Jon Bryant; (top right) Jon Bryant; (bottom) Jon Bryant; page 300 © (top) Esiksandora | Dreamstime.com; (bottom) Jon Bryant; page 304 © (top left) Jon Bryant; (top right) Ruby Soames; (bottom left) Jon Bryant; (bottom right) Jon Bryant; page 308 © (top) Brasilnut | Dreamstime.com; (left middle) Ruby Soames; (right middle) Jon Bryant; (bottom) Jon Bryant; page 312 © Jon Bryant; page 313 © (top left) Jon Bryant; (top right) Aletheia97 | Dreamstime.com; page 322 © (top) Jon Bryant; (bottom) Jon Bryant; page 323 © Andersastphoto | Dreamstime.com; page 325 © Deymos | Dreamstime.com; page 328 © Jon Bryant; page 333 © Jon Bryant; page 338 © (top left) Jon Bryant; (top right) Jon Bryant; (bottom) Jon Bryant; page 340 © (top left) Jon Bryant; (top right) Jon Bryant; (bottom) Jon Bryant; page 343 © Jlf4646 | Dreamstime.com; page 346 © Jon Bryant; page 352 © Jon Bryant; page 353 © Jon Bryant; page 354 © (top left) Jon Bryant; (top right) Billy Potočnik; (bottom) Jon Bryant; page 362 © (top) Denis Tikhomirov | Dreamstime.com; (left middle) Jon Bryant; (right middle) Jon Bryant; (bottom) Richard Banary | Dreamstime.com; page 368 © Jon Bryant; page 369 © Jon Bryant; page 372 © (top left) Jon Bryant; (top right) Jon Bryant; (bottom) Jon Bryant; page 374 © Jon Bryant; page 376 © Annedave | Dreamstime.com; page 380 © (top left) Jon Bryant; (top right) Jon Bryant; (bottom left) Ruby Soames; (bottom right) Jon Bryant; page 384 © (top) Barmalini | Dreamstime.com; (left middle) Zmphoto24 | Dreamstime.com; (right middle) Jon Bryant; (bottom) Jon Bryant; page 387 © Jon Bryant; page 390 © Jon Bryant; page 392 © Tagor784 | Dreamstime.com; page 394 © Jon Bryant; page 395 © (top left) Lyundovskaya | Dreamstime.com; (top right) Stevanzz | Dreamstime.com; page 403 © Pitchathorn Chitnelawong | Dreamstime.com; page 405 © (top) Ruby Soames; (left middle) Jon Bryant; (right middle) Jon Bryant; (bottom) Arsty | Dreamstime.com; page 411 © (top left) Ruby Soames; (top right) Jon Bryant; (bottom) Jon Bryant; page 413 © Spazgenev | Dreamstime.com; page 418 © Rglinsky | Dreamstime.com; page 420 © Elinaxx1vcom | Dreamstime.com; page 423 © Chdecout | Dreamstime.com; page 426 © (top left) Jon Bryant; (top right) Jon Bryant; (bottom) Jon Bryant; page 428 © Rglinsky | Dreamstime.com; page 430 © Jon Bryant; page 433 © Jon Bryant; page 435 © (top) Jon Bryant; (bottom) Jon Bryant; page 437 © Sam74100 | Dreamstime.com; page 452 © VVShots | Dreamstime.com; page 454 © Ryhor Bruyeu | Dreamstime.com

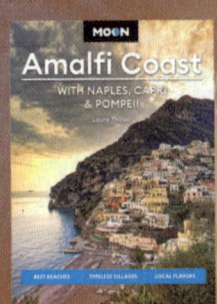
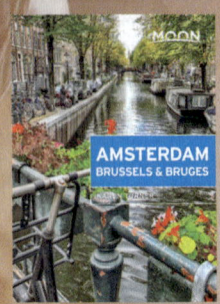
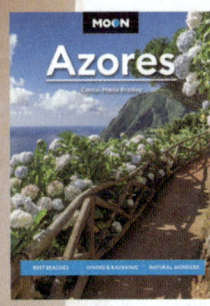
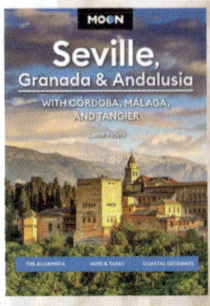

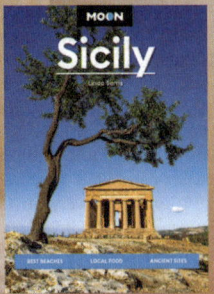
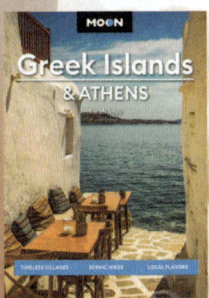

More European Travel Guides from Moon

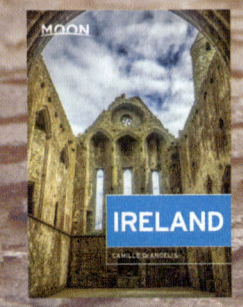

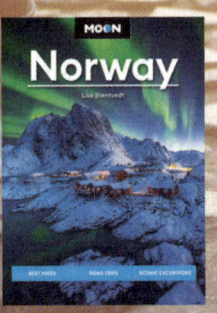
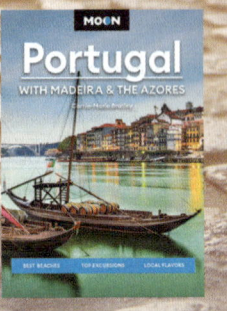
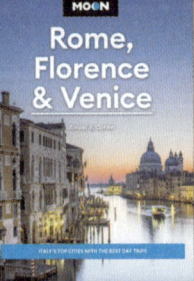
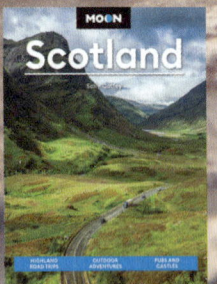
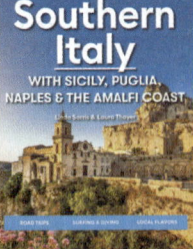